NUCLEAR ACTIVITY IN GALAXIES ACROSS COSMIC TIME
IAU SYMPOSIUM 356

COVER ILLUSTRATION:

AGN over Ethiopia courtesy of Habtamu Tadese

IAU SYMPOSIUM PROCEEDINGS SERIES

Chief Editor
MARIA TERESA LAGO, IAU General Secretary
Universidade do Porto
Centro de Astrofísica
Rua das Estrelas
4150-762 Porto
Portugal
mtlago@astro.up.pt

Editor
JOSÉ MIGUEL RODRÍGUEZ ESPINOSA, IAU Assistant General Secretary
Instituto de astrofísica de Canarias
La Laguna
38205 Tenerife
Spain
jre@iac.es

INTERNATIONAL ASTRONOMICAL UNION

UNION ASTRONOMIQUE INTERNATIONALE

NUCLEAR ACTIVITY IN GALAXIES ACROSS COSMIC TIME

PROCEEDINGS OF THE 356th SYMPOSIUM OF THE INTERNATIONAL ASTRONOMICAL UNION HELD IN ADDIS ABABA, ETHIOPIA 7–11 OCTOBER, 2019

Edited by

MIRJANA POVIĆ
Ethiopian Space Science and Technology Institute, Ethiopia
Instituto de Astrofísica de Andalucía - CSIC, Spain

PAOLA MARZIANI
National Institute for Astrophysics, Italy

JOSEFA MASEGOSA
Instituto de Astrofísica de Andalucía - CSIC, Spain

HAGAI NETZER
Tel Aviv University, Israel

SEBLU H. NEGU
Ethiopian Space Science and Technology Institute, Ethiopia

and

SOLOMON B. TESSEMA
Ethiopian Space Science and Technology Institute, Ethiopia

CAMBRIDGE UNIVERSITY PRESS
University Printing House, Cambridge CB2 8BS, United Kingdom
1 Liberty Plaza, Floor 20, New York, NY 10006, USA
10 Stamford Road, Oakleigh, Melbourne 3166, Australia

© International Astronomical Union 2021

This book is in copyright. Subject to statutory exception
and to the provisions of relevant collective licensing agreements,
no reproduction of any part may take place without
the written permission of the International Astronomical Union.

First published 2021

Printed in the UK by Bell & Bain, Glasgow, UK

Typeset in System LaTeX 2ε

A catalogue record for this book is available from the British Library Library of Congress Cataloguing in Publication data

This journal issue has been printed on FSC^{TM}-certified paper and cover board. FSC is an independent, non-governmental, not-for-profit organization established to promote the responsible management of the world's forests. Please see www.fsc.org for information.

ISBN 9781108492010 hardback
ISSN 1743-9213

This book is dedicated to all of the people who in one way or another contributed to the development of astronomy and space science in Ethiopia and Africa, and to all of those who will continue doing so in the future.

Table of Contents

Preface xvii

The Organising Committee and Acknowledgments xix

Opening Ceremony and Welcome Notes xxii

List of Participants xxxi

Scientific Program Overview xxxvi

Summary of activities organised during the symposium xxxviii

Introduction

Development in astronomy in Ethiopia and East-Africa through
nuclear activity in galaxies 3
    Mirjana Pović

Chapter I. Multiwavelength AGN surveys: past, present, and future

Multiwavelength surveys for Active Galactic Nuclei 11
    William Nielsen Brandt

Dissecting quasars with the J-PAS narrow-band photometric survey 12
    *Silvia Bonoli, Giorgio Calderone, Raul Abramo, Jailson Alcaniz,
Narciso Benitez, Saulo Carneiro, Javier Cenarro, David
Cristóbal-Hornillos, Renato Dupke, Alessandro Ederoclite,
Carlos López San Juan, Antonio Marín-Franch,
Claudia Mendes de Oliviera, Mariano Moles, Vinicius Placco,
Laerte Sodré Jr., Keith Taylor, Jesús Varela, Héctor Vázquez Ramió
and the J-PAS collaboration*

Unveiling the physical processes that regulate galaxy evolution with
SPICA observations 17
    Luigi Spinoglio, Juan A. Fernández-Ontiveros and Sabrina Mordini

Search for high-redshift blazars with Fermi-LAT 23
    Michael Kreter

A uniformly selected, all-sky, optical AGN catalogue 24
    Ingyin Zaw

Star formation and AGN activities in selected nearby HII galaxies in the
LeMMINGS survey 25
    Ikechukwu Obi

Panchromatic characterisation of accreting black holes in dusty
star-forming galaxies 26
    Gabriela Calistro Rivera

Chapter II. AGN types and unification model

AGN types and unification model 29
    Luigi Spinoglio and Juan Antonio Fernández-Ontiveros

Hypercat - hypercube of AGN tori 44
    Robert Nikutta, Enrique Lopez-Rodriguez, Kohei Ichikawa,
    Nancy A. Levenson and Christopher C. Packham

Towards a new paradigm of dust structure in AGN: Dissecting the mid-IR
emission of Circinus galaxy 50
    Marko Stalevski, Daniel Asmus and Konrad R. W. Tristram

Predicting the emission profile and estimation of model parameters for some
nearby LLAGN using accretion and jet models 56
    Bidisha Bandyopadhyay, Fu-Guo Xie, Neil M. Nagar,
    Dominik R. G. Schleicher, Venkatessh Ramakrishnan, Patricia Arévalo,
    Elena López and Yaherlyn Diaz

A new tool to derive chemical abundances in type-2 active galactic nuclei 61
    Rubén García-Benito, Enrique Pérez-Montero, Oli L. Dors,
    José M. Vílchez, Monica V. Cardaci and Guillermo F. Hägele

The quasar main sequence and its potential for cosmology 66
    Paola Marziani, Deborah Dultzin, Ascensión del Olmo, Mauro D'Onofrio,
    José A. de Diego, Giovanna M. Stirpe, Edi Bon, Natasa Bon,
    Bożena Czerny, Jaime Perea, Swayamtrupta Panda,
    Mary Loli Martinez-Aldama and C. A. Negrete

Spectroscopic properties of radio-loud and radio-quiet quasars 72
    Avinanda Chakraborty and Anirban Bhattacharjee

Fe II emission in NLS1s – originating from denser regions with
higher abundances? 77
    Swayamtrupta Panda, Paola Marziani and Bożena Czerny

The role of failed accretion disk winds in active galactic nuclei 82
    Margherita Giustini and Daniel Proga

Optical spectroscopy of nearby type1-LINERs 87
    Sara Cazzoli, Isabel Márquez, Josefa Masegosa, Ascensión del Olmo,
    Mirjana Pović, Omaira González-Martín, Barbara Balmaverde,
    Lorena Hernández-García and Santiago García-Burillo

Study of the diversity of AGN dust models 93
    Omaira González-Martín

AGN evolution as seen in spectral lines: The case of narrow-line Seyfert 1s 94
    Marco Berton

The properties of the dusty inner regions of nearby QSOs 95
    Itziar Aretxaga

AGN1 vs AGN2 dichotomy as seen from the point of view of ionized outflows . . 96
    Eleonora Sani

Elusive accretion discs in low luminosity AGN 97
    M. Almudena Prieto and Juan A. Fernandez-Ontiveros

Chapter III. Variability

A status report on AGN variability 101
    Paulina Lira

Recent results of measuring black hole masses via reverberation mapping 116
    Shai Kaspi

Revisiting old (AGN) friends - what's changed in their spectral looks 122
    Hartmut Winkler

Discovery of new changing look in NGC 1566 127
    Victor L. Oknyansky, Sergey S. Tsygankov, Vladimir M. Lipunov,
    Evgeny S. Gorbovskoy and Nataly V. Tyurina

The study of variability of 8 blazar candidates among the *Fermi*-LAT
unidentified gamma-ray sources 132
    Pheneas Nkundabakura, Jean D'amour Kamanzi,
    Jean D. Mbarubucyeye and Tom Mutabazi

High-resolution radio astronomy: An outlook for Africa 137
    Leonid I. Gurvits, Robert Beswick, Melvin Hoare, Ann Njeri,
    Jay Blanchard, Carla Sharpe, Adrian Tiplady and Aletha de Witt

X-ray properties of reverberation-mapped AGNs with super-Eddington
accreting massive black holes 143
    Jaya Maithil, Michael S. Brotherton, Bin Luo, Ohad Shemmer,
    Sarah C. Gallagher, Du Pu, Hu Chen, Jian-Min Wang and Yan-Rong Li

The BLR physics from the long-term optical monitoring of type-1 AGN 144
    Dragana Ilić

Chapter IV. Properties of AGN host galaxies

Properties of X-ray detected far-IR AGN in the green valley 147
    Antoine Mahoro, Mirjana Pović, Petri Väisänen, Pheneas Nkundabakura,
    Beatrice Nyiransengiyumva and Kurt van der Heyden

Properties of green valley galaxies in relation to their selection criteria 152
    Beatrice Nyiransengiyumva, Mirjana Pović, Pheneas Nkundabakura and
    Antoine Mahoro

Study of AGN contribution on morphological parameters of their
host galaxies 158
    Tilahun Getachew-Woreta, Mirjana Pović, Josefa Masegosa,
    Jaime Perea, Zeleke Beyoro-Amado and Isabel Márquez

Galaxy evolution studies in clusters: the case of Cl0024+1652 cluster
galaxies at $z \sim 0.4$ 163
    Zeleke Beyoro-Amado, Mirjana Pović, Miguel Sánchez-Portal,
    Solomon Belay Tessema, Tilahun Getachew-Woreta and the GLACE team

AGN, host galaxies, and starbursts 169
    Petri Väisänen

Quasar host galaxies and environments in the GAMA survey 170
    Jari Kotilainen

Role of environment on AGN activity 171
    Amirnezam Amiri

Hosts of jetted narrow-line Seyfert 1 galaxies in near-infrared 172
    Emilia Järvelä

Dirt-cheap gas scaling relations: Using dust attenuation and galaxy radius to predict gas masses for large samples of AGNs 173
    Hassen Yesuf

Chapter V. Triggering, feedback, and shutting off of AGN activity

AGN fueling and feedback 177
    Francoise Combes

Feedback from supermassive and intermediate-mass black holes at galaxy centers using cosmological hydrodynamical simulations 184
    Paramita Barai

Radiative feedback of low-L_{bol}/L_{Edd} AGNs 189
    Fu-Guo Xie

The impact of AGN on the life of their host galaxies at $z \sim 2$ 194
    Chiara Circosta

Do AGN really suppress star formation? 199
    Chris M. Harrison, David M. Alexander, Dalton J. Rosario, Jan Scholtz and Flora Stanley

Accretion and star formation in 'radio-quiet' quasars 204
    Sarah V. White (SVW), Matt J. Jarvis, Eleni Kalfountzou, Martin J. Hardcastle, Aprajita Verma, José M. Cao Orjales and Jason Stevens

MUSE-adaptive optics view of the starburst-AGN connection: NGC 7130 209
    Johan H. Knapen, Sébastien Comerón and Marja K. Seidel

The relationship between black hole accretion rate and gas properties at the Bondi radius 214
    De-Fu Bu

Tracing AGN feedback, from the SMBH horizon up to cluster scales 218
    Francesco Tombesi

Feedback and star formation in AGNs 223
    Luis C. Ho

The role of AGN in galaxy star formation: A case study of
a radio galaxy at z = 2.6 224
    Allison Man

Ionized AGN outflows are less powerful than assumed: A multi-wavelength
census of outflows in type II AGN 225
    Dalya Baron

Clustering dependence of Chandra COSMOS Legacy AGN on host galaxy
properties 226
    Viola Allevato

Chapter VI. Jets and environment

Radio jets: Properties, life and impact 229
    Raffaella Morganti

The origin of X-ray emission in Low-Excitation Radio Galaxies 243
    Shuang-Liang Li

Understanding the origin of radio outflows in Seyfert galaxies using radio
polarimetry 247
    *Biny Sebastian, Preeti Kharb, Christopher P. O' Dea, Jack F. Gallimore,
Stefi A. Baum and Edward J. M. Colbert*

Role of active galactic nuclei and flow of relativistic jets 252
    Abdissa Tassama and Tolu Biressa

Feedback from quasars: The prevalence and impact of radio jets 253
    Miranda Jarvis

Physics of SMBH in nearby AGNs 254
    Venkatessh Ramakrishnan

The spins of supermassive black holes 255
    Ranga-Ram Chary

The space VLBI mission RadioAstron: AGN results 256
    Yuri Kovalev

Observing AGN sources with the Event Horizon Telescope 257
    Maciek Wielgus

Chapter VII. The youngest AGN and AGN evolution

What do observations tell us about the highest-redshift supermassive
black holes? 261
    Benny Trakhtenbrot

AGN astrometry: A powerful tool for galaxy kinematic studies 276
    Naftali Kimani, Andreas Brunthaler and Karl M. Menten

The role of AGN activity in the building up of the BCG $z \sim 1.6$ 280
  Angela Bongiorno and Andrea Travascio

The role of LoBALs in quasar evolution 285
  Clare Wethers, Jari Kotilainen, Malte Schramm and Andreas Schulze

Luminosity functions and quasar lifetimes in a sample of mid-IR
selected quasars 290
  Susan E. Ridgway

Galactic Mergers at redshift $z \sim 5$, a sample of fast growing QSOs 291
  Nathen Nguyen

Supermassive black hole seed formation and the impact on black hole
populations across cosmic time 292
  Colin DeGraf

Chapter VIII. AGN posters

Multiwavelength morphological study of active galaxies 295
  *Betelehem Bilata-Woldeyes, Mirjana Pović, Zeleke Beyoro-Amado,
  Tilahun Getachew-Woreta and Shimeles Terefe*

The role of active galactic nuclei in galaxy evolution in terms of
radial pressure 299
  Biressa Tolu and Abate Feyissa

Optical variations in changing-look AGNs selected at X-rays 302
  *Sara Cazzoli, Josefa Masegosa, Isabel Márquez, Lorena Hernández-García,
  A. Álvarez-Hernández and Laura Hermosa-Muñoz*

NGC7469 as seen by MEGARA at the GTC 306
  *Sara Cazzoli, Armando Gil de Paz, Isabel Márquez, Josefa Masegosa,
  Jorge Iglesias, Jesus Gallego, Esperanza Carrasco, Raquel Cedazo,
  María Luisa García-Vargas, África Castillo-Morales, Sergio Pascual,
  Nicolás Cardiel, Ana Pérez-Calpena, Pedro Gómez-Alvarez,
  Ismael Martínez-Delgado and Laura Hermosa-Muñoz*

Optical spectral properties of radio loud quasars along the main sequence 310
  *Ascensión del Olmo, Paola Marziani, Valerio Ganci, Mauro D'Onofrio,
  Edi Bon, Natasa Bon, and Alenka C. Negrete*

Accretion rate in AGN and X-ray-to-optical flux ratio at $z \leqslant 0.2$ 314
  Asrate Gaulle, Mirjana Pović and Dejene Zewdie

Optical spectroscopy of type-2 LINERs 317
  Laura Hermosa Muñoz, Sara Cazzoli, Isabel Márquez and Josefa Masegosa

Jet opening angle and linear scale of launch region of blazars 320
  Xiang Liu, Pengfei Jiang and Lang Cui

Detailed characterisation of LINERs and retired galaxies in
the local universe 323
  Daudi T. Mazengo, Mirjana Pović, Noorali T. Jiwaji and Jefta M. Sunzu

Multiwavelength study of potential blazar candidates among *Fermi*-LAT
unidentified gamma-ray sources 326
    Jean Damascène Mbarubucyeye, Felicia Krauß and Pheneas Nkundabakura

A search for new γ-ray blazars from infrared selected candidates 329
    *Blessing Musiimenta, Bruno Sversut Arsioli, Edward Jurua and
Tom Mutabazi*

Fe II strength in NLS1s – dependence on the viewing angle and FWHM(Hβ) ... 332
    Swayamtrupta Panda, Paola Marziani and Bożena Czerny

Discovering exotic AGN behind the Magellanic Clouds 335
    *Clara M. Pennock, Jacco Th. van Loon, Cameron P. M. Bell,
Miroslav D. Filipović, Tana D. Joseph and Eleni Vardoulaki*

Environmental effects on star formation main sequence in
the COSMOS field 339
    Solohery M. Randriamampandry, Mattia Vaccari and Kelley M. Hess

A dying radio AGN in the ELAIS-N1 field 342
    Zara Randriamanakoto

Understanding galaxy mergers and AGN feedback with UVIT 345
    Khatun Rubinur, Mousumi Das, Preeti Kharb and P. T. Rahne

Exploring the X-ray universe via timing: mass of the active galactic nucleus
black hole XMMUJ134736.6+173403 348
    *Eva Šrámková, K. Goluchová, G. Török, Marek A. Abramowicz,
Z. Stuchlík, and Jiří Horák*

Dichotomy of radio loud and radio quiet quasars in four dimensional
eigenvector one (4DE1) parameter space 351
    Shimeles Terefe, Ascensión Del Olmo, Paola Marziani and Mirjana Pović

Determination of K4000 of potential blazar candidates among EGRET
unidentified gamma-ray sources 355
    Emmanuel Uwitonze, Pheneas Nkundabakura and Tom Mutabazi

The properties of inside-out assembled galaxies at $z < 0.1$ 358
    *Dejene Zewdie, Mirjana Pović, Manuel Aravena, Roberto J. Assef and
Asrate Gaulle*

Catalogue with visual morphological classification of 32,616 radio galaxies with
optical hosts 361
    Natalia Żywucka, Dorota Koziel-Wierzbowska and Arti Goyal

Extragalactic background light inhomogeneities and Lorentz-Invariance
Violation in gamma-gamma absorption and Compton scattering 364
    Hassan Abdalla

Probing black hole - host galaxy scaling relations with obscured
type II AGN 365
    Dalya Baron

Peering into the heart of darkness: Radio VLBI survey of
the NEP deep field 366
    Joseph Gelfand

X-ray variability plane revisited: Role of obscuration 367
    Omaira González-Martín

Optical variability of faint quasars 368
    Endalamaw Ewnu Kassa

Search for binary black holes in 10 years of Fermi LAT data with
information field theory 369
    Michael Kreter

Stellar populations and ages of ultra-hard X-ray AGN in the BASS survey 370
    Mirjana Pović

Adaptive optics imaging and spectroscopy of the radio galaxy 3C294 371
    Andreas Quirrenbach

GeMS/GSAOI near-infrared imaging of $z \sim 0.3$ BL Lacs 372
    Susan Ridgway

The circum-nuclear environment and jets of active galaxies at $z \sim 0$:
Recent results from a multi-frequency investigation 373
    Prajval Shastri

Spectral energy distribution of blazars 374
    Prospery Simpemba

The GLEAM 4-Jy Sample: The brightest radio-sources in the southern sky . . . 375
    Sarah White

An accreting $< 10^5 M\odot$ black hole at the center of dwarf galaxy IC750 376
    Ingyin Zaw

Chapter IX. Non-AGN posters

Science strategy of the African Astronomical Society (AfAS): *An outcome of the
Science Business held in synergy with the IAUS 356* 379
    *Lerothodi L. Leeuw, Kevin Govender, Charles M. Takalana,
    Zara Randriamanakoto, and Alemiye Mamo*

Mass-radius relation of compact objects 383
    Seman Abaraya and Tolu Biressa

Non-thermal radio emission from dark matter annihilation processes in
simulated Coma like galaxy clusters 385
    Fitsum Woldegerima Beyene and Remudin Reshid Mekuria

Star formation efficiency of magnetized, turbulent and
rotating molecular cloud 388
    Gemechu M. Kumssa and Solomon Belay Tessema

Multi-wavelength emission from dark matter annihilation processes in galaxy clusters and dark matter sub-halos 390
   Remudin Reshid Mekuria

14 years of 6.7 GHz periodic methanol maser observations towards G188.95+0.89 393
   Martin M. Mutie, Paul Baki, James O. Chibueze and Khadija El Bouchefry

Testing stellar evolution models 395
   Seblu H. Negu and Solomon Belay Tessema

Accelerating universe in modified teleparallel gravity theory 397
   Shambel Sahlu, Joseph Ntahompagaze, Amare Abebe and David F. Mota

The impact of CMEs on the critical frequency of F2-layer ionosphere (foF2) . . . 400
   Alene Seyoum, Nat Gopalswamy, Melessew Nigussie and Nigusse Mezgebe

Mass-loss varying luminosity and its implication to the solar evolution 403
   Negessa Tilahun Shukure, Solomon Belay Tessema and Endalkachew Mengistu

Howusu Metric Tensor - problems and prospects 405
   Obini Ekpe Ekpe

Progress of astronomy in Tanzania 406
   Noorali Jiwaji

The study of short orbital period of delta scuti pulsating variable stars 407
   Abduselam Mohammed

Design and development of a two-dish interferometer 408
   Dorcus Mulumba

Electron-proton interaction in radio sources 409
   Halima Ugomma Obini

Author Index 411

Preface

We are very pleased to introduce the proceedings of the 356th International Astronomical Union (IAU) Symposium titled *"Nuclear Activity in Galaxies Across Cosmic Time"* held in Addis Ababa, Ethiopia, from 7th to 11th of October, 2019. This was the third symposium organised in Africa in the past 100 years since the establishment of the IAU, and only the first one organised in Ethiopia and East-Africa. We hope that this meeting will inspire many of our colleagues, and that in future we will have more meetings organised in Africa. With this symposium we wanted to achieve two main goals.

First, we wanted to provide a general overview of recent findings and progress in observations, simulations, and theory of active galactic nuclei (AGN) from the local universe up to high redshifts. AGN play an important role in many aspects of modern cosmology. They are fundamental for understanding galaxy formation and evolution, black hole formation and growth, and the connection between the two. Multiwavelength studies by Herschel, ALMA, Planck, NuSTAR, HST and more, combined with larger and deeper AGN samples, resulted in better understanding of nuclear activity in galaxies and AGN physics across cosmic time. However, we are still far from understanding the full physics behind AGN activity and their role in galaxy formation and evolution. The timing of IAUS 356 was appropriate since we are now witnessing a rapid growth of results based on large data-sets at all wavelengths, and since we need to formulate the questions for several new missions such as eROSITA, JWST, Euclid, CTA, SKA, E-ELT, and Athena. The key topics highlighted in the symposium and covered in the proceedings encompass multiwavelength AGN surveys, AGN types and unification, AGN variability, active black holes and their host galaxies, triggering, feedback and shutting off AGN activity, jets and environment, highest redshift AGN and AGN evolution.

Second, we wanted to bring for the first time world experts in our field to the East-African region and contribute to the development of science in Ethiopia and Africa. We also wanted to call the attention of the international scientific community to the new activities and development of astrophysics in Sub-Saharan countries. For most of the international community South Africa is still the major player regarding scientific activities in astronomy and space science in Africa. However, over the past years many other countries (e.g., Egypt, Ghana, Kenya, Morocco, Namibia, Rwanda, Sudan, Uganda, etc.) have begun research activities, opened new MSc and PhD programs at their universities, and started developing new research centers and technological facilities for improving their socio-economical challenges. Many achievements have been realized, but still there are many difficulties, challenges, and needs. At the same time, the Ethiopian Space Science and Technology Institute (ESSTI) is a new research center established under the Ethiopian Ministry of Innovation and Technology in Nov 2016. It is the first center of this kind in Ethiopia and one of the first in all East-African region. Organizing this international conference in Ethiopia has shown to be an important contribution and a sign of international support to the new development of science in the country and Africa. We hope the conference provided important motivation for African scientists and students.

To take full advantage of the symposium and to benefit our society, different activities have been carried out for graduate students, young researchers, teachers, teachers trainers, school children, and general public, before, during, and after the symposium. Training for undergraduate, MSc and PhD students was organized before (5–6 Oct) and training for teachers after (12–13 Oct) the symposium. To promote science in general among the public and school children, one whole afternoon was dedicated to two public talks at Addis Ababa University and outreach activities that have been carried out in 8 public schools. A lunch session about *"Development in astronomy and space science in Africa"* was organized in parallel to the science-business meeting of the African Astronomical Society (AfAS) that has been held at the same venue on 10 and 11 of Oct. Finally, NOAO Data Lab training was also organised for interested participants. All education and outreach activities have been carried out voluntarily by our participants, both international and local. More details on all organised activities can be seen in section 'IAUS356 for society'.

More than 160 people from 30 countries and 5 continents participated in IAUS 356. The papers presented in this proceedings are only part of the contributions that have been presented in the symposium (talks and posters). We were able to enjoy high level presentations and fruitful discussions that have been organised after each session. In total we had 7 invited talks (4 males and 3 females, from 7 countries and 5 continents), 60 contributed talks (53% male and 47% female speakers), and 33 poster presentations in the AGN field. We also wanted to motivate our students and young researchers in general to present their work and we had in addition 17 poster presentations out of AGN field. We put a lot of efforts to take care of geographical distribution and gender balance and to achieve approximately 50%–50% of male-female participation in different aspects of symposium organisation (as can be seen with selection of invited speakers, contributed talks speakers, or SOC members). Participants had a chance to learn more about Ethiopian history and culture, by visiting the National Museum, and by enjoying traditional food, music and cultural dance, and to learn more about establishment of the ESSTI by visiting Entoto Observatory out of Addis Ababa.

In summary, we hope that this conference will contribute to our better understanding of nuclear activity in galaxies across cosmic time. We aimed at clarifying the role of nuclear activity in the broader context of galaxy evolution, outlining major unsolved problems and observational and theoretical strategies to solve them. We also hope that it will help to strengthen the development of science in Ethiopia and Africa and open new space for collaborations between Africa and other parts of the globe for the benefit of all.

This symposium and all organised activities would not be possible without huge support of the IAU and ESSTI, all SOC and LOC members, all our sponsors, invited speakers, all trainers and volunteers, and all IAUS 356 presenters and participants. Thank you everybody for all your contribution and support.

Mirjana Pović, Paola Marziani, Josefa Masegosa,
Hagai Netzer, Seblu H. Negu, and Solomon B. Tessema

Editors

Mirjana Pović
Ethiopian Space Science and Technology Institute, Ethiopia
Instituto de Astrofísica de Andalucía - CSIC, Spain

Paola Marziani
National Institute for Astrophysics, Italy

Josefa Masegosa
Instituto de Astrofísica de Andalucía - CSIC, Spain

Hagai Netzer
Tel Aviv University, Israel

Seblu H. Negu
Ethiopian Space Science and Technology Institute, Ethiopia

Solomon B. Tessema
Ethiopian Space Science and Technology Institute, Ethiopia

Organizing Committee
Scientific Organizing Committee

Mirjana Pović	(ESSTI, Ethiopia) - chair
Paola Marziani	(INAF, Italy) - co-chair
Hagai Netzer	(Tel Aviv University, Israel) - co-chair
Pheneas Nkundabakura	(University of Rwanda, Rwanda) - co-chair
Itziar Aretxaga	(INAOE, México)
Bililign T. Dullo	(Universidad Compultense de Madrid, Spain)
Andrew C. Fabian	(University of Cambridge, United Kingdom)
Lisa J. Kewley	(Australian National University, Australia)
Vincenzo Mainieri	(ESO, Germany)
Isabel Márquez	(IAA-CSIC, Spain)
Bradley M. Peterson	(STScI, USA)
Mara Salvato	(MPE, Germany)
Miguel Sánchez Portal	(IRAM, Spain)
Nebiha Shafi	(HartRAO, South Africa)
Prajval Shastri	(Indian Institute of Astrophysics, India)
Sylvain Veilleux	(University of Maryland, USA)
Jian-Min Wang	(Institute of High Energy Physics - CAS, China)

Local Organizing Committee

Mirjana Pović	(ESSTI) - chair
Alemiye Mamo Yacob	(ESSTI) - co-chair
Solomon Tessema Belay	(ESSTI) - co-chair
Feleke Zerihun	(ESSTI) - co-chair
Jerusalem Tamirat	(ESSTI) - co-chair
Josefa Masegosa	(IAA-CSIC, Spain) - co-chair
Biruk Abrham	(ESSTI)
Dugasa Belay	(ESSTI)
Zeleke A. Beyoro	(ESSTI and Kotebe Metropolitan University)
Tolu Biressa	(Jimma University)
Solomon Cherie	(ESSTI)
Eyoas Ergetu	(ESSTI)
Daniel Fekadu	(ESSTI)
Getinet Feleke	(Kotebe Metropolitan University)
Getnet Gebereegziabher	(ESSTI)
Etsegenet Getachew	(ESSTI)
Christopher D. Impey	(University of Arizona, USA)
Fraol Lenjisa	(ESSTI)
Getachew Mekonnen	(ESSTI)
Remudin R. Mekuria	(Addis Ababa University)
Seblu H. Negu	(ESSTI)
Sebhat Tadesse	(ESSTI)
Mekbeb Tamrat	(ESSTI)
Getachew Wollel	(ESSTI)
Betelehem B. Woldeyes	(ESSTI)
Tilahun G. Woreta	(ESSTI and Bule Hora University)

Figure 1. IAUS 356 LOC and supportive staff. *Credits. A. Solomon*

Acknowledgements

We highly appreciate the support of all institutions and people who in one way or another contributed to the realisation of this symposium, from the very initial stages of proposal writing to the final stage of all organisation. In particular, our gratitude goes to the IAU for giving us the opportunity and financial support to host for the very first time this symposium in Ethiopia and East-Africa. We also extend our thanks to the ESSTI and MInT for all the local support in terms of human and financial resources. This symposium would not be possible without huge support of SOC and LOC members, all our sponsors, invited speakers, all trainers and volunteers, and IAUS 356 participants. We hope that this symposium will contribute significantly to development of astronomy and science in Ethiopia and Africa.

The symposium is sponsored and supported by the IAU Divisions J (Galaxies and Cosmology), C (Education, Outreach and Heritage), and D (High Energy Phenomena and Fundamental Physics), by the IAU Commissions C.X1 (Inter-Division D-J Commission Supermassive Black Holes, Feedback and Galaxy Evolution), B2 (Data and Documentation), and C1 (Astronomy Education and Development), and by four African Regional Nodes of the Office of Astronomy for Development (IAU-OAD). We acknowledge the Cambridge University Press team for all the support offered during preparation and production of the IAU 356 book of proceedings.

The Local Organizing Committee operated under the auspices of the
Ethiopian Space Science and Technology Institute (ESSTI).

Funding and support by the

International Astronomical Union (IAU),

Ethiopian Space Science and Technology Institute (ESSTI),

Entoto Observatory and Research Centre (EORC),

Ethiopian Ministry of Innovation and Technology (MInT),

International Science Programme (ISP),

Development in Africa with Radio Astronomy (DARA),

Instituto de Astrofísica de Andalucía (IAA-CSIC) and its Severo Ochoa program SEV-2017-0709

Spanish Astronomical Society (SEA),

UK Science and Technology Facilities Council (STFC-UKRI),

Ethiopian Space Science Society (ESSS),

East-African Regional Office of Astronomy for Development (EA-ROAD),

Addis Ababa University (AAU),

and Nature Astronomy

is gratefully acknowledged and much appreciated.

Thank you.

Opening Ceremony and Welcome Notes

Address by the IAU 356 symposium chair and ESSTI Astronomy and Astrophysics Department Head, Dr. Mirjana Pović

His Excellency Dr. Ing. Getahun Mekuria State Minister of Ministry of Innovation and Technology, His Excellency Mr. Tefera Walwa Former Deputy Prime Minister and Patron of Ethiopian Space Science Society, Guests of Honor,

Dear Colleagues and Friends,

Welcome all to our symposium *'Nuclear Activity in Galaxies Across Cosmic Time'*. It is the 356th symposium of the International Astronomical Union (IAU), but only the third one to be organised in Africa in the past 100 years, and the first one in Ethiopia and East-Africa. We are very grateful to the IAU and all people who put their trust and confidence in us and gave us the opportunity to organise this symposium that is so important for recently established Ethiopian Space Science and Technology Institute (ESSTI) and our Astronomy and Astrophysics Department, astronomy in Ethiopia and Africa, and astronomy in general. With this meeting we hope to achieve some of the main proposed goals.

We hope that all together we will have very prospective and fruitful discussions and achieve better understanding of some of important points of AGN physics and observations, including: AGN multiwavelength surveys, AGN types and unification model, variability, properties of AGN host galaxies, triggering, feedback, and shutting off of AGN activity, jets and environment, and properties of youngest AGN and AGN evolution.

Beside that, this symposium goes much beyond a regular scientific meeting. Thanks to huge efforts of many of our African and international colleagues, many of African countries started with astronomy and space science developments over the past recent years, including Ethiopia. With this symposium we want to support those activities and contribute to these developments on national and continental level. We hope to give more visibility to the ESSTI, and astronomy and space science developments in Ethiopia and Africa, we want to inspire and motivate our first generation of MSc and PhD students by having all of you here, we want to benefit our society by organising in collaboration with all of you different education and outreach activities, and we wish to use this meeting for strengthening international collaborations on longer-term for the benefit of all.

All of this would not be possible without huge support of the IAU, ESSTI, Entoto Observatory and Research Center (EORC) and our Ministry of Innovation and Technology to whom we are very grateful. I want to thank all other institutions that supported organisation of this symposium, including the International Science Programme (ISP), Development in Africa with Radio Astronomy (DARA), Instituto de Astrofísica de Andalucía (IAA-CSIC) and and its Severo Ochoa program SEV-2017-0709, Spanish Astronomical Society (SEA), UK Science and Technology Facilities Council (STFC-UKRI), Ethiopian Space Science Society (ESSS), East-African Regional Office of Astronomy for Development (EA-ROAD), Addis Ababa University (AAU), and Nature Astronomy. I would also like to acknowledge amazing efforts that LOC and SOC members made, together with all invited speakers, all outreach and education trainers and volunteers for making possible this meeting and all activities organised for society.

As a main organiser I have to say that I was touched with the level of humanity that I found and felt every time that I contacted somebody asking for support with organisation of IAUS 356. It clearly showed me that there are many of us who believe that together through education and science we can fight poverty on a long-term and hopefully make this world a better and more fair place in future for everybody. Thank you all very much for coming and for helping us in reaching the goals of this IAU symposium.

I wish you to have fruitful meeting, and happy and unforgettable stay in Addis and Ethiopia. Thank you!

Mirjana Pović
Addis Ababa, 7 October 2019

Address by the ESSTI General Director, Dr. Solomon B. Tessema

His Excellency Dr. Ing. Getahun Mekuria State Minister of Ministry of Innovation and Technology and Guest of Honor, His Excellency Mr. Tefera Walwa Former Deputy Prime Minister and Patron of Ethiopian Space Science Society, SOC members, AfAS Executive Committee Members, LOC members, Dear colleagues,

As Director General of Ethiopian Space Science and Technology Institute (ESSTI), it is a great honor and pride to have the opportunity to say a few words before starting this symposium. First of all, on behalf of ESSTI, the government of Ethiopia and myself, I would like to express our great pleasure in welcoming all attendees of the International Astronomical Union (IAU) symposium 356 entitled 'Nuclear Activity in Galaxies Across Cosmic Time' to be held from 7th until 11th of October 2019 in Addis Ababa, Ethiopia, political capital of Africa and place of the origin of human being. I especially welcome those who traveled long distance to our lovely city. This is the third IAU symposium organized in Africa in the past 100 years, and the first one to be organized in Ethiopia and East- and Central-African region. It has been organised under the ESSTI Astronomy and Astrophysics Research and Development Department.

ESSTI was established by government of Ethiopia in 2016 by Regulation 393/2016. Since then some of the most important responsibilities and functions assigned to this Institute have been to promote, initiate, lead and coordinate basic and applied research activities in astronomy and astrophysics, space science, remote sensing, atmospheric and climate science, space geodesy, satellite development and operation, infrastructure development, capacity building in space science and technology, licensing, monitoring, controlling and authorizing space activities, strengthening international relations and collaborations, in collaboration with scientists, governmental and non-governmental organizations throughout our country as well as abroad. In view of the rapidly advancing frontiers of space science and technology, and the increasing importance of international collaborations, I strongly feel that our institute should play a leading role in promoting scientific activities in Africa as well as worldwide. This is not only a give-and-take information exchange, knowledge and technology transfer with the outside world, but also we intend to promote harmony between different scientific cultures in space science and technology through the establishment of different research and technology hubs at our institute.

Entoto Observatory and Research Center (EORC) has been established in 2013 and is currently one of the subordinates of ESSTI with two 1 meter telescopes able to connect Ethiopian astronomers with the rest of the world. In the past three years ESSTI has published more than 70 scientific papers in peer review journals. Presently, Ethiopia is hosting East African Office of Astronomy for Development and is working hard on the development of astronomy and astrophysics in the region. In addition to research and technology development, EORC has been giving trainings in both MSc and PhD in collaboration with Addis Ababa University in astronomy and astrophysics, space science, remote sensing, and space geodesy, including East-African students, with goal of being regional center of excellence in the above fields of study.

One activity among many is the global promotion of our research, technology and training, and hence we have decided to host different continental and international symposia on astronomy and space science in various topics in fields of our prioritized interest. The present IAU 356 symposium is one of the international conferences that have been conducted in past ten years. Since 2009, EORC and ESSTI have organized international and continental conferences and workshops, such as the two East African

Astronomical Society workshops, American Geophysical Union International Conference in Space Science, Middle East and Africa IAU Regional Meeting, annual ESSTI conference, and Africa Initiative for Planetary and Space Science workshop. The past conferences and the current symposium have been chosen based on the size, priority, national interest and knowledge and technology transfer mechanism. In addition to promoting exchange of expert insights, knowledge and technology transfer, we would like to encourage particularly young scientists to present papers in each symposium/ conference/ workshop on their new results from the astronomy, space science and technology areas and related fields, and to help them get an overview of fields they are involved in.

On behalf of ESSTI and myself, I assure you that such types of symposia, conferences and workshops are welcome and we are committed to support them at continental and international level. As the hosting country of African Union, Ethiopian government is committed to organize and lead research in astronomy and space science in both theoretical and observational aspects to support African development strategy through the ESSTI. In this symposium more than 120 papers and posters will be presented and known scientists as well as the youth, students and women will present their research findings, as well as panel discussions, invited talks, and public talks will be conducted. The current research development and future direction of extragalactic astronomy research will be forwarded and the next hosting countries will be proposed. Even though, Ethiopia is an emerging country in astronomy and space technology research, it has taken responsibility to support our African countries and widening to international level and strengthening scientific collaboration with the astronomical community in the world.

Finally, I would like to thank local and scientific organizing committees and ESSTI staff members for their hard work to organize this symposium. My thanks extend to IAU, Ministry of Innovation and Technology, International Science Programme (ISP), Development in Africa with Radio Astronomy (DARA), Instituto de Astrofísica de Andalucía (IAA-CSIC) and its Severo Ochoa program SEV-2017-0709, Spanish Astronomical Society (SEA), UK Science and Technology Facilities Council (STFC-UKRI), Ethiopian Space Science Society (ESSS), East-African Regional Office of Astronomy for Development (EA-ROAD), Addis Ababa University (AAU), and Nature Astronomy for all financial and organizational support.

We wish you all a very constructive and pleasant stay.

Solomon B. Tessema
Addis Ababa, 7 October 2019

Address by His Excellency Former Deputy Prime Minister and Patron of Ethiopian Space Science Society, Mr. Tefera Walwa

Dear colleagues,

My welcoming speech is not the usual welcome protocol. It is to present to you most serious points of mine as African. Main issues I want to raise are:
1. Is space science and technology important or imperative for development?
2. Should we limit to collaboration or should we come to unified efforts?
3. Who should be in charge?

1. Important or imperative?
Development is the result of imagination. What does space science and technology do? Its power of inspiration to release the power of imagination is its key role. Forget its being the first science. It is the center piece for all or I better say, it is gravity for all other sciences and technology. That is why it is imperative for development. Important makes optional. That is why I don't buy important. It is not me to talk to you on this. I better make world renowned scientists talk to you. For now, only one, out of very many.

"All the progress of human civilization, from the invention of the first tools to our nascent quantum technologies, is the result of the disciplined application of the imagination. ... But if we hadn't descended from people who, hundreds of thousands of years ago, imagined ways to harness fire, we would still be prey."

Time reborn, by Lee Smolin, p. 252

"Imagination enabled us to turn change and surprise into opportunities to extend our domain across the planet. Some 12,000 years ago, we adapted our environments to ourselves, becoming farmers rather than opportunistic hunter-gatherers. Since then, our footprint has extended to the point where our impositions on the Earth's natural systems threaten to cause us great harm. Because imagination is our game, and imagination got us here, only imagination can provide the new ideas that will take us safely through the surprises to come."

Ibid, p. 253

"Our way is to aspire always to more than and other than what we have. To be human is to imagine what is not, to seek beyond the limits, to test the constraints, to explore and rush and tumble across the intimidating boundaries of our known world."

Ibid, p. 254

2. To collaborate or unit?
Collaboration is the list option. It is duplication of effort and waste of many things (meager resources of the poor people of Africa, if not the whole humanity). Unity speeds up in all manner, by uniting the human resource of the field, the material wealth ..... etc. What not?! Unity means? One policy, strategy, organization, leadership and resource allocation, etc. means one school of thought leads and governs it. This is to the benefit of the whole humanity; it is not limited to Africa alone.

3. Who should be in charge?

The youth, if not the teenagers. The simple reason could be the next generation is better than the past and the current, undoubtedly. But many reasons more than this simple reason. Let's talk to Prof. Stephen Hawking and some other one.

"But what lies ahead for those who are young now? I can say with confidence that their future will depend more on science and technology than any previous generation's has done. They need to know about science more than any before them because it is part of their daily lives in an unprecedented way."

Brief answers to the big questions, by Stephen Hawking, p.203

"There is so much more to come and I hope that this prospect offers great inspiration to schoolchildren today. But we have a role to play in making sure this generation of children have not just the opportunity but the wish to engage fully with the study of science at an early level so that they can go on to fulfill their potential and create a better world for the whole human race. ..... And I believe the future of learning and education is the Internet. People can answer back and interact. In a way, the Internet connects us all together like the neurons in a giant brain. And with such an IQ, what cannot we be capable of?"

Ibid, pp. 207 - 208

Let me now take you to another one:

"Apollo inspired generations, including people who are now the explorers working to take us back to the Moon, on to Mars and beyond. ..... Their work is taking space exploration to new heights. But who will actually put new footprints on the Moon and take the first step on Mars? Who will take us beyond that goal? Who is responsible for the future of space exploration? Students. Today's students are that future-and it is our job to inspire them and help them see themselves as those leaders. ..... To these students, this is not hard to imagine. It is not a dream world or something from a storybook. Setting foot on Mars, going back to the Moon and exploring beyond is not just an idea; it is their reality. Today's students have a 'no limits, anything is possible' mindset-the type needed for space exploration. ...... While adults talk about technology as a thing, students simply experience it as part of their lives......, in your own way, you can inspire the next generation. Today's students have the mindset to transform the future of space exploration into something we cannot even imagine....."

Aviation Week and Space Technology, July 4 2019, by Lance Bush

Lance Bush is president and CEO of the Challenger Center, a non-profit formed by the families of the crew members who perished when the space shuttle Challenger broke apart.

Tefera Walwa
Addis Ababa, 7 October 2019

Address by East African Regional Office of Astronomy for Development Director, Mr. Alemiye Mamo

Your Excellency Dr. Ing. Getahun Mekuria State Minister of Ministry of Innovation and Technology, His Excellency Mr. Tefera Walwa Former Deputy Prime Minister and Patron of Ethiopian Space Science Society, distinguished invited speakers and guests, symposium participants, ladies and gentlemen,

On behalf of the East African Regional Office of Astronomy for Development (EA-ROAD), as co-organizer of the symposium, I would like to pass my sincere gratitude and thanks for coming and to welcome you all to the symposium.

The International Astronomical Union (IAU) Office of Astronomy for Development (OAD) has 10 regional offices across the world. EA-ROAD is one of this offices opened in Ethiopia in 2014 as an East-Africa regional node. The office is hosted at ESSTI under the supervision of Ministry of Innovation and Technology.

The main objectives of EA-ROAD are to further use astronomy as a tool for development by mobilizing international collaborations, human and financial resources to realize the scientific, technological and cultural benefits to society, and by implementing the IAU strategic plan and missions of OAD.

For the past one decade the level of astronomy and space science development in East-African region has been gradually improved. The establishment of space agencies and institutes, formulation of space policies, introduction of astronomy education in the curriculum at both undergraduate and postgraduate levels, establishment of East-Africa Astrophysics Research Network (EAARN) and establishing of ROAD have contributed a lot to the development of astronomy and space science in the region. All this gradual development shows the existence of a promising and fertile landscape to flourish astronomy and space science in the region. In addition, there are green lights that encourage investment and commitment at all levels. However, there are also bottlenecks in coordinating and finding a synergy in the region to reach to the point where the development of astronomy should be.

Thus, it needs a collaborative approach that engages government and policy makers, science educators, advocates and professional societies to realize and advocate for the role of astronomy for development and to inspire and attract young people to the field. I, therefore, take this opportunity to pledge the respective authorities and scientific community to support and contribute for the development of astronomy in East-Africa as well as the whole Africa in terms of leadership, finance and scientific contributions and collaborations.

Finally, I wish you to have a fruitful deliberation and successful symposium.

Thank you!

Alemiye Mamo
Addis Ababa, 7 October 2019

Address by African Astronomical Society Vice-President, Dr. Lerothodi Leeuw

His Excellency the former Deputy Prime Minister of Ethiopia, his Excellency the Minister of Innovation and Technology of Ethiopia, honorable guests, invited speakers, presenters and participants at the International Astronomical Union (IAU) Symposium 356, good morning. Good morning.

I speak this morning on behalf of the very young and the very ancient. Who, you are welcome to ask is that. The young is the African Astronomical Society (Buckley *et al.* 2019, AfAS2019), that was just recently re-organized and re-launched; and, I come before you as its Vice-President. The ancient is the African continent, a continent that is the cradle of humanity and its development, and indeed where ancient humans distinguished themselves, mastering the science of making fire and using tools for their development and prosperity. I draw on this youthful African Astronomical Society and ancient history and wisdom of Africa to welcome you here this week, to the International Astronomical Union Symposium 356 titled 'Nuclear Activity in Galaxies Across Cosmic Time'.

The Two Goals of the Symposium: Young and Ancient

The symposium has two goals, and to be consistent with my opening, please allow me to cast them as young and ancient. The young goal, in the words of the organizers, is "strengthening the development of science in Ethiopia and Africa and opening new space for collaborations between Africa and other parts of the globe for the benefit of all." Here the description young is especially relevant when having in mind the proposed new developments and collaborations, that it is hoped will come and grow from this symposium. The ancient goal is the deepening of "our understanding of nuclear activity in galaxies across cosmic time", and for the cosmic objects that are the subject of this meeting, ancient indeed goes deeper and beyond the time of humans here in Africa and on Earth. These goals are deep, expansive and non-trivial, and rightfully ambitious and important; and, on behalf on the African Astronomical Society I proudly commend you in your efforts to be here and to tackle them. I do that with strong conviction that you will indeed make progress in advancing them.

The African Astronomical Society (AfAS) and the IAUS 356

The African Astronomical Society (AfAS) will be running a two-day meeting on the 10th and 11th of October, parallel to this IAU Symposium. The goal of that meeting is to develop the science strategy of the society and deliberate its implementation. The society will welcome input from attendees of the symposium, in addition of the participants that will be gathering for that particular meeting. We hope that beyond the programs that have been formally organized for this IAU Symposium and the AfAS meeting, there will be opportunities for the respective participants to network and develop synergies to both tackle science and the development of Astronomy in Africa, that are the goals of this symposium. The AfAS Science Business Meeting is organized by the society together with the Ethiopian Space Science and Technology Institute (ESSTI) and partners; and, we are grateful to the amazing work that ESSTI has put in organizing these meetings.

The last time we were at this venue for an international astronomy meeting, it was two years ago for the Middle East and Africa Region IAU Meeting IV (MEARIM IV) of the IAU. In addition to the science discussed at that meeting, there was also a session dedicated to the revitalizing of the AfAS, that at the time was dormant. Rigorous debate ensued in the meeting and following it, a resolution was taken to revitalize and give new life to the AfAS. I'm proud to say, that meeting led to the successful revival of the society; and, I stand here today as its vice-president as a result of the first steps from that meeting. In welcoming you here and opening this symposium officially, on behalf of our society, I hope that similar concrete and significant deliberations and actions will come from this meeting. On behalf of the African Astronomical Society, I welcome you all to Africa and to this meeting.

Lerothodi Leeuw
Addis Ababa, 7 October 2019

Acknowledgements: Lerothodi Leeuw (LL) acknowledges funding from the South African National Foundation (NRF) and Department of Science and Innovation (DSI) to the African Astronomical Society (AfAS) and the Ethiopian Space Science and Technology Institute (ESSTI) and partners, to attend this meeting. Further, LL acknowledges the organizers of this symposium for the opportunity to present here.

List of Participants

Name & Institution	Email
Seman Abaraya, Jimma University, Ethiopia	semanabaraya06@gmail.com
Hassan Abdalla, North-West University, South Africa	hassanahh@gmail.com
Biruk Abrham, ESSTI, Ethiopia	bab3590@gmail.com
Admasu Abwari, Addis Ababa University, Ethiopia	
Etsegenet Alemu, UCT/SAAO, South Africa and ESSTI, Ethiopia	tsegiastro@gmail.com
Abdu Ali Amede, Woldia University, Ethiopia	abduali2009@gmail.com
Viola Allevato, Scuola Normale Superiore, Italy	viola.allevato@sns.it
Amirnezam Amiri, IPM, Iran	amirnezamamiri@gmail.com
Fentanesh Anley, Addis Ababa University, Ethiopia	
Wubsera Anley Yihunie, Woldia University, Ethiopia	wubseraanley@gmail.com
Ghion Ashenafi Getahun, ESSTI, Ethiopia	astroghi@gmail.com
Kumera Assefa, Addis Ababa University, Ethiopia	assefakume@gmail.com
Bidisha Bandyopadhyay, Universidad de Concepción, Chile	bidisharia@gmail.com
Paramita Barai, Universidade de São Paulo, Brasil	paramita.barai@iag.usp.br
Dalya Baron, Tel Aviv University, Israel	dalyabaron@gmail.com
Meron Bekabil, Wolkite University, Ethiopia	meronbekabil@gmail.com
Solomon Belay Tessema, ESSTI, Ethiopia	tessemabelay@gmail.com
Dugasa Belay Zeleke, ESSTI, Ethiopia	dugasa32@gmail.com
Marco Berton, University of Turku, Finland	marco.berton@utu.fi
Fitsum Beyene Woldegerima, Wachamo University, Ethiopia	fitsewgerima@gmail.com
Zeleke Beyoro Amado, ESSTI, Ethiopia	zbamado@gmail.com
Betelehem Bilata, ESSTI, Ethiopia	betelehem.bilata@gmail.com
Tolu Biressa, Jimma University, Ethiopia	tolu_biressa@yahoo.com
Sena Bokona, Jimma University, Ethiopia	nafyadibokona450@gmail.com
Angela Bongiorno, INAF-Observatory of Rome, Italy	angela.bongiorno@inaf.it
Silvia Bonoli, Donostia International Physics Center, Spain	sb.bonoli@gmail.com
William N. Brandt, Pennsylvania State University, USA	wnbrandt@gmail.com
De-Fu Bu, Shanghai Astronomical Observatory, China	dfbu@shao.ac.cn
Gabriela Calistro Rivera, Leiden Observatory/ESO, The Netherlands	gcalistrorivera@gmail.com
Sara Cazzoli, IAA-CSIC, Spain	sara@iaa.es
Avinanda Chakraborty, Presidency University, India	avinanda94@gmail.com
Ranga-Ram Chary, IPAC/Caltech, USA	rchary@caltech.edu
Liang Chen, Shanghai Astronomical Observatory, China	chenliang@shao.ac.cn
Chiara Circosta, ESO, Germany	ccircost@eso.org
Francoise Combes, Observatoire de Paris/LERMA, France	francoise.combes@obspm.fr
Jean Damascene Mbarubucyeye, University of Rwanda, Rwanda	mbjdamas@gmail.com
Yusuf Debela, Gunfi Secondary School, Ethiopia	yusufdebela2017@gmail.com
Colin DeGraf, University of Cambridge, United Kingdom	cdegraf@ast.cam.ac.uk
Terefe Demessa, Jimma University, Ethiopia	terefedemessa85@gmail.com
Obini Ekpe Ekpe, Michael Okpara University, Nigeria	ekpe.obini@mouau.edu.ng
Endalamaw Ewnu Kassa, Woldia University, Ethiopia	alexewnu4@gmail.com
Sisay Fantahun, ESSS, Ethiopia	sisayfantahun4@gmail.com
Lamessa Fekede, Dambe Dollo University, Ethiopia	lamefeke@gmail.com
Getinet Feleke, Kotebe Metropolitan University, Ethiopia	getinet13@gmail.com
Daniel Fikadu, ESSTI, Ethiopia	danielfikadu@gmail.com
Balina Galata, Goverment Preparatory School, Ethiopia	balinagalata07@gmail.com
Rubén García-Benito, IAA-CSIC, Spain	rgb@iaa.es
Haregeweyn Gashaw, ESSS, Ethiopia	asegedgashaw@gmail.com
Asrate Gaulle Weldegebreal, Dilla University, Ethiopia	asrieguale@gmail.com
Getnet Gebereegziabher, ESSTI, Ethiopia	getnetg1@gmail.com
Joseph Gelfand, New York University Abu Dhabi, United Arab Emirates	jg168@nyu.edu
Tilahun Getachew Woreta, ESSTI, Ethiopia	tilahun85@gmail.com
Omaira González Martín, IRyA/UNAM, Mexico	o.gonzalez@irya.unam.mx

List of Participants

Name & Institution	Email
Leonid Gurvits, JIVE & Delft University of Technology, The Netherlands	lgurvits@jive.eu
Chris Harrison, Newcastle University, United Kingdom	christopher.harrison@newcastle.ac.uk
Laura Hermosa Muñoz, IAC - CSIC, Spain	lhermosa@iaa.es
Luis Ho, Kavli Institute for Astronomy and Astrophysics, China	lho.pku@gmail.com
Seblu Humne Negu, ESSTI, Ethiopia	seblu1557@gmail.com
Dragana Ilić, University of Belgrade, Serbia	dilic@matf.bg.ac.rs
Chris Impey, University of Arizona, USA	cimpey@as.arizona.edu
Emilia Järvelä, University of California, USA	ejarvela@ucsb.edu
Miranda Jarvis, ESO, Germany	mjarvis@eso.org
Noorali Jiwaji, Open University of Tanzania, Tanzania	ntjiwaji@yahoo.com
Seid Kasim Geriyo, Werabe University, Ethiopia	seidkasimgeriyo2018@gmail.com
Shai Kaspi, Tel Aviv University, Israel	shai@wise.tau.ac.il
Naftali Kimani, Kenyatta University, Kenya	kimani.naftali@ku.ac.ke
Johan Knapen, IAC, Spain	johan.knapen@iac.es
Jari Kotilainen, University of Turku, Finland	jarkot@utu.fi
Michael Kreter, North West University, South Africa	michael@kreter.org
Gemechu Kumssa, ESSTI, Ethiopia	gemechumk@gmail.com
Lerothodi Leeuw, UNISA, South Africa	Lerothodi@alum.mit.edu
Wogayehu Legese, ESSTI, Ethiopia	wogmet@gmail.com
Fraol Lenjisa, ESSTI, Ethiopia	fraol.lenjisa@gmail.com
Shuang-Liang Li, Shanghai Astronomical Observatory, China	lisl@shao.ac.cn
Paulina Lira, Universidad de Chile, Chile	plira@das.uchile.cl
Xiang Liu, CAS, China	liux@xao.ac.cn
Antoine Mahoro, SAAO, South Africa	antoine@saao.ac.za
Jaya Maithil, University of Wyoming, USA	jmaithil@uwyo.edu
Alemiye Mamo Yacob, ESSTI, Ethiopia	malemiye@gmail.com
Allison Man, Dunlap Institute, University of Toronto, Canada	allison.man@dunlap.utoronto.ca
Feven Markos, ESSTI, Ethiopia	fevenmarkos06@gmail.com
Paola Marziani, Observatory of Padova-INAF, Italy	paola.marziani@inaf.it
Daudi Mazengo, University of Dodoma, Tanzania	mazengod@yahoo.co.uk
Kirubel Menberu, ESSS, Ethiopia	kirubelmenberu5@gmail.com
Alene Mitiku, ESSTI, Ethiopia	aleneseyoum4@gmail.com
Abduselam Mohammed, Woldia University, Ethiopia	abdiphy@gmail.com
Nebiyu Mohammed, Space Generation Advisory Council/ESSS, Ethiopia	hellonebo@gmail.com
Nuru Mohammed, Woldia University, Ethiopia	rawuda1989@gmail.com
Raffaella Morganti, ASTRON/Kapteyn Institute, The Netherlands	morganti@astron.nl
Martin Mule Mutie, Technical University of Kenya, Kenya	martmulesh@gmail.com
Dorcus Mulumba, Kenyatta University, Kenya	nthokidorcus@gmail.com
Blessing Musiimenta, MUST, Uganda	mblessing78@gmail.com
Hagai Netzer, Tel Aviv University, Israel	hagainetzer@gmail.com
Robert Nikutta, NSF's NOIRLab, USA	nikutta@noao.edu
Pheneas Nkundabakura, University of Rwanda, Rwanda	nkundapheneas@yahoo.fr
Beatrice Nyiransengiyumva, University of Rwanda, Rwanda	beatny1990@gmail.com
Victor Oknyansky, SAI MSU, Russia	oknyan@mail.ru
Ascensión del Olmo, IAA-CSIC, Spain	chony@iaa.es
Tom Oosterloo, ASTRON/Kapteyn Institute, The Netherlands	oosterloo@astron.nl
Swayamtrupta Panda, Center for Theoretical Physics, Nicolaus Copernicus Astronomical Center, Poland	panda@cft.edu.pl
Clara Marie Pennock, Keele University, United Kingdom	c.m.pennock@keele.ac.uk
Mirjana Pović, ESSTI, Ethiopia	mpovic@iaa.es

List of Participants

Name & Institution	Email
Almudena Prieto, IAC, Spain	aprieto@iac.es
Andreas Quirrenbach, Landessternwarte, University of Heidelberg, Germany	A.Quirrenbach@lsw.uni-heidelberg.de
Venkatessh Ramakrishnan, Universidad de Concepción, Chile	vramakrishnan@udec.cl
Solohery Randriamampandry, SAAO, South Africa	soloherymampionona@gmail.com
Zara Randriamanakoto, SAAO, South Africa	zara@saao.ac.za
Remudin Reshid Mekuria, Addis Ababa University, Ethiopia	remudin.mekuria@gmail.com
Susan Ridgway, NOAO, USA	ser@noao.edu
Khatun Rubinur, National Centre for Radio Astrophysics - Tata Institute of Fundamental Research, India	rubi.khatun35@gmail.com
Shambel Sahlu, ESSTI, Ethiopia	sahlushambel@gmail.com
Eleonora Sani, ESO, Chile	esani@eso.org
Biny Sebastian, NCRA-TIFR, India	biny@ncra.tifr.res.in
Alazar Seyoum, ESSTI, Ethiopia	alazarseyoum12@gmail.com
Prajaval Shastri, Indian Institute of Astrophysics, India	prajval.shastri@gmail.com
Prospery Simpemba, SA-ROAD, Copperbelt University, Zambia	pcs200800@gmail.com
Samuel Siyum, Woldia University, Ethiopia	samisiym@gmail.com
Luigi Spinoglio, IAPS-INAF, Italy	luigi.spinoglio@inaf.it
Eva Sramkova, Silesian University in Opava, Czech Republic	sram_eva@centrum.cz
Marko Stalevski, Astronomical Observatory of Belgrade, Serbia	marko.stalevski@gmail.com
Sebhat Tadesse, ESSTI, Ethiopia	sebhattad@gmail.com
Charles Takalana, Department of Science and Innovation, South Africa	charles.takalani@dst.gov.za
Jerusalem Tamirat Teklu, ESSTI, Ethiopia	jerrytamirat21@gmail.com
Abdissa Tassama, Jimma University, Ethiopia	yomiyuglt@gmail.com
Biruk Terefe, ESSS, Ethiopia	b.boo.tf@gmail.com
Beza Tesfaye, ESSS, Ethiopia	zionbez@gmail.com
Habtamu Tesfaye, Woldia University, Ethiopia	habtamutesfaye95@gmail.com
Ephrem Teshome Bogale, Addis Ababa University, Ethiopia	Ephphy@gmail.com
Nuredin Teshome Abegaz, ESSTI, Ethiopia	
Negessa Tilahun, ESSTI, Ethiopia	nagessa2006@gmail.com
Francesco Tombesi, University of Rome, Italy	francesco.tombesi@roma2.infn.it
Benny Trakhtenbrot, Tel Aviv University, Israel	benny.trakht@gmail.com
Halima Ugomma Obini, Michael Okpara University, Nigeria	obiniekpe@yahoo.com
Emmanuel Uwitonze, MUST, Uganda	uwitonze_emmanuel@yahoo.com
Getachew Wollel, ESSTI, Ethiopia	g.w.tiru@gmail.com
Petri Väisänen, SAAO, South Africa	petri@saao.ac.za
Clare Wethers, University of Turku, Finland	clweth@utu.fi
Sarah White, SARAO/Rhodes University, South Africa	sarahwhite.astro@gmail.com
Maciek Wielgus, Black Hole Initiative at Harvard University, United States	maciek.wielgus@gmail.com
Hartmut Winkler, University of Johannesburg, South Africa	hwinkler@uj.ac.za
Dejene Woldeyes, Universidad Diego Portales, Chile	dzewdie12@gmail.com
Wudu Worku, ESSTI, Ethiopia	wuduwork@gmail.com
Fu-Guo Xie, Shanghai Astronomical Observatory, China	fgxie@shao.ac.cn
Ayenew Yehualaw Gebreyes, Kotebe Metropolitan University, Ethiopia	yehualawayenew@gmail.com
Hassen Yesuf, Kavli PKU/IPMU, China	myesuf@pku.edu
Ingyin Zaw, New York University Abu Dhabi, United Arab Emirates	iz6@nyu.edu
Feleke Zerihun, ESSTI, Ethiopia	felekez@essti.gov.et

Plus the ESSTI and ESSS supportive staff members.

List of Participants

We would like to give our thanks to all participants for being a part of the IAUS 356.

Figure 1. Group photos from the opening session, National Museum visit, Entoto Observatory visit, and conference dinner. *Credits. A. Solomon and V. Oknyansky.*

Figure 2. IAUS 356 photos. *Credits. A. Solomon*

Scientific program overview

IAUS 356 hosted in total 7 Sessions on different aspects of AGN physics and observations. Each Session consisted of one invited talk, number of contributed talks, and discussion. Each Session is related with one book Chapter and is briefly described here.

Session 1. Multiwavelength AGN surveys: past, present, and future

This session focuses on photometric, spectroscopic and polarimetric observations of AGN over the entire electromagnetic spectrum including past and present multi-wavelength surveys. Expectations from future missions are also discussed.

Session 2. AGN types and unification

Classification schemes in different wavelength bands with emphasis on new results from MIR, FIR and X-ray surveys and comparison between optical and radio classification are discussed. The session also includes new ideas about type 1, type 2 and type 1.9 AGN and discussion about torus observations and models.

Session 3. Variability

This session includes broad band and spectroscopic studies of variable AGN. The discussion is focused on what we know about the duration and origin of variations in different parts of SED, and on new theoretical improvements. We will address the status of reverberation mapping and its use for determining black hole mass, and new observations and theoretical models regarding continuum variations and accretion disk properties. The session includes also a discussion about the role of large facilities (e.g., eROSITA, LSST, and SKA) in variability studies.

Session 4. Active black holes and their host galaxies

New observational, theoretical, and numerical results about the connections between SMBHs and their host galaxies are presented. This includes models of SMBH growth, and stellar mass growth via star formation, at different epochs, and the various scaling relationships between the two. AGN morphology and types with regard to the growth scenarios, AGN in blue and red galaxies, and the role of environment, are also part of this session.

Session 5. Triggering, feedback and shutting off AGN activity

The status of AGN theory and numerical simulations regarding triggering of AGN activity, and shutting-off black hole accretion and star formation via AGN feedback is discussed. This includes major and minor galaxy mergers, secular evolution and feeding from the halo. Special attention is given to differences in those mechanisms at different epochs.

Session 6. Jets and environment

New observational, theoretical, and numerical results regarding relativistic and non-relativistic jets are presented. This includes modeling jet morphologies, propagation, and stability and the origin of AGN jets. We aim to understand better the evolution of astrophysical jets at different redshifts and the influence of jets on their environment.

Session 7. The Highest redshift AGN and AGN evolution

AGN evolution from the early seed black holes until the present time is discussed. Observations and modeling of the first AGN and plans to observe such objects at redshifts larger than 7.5 - the present record.

Invited and contributed talks were supported with significant number of poster presentations that in this book are presented under Chapter 8 (AGN posters). Finally, we had a significant number of poster presentations out of the field of AGN that are given in Chapter 9 (non-AGN posters).

Summary of activities organised during the symposium

We are pleased to introduce here different outreach and education activities that have been carried out along the IAU 356 symposium for benefiting our society. These activities would not be possible without the IAU and ESSTI, all our sponsors, IAUS 356 participants who voluntarily did education and outreach activities, SOC and LOC members, ESSS members and volunteers, and active participants who directly benefited from all activities. Huge efforts have been made by all parties to organise all activities summarised below. We are very grateful for all received help and support, and we deeply believe in importance that such activities can have for science and education development and long-term benefits of our society.

Training for MSc/PhD students and young researchers (5–6 of October, 2019)

This training was organised before the IAU 356 symposium, and was held in one of the halls of Addis Ababa University in collaboration with Physics Department. Its aim was to improve the skills of our MSc/PhD students and young researchers who are already attached to different universities and research centres. We had 45 participants in total, from different African countries, including Ethiopia (great majority of participants), Kenya, Nigeria, Rwanda, South Africa, Tanzania, and Uganda. In case of Ethiopian participants we supported colleagues coming from more than 10 different universities across the country. The training was planned for benefiting broader community, and not necessarily people involved in astronomy. Therefore, one part of the training (10 hours) was focused on *'Introduction to python'* programming given by Dr. Rubén García Benito (IAA-CSIC, Spain), while another one (6 hours) was related with tips on *'CV, motivation letter, proposals, and research papers writing'* given by Dr. Allison Man (University of Toronto, Canada) and Dr. Johan Knapen (IAC, Spain). Colleagues from astronomy, space physics, physics in general, and engineering attended the training. We got very positive feedback from both participants and trainers. Our appreciation goes to all facilitators, without whom this training would not be possible. Picture from training is shown in Fig. 1 (left plot).

Public talks and outreach activities at schools (8 of October, 2019)

During one afternoon we organised in collaboration with the Ethiopian Space Science Society (ESSS) 2 public talks at Addis Ababa University, given by Prof. Christopher Impey (University of Arizona, USA) about *'Black holes and nuclear activity in galaxies'* and Prof. Petri Vaisanen (SAAO, South Africa) about *'South African astronomy in 2020s'*. In parallel, outreach activities were organised in 8 public schools where more than 40 symposium participants interacted with our primary and secondary school children and their teachers, and promoted during several hours astronomy and science. In total, with all activities we reached between 700 and 800 school children, students, and general public. We are grateful to all facilitators for their volunteer work done during all activities. Attached in Fig. 1 (right plot) and Fig. 2 are the poster announcing public talks and several pictures from the outreach events in schools.

Figure 1. Left: Group picture from MSc/PhD students and young researchers training. *Credits: M. Pović.* **Right:** Poster with announced public talks. *Credits: ESSS.*

Figure 2. Top left: Group picture from AAU during public talks. *Credits: M. Pović.* **Top right and bottom:** Outreach activities in schools. *Credits: S. Ridgway, C. Harrison, and Z. Beyoro-Amado, respectively.*

The Astro Data Lab - An Open-Data, Open-Access Science Platform (10 of October, 2019)

During the afternoon of 10 of October, after all talks, practical session was organised for our MSc/PhD students and other IAUS participants on *'The NOAO Data Lab - An Open-Data, Open Access Science Platform'* given by Dr. Robert Nikutta (NOIRLab, USA). We are very thankful to Robert for his support and time.

Lunch session on Astronomy in Africa (10 of October, 2019)

In parallel with the IAUS 356, there was a 2 days scientific-business meeting of recently re-established African Astronomical Society (AfAS), on 10–11 of Oct. We used this opportunity to organise a lunch session for presenting AfAS to our participants and for discussing about the status of astronomy in Africa. We had a very fruitful discussion with approximately 70 participants attending the session. Few pictures are attached in Fig. 3.

Figure 3. Left: NOAO data Lab training. *Credits: M. Pović.* **Middle:** Kevin Govender, Office of Astronomy for Development (IAU-OAD) Director chairing 'Astronomy in Africa' session. *Credits: A. Solomon.* **Right:** participants attending 'Astronomy in Africa session'. *Credits: A. Solomon.*

Training in practical astronomy for teachers and teachers trainers (12–13 of October, 2019)

This training was carried out after the symposium at the ESSTI for 44 participants, who were public school teachers, teacher trainers, and ESSTI and ESSS staff members and volunteers actively involved in astronomy/science outreach activities in Ethiopia. We also had 2 colleagues from Tanzania and Zambia as participants. On Saturday morning the training was given by 4 ESO colleagues who developed and brought different outreach materials that were used during the workshop. We are very much grateful to Dr. Chris Harrison, Ms. Miranda Jarvis, Dr. Gabriela Calistro Rivera, and Dr. Chiara Circosta for giving the training and for donating all their materials to the ESSTI. We are also grateful to Prof. Prajval Shastri who spent Saturday morning with us and shared with teachers some of her experiences.

Saturday afternoon and Sunday the training was done in collaboration with the Network for Astronomy School Education (NASE), where experiments have been constructed using recycled and easily accessible materials for showing different physical and astrophysical laws. Three workshops were given by Dr. Mirjana Pović (ESSTI), related with solar spectrum and light, stellar properties and stellar lives, and expansion of the universe. Introduction to the virtual planetarium software 'Stellarium' and stargazing was given by Mr. Alemiye Mamo (ESSTI). Few pictures from the training can be seen in Fig. 4.

Figure 4. Pictures from school teachers training. *Credits: A. Mamo and M. Pović.*

INTRODUCTION

INTRODUCTION

Nuclear Activity in Galaxies Across Cosmic Time
Proceedings IAU Symposium No. 356, 2019
M. Pović, P. Marziani, J. Masegosa, H. Netzer, S. H. Negu & S. B. Tessema, eds.
doi:10.1017/S174392132000246X

Development in astronomy in Ethiopia and East-Africa through nuclear activity in galaxies

Mirjana Pović[1,2]

[1] Astronomy and Astrophysics Research and Develoment Division, Etiopian Space Science and Technology Institute (ESSTI), Addis Ababa, Ethiopia

[2] Instituto de Astrofísica de Andalucía (IAA-CSIC), Granada, Spain

Abstract. In this paper we summarise the research that is currently going on in Ethiopia and East-Africa in extragalactic astronomy and physics of active galaxies and active galactic nuclei (AGN). The study is focused on some of the still open questions such as: what are the stellar ages and populations of ultra hard X-ray detected AGN and connection between AGN and their host galaxies?, what are the properties of AGN in galaxy clusters and the role that environment has in triggering nuclear activity?, what are the morphological properties of AGN and how precisely we can deal with morphological classification of active galaxies?, what are the properties of galaxies in the green valley and the role of AGN in galaxy evolution?, and what are the properties of radio-loud and radio-quiet quasars (QSO) and dichotomy between the two?. Each of these questions has been developed under one specific project that will be briefly introduced. These projects involve 6 PhD and 3 MSc students and collaborations between Ethiopia, Rwanda, South Africa, Uganda, Tanzania, Spain, Italy, and Chile. With all projects we aim: first, to contribute to our general knowledge about AGN, and second, to contribute to the development in astronomy and science in Ethiopia and East-Africa.

Keywords. galaxies: active; galaxies: main properties; astronomy development in Africa

---

1. Introduction

How galaxies form and evolve is still one of the fundamental questions in modern cosmology. Many properties of different types of galaxies are still fairly understood, especially when going to higher redshifts. In particular, active galaxies, having active galactic nuclei (AGN) in their centre, play an important role in understanding galaxy formation and evolution, supermassive black hole (SMBH) formation and growth, star formation (SF) in galaxies, morphological transformation and role that AGN may have in moving galaxies from late- to early-types, or how the Universe was in its early stage (e.g., Heckman & Best 2014; Netzer 2015; Hickox & Alexander 2018). In the following we describe briefly several projects that are related with some of the still open questions in the field of AGN, and in the same time are important for human capacity building and astronomy development in Ethiopia and East-Africa.

2. Stellar ages and morphologies of ultra-hard AGN

Connection between SF and AGN activity was studied widely over the past years, and shown to be very important for understanding the role of AGN in galaxy evolution (e.g., Pović *et al.* 2013; Shimizu *et al.* 2017; Masoura *et al.* 2018). What are the stellar ages and average stellar populations of AGN host galaxies, and if there are differences depending on AGN type, are still some of open questions. The AGN sample detected in the ultra-hard

© The Author(s), 2021. Published by Cambridge University Press on behalf of International Astronomical Union

X-rays (14 - 195 keV) by the Swift BAT telescope is not affected by obscuration nor it is contaminated by stellar emission, and therefore presents some of the most unbiased samples. Therefore, the Swift-BAT AGN Spectroscopic Survey (BASS†; Koss et al. 2017) gives us an unique opportunity to understand connection between AGN and their host galaxies by studying the SMBH mass and accretion rate (Koss et al. 2017; Ricci et al. 2017) in relation to stellar properties, metallicities, and morphologies of AGN hosts. It is for the first time that this kind of study will be carried out for a complete sample of ultra-hard X-ray detected AGN.

Using the optical spectra and BASS DR1 data (Koss et al. 2017) we are carrying out spectral energy distribution (SED) fitting by using the STARLIGHT code (Cid Fernandes et al. 2004), with aim to study stellar ages and populations of ultra-hard X-ray AGN. For fitting the type-2 AGN, we are following the same procedure as in (Pović et al. 2016), and testing templates with different metallicities and stellar ages Bruzual & Charlot (2003). Using the obtained emission spectra we are measuring properties of all emission lines (integrated fluxes and equivalent widths), again using the same methodology as in (Pović et al. 2016). Metallicities will be measured using the new Bayesian-like approach of Pérez-Montero et al. (2019) that has been tested on type-2 AGN. This study will be combined by morphological analysis of the same sample. Visual multiwavelength classification in optical, radio, and X-rays was carried out (see Bilata-Woldeyes et al. 2020, poster paper in Chapter 8), where we obtained that most of ultra hard X-ray detected AGN are hosted by spirals in optical, are radio-quiet, and have compact morphologies in X-rays. We are finally planning to study in more details how/if X-ray luminosities, SMBH masses, and accretion rates are correlated with obtained stellar ages and populations, metallicities, and morphology, for understanding better connection between the ultra-hard AGN and their host galaxies.

This project is carried out as a collaboration between Ethiopia and Spain. Morphological analysis resulted in MSc degree of Betelehem Bilata, one of our few female students.

3. Properties of galaxies in galaxy clusters up to z ∼ 1.0

The study of properties of galaxies inside galaxy clusters represents one of the main steps in understanding the formation and evolution of the Universe. In particular, it is important to understand how galaxies transform inside the clusters and how they change their properties as a function of redshift and environment. The research case proposed here has been carried out at the ESSTI under the GLACE collaboration (GaLAxy Cluster Evolution survey; Sánchez-Portal et al. 2015), with general aim to study the properties and evolution of galaxies in galaxy clusters up to z ∼ 1. Our main objective is to better understand some of the still open questions such as: the role of AGN in galaxy clusters, metallicity variability, and galaxy transformation and evolution within clusters at different cosmic times.

Two GLACE clusters, RXJ1257.2+4738 at z = 0.866 and ZwCl0024.0+1652 at z = 0.395, have been analysed using tunable filters (TF) data available from the OSIRIS instrument at the GTC 10m telescope and public data. For the RXJ1257 cluster we carried out morphological classification and analysis (Pintos-Castro et al. 2016) using the non-parametric methods based on the galSVM code (Huertas-Company et al. 2008). Regarding ZwCl0024 cluster, emission line galaxies have been previously selected and analysed in H_α and [NII] lines (Sánchez-Portal et al. 2015). We carried out morphological classification of galaxies in ZwCl0024, in a consistent way as in RXJ1257, by classifying all galaxies as early- or late-types, and obtaining the most detailed catalogue

† https://www.bass-survey.com/

up to date up to a clustercentric distance of 1 Mpc (Beyoro-Amado et al. 2019). Data reduction and analysis of H$_\beta$ and [OIII] lines in the ZwCl0024 cluster has been finalised. Obtained emission line fluxes and luminosities have been tested and compared with those of H$_\alpha$ and [NII] (Beyoro-Amado et al. 2020, in prep). We obtained pseudo-spectra of both lines and selected possible emission line galaxies after inspection of pseudo-spectra and HST/ACS images. We are studying the nuclear activity in this cluster using the BPT diagram and 4 emission lines from TF observations. Metallicity estimates will be carried out using [NII] line, SFR using H$_\alpha$ line, and extinctions using H$_\alpha$ and H$_\beta$ lines. All properties including morphology will be analysed in relation to local density and clustercentric distance. Finally, we will provide a more global analysis where properties measured for RXJ1257 and ZwCl0024 clusters will be compared with those obtained for local Virgo cluster using public data. For more information about this work see Beyoro-Amado et al. (2020) paper in Chapter 4.

This project is related to PhD thesis of Zeleke Beyoro-Amado, and forms a part of collaboration between Ethiopia and Spain.

4. Morphological properties of active galaxies

Morphology is the most accessible indicator of galaxy physical structure, being crucial for understanding the formation of galaxies throughout cosmic time and for providing answers to some of still open questions mentioned above. In case of active galaxies there are still many inconsistencies between the results obtained and their interpretation regarding morphology (e.g., Pierce et al. 2007; Georgakakis et al. 2008; Gabor et al. 2009; Pović et al. 2009, 2012; Mahoro et al. 2019), and how the AGN affects morphological classification of its host galaxies (Gabor et al. 2009; Cardamone et al. 2010; Pierce et al. 2010). In particular, the interpretation of morphology still remains a problem in the framework of galaxy evolution. Since measured morphology depends strongly on image resolution, classification in deep surveys remains very difficult, especially when dealing with faint and high redshifts sources (Pović et al. 2015). Still detailed morphological analysis is missing in the field of AGN, especially at higher redshifts.

In this work we are carrying out a detailed study on how the AGN contribution affects morphological classification of active galaxies. To do so, we are applying a similar method as used in Pović et al. (2015). Using a local sample of $\sim$ 2000 visually classified SDSS inactive galaxies (Nair et al. 2010) we simulated their images by adding in the centre different AGN contributions from 5% up to 75% of the total flux. By running the galSVM code (Huertas-Company et al. 2008), we are measuring six commonly used morphological parameters and are evaluating how well the morphological classification of active galaxies can be determined at z $\sim$ 0 and at higher redshift in COSMOS conditions (Getachew et al. 2020a,b, in prep.). Secondly, we are planning to estimate the possible AGN contribution of COSMOS galaxies by using GALFIT code (Peng et al. 2002). Finally, taking into account the previous results we will re-classify active galaxies in the COSMOS field and understand better how they evolved across cosmic time. For more information see Getachew-Woreta et al. (2020) paper in Chapter 4.

This project is related to PhD thesis of Tilahun Getachew-Woreta, and is carried out between Ethiopia and Spain.

5. Properties of green valley galaxies and role of AGN in galaxy evolution

On the colour-magnitude or colour-stellar mass diagrams green valley (GV) galaxies are located between the 'red sequence' and 'blue cloud' which are mainly populated by early- and late-type galaxies, respectively (e.g., Pović et al. 2013; Schawinski et al. 2014; Salim 2014; Bremer et al. 2018). Different studies suggested that morphological transformation of galaxies happens in the GV during different timescales (Schawinski et al. 2014; Smethurst et al. 2015; Trayford et al. 2016; Bremer et al. 2018), being also dependent on morphology (Nogueira-Cavalcante et al. 2018). Study of GV galaxies is therefore crucial for understanding the process of SF quenching, how galaxies transform from late- to early-types, and what is the role of AGN in morphological transformation of galaxies and their evolution.

To understand better the properties of GV sources and connection between inactive and active galaxies we selected a large sample of sources and studied their properties such as SFRs, stellar masses, sizes, morphologies, stellar populations, and ages. We are carrying out this study at low, intermediate, and higher redshifts using public SDSS, GAMA, and COSMOS data, respectively. In addition, for a smaller number of sources we are using our own spectroscopic data from SALT telescope. We measured SFRs in COSMOS through the FIR Herschel/PACS data and SED fitting, and we observed the location of active and inactive galaxies on the main-sequence (MS) of SF. We found that most of our GV X-ray detected AGN with far-IR emission have SFRs higher than the ones of inactive galaxies at fixed stellar mass ranges. Therefore, they do not show signs of SF quenching, as shown in most of previous optical studies (e.g., Nandra et al. 2007; Pović et al. 2012; Leslie et al. 2016), but rather its enhancement. Our results may suggest that for X-ray detected AGN with FIR emission if there is an influence of AGN feedback on SF in the GV the scenario of AGN positive feedback seems to take place, rather than the negative one (Mahoro et al. 2017). In the second paper published recently, we studied morphological properties and found that a significant number of our AGN (38%), but not the majority, are related to interactions and mergers (Mahoro et al. 2019). At low redshifts using SDSS optical and GALEX UV data we analysed six criteria commonly used for selecting GV galaxies. We observed that depending on criteria different population of galaxies are selected, bringing therefore to different GV results (Nyiransengiyumva et al. 2020, in prep.). For understanding better the role of AGN in SF quenching in the GV, we are now measuring stellar ages, average stellar populations, and metallicities of selected active and inactive galaxies at both lower and higher redshifts. For more information see papers of Mahoro et al. (2020) and Nyiransengiyumva et al. (2020) in Chapter 4.

This project is related to PhD dissertations of Antoine Mahoro and Betrice Nyiransengiyumva from Rwanda, and is carried out as a collaboration between Rwanda, Ethiopia, South Africa, Uganda, and Spain.

6. Dichotomy of radio-loud and radio-quiet quasars

Quasars (QSOs) were discovered more than 50 years ago. Being some of the most luminous sources in the Universe they are fundamental for cosmological studies (see paper of Marziani et al. 2020, in Chapter 2). Most of the QSOs in the local universe are radio-quiet (RQ). It is still under debate if there is an evidence for a continuity or physical dichotomy between RL and RQ QSOs, what is the origin of the powerful relativistic jets, and the effect that they have on the surrounding medium. Previous studies suggested the existence of two Populations (A and B) of QSOs in the 4D Eigenvector 1 (4DE1) plane defined by the FWHM of Hβ and the strength of the optical FeII blend at 4570Å normalised by the intensity of Hβ line (RFe), and the existence of 'QSO main sequence' (Sulentic et al. 2000; Shen & Ho 2014; Marziani et al. 2018). It has been shown that RL QSOs are strongly concentrated in the Population B (Zamfir et al. 2008). In this work we are analysing the dichotomy of RL and RQ QSOs and the effect of the relativistic radio jets on the gas in the broad line emission region.

We are using our spectroscopic data from CAHA 3.5m and GTC 10m telescopes of ~ 60 RL QSOs. We want to quantify broad emission lines differences between RL and RQ sources, exploiting larger and more complete samples of QSOs with spectral coverage in Hβ, FeII, MgII and CIV emission lines than it was done previously. We are using SPECFIT code to determine the main parameters of each component. This will allow us to verify whether the larger values of FWHM(Hβ) among RL sources can be due to orientation, and study the wind properties affecting the profiles of the CIV and MgII lines. Currently we are analysing a sample of 12 RL QSOs using our CAHA spectra. We will focus our comparisons on RQ and RL sources that have the same mass and L/L$_{Edd}$ ranges. For more information see Terefe et al. (2020) paper in Chapter 8.

This project is related with PhD thesis of Shimeles Terefe from Ethiopia, and is carried out in collaboration between Ethiopia, Spain, and Italy.

7. Conclusions

This paper describes some of the main projects that are running under the extragalactic astronomy group at the ESSTI. Several other works have been conducted related with: 'Properties of inside-out assembled galaxies at z < 0.1' (see Zewdie et al. 2020a,b, in Chapter 8), 'Testing the alternative method to measure the accretion rate in galaxies' (Gaulle et al. 2020, in Chapter 8), and 'Characterisation of LINERs and retired galaxies at z < 0.1' (Mazengo et al. 2020, in Chapter 8). These three projects resulted in three MSc dissertations in Ethiopia and Tanzania. Beside contributing to our general knowledge about galaxies and AGN, all mentioned projects contributed to development in astronomy and science in Ethiopia, Rwanda, Uganda, South Africa, and Tanzania. We managed to give more visibility to astronomy in Africa, to contribute to the institutional development of ESSTI and partner institutions, and to strengthen international collaborations. We are contributing to human capacity building of our first MSc and PhD students in the field, and we are also inspiring many other young people who went through different trainings and education and outreach activities that have been organised in Ethiopia and East-Africa over the past recent years.

Acknowledgements

Support of the Ethiopian Space Science and Technology Institute (ESSTI) under the Ethiopian Ministry of Innovation and Technology (MInT), Spanish MEC under AYA2016-76682-C3-1-P and Center of Excellence Severo Ochoa award for the IAA (SEV-2017-0709) are gratefully acknowledged. This proceedings paper would not be possible without the IAUS 356 support of the IAU, ESSTI, EORC, MInT, ISP, IAA-CSIC, SEA, STFC-UKRI, DARA, ESSS, EA-ROAD, AAU, and Nature Astronomy.

References

Beyoro-Amado, Z. et al. 2019, MNRAS, 485, 1528
Beyoro-Amado, Z. et al. 2020, IAUS 356 proceedings, Cambridge University Press, in press (arXiv:2004.01892)
Bilata-Woldeyes, B. et al. 2020, IAUS 356 proceedings, Cambridge University Press, in press (arXiv:2003.12416)
Bremer, M. N. et al. 2018, MNRAS, 476, 12
Bruzual, G. & Charlot, S. 2003, MNRAS, 344, 1000
Cardamone, C. N. et al. 2010, ApJ, 721, L38
Cid Fernandes, R. et al. 2004, ApJ, 605, 105
Gabor, J. M. et al. 2009, ApJ, 691, 705
Gaulle, A. et al. 2020, IAUS 356 proceedings, Cambridge University Press, in press (arXiv:2003.13487)
Georgakakis, A. et al. 2008, MNRAS, 385, 2049
Getachew-Woreta, T. et al. 2020, IAUS 356 proceedings, Cambridge University Press, in press (arXiv:2004.02250)

Heckman, T. M. & Best, P. N. 2014, *ARA&A*, 52, 589
Hickox, R. C. & Alexander, D. M. 2018, *ARA&A*, 56, 625
Huertas-Company, M. *et al.* 2008, *A&A*, 478, 971
Ilbert, O. *et al.* 2016, *ApJ*, 690, 1236
Koss, M. *et al.* 2017, *ApJ*, 850, 74
Leslie, S. K. *et al.* 2016, *MNRAS*, 455, 82
Mahoro, A., Pović, M., & Nkundabakura, P. 2017, *MNRAS*, 471, 3226
Mahoro, A. *et al.* 2019, *MNRAS*, 485, 452
Mahoro, A. *et al.* 2020, IAUS 356 proceedings, Cambridge University Press, in press (arXiv:2003.12033)
Marziani, P. *et al.* 2018, *FrASS*, 5, 6
Marziani, P. *et al.* 2020, IAUS 356 proceedings, Cambridge University Press, in press (arXiv:2002.07219)
Masoura, V. A. *et al.* 2018, *A&A*, 618, 31
Mazengo, D. *et al.* 2020, IAUS 356 proceedings, Cambridge University Press, in press
Nair, P. B. & Abraham, R. G. 2010, *ApJS*, 186, 427
Nandra, K. *et al.* 2007, *ApJ*, 660, 11
Netzer, H. 2015, *ARA&A*, 53, 365
Nogueira-Cavalcante, J. P. *et al.* 2018, *MNRAS*, 473, 1346
Nyiransengiyumva, B. *et al.* 2020, IAUS 356 proceedings, Cambridge University Press, in press (arXiv:2004.01104)
Peng, C. Y. *et al.* 2002, *AJ*, 124, 266
Pérez-Montero, E. *et al.* 2019, *MNRAS*, 483, 3322
Pierce, C. M. *et al.* 2007, *ApJ*, 660, L19
Pierce, C. M. *et al.* 2010, *MNRAS*, 408, 139
Pintos-Castro, I. *et al.* 2016, *A&A*, 592, 108
Pović, M. *et al.* 2009, *ApJ*, 706, 810
Pović, M. *et al.* 2012, *A&A*, 541, A118
Pović, M. *et al.* 2013, *MNRAS*, 435, 3444
Pović, M. *et al.* 2015, *MNRAS*, 453, 1644
Pović, M. *et al.* 2016, *MNRAS*, 468, 2878
Ricci, C. *et al.* 2017, *ApJS*, 233, 17
Salim, S. 2014, *Serbian Astronomical Journal*, 189, 1
Sánchez-Portal, M. *et al.* 2015, *A&A*, 578, 30
Schawinski, K. *et al.* 2014, *MNRAS*, 440, 889
Shen, Y. & Ho, L. C. 2014, *Nature*, 513, 210
Shimizu, T. T. *et al.* 2017, *MNRAS*, 466, 3161
Smethurst, R. J. *et al.* 2015, *MNRAS*, 450, 435
Sulentic, J. W. *et al.* 2000, *ApJ*, 536, 5
Terefe, S. *et al.* 2020, IAUS 356 proceedings, Cambridge University Press, in press (arXiv:2003.12736)
Trayford, J. V. *et al.* 2016, *MNRAS*, 460, 3925
Zamfir, S., Sulentic, J. W., & Marziani, P., 2008, *MNRAS*, 387, 856
Zewdie, D. *et al.* 2020a, IAUS 356 proceedings, Cambridge University Press, in press (arXiv:2004.00718)
Zewdie, D. *et al.* 2020b, *MNRAS*, 498, 4345

CHAPTER I. Multiwavelength AGN surveys: past, present, and future

Multiwavelength surveys for Active Galactic Nuclei

William Nielsen Brandt

Department of Astronomy & Astrophysics, 525 Davey Lab,
The Pennsylvania State University, University Park, PA 16802, USA
email: wnbrandt@gmail.com

Abstract. Most of what we know about active galactic nuclei (AGNs) has been driven, or at least strongly shaped, by our methods for finding them, and multiwavelength AGN surveys have achieved remarkable successes in recent decades. I will present a broad, and thus necessarily shallow, review of such multiwavelength AGN surveys. I will first present some brief introductory points on, e.g., general survey approaches, AGN luminosities, host galaxies, and anisotropic emission/obscuration. I will then review many of the key current surveys and their results, separating these into ground-based and space-based surveys. Finally, I will discuss some future prospects including essential remaining questions and "discovery space" considerations.

Keywords. galaxies: active, galaxies: nuclei, black hole physics, surveys

1. Introduction and Summary

The topic of multiwavelength surveys for AGNs is an enormous one, and indeed entire conferences have been held on this topic (e.g., IAU Symposium 304 in Yerevan, Armenia). This broad topic cannot be reviewed well in a brief proceedings article. Thus, I have prepared an expanded one-hour version of my conference talk and placed it on YouTube as https://www.youtube.com/watch?v=jnItH5jxmlA, since this allows a more complete presentation. As per agreement with the editors, this will serve as my conference proceedings contribution, and I hope this can be of some use for future generations of Ethiopian students.

Acknowledgments

I thank the organizers for a stimulating and enjoyable symposium. I acknowledge support from the Chandra X-ray Center grant GO8-19076X and the NASA ADAP grant 80NSSC18K0878.

Dissecting quasars with the J-PAS narrow-band photometric survey

Silvia Bonoli[1,2,3], Giorgio Calderone[4], Raul Abramo[5],
Jailson Alcaniz[6], Narciso Benitez[7], Saulo Carneiro[8], Javier Cenarro[3],
David Cristóbal-Hornillos[3], Renato Dupke[6], Alessandro Ederoclite[9],
Carlos López San Juan[3], Antonio Marín-Franch[3],
Claudia Mendes de Oliviera[9], Mariano Moles[7], Vinicius Placco[10],
Laerte Sodré Jr.[9], Keith Taylor[11], Jesús Varela[3],
Héctor Vázquez Ramió[3] and the J-PAS collaboration

[1]Donostia International Physics Centre (DIPC), Paseo Manuel de Lardizabal 4, 20018 Donostia-San Sebastian, Spain
Email: silvia.bonoli@dipc.org

[2]IKERBASQUE, Basque Foundation for Science, E-48013, Bilbao, Spain

[3]Centro de Estudios de Física del Cosmos de Aragón, Unidad Asociada al CSIC, Plaza San Juan 1, 44001 Teruel, Spain

[4]INAF Trieste, Via Giambattista Tiepolo, 11, 34131 Trieste, Italy

[5]Departamento de Física Matemática, Instituto de Física, Universidade de Sao Paulo, SP, Rua do Matao 1371, Sao Paulo, Brazil

[6]Observatorio Nacional, Rio de Janeiro, 20921-400, RJ, Brazil

[7]Instituto de Astrofísica de Andalucía (IAA-CSIC), Glorieta de la Astronomía, E-18008, Granada, Spain

[8]Instituto de Física, Universidade Federal da Bahia, 40210-340, Salvador, BA, Brazil

[9] Departamento de Astronomia, Instituto de Astronomia, Geofisica e Ciências Atmosféricas da USP, Cidade Universitária, 05508-900, Sao Paulo, SP, Brazil

[10]Department of Physics, University of Notre Dame, Notre Dame, IN 46556 USA

[11]Instruments4

Abstract. The J-PAS survey will soon start observing thousands of square degrees of the Northern Sky with its unique set of 56 narrow band filters covering the entire optical wavelength range, providing, effectively, a low resolution spectra for every object detected. Active galaxies and quasars, thanks to their strong emission lines, can be easily identified and characterized with J-PAS data. A variety of studies can be performed, from IFU-like analysis of local AGN, to clustering of high-z quasars. We also expect to be able to extract intrinsic physical quasar properties from the J-PAS pseudo-spectra, including continuum slope and emission line luminosities. Here we show the first attempts of using the QSFit software package to derive the properties for 22 quasars at $0.8 < z < 2$ observed by the miniJPAS survey, the first deg^2 of J-PAS data obtained with an interim camera. Results are compared with the ones obtained by applying the same software to SDSS quasar spectra.

Keywords. quasars: general, quasars: emission lines, galaxies: nuclei, galaxies: active, surveys, techniques: photometric

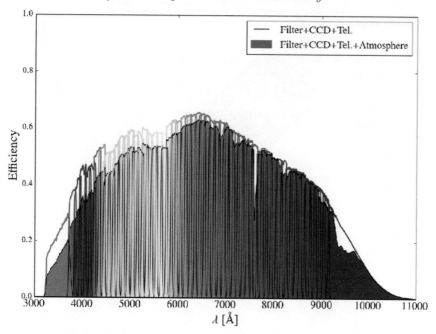

Figure 1. The measured transmission curves for the J-PAS filter set before and after accounting for sky absorption. Effects of the CCD quantum efficiency and the entire optical system of the JST/T250 telescope are included.

1. Introduction: the J-PAS survey

The Javalambre Physics of the Accelerating Universe Astrophysical Survey (J-PAS) is an ambitious photometric survey which will observe thousands of square degrees of the northern sky starting in mid-2020 Benítez *et al.* (2014). The novelty of the survey is in its filter system: 54 narrow band (FWHM$\sim 145 \text{Å}$) covering from $\sim 3800 \text{Å}$ to $\sim 9000 \text{Å}$ and two medium bands extending to $\sim 3000 \text{Å}$ in the blue side and to $\sim 10,000 \text{Å}$ in the red side of the spectrum (see Figure 1).

The survey will be carried out from a new wide field-of-view telescope, the "JST/T250", the largest telescope of the OAJ† observatory (Moles *et al.* 2010; Cenarro *et al.* 2014), located in continental Spain. Once equipped with the JPCam camera (1.2 Gpixel, 0.23''/pixel), the telescope will start observing 14 filters at the same time, covering a total of 4.2 deg^2/pointing Marín-Franch *et al.* (2017).

The unique filter system of J-PAS will effectively produce a low-resolution spectrum for every object observed, allowing to perform a wide variety of scientific projects, from Milky Way stellar population to galaxy evolution studies. The exquisite, sub percent, photometric redshifts Benítez *et al.* (2009) will also allow to perform cosmological experiments, from a precise mapping of the cosmic structure to galaxy clusters detection down to group masses.

Before JPCam is being installed on the telescope, the JST/T250 has been equipped with the *Pathfinder* camera, a single-CCD, 9.2k × 9.2k pixel camera with an effective field of view of 0.27 deg^2. The *Pathfinder* camera has been used to perform the commissioning of the JPCam Actuator System and to start the scientific operation of the JST/T250. In particular, it carried out the JPAS-*Pathfinder* survey, also called

† http://www.oaj.cefca.es

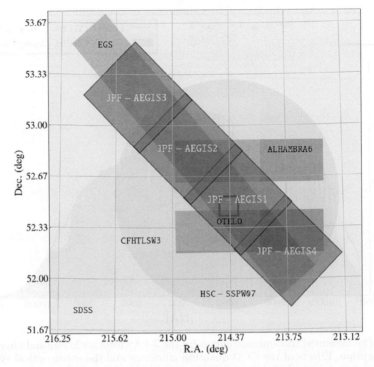

Figure 2. Footprint of the miniJPAS field, with individual pointings shown as red squares. The footprints of other projects are also shown.

"miniJPAS", a 1 deg² along the Extended Gorth Strip observed to full depth[†] with all J-PAS filters (see the footprint in Figure 2. The miniJPAS data have been recently made public to the whole astrophysical community (http://www.j-pas.org/datareleases/minijpas_public_data_release_pdr201912) and are been used to test the capabilities of J-PAS data (Bonoli *et al.*, in prep).

2. AGN and Quasar studies with the J-PAS narrow bands

The narrow bands of the J-PAS filter system make J-PAS data particularly suited to study and characterize emission line objects, active galaxies with broad emission lines in particular. In Figure 3 we show two quasars observed in miniJPAS, with the photometric data (red points) compared to the SDSS spectroscopic data (gray line).

It is clear visually how the strong emission features typical of quasars can be properly identified in the J-PAS pseudo-spectra. Indeed, quasars can be identified and their photo-z estimated to sub-percent precision (Bonoli *et al.*, in prep.).

To test the full capabilities of J-PAS data to estimate quasar spectral properties, we used the QSFit[‡] software package to analyze the low-resolution miniJPAS spectra, and compare the results with the corresponding analysis carried out on SDSS data. QSFit is a software aimed to perform automatic spectral analysis of AGN optical/UV spectra Calderone *et al.* (2017), and its fitting procedure takes into account several components simultaneously, providing estimates of broad band continuum luminosities and slopes, emission lines luminosities and width, etc. In order to compare the spectral quantities measured on the J-PAS and SDSS spectra we chose to limit the analysis in the redshift range $0.8 < z < 1.2$, since in this range the several broad band components building up the

[†] mag$_{AB}$ between 22 and 22.5 for the narrow bands (depth at 5σ in a $3''$ aperture).
[‡] https://qsfit.inaf.it/

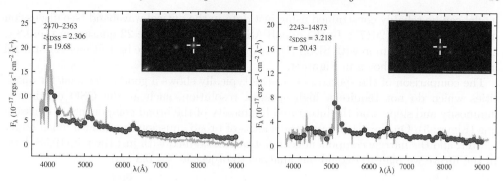

Figure 3. Two examples of SDSS quasar spectra (gray lines) as observed with miniJPAS (red points).

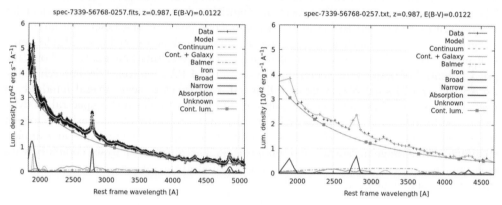

Figure 4. An example of the best fit spectral model obtained running the QSFit software on a SDSS data (left panel) and on the corresponding low-resolution miniJPAS spectrum (right panel) of a $z \sim 1$ quasar. The data (black points) and the overall model (orange lines) are shown, together with the contributions of the different components, from the continuum (dashed red lines) to the broad line components (blue lines), as reported in the legend.

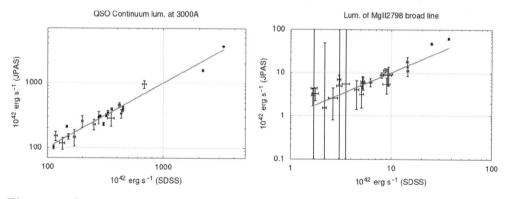

Figure 5. Comparison between the spectral quantities estimated on the SDSS spectra and on the correspoding miniJPAS pseudospectra, as obtained with the QSFit software. Left panel: QSO continuum luminosity at 3000Å. Right: integrated luminosity of the MgII emission line.

model can be reliably constrained using just data in the optical waveband (see discussion in Calderone *et al.* (2017)). In the miniJPAS footprint there are 22 quasars in the $0.8 < z < 1.2$ range in common with SDSS (DR14Q). An example of the best fit model obtained in the two cases is shown in Figure 4.

The comparison of the spectral quantities typically shows a good agreement for quantities which do not require a high spectral resolution, such as the QSO continuum luminosity and slope, and the integrated luminosity of the broad components of emission line (Figure 5). On the other hand, the widths of broad emission line show a much larger scatter, and the narrow components can not be constrained with just the miniJPAS data. In the future we plan to extend our analysis to object at $z < 0.8$ and $z > 2$.

3. Conclusions

We ran the QSFit spectral analysis software on a small sample of 22 quasars with $0.8 < z < 1.2$ selected from the miniJPAS survey and with corresponding SDSS spectra. The comparison of the resulting spectral quantities shows a very good agreement for the quantities associated to broad band component (such as the QSO continuum) or integrated quantities (such as the luminosities of broad emission lines). While being still preliminary and based on a small number of objects, this first analysis shows the great potential of performing spectral analysis on the J-PAS photometric data. Once the J-PAS survey will start observing at full speed, we expect to scan several hundreds of $\deg^2$/yr, thus detecting hundreds of thousands of quasars to which this spectral analysis can be performed. Although possibly inaccurate on a object-by-object basis, we expect to be able to statistically derive intrinsic physical properties of large quasar samples, spanning a wide range in redshift and luminosities. This spectral analysis, combined with environment and clustering studies, will provide important information on the redshift evolution of active galaxies.

References

Benitez, N., Moles, M., Aguerri, J. A. L., *et al.* 2009, *ApJ*, 692, L5
Benitez, N., Dupke, R., Moles, M., *et al.* 2014, arXiv:1403.5237
Bonoli, S., Dupke, R., Cenarro, A. J., *et al. in prep.*
Calderone, G., Nicastro, L., Ghisellini, G., *et al.* 2017, *MNRAS*, 72, 4
Cenarro, A. J., Moles, M., Marín-Franch, A., *et al.* 2014, *Society of Photo-Optical Instrumentation Engineers (SPIE) Conference Series*, Vol. 9149
Marín-Franch, A., Taylor, K., Santoro, F. G., *et al.* 2017, *Highlights on Spanish Astrophysics IX*, 670
Moles, M., Sánchez, S. F., Lamadrid, J. L., *et al.* 2010, *PASP*, 122, 363

Unveiling the physical processes that regulate galaxy evolution with SPICA observations

Luigi Spinoglio, Juan A. Fernández-Ontiveros and Sabrina Mordini

Istituto di Astrofisica e Planetologia Spaziali - INAF, Rome,
Via Fosso del Cavaliere 100, 00133, Roma, Italia
emails: `luigi.spinoglio@iaps.inaf.it`, `j.a.fernandez.ontiveros@gmail.com`,
`sabrina.mordini@uniroma1.it`

Abstract. To study the dust obscured phase of the galaxy evolution during the peak of the Star Formation Rate (SFR) and the Black Hole Accretion Rate (BHAR) density functions ($z = 1-4$), rest frame mid-to-far infrared (IR) spectroscopy is needed. At these frequencies, dust extinction is at its minimum and a variety of atomic and molecular transitions, tracing most astrophysical domains, occur. The future IR space telescope mission, *SPICA*, fully redesigned with its 2.5 m mirror cooled down to $T < 8\,\mathrm{K}$, will be able to perform such observations. With *SPICA*, we will: 1) obtain a direct spectroscopic measurement of the SFR and of the BHAR histories, 2) measure the evolution of metals and dust to establish the matter cycle in galaxies, 3) uncover the feedback and feeding mechanisms in large samples of distant galaxies, either AGN- or starburst-dominated, reaching lookback times of nearly 12 Gyr. *SPICA* large-area deep surveys will provide low-resolution, mid-IR spectra and continuum fluxes for unbiased samples of tens of thousands of galaxies, and even the potential to uncover the youngest, most luminous galaxies in the first few hundred million years. In this paper a brief review of the scientific preparatory work that has been done in extragalactic astronomy by the *SPICA* Consortium will be given.

Keywords. telescopes, galaxies: evolution, galaxies: active, galaxies: starburst, quasars: emission lines, galaxies: ISM, galaxies: abundances, galaxies: high-redshift, infrared: galaxies

1. Introduction

The bulk of the star formation and supermassive black hole (SMBH) accretion in galaxies took place more than six billion years ago, with a sharp drop to the present epoch (e.g., Madau & Dickinson 2014, Fig. 1). Since most of the energy emitted by stars and accreting SMBHs is absorbed and re-emitted by dust, understanding the physics of galaxy evolution requires infrared (IR) observations of large, unbiased galaxy samples spanning a range in luminosity, redshift, environment, and nuclear activity. From *Spitzer* and *Herschel* photometric surveys the Star Formation Rate (SFR) and Black Hole Accretion Rate (BHAR) density functions have been *estimated* through measurements of the bolometric luminosities of galaxies (Le Floc'h *et al.* 2005; Gruppioni *et al.* 2013; Delvecchio *et al.* 2014). However, such integrated measurements could not separate the contribution due to star formation from that due to BH accretion (see, e.g., Mullaney *et al.* 2011). This crucial separation has been attempted so far through modelling of the spectral energy distributions and relied on model-dependent assumptions and local templates, with large uncertainty and degeneracy. On the other hand, determinations from UV (e.g. Bouwens *et al.* 2007) and optical spectroscopy (e.g., from the *Sloan* Digital Sky Survey, Eisenstein *et al.* 2011) track only marginally ($\sim 10\%$) the total integrated light (Fig. 1). X-ray analyses of the BHAR, in turn, are based on large extrapolations

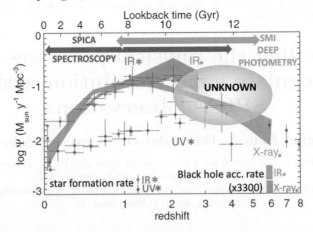

Figure 1. Estimated star-formation rate densities from the far-ultraviolet (blue points) and far-IR (red points) photometric surveys (figure adapted from Madau & Dickinson 2014). The estimated BHAR density, scaled up by a factor of 3300, is shown for comparison (in green shading from X-rays and light blue from the IR).

and possibly miss a large fraction of obscured objects. Furthermore, the SFR density at $z > 2-3$ is very uncertain, since it is derived from UV surveys, highly affected by dust extinction. As opposite, through IR emission lines, the contributions from stars and gravity can be separated. *SPICA* spectroscopy will allow us to directly measure redshifts, SFRs, BHARs, metallicities and dynamical properties of gas and dust in galaxies at lookback times up to about 12 Gyrs. *SPICA* spectroscopic observations will allow us for the first time to redraw the SFR rate and BHAR functions (Fig. 1) in terms of measurements directly linked to the physical properties of the galaxies.

2. Why IR spectroscopy

The mid- to far-IR spectral range hosts a suite of atomic and ionic transitions, covering a wide range of excitation, density, and metallicity, directly tracing the physical conditions in galaxies, which are typically obscured during most of their evolution. Ionic fine structure lines (e.g. [NeII], [SIII], [OIII]) probe HII regions around hot young stars, providing a measure of the SFR and the gas density. Lines from highly ionized species (e.g. [OIV], [NeV]) trace the presence of AGN and can measure the BHAR. Photo-dissociation regions (PDR), the transition between young stars and their parent molecular clouds, can be studied via the strong [CII] and [OI] lines (Spinoglio & Malkan 1992, Fig. 2-a).

Through line ratio diagrams, like the *new IR BPT diagram* (Fernández-Ontiveros *et al.* 2016, Fig. 2-b), IR spectroscopy can separate the galaxies in terms of both the source of ionization – either young stars or AGN excitation – and the gas metallicity, during the dust-obscured era of galaxy evolution ($0.5 < z < 4$).

3. SPICA observations of galaxy evolution

The SPace Infrared telescope for Cosmology and Astrophysics (*SPICA*) will combine a 2.5 m mirror cooled to below 8 K with instruments employing state-of-the-art detectors, becoming the first large telescope cooled in space using mechanical cryo-coolers instead of liquid cryogen. The low telescope background and the new generation of detectors will provide about two orders of magnitude sensitivity improvement with respect to previous missions. *SPICA* instruments will provide a spectral resolving power ranging from $R = 50-120$ to 11000 in the $17-230\,\mu$m domain as well as $R \sim 28000$ between $12-18\,\mu$m. The Transition Edge Superconductor detectors in the SAFARI spectrometer

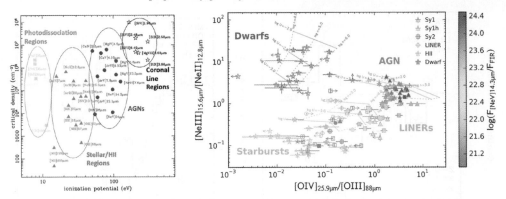

Figure 2. Left: the ionization density diagram of the IR fine structure lines (Spinoglio & Malkan 1992). **Right:** Observed line ratios of [NeIII]15.6μm/[NeII]12.8μm vs. [OIV]26μm/[OIII]88μm for AGN, LINER, starburst and dwarf galaxies in the local Universe. The active galaxies symbols have been color-coded from blue to red, according to the value of the ratio of the [NeV14.3]μm line to the FIR continuum as measured with *Herschel-PACS* spectra at ∼ 160μm, measuring the AGN dominance, taken from Fernández-Ontiveros *et al.* (2016).

(35 − 230 μm) will reach an unprecedented sensitivity of ∼ 7 × 10^{-20} W m^{-2} (5σ/1 hr). Thanks to its large field of view of 10′ × 12′, the SMI imager and spectrometer will deliver simultaneous spectroscopy (17 − 37 μm) and photometry (34 μm) mapping for large areas in the sky either. Additionally the B-BOP instrument will provide accurate polarimetric imaging at 70, 220 and 350 μm (André *et al.* 2019). A full description of the mission, the telescope and the focal plane instruments can be found in Roelfsema *et al.* (2018).

In the field of galaxy evolution, *SPICA* will: 1) obtain the first spectroscopic characterization of the SFR and the BHAR histories (Spinoglio *et al.* 2017); 2) measure the evolution of metals and dust and establish the matter cycle in galaxies (Fernández-Ontiveros *et al.* 2017); 3) uncover the feedback and feeding mechanisms in large samples of distant galaxies, either AGN- or starburst-dominated (González-Alfonso *et al.* 2017); 4) provide low-resolution, mid-IR spectra and continuum fluxes for deep unbiased samples of tens of thousands of galaxies, and even the potential to uncover the youngest, most luminous galaxies in the first few hundred million years (Gruppioni *et al.* 2017; Kaneda *et al.* 2017); 5) probe the spectra of hyper-luminous IR galaxies at redshift $z = 5 - 10$, allowing us to characterize their main physical properties (Egami *et al.* 2018).

3.1. *Feedback from powerful AGN*

The correlations between the SMBH mass and the velocity dispersion, stellar mass, and luminosity of galaxies in the local Universe (Magorrian *et al.* 1998; Ferrarese & Merrit 2000) suggests a link between the growth of the BH and the stellar population in its host galaxy. The bimodal color distribution observed in local galaxies (Strateva *et al.* 2001; Baldry *et al.* 2004) points to an scenario where massive red-and-dead galaxies finished their evolution on very short timescales, while the evolution of low-mass blue galaxies is still ongoing (Hopkins *et al.* 2006; Schawinski *et al.* 2014). But even more relevant for our current (poor) knowledge of galaxy evolution would be to reconcile the shape of the observed luminosity function of galaxies with that of the theoretical halo mass function in CDM. Feedback from supernovae can match observations with theory at low masses, while AGN feedback would be needed to make them agree in the high mass end. In this self-regulated feedback model, the funneling of large amounts of gas into the nuclear region generates both a nuclear starburst (SB) and drives the accretion onto the SMBH. The latter eventually reaches a critical mass/luminosity when

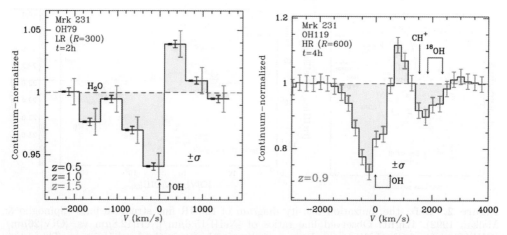

Figure 3. Left: The OH 79µm P Cygni profile observed by *Herschel* in Mrk231 (Fischer et al. 2010) and simulated with the SAFARI low resolution mode at z = 0.5, 1.0, 1.5 (see errorbars) (González-Alfonso et al. 2017). **Right:** Same, but simulated with the SAFARI FTS high resolution mode at redshift of z = 0.9 (González-Alfonso et al. 2017).

the energy and momentum released couples with the surrounding interstellar medium (ISM), limiting the accretion onto the SMBH and quenching the SBs via injection of turbulence, through a fast sweeping out of the ISM gas reservoir, or by heating the circumgalactic gas and preventing further accretion of gas onto the galaxy (negative feedback), ultimately yielding the MBH-σ relationship (Silk & Rees 1998; di Matteo et al. 2005; Springel et al. 2005; Murray et al. 2005; Hopkins et al. 2006).

With *SPICA* we will be able to address the following questions: is AGN feedback responsible for the decline of SF in the last 7 Gyr, driving massive galaxies into the red-and-dead sequence? What physical processes – mechanical or thermal energy injection – drive molecular outflows? *SPICA* will be able to detect P-Cygni profiles in the OH lines of powerful ultra-luminous galaxies up to a redshift of $z \sim 2$ in low resolution mode, and up to $z \sim 1$ at high spectral resolution, as shown in Fig. 3.

3.2. Measuring metallicity evolution with SPICA

To study the chemical evolution of galaxies, especially during the dust-obscured era across the peak of the SFR density ($1 < z < 3$), the use of metallicity tracers almost independent of the dust extinction, radiation field, and of the gas density are crucial. Optical nebular lines are likely probing just the most external (unobscured) regions of these galaxies, as suggested by the order of magnitude discrepancy found between the dust content of high redshift sub-millimetre galaxies (SMG) seen by *Herschel* and the metallicity measurements based on optical lines (Santini et al. 2010). This is likely caused by dust obscuration, since the ISM of SMG is optically thick at visual wavelengths, thus the optical emission in these galaxies comes probably from the outer parts which would be poorly enriched with heavy elements.

Gas metallicities can be determined in dust-obscured regions and galaxies using the diagnostic shown in Fig. 4-a, up to $z \sim 1.6$, where the [OIII]88µm line would still fall within the spectral range of SAFARI. A diagnostic based on the [OIII]52µm/[NIII]57µm ratio (see Nagao et al. 2011) is also a metallicity tracer up to $z \sim 3$ if the density is constrained through other line ratios. Above $z > 0.15$ and 0.7, the [SIV]10.5µm line would enter in the SMI/HR and MR ranges, respectively, enabling an additional indirect

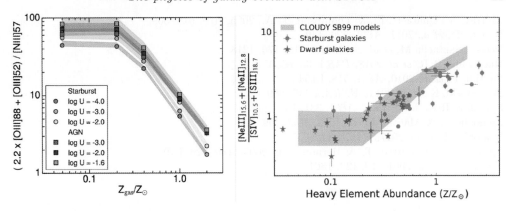

Figure 4. Left: AGN and starburst models for the metallicity sensitive $(2.2\times$ [OIII]88μm+ [OIII]52μm)/[NIII]57μm line ratio as a function of the gas-phase metallicity (Pereira-Santaella et al. 2017). **Right:** The ([NeII]12.8μm + [NeIII]15.6μm) to ([SIII]18.7μm + [SIV]10.5μm) line ratio from Spitzer/IRS observations of local starburst galaxies vs. indirect gas-phase metallicity determined from strong optical lines (Moustakas et al. 2010; Pilyugin et al. 2014). Cloudy simulations including sulphur stagnation above $Z > 1/5\, Z_\odot$ are in agreement with the observed increase of this line ratio (Fernández-Ontiveros et al. 2016; Fernández-Ontiveros et al. 2017).

abundance diagnostic – based on the calibration of metallicity-sensitive line ratios – using the ([NeII]12.8μm + [NeIII]15.6μm) to ([SIII]18.7μm + [SIV]10.5μm) line ratio (Fig. 4-b; Fernández-Ontiveros et al. 2016).

Acknowledgements

We acknowledge the whole SPICA Collaboration Team, as without its multi-year efforts and work this paper could not have been possible. LS and JAFO acknowledge financial support by the Agenzia Spaziale Italiana (ASI) under the research contract 2018-31-HH.0.

References

André, Ph. et al. 2019, *PASA*, 36, e029
Baldry, I. K. et al. 2004, *ApJ*, 600, 681
Bouwens, R. R. et al. 2007, *ApJ*, 670, 928
Delvecchio, I. et al. 2014, *MNRAS*, 439, 2736
di Matteo, T., Springel, V., & Hernquist, L. 2005, *Nat.*, 433, 604
Egami, E. et al. 2018, *PASA*, 35, e048
Eisenstein, D. J. et al. 2011, *AJ*, 142, 72
Fernández-Ontiveros, J. A. et al. 2016, *ApJS*, 226, 19
Fernández-Ontiveros, J. A. et al. 2017, *PASA*, 34, e053
Ferrarese, L. & Merritt, D. 2000, *ApJ*, 539, L9
Fischer, J. et al. 2010, *A&A*, 518, L41
González-Alfonso, E. et al. 2017, *PASA*, 34, e054
Gruppioni, C. et al. 2013, *MNRAS*, 432, 23
Gruppioni, C. et al. 2017, *PASA*, 34, e055
Hopkins, P. F., et al. 2006, *ApJS*, 163, 50
Kaneda, H. et al. 2017, *PASA*, 34, e059
Le Floc'h, E. et al. 2005, *ApJ*, 632, 169
Madau, P. & Dickinson, M. 2014, *ARA&A*, 52, 415
Magorrian, J. et al. 1998, *AJ*, 115, 2285
Mullaney, J. R. et al. 2011, *MNRAS*, 414, 1082

Murray, N., Quataert, E., & Thompson, T. A., 2005, *ApJ*, 618, 569
Nagao, T., *et al.* 2011, *A&A*, 526, 149
Pereira-Santaella M. *et al.* 2017, *MNRAS*, 470, 1218
Roelfsema, P. R. *et al.* 2018, *PASA*, 35, e030
Santini, P. *et al.* 2010, *A&A*, 518, L154
Schawinski, K. *et al.* 2014, *MNRAS*, 440, 889
Silk, J. & Rees, M. J. 1998, *A&A*, 331, L1
Spinoglio, L. & Malkan, M. A. 1992, *ApJ*, 399, 504
Spinoglio, L. *et al.* 2017, *PASA*, 34, e057
Springel, V., Di Matteo, T., & Hernquist, L. 2005, *ApJ*, 620, L79
Strateva, I. *et al.* 2001, *AJ*, 122, 1861

Search for high-redshift blazars with Fermi-LAT

Michael Kreter

Centre for Space Research, North-West University, Private Bag X6001, Potchefstroom 2520, South Africa

Abstract. High-redshift blazars ($z \geqslant 2.5$) are one of the most powerful classes of gamma-ray sources in the Universe. These objects posses the highest jet powers and luminosities and have black-hole masses often in excess of 10^9 solar masses. In addition, high-redshift blazars are important cosmological probes and serve as test objects for blazar evolution models. Due to their large distance, their high-energy emission peak is downshifted to energies below the GeV range, which makes them difficult to study with Fermi/LAT and only the very brightest objects are detectable. Hence, only a small number of high-redshift blazars could be detected with Fermi/LAT so far. In this work, we present a strategy to significantly increase the detection statistics at redshift $z \geqslant 2.5$ via a search for flaring events in high-redshift gamma-ray blazars whose long-term flux remains below the sensitivity limit of Fermi/LAT. Seven previously GeV undetected high-redshift blazars have been identified from their bright monthly outburst periods, while more detections are expected in the future.

Keywords. galaxies: active, galaxies: blazars, surveys: gamma-rays

A uniformly selected, all-sky, optical AGN catalogue

Ingyin Zaw

New York University Abu Dhabi, Abu Dhabi, United Arab Emirates

Abstract. We have constructed an all-sky AGN catalogue, based on optical spectroscopy, from the parent sample of galaxies in the 2MASS Redshift Survey (2MRS), a near-complete census of the nearby ($z < 0.09$) universe. In addition to identifying the 8491 AGNs and providing line measurements for all the emission line galaxies so that the users can customise the selection criteria, we assess the affects of spectral quality on AGN identification. We find that spectral signal-to-noise and resolution affect not only the overall AGN detection rates but also the broad-line to narrow-line AGN ratios. These systematic effects must be taken into account when using any optical AGN catalogue and in comparing the results from different catalogues. We develop a way to account for the inhomogeneities by parametrizing the AGN detection rates as a function of the spectral signal-to-noise, making our catalogue suitable for statistical analyses. We will also present cross-correlation studies between this catalogue and all-sky catalogs at other wavelengths to better understand the different physical processes which lead to the emission at different wavelengths.

Keywords. galaxies: active, surveys: optical catalogue

Star formation and AGN activities in selected nearby HII galaxies in the LeMMINGS survey

Ikechukwu Obi

NASRDA - Centre for Basic Space Science (CBSS), University of Nigeria Nsukka, Nigeria

Abstract. We investigate the relative fraction of the emission generated by star formation and nuclear activities in 6 nearby HII galaxies selected from the first high resolution radio data release of LeMMINGS, the Legacy e-MERLIN Multi-band Imaging of Nearby Galaxies Survey. These galaxies are supposed to be powered solely by star formation according to the BPT diagram but exhibit jetted morphologies on parsec scales indicating the presence of a low luminosity AGN. We further carried out a multi-wavelength SED fiiting and analysis using the CIGALE code, estimating stellar masses and star formation rates.

Keywords. galaxies: active, galaxies: star formation

Panchromatic characterisation of accreting black holes in dusty star-forming galaxies

Gabriela Calistro Rivera

Leiden Observatory, Leiden, Netherlands

Abstract. Although AGN do not typically dominate the bolometric emission of dusty star forming galaxies, large AGN fractions (sometimes > 40%) have been observed in various sub-millimeter surveys. These diagnostics have been however mostly based on X-ray counterpart selections and a complete multiwavength census of the fraction of AGN hosts is needed. I will present new advances in the modelling of panchromatic spectral energy distributions (SEDs) of active galactic nuclei (AGN), based on our publicly available code AGNfitter (Calistro-Rivera *et al.* 2016). AGNfitter implements a fully Bayesian Markov Chain Monte Carlo method to fit the spectral energy distributions of AGNs pushing the wavelengths frontiers from the radio to the X-rays. I will present a recent application of AGNfitter on dusty star forming galaxies in the ALESS sub-millimeter survey to obtain an unbiased multiwavelength characterisation of the nuclear activity buried in dusty star formation. Our method reveals a significantly larger contribution of AGN activity to the emission in these galaxies than previously observed based on X-rays diagnostics. Our method represents a unique tool to potentially characterise an unbiased accretion history of the Universe when applied to larger populations of star-forming galaxies.

Keywords. galaxies: active, galaxies: spectral energy distributions, galaxies: star formation

Reference

Calistro-Rivera, G., *et al.* 2016, *ApJ*, 833, 98

CHAPTER II. AGN types and unification model

AGN types and unification model

Luigi Spinoglio and Juan Antonio Fernández-Ontiveros

Istituto di Astrofisica e Planetologia Spaziali - INAF, Rome,
Via Fosso del Cavaliere 100, 00133, Roma, Italia
emails: luigi.spinoglio@iaps.inaf.it, j.a.fernandez.ontiveros@gmail.com

Abstract. The motivation of the "unified model" is to explain the main properties of the large zoo of active galactic nuclei with a single physical object. The discovery of broad permitted lines in the polarized spectrum of type 2 Seyfert galaxies in the mid 80's led to the idea of an obscuring torus, whose orientation with respect to our line of sight was the reason of the different optical spectra. However, after many years of observations with different techniques, including IR and mm interferometry, the resulting properties of the observed dust structures differ from the torus model that would be needed to explain the type 1 vs type 2 dichotomy. Moreover, in the last years, multi-frequency monitoring of active galactic nuclei has shown an increasing number of transitions from one type to the other one, which cannot be explained in terms of the simple orientation of the dusty structure surrounding the active galactic nucleus (AGN). The interrelations between the AGN and the host galaxy, as also shown in the Magorrian relation, suggest that the evolution of the host galaxy may also have an important role in the observed manifestation of the nuclei. As an example, the observed delay between the maximum star formation activity and the onset of the AGN activity, and the higher occurrence of type 2 nuclei in star forming galaxies, have suggested the possible evolutionary path from, e.g., H II → AGN2 → AGN1. In the next years the models of unification need to also consider this observational framework and not only simple orientation effects.

Keywords. accretion, galaxies: active, galaxies: nuclei, galaxies: Seyfert, ISM: dust, galaxies: evolution

1. Introduction

Since their discovery in 1943, Seyfert galaxies appeared to show different optical spectra, as already described in the paper by Seyfert (1943): "The observed relative intensities of the emission lines exhibit large variations from nebula to nebula". As an example of this variety, the maximum observed widths of hydrogen recombination lines were ranging from 3600 to 8500 $\rm km\,s^{-1}$. In a following paper, Seyfert (1946) showed that "the hydrogen lines in NGC 4151 and NGC 7469 are of unusual interest, being composed of relatively narrow cores ($1100\,\rm km\,s^{-1}$) superposed on very wide wings ($7500\,\rm km\,s^{-1}$)".

The main difference between type 1 and type 2 Seyfert galaxies come from the optical and UV spectroscopy: *i)* type 1 have broad permitted lines (i.e., hydrogen recombination lines, intercombination lines – e.g., C IV 1459 Å, Mg II 2798 Å – and semi-forbidden lines, e.g., C III] 1909 Å) with line widths of FWHM $\sim 1000 - 10000\,\rm km\,s^{-1}$; *ii)* both type 1 and type 2 have narrow permitted and forbidden lines: FWHM $\sim 500 - 1000\,\rm km\,s^{-1}$.

The complexity and variety of Active Galactic Nuclei (AGN), which extend the class of Seyfert galaxies to higher luminosities, has been acknowledged in the following decades as the observational material were accumulated. The active galactic nuclei *zoo* has been presented in a coherent framework in the review by Heckman & Best (2014, see their Fig. 4): the main parameter is the total luminosity in Eddington units, which is linked to

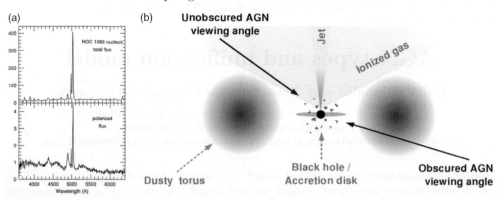

Figure 1. Left: (a) Spectropolarimetry of NGC 1068, showing (above) the flux spectrum and below the polarized flux, which is indistinguishable from the flux spectra of Type 1 spectra. Figure from Antonucci (1993). **Right: (b)** Schematic view of the torus model. Figure from Gandhi (2005).

the accretion rate of the supermassive black holes (SMBH) and sets the transition from the jet mode to the radiative mode.

In the mid 1980s, Antonucci & Miller (1985) first reported that the optical polarization spectrum of the prototype Type 2 Seyfert galaxy NGC 1068 showed very broad symmetric Balmer lines ($\sim$7500 km s^{-1} FWZI) and also permitted Fe II, indicating a very close similarity with Type 1 Seyfert galaxies (see Fig. 1-a). In their paper, they say "We favor an interpretation in which the continuum source and broad line clouds are located inside a thick disk, with electrons above and below the disk scattering continuum and broad line photons in to the line of sight". This paper coincides with the beginning of the "Unification Era", i.e. the attempt to explain the different spectral manifestations of AGN with a single physical object, viewed from different angles. In his review, Antonucci (1993) introduced what he named the "straw person model", for which there are only two types of AGN: those radio quiet and those radio loud. Considering only radio quiet objects, this statement means that "all properties such as spectroscopic classification [...] are ascribed to orientation" (Antonucci 1993).

We can anticipate here that the situation, after three decades of observations, more and more evidence has shown that the simple unification scheme proposed by Antonucci (1993) is far too simplistic to account for them: if on the one hand it can be recognised that some unification is well proven by observations, i.e. the so-called "Hidden Broad-Line Region Galaxies" (HBLR) are the same as the type 1 AGN, on the other hand, the occurrence of other intrinsic factors in the nature and appearance of AGN are present. Among these: the effect of evolution of the accretion disk, which is not a stable physical system; the effect of the host galaxy, which is not negligible. Moreover, the simple unification scheme relies on the presence of a geometrically thick obscuring structure, the so-called dusty torus, whose robust observational evidence is still not available.

This review is organized as follows: Section 2 introduces the concept of the hypothetical torus which would be needed to demonstrate the unification between type 1 and type 2 AGN; Section 3 describes some of the observational material which has been collected so far with different techniques aiming to the detection of the molecular tori around SMBHs; Section 4 describes the unification which has been already reached for low luminosity AGN; Section 5 tries to make an observational review, without any ambition to be complete, of the data on the so-called *changing look* AGN, i.e. those AGN which have been transited from one type to another one and may question the simple unification model. Section 6 attempts to assess the role of evolution and of the host galaxy in the AGN appearance. Finally, in Section 7 we present our conclusions.

2. The needed molecular torus

A large number of works have been published on models of molecular tori, either theoretical and phenomenological. A comprehensive review of what has been done up to the mid 2010's is that by Netzer (2015), to which we refer the reader to obtain a full comprehension of the various approaches used in the literature to investigate AGN tori, an essential work to assess the unification model. Other relevant reviews on obscuration around AGN and obscured AGN are those of Ramos Almeida & Ricci (2017), Hickox & Alexander (2018), and the updated review by Antonucci (2012) respectively. In this review, we will limit our attention here to understand what characteristics a torus needs to have in order to be compliant with the unification model. We show, in Fig. 1-b, the schematic view of the torus model from Gandhi (2005). The expected characteristics and the physical role of the torus are:

• It needs to block the Broad Line Region (BLR) radiation. This is the main requirement of the torus to allow for the main observational difference between type 1 and type 2 AGN, which is the appearance of the optical and ultraviolet emission line spectra. A strong observational constraint on this can be derived from the statistics of the fraction of type 1 versus type 2 AGN seen, e.g., in the Local Universe. This fraction, however, strongly depends on the AGN luminosity. At the typical Seyfert galaxy luminosities of $L_{bol} \leqslant 10^{44}$ erg s^{-1} the Type 1 fraction is $\sim$0.3, increasing to $\sim$0.7 at $L_{bol} \leqslant 10^{46}$ erg s^{-1} (Assef et al. 2013). Recent hard X-ray surveys have significantly improved our understanding of AGN obscuration, showing that 70% of all local AGN are obscured (Burlon et al. 2011; Ricci et al. 2015). If the opening angle of the hypothetical torus (shaded angle in Fig. 1-b) is, e.g., 120°, then the surface area with BLR visibility will intersect exactly half of the hemisphere visible from an observer, which will correspond to an equal fraction of types 1 and types 2. However, if this angle is as narrow as 60°, then the fraction of type 1 should be smaller than 15%.

• It needs to collimate the AGN ionizing radiation and induce the biconical shape of the Narrow Line Region (NLR), i.e the so-called ionization cones: studies on the NLR morphology are relevant in this respect.

• It needs to allow the feeding of the Black-Hole through accretion, and act at the same time as a gas reservoir.

3. Observations of the "torus"

3.1. Mid-IR interferometry

We describe in this Section the most relevant "direct" observations of the torus, through interferometric techniques, either in the mid-IR and in the millimeter ranges. The first solid claim of the detection of a torus has been reported by Jaffe et al. (2004) for the prototype Seyfert type 2 galaxy NGC 1068. It is not surprising that this galaxy was chosen, because of the discovery of broad hydrogen lines in the polarized optical spectrum by Antonucci & Miller (1985) and because of its vicinity (only 14.4 Mpc from us).

Jaffe et al. (2004) report interferometric mid-IR observations with the MIDI (Leinert et al. 2003) instrument at the focal plane of the ESO-VLTI (Glindemann et al. 2003) interferometer, reaching a spatial resolution of $\sim$10 mas at $\lambda = 10\,\mu$m, that spatially resolve the mid-IR brightness distribution in the nucleus of NGC 1068. These observations were apparently consistent with a warm (320 K) dust structure with a size of of $2.1 \times 3.4\,\text{pc}^2$, enclosing a smaller and hotter structure (Fig. 2-a). However, these results were based on a relatively poor coverage or spatial frequencies in the uv-plane and therefore the

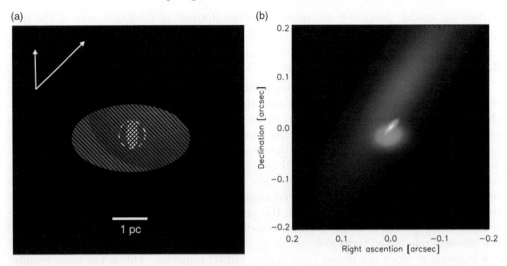

Figure 2. Left: (a) Model dust structure in the nucleus of NGC 1068, showing a central hot component (dust temperature T > 800 K, light) marginally resolved along the source axis. The much larger warm component (T = 320 K, dark shade) is well resolved. Single hatching represents the averaged optical depth in the silicate absorption, $\tau_{SiO} = 0.3$, while the higher value $\tau_{SiO} = 2.1$ is found towards the hot component (cross-hatched). Figure from Jaffe et al. (2004). **Right: (b)** Image of the three component model 2 of López-Gonzaga et al. (2014) for the mid-infrared emission at 12μm of the nuclear region of NGC 1068. Figure from López-Gonzaga et al. (2014).

modeling of the observed visibilities needed to reconstruct the torus geometry has to be taken with extreme caution. As a matter of fact, the uv-coverage was improved later by López-Gonzaga et al. (2014), with the same instrument, but including longer baselines thanks to the Auxiliary Telescopes, allowing to measure also the 5−10 pc scales. Their results show that most of the mid-IR emission originates from the large-scale structures and is associated with warm dust distributed in two major components, one close to the center and one at a distance of more than 80 mas, close to 16°−18° in the NW direction (Fig. 2-b). While the central warm region is interpreted as an extension of the hot emission region, the offset region is attributed to dusty clouds close to the northern ionization cone. Models infer a size of 14 pc, strong elongation along a position angle of PA $\sim -35°$, and three times more 12 μm flux than that of the central hot region.

The warm component, named 3 in López-Gonzaga et al. (2014), located $\sim$7 pc north of the hotter nuclear disk, apparently intercepts a large fraction of the nuclear UV emission. Thus there are several obscuring components at different disk latitudes that can cause type 2 appearance in AGN. The volume that is heated by this emission is quite narrow; the viewing angles from which this galaxy would be classified as Seyfert type 1 cover only $\sim$10% of the sky. Even if the spatial frequency sampling of these data is very good, compared to the previous work, however the best model does not appear to have the characteristics of the needed torus.

Tristram et al. (2014) collected extensive mid-IR interferometric data of the closest Seyfert type 2 galaxy Circinus (at $\sim$4 Mpc) with MIDI at the VLTI during many observing runs from 2008 to 2011. Besides confirming the presence of two components in the dust distribution, one inner dense disk component and an extended emission region, they analyse with detailed models these components. For a better understanding, we report their figure here (Fig. 3).

The disk-like component is highly elongated (along PA $\simeq 46°$) with a size of $\sim 0.2 \times 1.1$ pc^2, aligned with the orientation of the nuclear maser disk and perpendicular to the

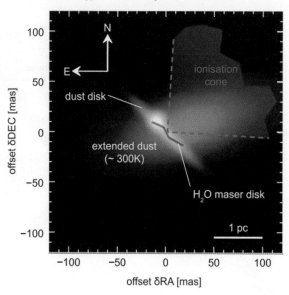

Figure 3. Image of the three-component model for the mid-IR emission of the nucleus of the Circinus galaxy. The gradient of the extended component due to the increase in the silicate depth towards the south-east is clearly visible. Despite the lower surface brightness, 80% of the emission originates from the extended component. The trace of the water maser disk is also plotted: the blue and red parts trace the approaching and receding sides of the maser disk respectively. Figure from Tristram *et al.* (2014), adapted by Hönig (2016).

ionisation cone and outflow (along PA $\sim -44°$). They interpret this component either as emission from material associated with the inner funnel of the torus directly above or below the disk or from the directly illuminated portion of a warped disk slightly oriented towards us.

The extended dust emission accounts for 80% of the mid-IR emission. It has a size of $\sim 0.8 \times 1.9\,\mathrm{pc}^2$ and is elongated along PA $\sim 107°$, that is roughly along the polar direction. It is interpreted as the emission from the inner funnel of a more extended dust distribution and especially as emission from the funnel edge along PA $\sim -90°$. Dense dusty material enters the ionisation cone primarily on this side of the funnel, which is also preferentially illuminated by the inclined accretion disk.

They find both emission components to be consistent with dust at $T \sim 300\,\mathrm{K}$, i.e. no evidence of an increase in the temperature of the dust towards the center, meaning that most of the near-IR emission probably comes from parsec scales as well. They further argue that *the disk component alone is not sufficient to provide the necessary obscuration and collimation of the ionising radiation and outflow*. The material responsible for this must instead be located on scales of $\sim 1\,\mathrm{pc}$, surrounding the disk. They associate this material with the dusty torus. However, they conclude that the presence of a bright disk-like component, polar elongated dust emission and the lack of a temperature difference are not expected for typical models of the centrally heated dust distributions of AGN. Moreover, they state that new sets of detailed radiative transfer calculations will be required to explain their observations and to better understand the three-dimensional dust morphology in the nuclei of active galaxies.

While initial observations using mid-IR interferometry initially confirmed the presence of a geometrically-thick obscuring structure in NGC 1068 (Jaffe *et al.* 2004), further observing campaigns on prototypical hidden Seyfert type 1 galaxies favor a

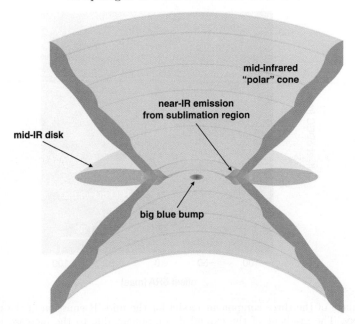

Figure 4. Schematic view of the pc-scale AGN infrared emission consisting of a geometrically-thin disk in the equatorial plane (light brown) and a hollow dusty cone towards the polar region (dark brown). The inner part of the disk (pink) emits the near-IR emission dominating the 3–5 μm bump. Figure from Hönig (2019).

geometrically-thin dust distribution, i.e. a disk instead of a torus. This casts doubts on the role that this structure plays on the obscuration and the type 1/type 2 dichotomy.

3.2. Redefining the torus

Molecular lines show large, massive disks while mid-IR observations are dominated by a strong polar component, which is interpreted as a dusty wind. A unifying view of AGN in the infrared (IR) and sub-mm has been recently proposed by Hönig (2019). His paper aims at using characteristics shared by AGN in each of the wavebands and a set of simple physical principles to form a unifying view of these seemingly contradictory observations:

• Dusty molecular gas flows in from galactic scales of ∼100 pc to the sub-pc environment via a disk with small to moderate scale height.

• The hot, inner part of the disk inflates due to IR radiation pressure and unbinds a large amount of the inflowing gas from the BH gravitational potential, providing the conditions to launch a wind driven by the radiation pressure from the AGN.

• The dusty wind feeds back mass into the galaxy at a rate of the order of ∼0.1−100 $M_\odot$ yr^{-1}, depending on AGN luminosity and Eddington ratio.

• Angle-dependent obscuration as required by AGN unification is provided by a combination of disk, wind, and wind launching region.

González-Martín et al. (2019a,b) recently made an investigation of the predictions of six dusty torus models of AGN, including either clumpy torus models, two phase models and also the wind/torus model (Hönig & Kishimoto 2017), all with available spectral energy distributions, aiming at exploring which model describes better the data and the resulting parameters. They show that different torus models explain the same energy distributions with very different geometries, such as the viewing angle and the covering factor. Being these two parameters among the most important physical quantities

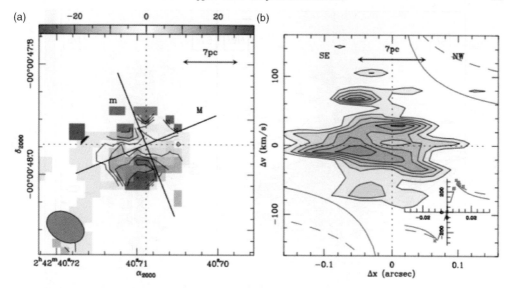

Figure 5. **Left:(a)** ALMA CO(6-5) mean-velocity map of the torus. The black lines labeled as "M" and "m" show, respectively, the orientations of the major and minor axes of the CO torus. **Right:(b)** CO(6-5) position-velocity diagram along the M axis. The inset shows the velocities relative to v_{sys} as a function of radius (in arcseconds) as derived for the H_2O megamaser spots (red markers) detected along $PA^{maser} = 140° \pm 5°$ (Greenhill et al. 1996). The dashed (solid) blue curve shows the best-fit sub-Keplerian (Keplerian) rotation curve $v_{rot} \propto r^{-a}$ of Greenhill et al. (1996) with $a = 0.31$ (0.50). Figure from Garcia-Burillo et al. (2016).

characterizing the AGN appearance, we conclude that the torus models in general do not appear to be robustly linked to the observed properties, making their scope to link theory to observations very difficult.

3.3. ALMA interferometry

González-Martín et al. (2016) used the Atacama Large Millimeter Array (ALMA) interferometer (Wootten & Thompson 2009) to map the NGC 1068 circumnuclear disk (CND), extended ∼300 pc, in the CO(6-5) transition and in the continuum at 432 μm at a spatial resolution of ∼4 pc.

The CND has been spatially resolved and its dust emission imaged, the molecular gas component shows a disk with a 7–10 pc diameter (Fig. 5-a). However, no clear rotation pattern is present in the data, even if this molecular disk might be a gas reservoir, it is hard to say that this is a stable structure. Moreover, the CO emission does not appear to be related to the H_2O masers orbiting the SMBH (Greenhill et al. 1996), indicating that the CO gas is not in Keplerian motion around the BH, as one would expect (Fig. 5-b). The dynamics of the molecular gas in the torus show instead strong non-circular motions and enhanced turbulence superposed on a surprisingly slow rotation pattern of the disk. How can we explain the BH feeding from such a turbulent structure? How universal is this picture?

Impellizzeri et al. (2019) mapped the inner region of tens of parsecs around the nucleus of NGC 1068 in the HCN (J = 3-2) transition at 256 GHz. They identify three kinematically distinct regions: (1) an outflow component in emission on the HCN position-velocity diagram and detected as a blueshifted wing in absorption against the nuclear continuum source, with projected outflow speeds approaching ∼450 km s^{-1}; (2) an inner disk with

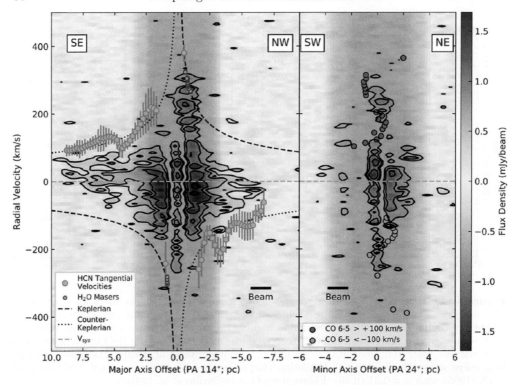

Figure 6. ALMA HCN (J = 3-2) position-velocity diagrams extracted along the HCN major axis (left panel) and minor axis (right panel) of the nuclear region of NGC 1068. On the major axis diagram, the tangential HCN velocities are marked by yellow dots with red errorbars, and, for comparison, the H$_2$O maser radial velocities along PA $\sim 131°$ are shown as filled blue circles (data from Greenhill & Gwinn 1997) corrected for the best-fit inclination, $i = 79°$. The best-fitting model for the H$_2$O masers, shown with a dashed line, is Keplerian rotation around a central mass of 1.66×10^7 M$_\odot$. Figure from Impellizzeri et al. (2019).

radius of ~1.2 pc; and (3) an outer disk extending to $r \sim 7$ pc. The two disks counter-rotate, and the kinematics of the inner disk agree with the H$_2$O megamaser disk mapped by the VLBA (Greenhill & Gwinn 1997). The outer disk shows a Keplerian rotation curve consistent with an extrapolation of the rotation curve of the inner disk. They also find that the HCN radial velocity field is more complex along the molecular outflow axis, which suggests detecting but not fully resolving HCN emission associated with the outflow (Fig. 6). They conclude that the molecular obscuring medium in NGC 1068 consists of counter-rotating and misaligned disks on parsec scales. It appears therefore that the observational situation is far more complicate than we would expect from the *simple* torus.

4. Low Luminosity AGN Unification

At low luminosities, AGN seem to depart from the classical unification scheme. The big blue bump – associated with the thermal emission from the accretion disk – is typically missing in these nuclei (Fig. 7), whose emission is usually dominated by a featureless, non-thermal power-law continuum (Ho et al. 1996; Ho 2008). Furthermore, some models predict the vanishing of the torus at low AGN luminosities ($\log(L_{\rm bol}/L_{\rm edd}) \lesssim -3$; $L_{\rm bol} \lesssim 10^{42}$ erg s^{-1}), when the radiation pressure from the nucleus can no longer sustain this structure (Elitzur & Shlosman 2006), following also the collapse of the BLR (Nicastro et al. 2003; Elitzur & Ho 2009). Radiatively inefficient accretion flow models (Narayan & Yi 1995) and jet outflow models (Falcke & Markoff 2000) were introduced to explain the

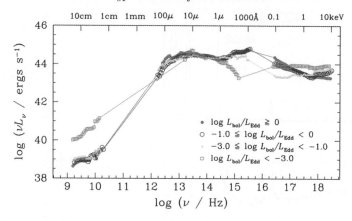

Figure 7. Composite SEDs for radio-quiet AGN binned by Eddington ratio. The SEDs are normalized at 1 μm. Figure from Ho (2008).

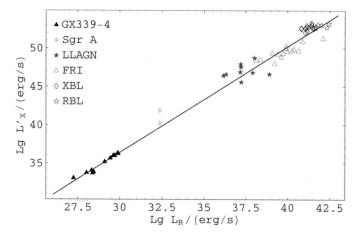

Figure 8. Radio/X-ray correlation for XRBs and AGN, where the X-ray flux of all AGN has been increased by a constant value of 10^7, corresponding to an average AGN mass of 3×10^9 M$_\odot$. Figure from Falcke et al. (2004).

disappearance of the accretion disk and ultimately the low radiative efficiency observed in a variety of sources across a wide range in BH masses, from X-ray binaries in the low-hard state (Markoff et al. 2001) to Sgr A* (Narayan et al. 1995; Falcke & Markoff 2000) and Low Luminosity AGN (LLAGN Yu et al. 2011; Markoff et al. 2008). In these models the accretion disk recedes at low accretion rates, and the innermost region is dominated by a geometrically-thick structure where the material is either advected towards the BH or ejected along the system axis forming a collimated jet or wind outflow. This causes the disappearance of the big blue bump, since viscous phenomena are not acting within the innermost and hottest disk radii, and the advected material does not radiate the energy gained through accretion. On the other hand, the higher radio activity observed in LLAGN is associated with the jet (Ho 2008).

This common framework to explain the physics of accretion at low luminosities have provided a wide unification for low-power BHs across the mass spectrum, which was confirmed by the discovery of the fundamental plane of BH accretion (Merloni et al. 2003; Falcke et al. 2004, Fig. 8). The latter is a tight correlation found between the

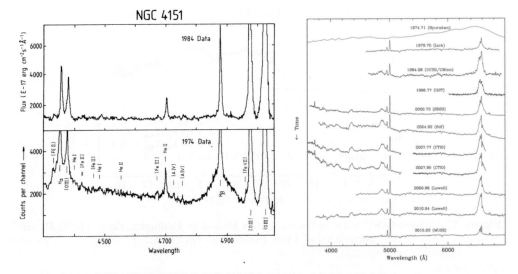

Figure 9. Left: fading out of the broad emission lines and the featureless blue continuum in the optical spectrum of NGC 4151 between 1974 and 1984 (lower and upper panels, respectively; Penston & Pérez 1984). **Right:** back and forth transition of Mrk 1018 from a type 2 in late 1979 to a type 1 in 2007 and back to the type 2 class after 2010 (Kim et al. 2018).

continuum emission at radio, the X-ray flux, and the BH mass, from stellar-mass BH to SMBHs, proving the common physics of the low-hard state in these systems over ∼8 orders of magnitude in mass.

5. The Mutant AGN

Variability is a well-know characteristic of active nuclei since their discovery (e.g. Peterson et al. 1982; Clavel et al. 1983), and has been successfully exploited to infer the size of the innermost components that cannot be spatially resolved with current facilities (e.g. Peterson et al. 1998). The first cases of spectral transitions in Seyfert galaxies were reported very early, even before the unification theory was settled in the field. For instance, the discovery of broad emission lines fading out from the optical spectra of NGC 4151 and 3C 390.3 within a timescale of 10 years (left panel in Fig. 9) was interpreted by Penston & Pérez (1984) as a possible evolutionary sequence in AGN. That is, the presence of the BLR in these nuclei would be linked to the activity of the central continuum source instead of the orientation with respect to the distribution of the obscuring material, thus disappearing when the continuum weakens. Broad emission lines emerging in type 2 and LINERs have also been detected, e.g. in NGC 1097 (Storchi-Bergmann et al. 1993), NGC 7582 (Aretxaga et al. 1999), and NGC 3065 (Eracleous & Halpern 2001), usually correlated with variations in the blue-featureless AGN continuum. Furthermore, back and forth transitions have also been detected for some nuclei that have returned to the initial spectral type after experiencing an earlier change, e.g. Mrk 1018 (Kim et al. 2018, right panel in Fig. 9), Mrk 590 (Mathur et al. 2018), NGC 1566 (Oknyansky et al. 2019; Parker et al. 2019), NGC 1365 (Risaliti et al. 2005). Recently, an increasing number of spectral transitions between type 1 and type 2 AGN have been reported, the so-called "changing-look" quasars (e.g. Matt et al. 2003; LaMassa et al. 2015), which can now be identified thanks to the wide sky monitoring surveys such as PanSTARSS (Chambers et al. 2016), CATALINA (Drake et al. 2009), etc.

In the context of unification such transitions were initially ascribed to the torus structure, and therefore the interpretation was made in terms of variations of the absorption column density. The latter could be caused by e.g. overdensities or clouds in the patchy torus that intercept our line of sight to the central engine, and thus to the BLR that originated the broad lines. However, variable absorption is not compatible with the mid-IR light curves of changing-look AGN, since the expected crossing time for the obscuring material is significantly longer than the observed mid-IR variability (Sheng et al. 2017). On the other hand, the low ($\lesssim 1\%$) UV polarisation observed in most of the 13 changing-look quasars observed by Hutsemékers et al. (2019) suggests that such transitions are not likely caused by changes the configuration of the dust obscuring structure. Therefore, it appears that the presence or absence of the BLR may be, at least in part, determined by the activity of the central continuum source, and not by the particular orientation of the system with respect to our line of sight.

6. The role of evolution and of the host galaxy

There is evidence accumulated in the last tens of years that nuclear activity is linked to the host galaxy population. This is primarily witnessed by the so-called *Magorrian relation* (Magorrian et al. 1998; Ferrarese & Merrit 2000) showing a correlation between SMBH masses and velocity dispersion, stellar mass and luminosity of their host galaxies in the local Universe and by the similar shapes, as a function of cosmic time, of the Star Formation (SF) density and BH Accretion (BHA) density, i.e. the so-called *AGN/SF co-evolution* (Madau & Dickinson 2014). Moreover, a possible evolution is envisaged between the various types of active galaxies. Two possible evolutionary progressions are HII → Seyfert type 2 (Storchi-Bergmann et al. 2001; Kauffmann et al. 2003), or a fuller scenario of HII→ Seyfert type 2 →Seyfert type 1 (Hunt & Malkan 1999; Levenson et al. 2001; Krongold et al. 2002). These predict that galaxy interactions, leading to the concentration of a large gas mass in the circumnuclear region of a galaxy, trigger starburst emission. Then mergers and bar-induced inflows can bring fuel to a central BH, stimulating AGN activity. While relatively young (<1 Gyr) stellar populations are found in more than half of Seyfert 2s (Schmitt et al. 1999; González Delgado et al. 2001; Raimann et al. 2003), they are also found in broad-lined AGN (Kauffmann et al. 2003). The photometric mid-IR studies of Edelson et al. (1987) and Maiolino et al. (1995) did indeed find that Seyfert 2s galaxies more often have enhanced star formation than Seyfert 1s.

A temporal link, if not a casual link, between SF and BHA has been demonstrated both observationally and theoretically. Wild et al. (2010) studied the growth of BHs, with masses of $10^{6.5}-10^{7.5}$ M$_\odot$ and AGN luminosities of $10^{42}-10^{44}$ erg s^{-1}, in 400 local galactic bulges which have experienced a strong burst of star formation in the past 600 Myr. During the first 600 Myr after a starburst, the BHs in the sample increase their mass by 5% on-average and the total mass of stars formed is about 10^3 times the total mass accreted onto the BH. This ratio is similar to the ratio of stellar to BH mass observed in present-day bulges. They also find that the average rate of accretion of matter onto the BH rises steeply roughly 250 Myr after the onset of the starburst (Fig. 10-a). Hopkins' (2012) simulations of AGN fueling by gravitational instabilities naturally produce a delay between the time when SFRs peak inside of a given annulus and the time when AGN activity peaks (Fig. 10-b). This offset scales as the gas consumption time, $\sim 10-100$ dynamical times. On small scales ($\lesssim 10$ pc), this is characteristically $\sim 10^7$ yr, rising to a few 10^8 yr on kpc scales. These offsets are similar to the magnitude of time offsets suggested by various observations on both small and large scales (Wild et al. 2010).

Furthermore, the evolution of AGN and their observational appearance is connected to the evolution of the host galaxy (Ballantyne et al. 2006). The ratio of obscured to unobscured AGN (R) increases with redshift, implying a change to the traditional unification model. Since the obscuring medium is changing with redshift, it must be influenced, e.g., by the cosmic star formation rate, which peaks at a very similar redshift as R does.

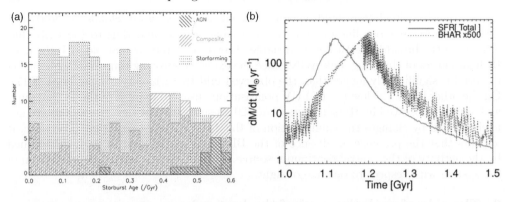

Figure 10. Left: (a) the number of star-forming (blue), composite-AGN (orange) and pure-AGN (red) in our sample as a function of time since the onset of the starburst. Figure from Wild et al. (2010). **Right:** (b) Galaxy merger simulation, near coalescence of the two galaxy nuclei (at t $\sim$ 1.1 Gyr). The total SFR peaks near this coalescence, as inflows first reach $\sim$kpc scales. The BHAR (here multiplied by 500 for ease of comparison) grows rapidly during this time, but sufficient gas remains to fuel BH growth until $\sim 10^8$ yr later. Figure from Hopkins' (2012).

As star formation increases in a galaxy, also the absorbing gas and dust increase, acting as an obscuring medium. The absorbing material would be located either close to the dust sublimation radius (and mimic some properties of the absorbing torus), or it could also be spread over most of the inner part of the galaxy (McLeod & Rieke 1995). The idea of an extended, more galactic-scale obscuring medium is consistent with IR observations, which have pointed out the remarkable similarity in the mid-to-far-IR emission between Seyfert 2s and 1s (Kuraszkiewicz et al. 2003; Lutz et al. 2004), in contrast to the predictions of the simple molecular torus model. Ballantyne et al. (2006) conclude that most of the accretion in the universe is obscured and that this obscuration evolves similar to the star formation rate. These facts deepen the connection between star formation and AGN fueling, as well as that between BH growth and galaxy evolution.

Buchanan et al. (2006), using the 12μm Seyfert galaxies sample (Spinoglio & Malkan 1989; Rush et al. 1993), find that the Seyfert 2 galaxies typically show stronger starburst contributions than the Seyfert 1 galaxies in the sample, contrary to what is expected based on the unified scheme for AGN. Tommasin et al. (2010), using the same local sample of AGN, found that the mid-IR emission properties characterize all the type 1 AGNs (which include both Seyfert type 1 and Hidden Broad Line galaxies) as a single family, with strongly AGN-dominated spectra. In contrast, the type 2 AGNs can be divided into two groups, the first one with properties similar to the type 1 AGNs except without detected broad lines, and the second with properties similar to the non-Seyfert galaxies, such as LINERs or starburst galaxies.

The combination of *Spitzer* and *Herschel* mid- and far-IR spectroscopy, respectively, of Seyfert galaxies, LINER and dwarf galaxies in the Local Universe, allowed us to define a line ratio diagram, the so-called *new IR BPT diagram* which can separate the various types of AGN through the ionized fine-structure lines (Fernández-Ontiveros et al. 2016). We report such diagram in Fig. 11, where we can easily see that the position of Seyfert type 1 galaxies is well displaced from that one of the Seyfert type 2 galaxies. Many of these latter galaxies lie closer to the Starburst galaxies region, while the LINER occupy a region which is intermediate between the Starburst galaxies and the Seyfert type 1 galaxies. We confirm here the result of Tommasin et al. (2010) that the "hidden broad-line region galaxies are indistinguishable form the Seyfert type 1's".

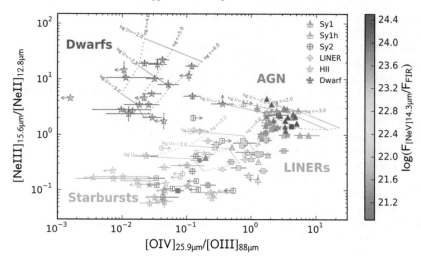

Figure 11. The [Ne III]$_{15.6\mu m}$/[Ne II]$_{12.8\mu m}$ line ratio *vs* the [O IV]$_{25.9\mu m}$/[O III]$_{88\mu m}$ ratio for AGN with different classifications based on the optical spectrum. Photoionisation models of AGN, LINER, starburst galaxies, and dwarf galaxies are shown as blue, green, yellow, and purple grids, respectively. The logarithmic values of the density (n_H) and ionisation potential (U) of the photoionisation models are indicated in the figures. Symbols are colour-coded according to their F$_{[Ne V]14.3\mu m}$/F$_{FIR}$ flux ratio, when available (see colour bar). Figure from Fernández-Ontiveros *et al.* (2016).

7. Conclusions

Our conclusions can be summarized as follows:

• The "unified model" has been conceived to explain the large *zoo* of different AGN with a single physical object; however, while it is well recognized that it works for unificating the so-called "Hidden Broad-Line Region Galaxies" (HBLR), i.e. those galaxies for which, either polarization spectra or near-IR ones, show broad lines, with the AGN type 1, it cannot be generalized to all AGN types.

• It seems so far that no strong observational evidence of tori with the needed characteristics to block BLR, collimate radiation and let AGN feeding has yet been found.

• In a large fraction of the changing-looking AGN, "transitions" can be explained by the AGN duty cycle.

• AGN and host galaxy interrelations may indicate an evolutionary path, e.g. HII → AGN2 → AGN1.

• In the next years the models of unification need to also consider this observational framework and not only simple orientation effects.

Acknowledgements

LS acknowledges Mirjana Povic for her kind invitation at the conferece in Addis Ababa, organized with the Ethiopian Space Science and Technology Institute (ESSTI). LS and JAFO acknowledge financial support by the Agenzia Spaziale Italiana (ASI) under the research contract 2018-31-HH.0.

References

Antonucci, R. R. J. 1993, *ARA&A*, 31, 473
Antonucci, R. 2012, *Astronomical and Astrophysical Transactions*, 27, 557
Antonucci, R. R. J. & Miller, J. S. 1985, *ApJ*, 297, 621

Aretxaga, I., Joguet, B., Kunth, D., et al. 1999, ApJL, 519, L123
Assef, R. J., Stern D., Kochanek C. S., et al. 2013, ApJ, 772, 26
Ballantyne, D. R. Everett, J. E., & Murray, N. 2006, ApJ, 639, 740
Buchanan, C. L., Gallimore, J. F., ODea, C. P., et al. 2006, AJ, 132, 401
Burlon, D., Ajello, M., Greiner, J., et al. 2011, ApJ, 728, 58
Chambers, K. C., Magnier, E. A., Metcalfe, N., et al. 2016, arXiv e-prints, arXiv:1612.05560
Clavel, J., Joly, M., Collin-Souffrin, S., et al. 1983, MNRAS, 202, 85
Drake, A. J., Djorgovski, S. G., Mahabal, A., et al. 2009, ApJ, 696, 870
Edelson, R. A., Malkan, M. A., Rieke, G. H., et al. 1987, ApJ, 321, 233
Elitzur, M. & Shlosman, I. 2006, ApJL, 648, L101
Elitzur, M. & Ho, L. C. 2009, ApJL, 701, L91
Eracleous, M. & Halpern, J. P. 2001, ApJ, 554, 240
Falcke, H. & Markoff, S. 2000, A&A, 362, 113
Falcke, H., Körding, E., & Markoff, S. 2004, A&A, 414, 895
Fernández-Ontiveros, J. A., Spinoglio, L., Pereira-Santaella, M., et al. 2016, ApJS, 226, 19
Ferrarese, L. & Merritt, D. 2000, ApJ, 539, L9
Gandhi, P. 2005, *Asian Journal of Physics*, 13, 90
García-Burillo, S., Combes, F., Ramos Almeida, C., et al. 2016, ApJ, 823, 12
Glindemann, A., Algomedo, J., Amestica, R., et al. 2003, SPIE, 4838, 89
González Delgado, R., Heckman, T., & Leitherer, C. 2001,ApJ, 546, 845
González-Martín, O., Masegosa, J., García-Bernete, I., et al. 2019, ApJ, 884, 10
González-Martín, O., Masegosa, J., García-Bernete, I., et al. 2019, ApJ, 884, 1
Greenhill, L. J., Gwinn, C. R., Antonucci, R., & Barvainis, R. 1996, ApJL, 472, L21
Greenhill, L. J. & Gwinn, C. R. 1997, Ap&SS, 248, 261
Heckman, T. M. & Best, P. N. 2014, ARA&A, 52, 589
Hickox, R. C. & Alexander, D. M. 2018, ARA&A, 56, 625
Ho, L. C., Filippenko, A. V., & Sargent, W. L. W. 1996, ApJ, 462, 183
Ho, L. C. 2008, ARA&A, 46, 475
Hönig, S. F. 2016, *Astronomy at High Angular Resolution*, 95
Hönig, S. F. & Kishimoto, M. 2017, ApJL, 838, L20
Hönig, S. F. 2019, ApJ, 884, 171
Hopkins, P. F. 2012, MNRAS, 420, L8
Hunt, L. K. & Malkan, M. A. 1999, ApJ, 516, 660
Hutsemékers, D., Agís González, B., Marin, F., et al. 2019, A&A, 625, A54
Impellizzeri, C. M. V., Gallimore, J. F., Baum, S. A., et al. 2019, ApJL, 884, L28
Jaffe W., Meisenheimer K., Röttgering H. J. A., et al. 2004, Nature, 429, 47
Kauffmann, G., Heckman, T. M., Tremonti, C. et al. 2003, MNRAS, 346, 1055
Kim, D.-C., Yoon, I., & Evans, A. S. 2018, ApJ, 861, 51
Krongold, Y., Dultzin-Hacyan, D., & Marziani, P. 2002, ApJ, 572, 169
Kuraszkiewicz, J. K. et al. 2003, ApJ, 590, 128
LaMassa, S. M., Cales, S., Moran, E. C., et al. 2015, ApJ, 800, 144
Leinert, Ch., Graser, U., Przygodda, F. et al. 2003, Ap&SS, 286, 73
Levenson, N. A., Weaver, K. A., & Heckman, T. M. 2001, ApJ, 550, 230
López-Gonzaga, N., Jaffe, W., Burtscher, L., et al. A&A, 565, A71
Lutz, D., Maiolino, R., Spoon, H. W. W., & Moorwood, A. F. M. 2004, A&A, 418, 465
Madau, P. & Dickinson, M. 2014, ARA&A, 52, 415
Magorrian, J. et al. 1998, AJ, 115, 2285
Maiolino, R. Ruiz, M., Rieke, G. H., Keller, L. D. 1995, ApJ, 466, 561
Markoff, S., Falcke, H., & Fender, R. 2001, A&A, 372, L25
Markoff, S., Nowak, M., Young, A., et al. 2008, ApJ, 681, 905
Mathur, S., Denney, K. D., Gupta, A., et al. 2018, ApJ, 866, 123
Matt, G., Guainazzi, M., & Maiolino, R. 2003, MNRAS, 342, 422
McLeod, K. K. & Rieke, G. H. 1995, ApJ, 441, 96
Merloni, A., Heinz, S., & di Matteo, T. 2003, MNRAS, 345, 1057

Narayan, R., Yi, I., & Mahadevan, R. 1995, *Nature*, 374, 623
Narayan, R. & Yi, I. 1995, *ApJ*, 452, 710
Netzer, H. 2015 *ARA&A*, 53, 365
Nicastro, F., Martocchia, A., & Matt, G. 2003, *ApJL*, 589, L13
Oknyansky, V. L., Winkler, H., Tsygankov, S. S., et al. 2019, *MNRAS*, 483, 558
Parker, M. L., Schartel, N., Grupe, D., et al. 2019, *MNRAS*, 483, L88
Peterson, B. M., Foltz, C. B., Byard, P. L., et al. 1982, *ApJS*, 49, 469
Peterson, B. M., Wanders, I., Bertram, R., et al. 1998, *ApJ*, 501, 82
Penston, M. V. & Pérez, E. 1984, *MNRAS*, 211P, 33
Raimann, D., Storchi-Bergmann, T., González Delgado, R. M., et al. 2003, *MNRAS*, 339, 772
Ramos Almeida, C. & Ricci, C. 2017, *Nat.As.*, 1, 679
Ricci, C., Ueda, Y., Koss, M. J. et al. 2015, *ApJ*, 815, L13
Risaliti, G., Elvis, M., Fabbiano, G., et al. 2005, *ApJL*, 623, L93
Rush, B., Malkan, M. A. & Spinoglio, L. 1993, *ApJSS*, 89, 1
Schmitt, H., Storchi-Bergmann, T., & Cid Fernandes, R. 1999, *MNRAS*, 303, 173
Seyfert, C. K. 1943, *ApJ*, 97, 28
Seyfert, C. K. 1946, *PAAS*, 10, 317
Sheng, Z., Wang, T., Jiang, N., et al. 2017, *ApJL*, 846, L7
Spinoglio, L. & Malkan, M. A. 1989, *ApJ*, 342, 83
Storchi-Bergmann, T., Baldwin, J. A., & Wilson, A. S. 1993, *ApJL*, 410, L11
Storchi-Bergmann, T., González Delgado, R. M., Schmitt, H. R., et al. 2001, *ApJ*, 559, 147
Tristram, K. R. W., Burtscher, L., Jaffe, W., et al. 2014, *A&A*, 563, A82
Tommasin, S., Spinoglio, L., Malkan, M. A., et al. 2010, *ApJ*, 709, 1257
Wild, V., Heckman, T., & Charlot, S. 2010, *MNRAS*, 405, 933
Wootten, A. & Thompson, A. R., 2009, *Proc. of the IEEE*, Vol. 97, Issue 8, 1463
Yu, Z., Yuan, F., & Ho, L. C. 2011, *ApJ*, 726, 87

Hypercat - hypercube of AGN tori

Robert Nikutta[1], Enrique Lopez-Rodriguez[2], Kohei Ichikawa[3,4], Nancy A. Levenson[5] and Christopher C. Packkham[6,7]

[1]NSF's National Optical-Infrared Astronomy Research Laboratory
950 N Cherry Avenue, Tucson, AZ 85719, USA
email: nikutta@noao.edu

[2]SOFIA Science Center, NASA Ames Research Center, Moffett Field, CA 94035, USA

[3]Frontier Research Institute for Interdisciplinary Sciences, Tohoku University, Sendai 980-8578, Japan

[4]Astronomical Institute, Tohoku University, Aramaki, Aoba-ku, Sendai, Miyagi 980-8578, Japan

[5]Space Telescope Science Institute, Baltimore, MD 21218, USA

[6]Department of Physics & Astronomy, University of Texas at San Antonio, One UTSA Circle, San Antonio, TX 78249, USA

[7]National Astronomical Observatory of Japan, 2-21-1 Osawa, Mitaka, Tokyo 181-8588, Japan

Abstract. We introduce HYPERCAT, a large set of 2-d AGN torus images computed with the state-of-the-art clumpy radiative transfer code CLUMPY. The images are provided as a 9-dimensional hypercube, in addition to a smaller hypercube of corresponding projected dust distribution maps. HYPERCAT also comprises a software suite for easy use of the hypercubes, quantification of image morphology, and simulation of synthetic observations with single-dish telescopes, interferometers, and Integral Field Units. We apply HYPERCAT to NGC 1068 and find that it can be spatially resolved in Near- and Mid-IR, for the first time with single-dish apertures, on the upcoming generation of 25–40m class telescopes. We also find that clumpy AGN torus models within a range of the parameter space can explain on scales of several parsec the recently reported polar elongation of MIR emission in several sources, while not upending basic assumptions about AGN unification.

Keywords. galaxies: active, galaxies: nuclei, galaxies: Seyfert, infrared: galaxies radiative transfer, instrumentation: high angular resolution, instrumentation: interferometers, methods: data analysis, techniques: high angular resolution

1. Introduction

Unification of active galactic nuclei (AGN) (e.g. Antonucci 1993; Urry & Padovani 1995) has explained much of the dichotomy between type 1 and type 2 AGN, by simply assuming an axially symmetric, anisotropic, dusty obscurer – a torus – and its orientation to the observer's line of sight (LOS). Complications were brought about by ever-improving observations, and these required modifications to the simple torus picture; the field has mostly converged on very compact (e.g. Jaffe *et al.* 2004; Gallimore *et al.* 2016) and clumpy (e.g. Nenkova *et al.* 2008a,b; Markowitz *et al.* 2014) distributions of dust and molecular gas surrounding the accreting supermassive black hole. The structure and dynamics of the molecular matter within the innermost few parsecs appear to be rather complex (Impellizzeri *et al.* 2019; Combes *et al.* 2019).

In the simple torus picture, most of the circum-nuclear dust is concentrated, by design, toward the torus equatorial plane, and that is where most of the dust emission is expected to emerge. Recent observation yet again are pushing the envelope; in several nearby sources, where high-spatial resolution interferometric or photometric observations are possible, the MIR emission shows clear signs of elongation along the system axis, not along its equatorial plane (e.g. Hönig et al. 2013; López-Gonzaga et al. 2016; Asmus 2019). At first this seems incompatible with torus-based unification. Modified models have been proposed to explain these observations e.g. as an inversion of the cloud distribution from toroidal to bi-conical (Hönig & Kishimoto 2017), or through interplays between a tilted accretion disk that anisotropically illuminates a hollow, bi-conical, dusty wind (Stalevski et al. 2017). Tristram et al. (2014) suggested for the case of Circinus that the elongation is due to seeing the inner wall of a torus funnel slightly tilted toward us.

Today's single-dish telescopes can not resolve the central tens of parsecs even in the closest sources. Second-generation IR interferometers such as *VLTI* combine light from only two telescopes, thus do not achieve phase closure. This fundamentally prevents image reconstruction. However, the observed visibilities as function of baseline, position angle, and wavelength can be compared to models of brightness distribution. The most common approach so far has been to model a synthetic brightness distribution with a linear combination of 2d Gaussian components. Even in its simplest incarnation, this approach required no fewer than 18 free parameters in the case of Circinus (Tristram et al. 2014). The upcoming class of 25–40m giant telescopes (GMT, TMT, ELT) will for the first time deliver model-free resolved imagery of the central parsecs in nearby AGN at IR wavelengths. It is clear that further progress must come from modeling the observed brightness distribution as the result of radiative transfer through the underlying physical, 3d, distribution of dust. Only inferences following this approach will be able to constrain the actual geometry and physics and parsec scales. HYPERCAT's image and dust cubes are developed precisely for that purpose.

2. Hypercubes of light and dust

Model SEDs of AGN emission are abundant and have been used with success in the literature, but studies using resolved model images remain in their infancy. First, because until recently no spatially resolved data were available and, second, because computation, storage, and processing of image sets with sufficient parameter coverage are prohibitively expensive. *VLTI* and *ALMA*, and upcoming facilities such as *TMT*, *GMT* and *ELT* have or will alleviate the first problem. HYPERCAT is designed to break the second barrier.

It is a user-friendly software that hides the complexity of handling very large multi-dimensional hypercubes of data. HYPERCAT can generate an image of the torus emission (or a dust map) for any combination of parameters via multilinear interpolation in fractions of a second. It can also simulate observations of these images with single-dish telescopes. These include PSF convolution (provided by the user or computed by HYPERCAT), image transformations (e.g. rotation, scaling, resampling), noise addition to a signal-to-noise ratio (SNR), and deconvolution. Interferometry can also be simulated for a set of (u,v) points. Furthermore, spatially resolved maps of spectral properties can be analyzed akin to integral-field-units (IFU). Finally, a module to compute image morphology quantifiers completes the HYPERCAT software. HYPERCAT is open-source, written in Python, and will be publicly available shortly[1]. We also release our image hypercubes[2] generated with CLUMPY, which we describe below.

2.1. Model hypercubes

CLUMPY models the distribution of dust clouds around an AGN as an axisymmetric function of cloud number density, requiring four free parameters – ratio of outer to inner

[1] Please visit https://clumpy.org/
[2] Available at: ftp://ftp.noao.edu/pub/nikutta/hypercat/

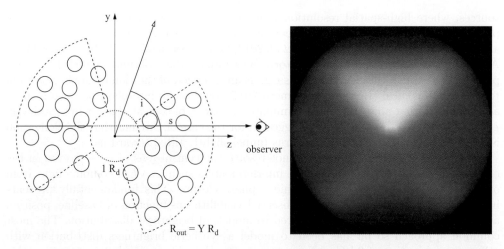

Figure 1. *Left:* Vertical cut through a schematic toroidal cloud distribution. The torus angular edge is soft, i.e. cloud number falls off from the mid-plane like a Gaussian. *Right:* What an observer in the left panel might see; composite of three CLUMPY brightness maps at 4.6, 10, 25 μm (blue, green, red). The relative contributions of all maps were equalized for clarity. The model has parameters $\sigma = 30$ deg, $i = 80$ deg, $Y = 20$, $\mathcal{N}_0 = 7$, $q = 0.1$, and $\tau_V = 75$. In this configuration the MIR emission occurs predominantly from polar directions, i.e. from the inner walls of the torus cone (green color).

torus radii, Y; mean number of clouds along a radial ray in the equatorial plane, $\mathcal{N}_0$; exponent q of the radial power-law profile $1/r^q$ (with r the distance); and torus angular width σ measured in degrees from the equatorial plane. In addition, the viewing angle i between the system axis and the observer's LOS, and the wavelength λ are required to quantify a model. Dust radiative transfer calculations performed by CLUMPY result in an image $I(x,y)$ discretized along axes x and y. Figure 1 shows on the left a schematic view of a cloud distribution and the propagation of photons along the LOS. On the right is what an observer will see – in this case a composite image of three distinct wavelengths.

A hypercube of all computed images is 9-dimensional. Our sampling of parameters yields 336,000 combinations, times 25 wavelengths between 1.2 and 945μm. Together with the ≈ 2000 times smaller corresponding dust maps, the hypercubes contain 2.45×10^{11} values (when storing half-images), at a resolution of six pixels per dust sublimation radius. This corresponds to 913/271 GB of raw/compressed storage. Computation consumed 3.2 years of CPU time. Downloading and storage are tractable with a broad-band internet connection and storage capacities on modern laptops, but handling a data hypercube of this size is not. HYPERCAT was designed to jump this entry barrier.

2.2. *Software suite*

The HYPERCAT software suite comprises Python modules, accessible through a high-level API suitable for most users. A low-level set of functions allows to operate on images more directly. A simple GUI is also provided, which allows image generation for any parameter combination, saving to a FITS file for further processing, or displaying of images in DS9[1] directly. A user manual and example Jupyter notebooks facilitate a quick start with HYPERCAT.

---

[1] SAO DS9: http://ds9.si.edu/

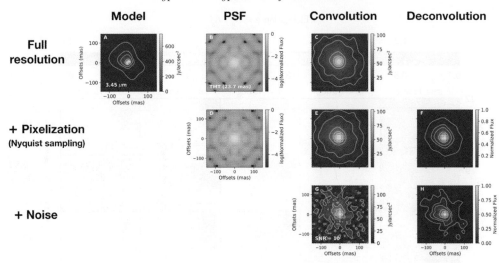

Figure 2. HYPERCAT step-by-step process to obtain synthetic observations of the 2-d dust emission distribution using CLUMPY models. *(A)* Dust emission image at 3.45 μm of NGC 1068 produced by a CLUMPY model using the best-fit parameters obtained via SED fitting by Lopez-Rodriguez et al. (2018) ($\sigma = 45°$, $Y = 18$, $\mathcal{N}_0 = 4$, $q = 0.08$, $\tau_V = 70$, $i = 75°$). *(B)* PSF at 3.45 μm estimated using the pupil image of the TMT at full resolution (1.21 mas). *(C)* Convolved image using the full resolution dust emission (A) and PSF (B) images. *(D)* PSF pixelated at the Nyquist sampling of the TMT at 3.45 μm. Pixel scale is 11.6 mas. *(E)* Pixelated dust emission image of NGC 1068 at same pixel scale as D. *(F)* Deconvolved dust emission image using D and E. *(G)* Gaussian noise with SNR = 10 at the peak pixel was applied to E. *(H)* Deconvolved image using the pixelated PSF D and noisy image G.

3. Application to NGC 1068

We applied HYPERCAT to NGC 1068, a nearby AGN with very good data available from many campaigns. Applying best-fit CLUMPY model parameters found by Lopez-Rodriguez et al. (2018) through detailed SED fitting, HYPERCAT interpolated the corresponding images at all wavelengths of interest. Figure 2 shows in panel (A) the brightness map at 3.45μm, and outlines how synthetic observations can be simulated step-by-step in HYPERCAT using this input image. The steps include convolution with a PSF (provided by user or computed by HYPERCAT from telescope pupil images), addition of noise to a specified SNR, and image deconvolution. A major result is that the upcoming generation of extremely large telescopes will indeed resolve the circum-nuclear dust emission in several AGN in the L, M and N bands, as is evident from e.g. panel (H) in Fig. 2.

HYPERCAT can also simulate simple interferometric observations, and analyze spectrally-resolved observations per-pixel, akin to IFUs. We will publish these results in a forthcoming paper (Nikutta et al., in prep.).

4. Morphology

We developed a mechanism based on image moments to quantify morphological properties of images generated by CLUMPY, which allow us to measure the extension of an emission region in x and y directions independently, to measure the extension direction and amplitude, and other useful morphological properties. We investigated for the case of NGC 1068 which parameter variations produce image elongations compatible with those reported in observations. Figure 3 shows the elongation e, defined as the ratio of

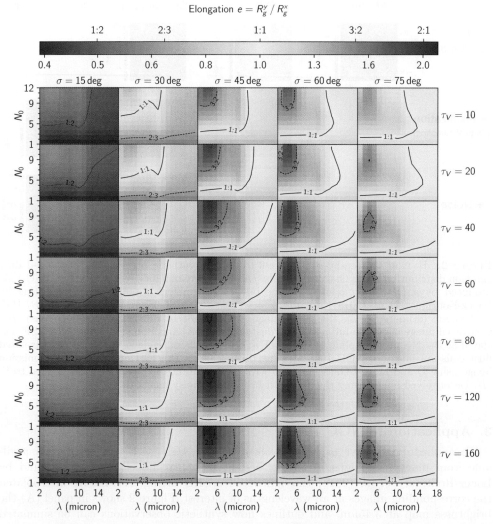

Figure 3. Image elongation e as function of varied parameters, shown as heatmaps. Red colors show elongation in y-direction (polar), blue colors in x-direction (equatorial). Some best-fit parameters values from SED fitting of NGC 1068 are held fixed, i.e. $i = 75$ degrees, $q = 0.08$, $Y = 18$, while all other parameters vary as indicated. All panels are normalized to the same range and are logarithmically stretched separately for values smaller/greater than 1.0 (see color bar). Contour lines show aspect ratios 1:1 (solid), 2:3 and 3:2 (dashed), 2:1 and 1:2 (dotted), also marked on the colorbar (upper tickmarks).

so-called gyration radii in both directions, as a function of model parameters and wavelengths. Dark-red regions in the graphs indicate parameter spaces that produce strong polar elongations of 2:1 and higher.

5. Summary

HYPERCAT provides *(i)* a very large hypercube of AGN torus images computed with CLUMPY, *(ii)* the corresponding dust maps, and *(iii)* a software suite to make easy use of the maps, and to simulate observations with single-dish telescopes, interferometers, and IFUs. It also comprises a module to estimate various image morphology parameters. We have applied HYPERCAT to the case of NGC 1068 in an attempt to derive the torus geometry, and to quantify the emerging morphology of the observed brightness distributions. We find that in this case, and for a range of model parameters, clumpy tori can indeed produce emission maps strongly elongated in the polar directions of the system, requiring only a small tilt of the torus axis toward the observer. This affords a view of the MIR emission coming off the inner torus walls. The elongations can reach up to several pc in real systems, which can explain much of the very-high resolution data from e.g. VLTI. It is likely not sufficient, by the nature of model sizes, to explain MIR elongations on scales of tens of pc; these are presumably caused by other mechanisms, e.g. by a dusty outflow.

References

Antonucci, R. 1993, *ARA&A*, 31, 473, doi: 10.1146/annurev.aa.31.090193.002353
Asmus, D. 2019, *MNRAS*, 2220, doi: 10.1093/mnras/stz2289
Combes, F., García-Burillo, S., Audibert, A., *et al.* 2019, *A&A*, 623, A79, doi: 10.1051/0004-6361/201834560
Gallimore, J. F., Elitzur, M., Maiolino, R., *et al.* 2016, *ApJL*, 829, L7, doi: 10.3847/2041-8205/829/1/L7
Hönig, S. F. & Kishimoto, M. 2017, *ApJL*, 838, L20, doi: 10.3847/2041-8213/aa6838
Hönig, S. F., Kishimoto, M., Tristram, K. R. W., *et al.* 2013, *ApJ*, 771, 87, doi: 10.1088/0004-637X/771/2/87
Impellizzeri, C. M. V., Gallimore, J. F., Baum, S. A., *et al.* 2019, *ApJL*, 884, L28, doi: 10.3847/2041-8213/ab3c64
Jaffe, W., Meisenheimer, K., Röttgering, H. J. A., *et al.* 2004, *Nature*, 429, 47, doi: 10.1038/nature02531
López-Gonzaga, N., Burtscher, L., Tristram, K. R. W., Meisenheimer, K., & Schartmann, M. 2016, *A&A*, 591, A47, doi: 10.1051/0004-6361/201527590
Lopez-Rodriguez, E., Fuller, L., Alonso-Herrero, A., *et al.* 2018, *ApJ*, 859, 99, doi: 10.3847/1538-4357/aabd7b
Markowitz, A. G., Krumpe, M., & Nikutta, R. 2014, *MNRAS*, 439, 1403, doi: 10.1093/mnras/stt2492
Nenkova, M., Sirocky, M. M., Ivezić, Ž., & Elitzur, M. 2008a, ApJ, 685, 147, doi: 10.1086/590482
Nenkova, M., Sirocky, M. M., Nikutta, R., Ivezić, Ž., & Elitzur, M. 2008b, *ApJ*, 685, 160, doi: 10.1086/590483
Stalevski, M., Asmus, D., & Tristram, K. R. W. 2017, *MNRAS*, 472, 3854, doi: 10.1093/mnras/stx2227
Tristram, K. R. W., Burtscher, L., Jaffe, W., *et al.* 2014, *A&A*, 563, A82, doi: 10.1051/0004-6361/201322698
Urry, C. M., & Padovani, P. 1995, *PASP*, 107, 803, doi: 10.1086/133630

Towards a new paradigm of dust structure in AGN: Dissecting the mid-IR emission of Circinus galaxy

Marko Stalevski[1], Daniel Asmus[2] and Konrad R. W. Tristram[3]

[1] Astronomical Observatory, Volgina 7, 11060 Belgrade, Serbia
email: mstalevski@aob.rs

[2] Dept. of Physics & Astronomy, University of Southampton,
SO17 1BJ, Southampton United Kingdom

[3] European Southern Observatory, Casilla 19001, Santiago 19, Chile

Abstract. Recent mid-infrared (MIR) observations of nearby active galactic nuclei (AGN), revealed that their dust emission appears prominently extended in the polar direction, at odds with the expectations from the canonical dusty torus. This polar dust, tentatively associated with dusty winds driven by radiation pressure, is found to have a major contribution to the MIR flux from a few to hundreds of parsecs. One such source with a clear detection of polar dust is a nearby, well-known AGN in the Circinus galaxy. We proposed a phenomenological model consisting of a compact, thin dusty disk and a large-scale polar outflow in the form of a hyperboloid shell and demonstrated that such a model is able to explain the peculiar MIR morphology on large scales seen by VLT/VISIR and the interferometric data from VLTI/MIDI that probe the small scales. Our results call for caution when attributing dust emission of unresolved sources entirely to the torus and warrant further investigation of the MIR emission in the polar regions of AGN.

Keywords. galaxies: individual (Circinus), galaxies: active, galaxies: nuclei, galaxies: Seyfert, infrared: galaxies, radiative transfer, radiation mechanisms: thermal

1. Introduction

In a widely-accepted picture, dust in active galactic nuclei (AGN) is contained around the equatorial plane in a roughly toroidal shape, nicknamed "the dusty torus" (Antonucci 1993; Netzer 2015). However, recent mid-infrared (MIR) interferometric observations revealed that thermal dust emission in AGN appears to be originating dominantly along the polar direction, following the orientation of the ionisation cone, jet, polarisation angle or the accretion disc rotational axis (Hönig et al. 2012; López-Gonzaga et al. 2016). Furthermore, polar dust emission found to extend out to tens or even hundreds of parsecs (Asmus et al. 2016). The orientation of the polar dust emission on large and small scales roughly match, indicating that both might have a same physical origin, namely a dusty wind driven by the radiation pressure on the dust grains close to the sublimation radius. These findings have a potential to lead us towards a new paradigm for the dust structure in AGN.

2. Data

We obtained Circinus images at 8.6 μm and 11.9 μm with the VISIR instrument mounted on the Very Large Telescope (VLT) as part of its science verification after the upgrade (Stalevski et al. 2017). We complemented them with archival 10.5 μm and

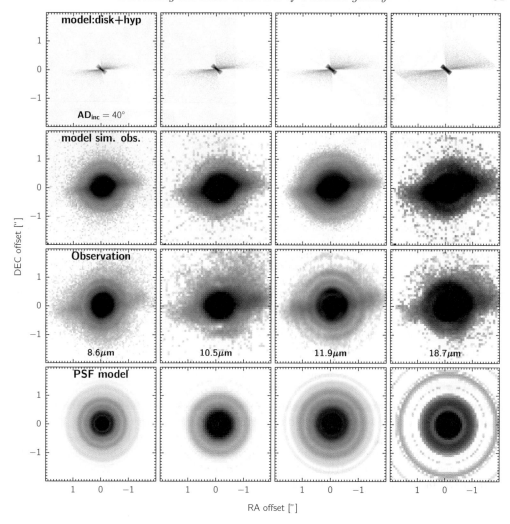

Figure 1. Comparison of the VISIR images of Circinus with the representative model, including foreground extinction. From top to bottom, the rows show: (1) the model images; (2) the model images as they would appear when observed with VISIR; (3) the images of Circinus acquired with VISIR; (4) our approximation of the observed PSF. Adapted from Stalevski et al. (2017)

18.7 μm VISIR images. The central 4" × 4" region of these images is shown in the third row of Fig. 1. A prominent bar-like feature is present in all of them, extending to ∼40 pc on both sides of the nucleus, and coinciding with the edge of the ionization cone seen at visible wavelengths on the Western side. Circinus was also observed with the MID-infrared Interferometric instrument (MIDI) at the VLT interferometer between 8 and 13 μm (Tristram et al. 2014). For comparison of the model and the data over a wider wavelength range, we also assembled the observed nuclear SED and spectra of Circinus from data available in the literature.

3. Model

The illumination pattern of the ionization cone (brighter toward the western edge) is indicative of the anisotropic emission pattern of the ionizing source. This is supported by the orientation of the inner part of the maser disk, which is roughly perpendicular to the

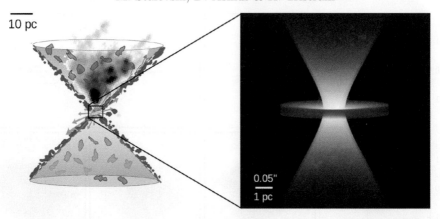

Figure 2. Left: Schematic of our model for the nuclear dust distribution in Circinus consisting of a compact dusty disk and a dusty hollow cone enveloping ionized gas seen in the optical (show in gray scale). The blue arrows illustrate the anisotropic emission pattern of the accretion disk, whose orientation matches the orientation of the inner part of the warped maser disk. Right: zoom into the central region, where our modeling of the MIDI data revealed a thin dusty disk and a dusty wind shaped like a hyperboloid; a colour composite image (in logarithmic scale) of our best model made by mapping the 8, 10 and 13 μm flux images obtained with radiative transfer simulations to the blue, green and red, respectively. Adapted from Stalevski et al. (2017) and Stalevski et al. (2019)

cone edge. An optically-thick, geometrically-thin accretion disk displays a $\cos\theta$ angular-dependent luminosity profile. Aligned with the inner part of the warped maser disk, such a disk will emit more strongly into or close to the western edge of the cone. If the cone wall is dusty, then the described setup could naturally produce the dusty bar seen in the VISIR image. The opposite side of the cone wall would remain cold and invisible, as the tilted anisotropic accretion disk would emit very little in that direction. Thus, we proposed a model consisting of a geometrically thin disc and a hollow cone (i.e. hyperboloid shell), as depicted in the schematic in Fig. 2. The dust composition and grain size distribution is a mixture of silicates (53%) and graphite (47%) in the disc and only graphite in the polar region, in both cases following the power-law size distribution ($\propto a^{-3.5}$) with grain sizes in the range of $a = 0.1-1$ μm. A foreground absorbing screen is applied to all the models to account for extinction by dust in the disc of the host galaxy. To calculate how the above described model would appear in the IR, we employed SKIRT†, a state–of–the–art 3D radiative transfer code based on the Monte Carlo technique (Baes et al. 2011; Baes & Camps 2015; Camps & Baes 2015). The photons are propagated through the simulation box following the standard Monte Carlo radiation transfer prescriptions that take into account all the relevant physical processes: anisotropic scattering, absorption and thermal re-emission. At the end of the simulation, images and SEDs of the model can be reconstructed for the comparison with the observations in the entire wavelength range of interest.

4. Results and conclusions

In Fig. 1, we compare the VISIR images of Circinus with the synthetic observations of our representative model, produce by convolution with the instrumental PSF and including background noise and foreground extinction. We see that the simulated model images provide a good match to the observed morphology at all wavelengths.

† http://www.skirt.ugent.be

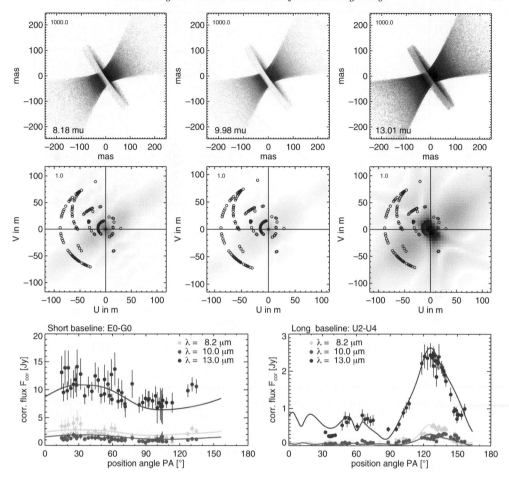

Figure 3. The interferometric diagnostic plots for the best disc+hyperboloid model. From top to bottom, the rows show: (1) The model images at 8, 10 and 13 μm. (2) The Fourier transform of the model surface brightness distribution in the uv plane with blue circles marking the positions of the interferometric measurements. (3) The correlated fluxes ($F_{\rm cor}$) of the interferometric measurements (dots) and of the model (lines) as a function of the position angle (PA) of the projected baseline lengths (BL). The signal seen in $F_{\rm cor}$ vs. PA is well reproduced by the model at both short and long baselines and at all wavelengths.

Fig. 3 represents the diagnostic plots we used to compare the model to the MIDI data. This figure contains our disc+hyperboloid model images (first row) and the Fourier transforms of the model surface brightness (second row), but the most important information is in the third row, featuring correlated flux as a function of the position angle, for short and long baselines and at three wavelengths. Here we show correlated fluxes for the two telescope combinations which probe different spatial scales alongn various directions. The correlated fluxes of the short baselines (third row, left panel) probe larger spatial scales. They show a wide minimum at PA $\sim 90°$ indicating extended emission in this direction, which is reproduced by the dusty hyperboloid shell of our model. The correlated fluxes of the long baselines (third row, right panel), which probe smaller scales, have a prominent peak at PA $\sim 125°$. This suggests a very elongated surface brightness of the source, which is in our model reproduced by the dusty disc seen almost edge-on.

In Fig. 4, we compare our VISIR-based model (solid black line) and the MIDI-based models (disc+hyperboloid in dash-dotted red line and clumpy version of the same model

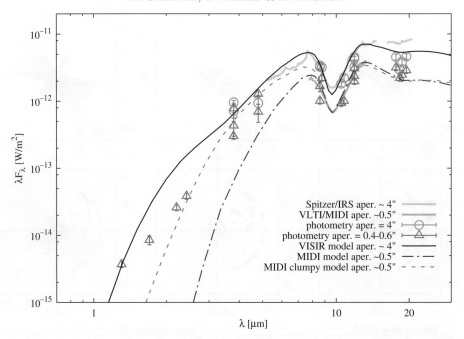

Figure 4. Comparison of the observed SED with the VISIR-based model SED from Stalevski et al. (2017) (solid black line), the MIDI-based models from Stalevski et al. (2019) (a smooth model in dash-dotted red line and a clumpy model in dashed blue line). Large-scale aperture photometry (4") is shown in green down-pointing triangles, while photometry extracted from apertures comparable or smaller than the resolution limit of VISIR (0.4−0.6") is marked by red up-pointing triangles. The aperture of the Spitzer/IRS spectrum ($\geqslant 3.6$") is comparable to the total aperture of VISIR in $5.2 - 14.5$ μm range, while significantly larger at longer wavelengths. The MIDI spectrum was extracted using a $\sim 0.54" \times 0.52"$ aperture and hence corresponds to the unresolved nucleus with VISIR. Adapted from Stalevski et al. (2019).

in dashed blue line) with the observed SED. For all the observations, we measured the fluxes consistently in a 4" diameter aperture (corresponding to the total VISIR aperture) and a 0.4−0.6" diameter aperture (corresponding to the unresolved core in VISIR and the MIDI total flux spectrum aperture). The clumpy disc+hyperboloid model accounts well for the entire SED, except at the shortest wavelength which are likely contamintated by the host galaxy.

We performed detailed modeling of the Circinus AGN, constructing a model based on observations across a wide range of wavelengths and spatial scales. The model consists of a compact dusty disc and a large-scale hollow dusty cone illuminated by a tilted accretion disc. We showed that this model is able to reproduce well the observed 40 pc scale MIR morphology of Circinus, its entire IR SED and the interferometric observations at all baselines (Stalevski et al. 2017 and Stalevski et al. 2019). Our results reinforce calls for caution when using the dusty tori models to interpret the IR data of AGN. Further study of polar dust emission in larger samples is necessary to constrain its properties and assess its ubiquity in the whole AGN population.

References

Antonucci, R. 1993, *ARA&A*, 31, 473
Asmus, D., Hönig, S. F., & Gandhi, P. 2016, *ApJ*, 822, 109
Baes, M., Verstappen, J., De Looze, I., et al. 2011, *ApJS*, 196, 22
Baes, M. & Camps, P. 2015, *Astronomy and Computing*, 12, 33

Camps, P., & Baes, M. 2015, *Astronomy and Computing*, 9, 20
Hönig, S. F., Kishimoto, M., Antonucci, R., et al. 2012, *ApJ*, 755, 149
López-Gonzaga, N., Burtscher, L., Tristram, K. R. W., et al. 2016, *A&A*, 591, A47
Netzer, H. 2015, *ARA&A*, 53, 365
Tristram, K. R. W., Burtscher, L., Jaffe, W., et al. 2014, *A&A*, 563, A82
Stalevski, M., Asmus, D., & Tristram, K. R. W. 2017, *MNRAS*, 472, 3854
Stalevski, M., Tristram, K. R. W., & Asmus, D. 2019, *MNRAS*, 484, 3334
Tristram, K. R. W., Burtscher, L., Jaffe, W., et al. 2014, *A&A*, 563, A82

Predicting the emission profile and estimation of model parameters for some nearby LLAGN using accretion and jet models

Bidisha Bandyopadhyay[1], Fu-Guo Xie[2], Neil M. Nagar[1], Dominik R. G. Schleicher[1], Venkatessh Ramakrishnan[1], Patricia Arévalo[3], Elena López[3] and Yaherlyn Diaz[3]

[1]Departamento de Astronomía, Facultad Ciencias Físicas y Matemáticas, Universidad de Concepción, Av. Esteban Iturra s/n Barrio Universitario, Casilla 160-C, Concepción, Chile
email: bidisharia@gmail.com

[2]Key Laboratory for Research in Galaxies and Cosmology, Shanghai Astronomical Observatory, Chinese Academy of Sciences, 80 Nandan Road, Shanghai 200030, China

[3]Instituto de Física y Astronomía, Facultad de Ciencias, Universidad de Valparaíso, Gran Bretana No. 1111, Playa Ancha, Valparaíso, Chile

Abstract. The Event Horizon Telescope (EHT) provides a unique opportunity to probe the physics of supermassive black holes through Very Large Baseline Interferometry (VLBI), such as the existence of the event horizon, the accretion processes as well as jet formation in Low Luminosity AGN (LLAGN). We build a theoretical model which includes an Advection Dominated Accretion Flow (ADAF) and a simple radio jet outflow. The predicted spectral energy distribution (SED) of this model can be compared to observations to get the best estimates of the model parameters. The model-predicted radial emission profiles at different frequency bands can be used to predict whether the inflow can be resolved by the EHT or other telescopes. We have applied this method to some nearby LLAGN such as M84, NGC 4594, NGC 4278 and NGC 3998. We also estimate the model parameters for each of them using high resolution data from different surveys.

Keywords. accretion, accretion disks, black hole physics, Galaxy: nucleus

1. Introduction

The detection of the photon ring around the black hole in the nucleus of M87 (EHT Collaboration 2019) with the Event Horizon Telescope (EHT) using very long baseline interferometric (VLBI) techniques has opened a new window to probe regions in the extreme proximity of supermassive black holes. This advancement in science and technology has not only enabled us to test Einstein's theory of General Relativity but also to probe regions in the accretion flow which were unresolvable before. It is thus of profound importance to investigate the various physical processes involved in the accretion flow to understand the powering source of such systems. It is hence also important to probe more such systems to enhance our knowledge about the physical processes. Besides the EHT, the global 3-mm VLBI array (GMVA), which operates at 86 GHz, imaging at high resolutions (few tens of microarcsecs), will enable to observe accretion regions as well as the jet base in nearby accreting supermassive black holes.

In this proceeding we present the primary results that we obtained in our recent work (Bandyopadhyay et al. 2019) where we compared our modeled spectral energy distribution (SED) to the observed high resolution multi wavelength (MW) data for some selected candidates to better estimate the model parameters. Most of these systems are expected to accrete at a sub-Eddington rate and are generally radiatively inefficient with an advection dominated accretion flow and hence are also known as radiatively inefficient accretion flows (RIAFs), a subgroup of the advection dominated accretion flows (ADAFs). Due to the presence of a magnetic field and low densities, the primary emission processes contributing to the SED from the ADAF are thermal synchrotron emission, bremsstrahlung emission and comptonised emission by the thermal electrons. Often a fraction of the thermal electrons is boosted to power-law electrons which too emit via synchrotron emission. At arcsec to sub-arcsec scales, the emission from the jet base may also contribute to the observed SED and thus it is important to include the jet emission (primarily synchrotron) to match the observed data set and have a better estimate on the model parameters. We then used the estimated parameters of the accretion flow to obtain a radial profile of the emission which allows us to predict which of the sources could possibly be resolved and detected by the EHT and the GMVA.

In the following sections, there is a brief description of the equations and the parameters involved in the accretion flow. We then test our model for M87 which has been extensively studied in literature (Nemmen et al. 2014; Li et al. 2016) and also for which we now have the EHT results. We then present the results for two of the sources which we expect to be resolvable by the EHT.

2. A brief description of the model

The steady state flow equations for an ADAF with a two temperature plasma can be written in terms of the four conservation equations of mass, radial momentum, angular momentum and energy and are expressed as follows:

$$\dot{M}(R) = \dot{M}(R_{tr}) \left(\frac{R}{R_{tr}}\right)^s = 4\pi \rho R H |v|. \tag{2.1}$$

$$v\frac{dv}{dR} - \Omega^2 R = -\Omega_K^2 R - \frac{1}{\rho}\frac{d}{dR}(\rho c_s^2). \tag{2.2}$$

$$\frac{d\Omega}{dR} = \frac{v\Omega_K(\Omega R^2 - j)}{\alpha R^2 c_s^2}. \tag{2.3}$$

$$\rho v \left(\frac{de_i}{dR} - \frac{p_i}{\rho^2}\frac{d\rho}{dR}\right) = (1-\delta)q^+ - q^{ie}.$$

$$\rho v \left(\frac{de_e}{dR} - \frac{p_e}{\rho^2}\frac{d\rho}{dR}\right) = \delta q^+ + q^{ie} - q^-. \tag{2.4}$$

Here R_{tr} is the truncation radius, s is the parameter that quantifies density reductions due to outflows, $H = c_s/\Omega_K$ is the scale height, ρ is the gas density, Ω is the angular momentum of the in-falling gas, Ω_K is the angular momentum of the Keplerian orbit, v is the radial velocity of the gas, c_s is the sound speed, j is the angular momentum at the gravitational radius R_g and is an eigenvalue for the system under consideration, α is the viscosity parameter, e_e and e_i are the specific internal energies of the electrons and ions, respectively, δ is the fraction of the viscous energy (q^+) that goes into heating the electrons (Xie & Yuan 2012; Chael et al. 2018), q^{ie} the energy that is exchanged between electrons and ions and q^- is the energy lost via radiation. In this work, we express the mass accretion rate $\dot{M}$ in terms of the Eddingtion accretion rate $\dot{M}_{\rm Edd}$ through the dimensionless parameter $\dot{m}$ as $\dot{M}(R) = \dot{m}(R)\dot{M}_{\rm Edd}$. This parameter at R_g is equal to the Eddington ratio ($L_{\rm Bol}/L_{\rm Edd}$) in case of thin disk accretion flows but is higher for ADAF. We use the Eddington ratio as a lower limit for $\dot{m}(R_g)$. We vary this parameter

Table 1. Model A and Model B parameter values for M87.

Model	$\dot{m}_{tr}$	δ	s	j	p_l	η	$\dot{m}_{jet}$	p_{jet}	ϵ_e	ϵ_b	ξ
Model A	4.2×10^{-4}	0.1	0.1	0.7999	3.0	0.015	1.0×10^{-8}	2.6	0.0009	0.0006	0.01
Model B	1.2×10^{-4}	0.5	0.3	1.8360	3.0	0.015	1.0×10^{-8}	2.5	0.0009	0.0006	0.01

by varying $\dot{m}(R_{\text{tr}})$ (from now on $\dot{m}_{\text{tr}}$) and s using eq. [2.1]. The pressure (p_i and p_e) in eq. [2.4] is the gas pressure ($p_{\text{gas}} = p_i + p_e$) expressed in terms of the total pressure ($p_{\text{tot}} = p_{\text{gas}} + p_{\text{magnetic}}$) as $p_{gas} = \beta p_{\text{tot}}$.

The accretion dynamics is more complex due to turbulence, the presence of magnetic fields, hot spots and outflows. Narayan & Yi 1994; Narayan & Yi 1995; Blandford & Begelman 1999 postulated that ADAFs should have strong winds followed by the formation of jets. In this work, we use a phenomenological model (Spada *et al.* 2001) to describe the jet, which is sufficient to model the SED. It is assumed to be composed of a normal plasma, consisting of electrons and protons, with velocities determined by a bulk Lorentz factor $\Gamma_j = 10$ (typical for jets in AGN as in Lister *et al.* 2016). In this model, a fraction ξ of the electrons is boosted to a power law (power law index p_{jet}) energy distribution due to internal shocks within the jet. Parameters defining the fraction of the shock energy that goes into electrons and magnetic fields, ϵ_e and ϵ_B respectively, are included. The mass loss rate $\dot{M}_{\text{jet}}$ is sensitively coupled with the jet outflow velocity V_{jet} (assumed to be constant for all our LLAGN) which controls the beaming effect and gas density.

Additional parameters are included when a fraction η of the thermal electrons is boosted to become power-law (p_l) electrons in the accretion flow and these electrons then emit via synchrotron emission. We then used these combined models to obtain the SED and cross match with the data.

3. Results

We initially applied our results to M87 which has a black hole mass of $6.5 \pm 0.7 \times 10^9 M_\odot$ and is at a distance of 16.8 Mpc. We used two set of model parmeter values (model A and model B) to fit to the high resolution data (Prieto *et al.* 2016) where the difference in the two models was primarily the choice of the accretion rate and the outflow parameter. We obtained a better fit to the data using the model A parameter values (see table [1] for the model parameter values) where the accretion rate agrees with the GRMHD simulation of Mościbrodzka *et al.* 2016. The SED fits with this model are shown in the left panel of fig. [1]. We then used these parameter values to obtain the radial emission profile from the ADAF at 230 GHz (EHT), 86 GHz (GMVA) and 22 GHz (EVN). The right panel of fig. [1] displays the radial profiles at the frequencies mentioned for the cases with and without the emission from non-thermal electrons in the ADAF.

We then applied our model to the high resolution data (see the Appendix in Bandyopadhyay *et al.* 2019) of the five sources (Cen A, M84, NGC 4594, NGC 3998 and NGC 4278) and obtained the best parameter value to our model to finally obtain the radial profile for these sources. Cen A which has a comparatively smaller black hole mass is also one of the nearest sources with a radio loud core. Although resolving the photon ring may not be possible, the derived radial profile of the emission suggests that a part of the accretion flow may be detectable with the EHT and also the GMVA as shown in fig. [2]. The other source whose accretion flow can be detectable by the EHT according to our analysis is NGC 3998. This source is well studied in the X-ray with various probes but is not quite radio loud. As can be seen in the right panel of fig. [2], a part of the inner accretion flow may be observable and detectable by the EHT with emission only from thermal electrons. Our analysis of all the five sources is briefly mentioned in table [3].

Table 2. List of the sources we studied based on their mass, distance, ring sizes and Eddington ratios.

Source	$log(M_{BH}/M_\odot)$	Distance (Mpc)	θ_{Ring} (μas)	Eddington Ratio (L_{Bol}/L_{Edd})
NGC 5128 (Cen A)	7.7	3.8	1.5	5.0×10^{-4}
NGC 4374 (M84)	8.9	17.1	4.8	5.0×10^{-6}
NGC 4594 (Sombrero, M 104)	8.5	9.1	3.6	1.5×10^{-6}
NGC 3998	8.9	13.1	6.2	1.0×10^{-4}
NGC 4278	8.6	14.9	2.7	5.0×10^{-6}

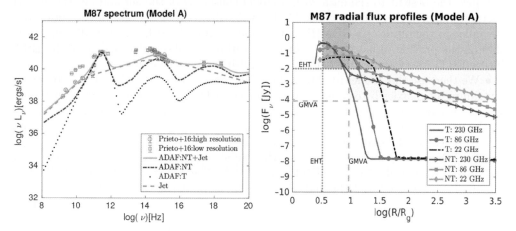

Figure 1. *Left*: SED model (Model A) fit to the high resolution data (Prieto *et al.* 2016). *Right*: The radial flux profile with the same model. The pink shaded region marks the region of detectability by the EHT only and the yellow by the GMVA while pink marks the common region of detectability.

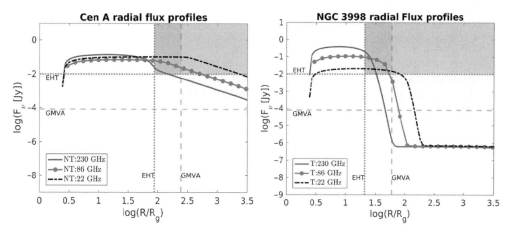

Figure 2. Radial flux profiles for Cen A and NGC 3998 respectively

4. Discussion and Conclusion

We summarize our main conclusions as follows:

• The framework described in section [2] was tested with the SED of M87 considering models A and B. The basic motivation to select these models was to consider different accretion rates and outflow parameters which compare to the values in literature (Nemmen *et al.* 2014; Li *et al.* 2016; Mościbrodzka *et al.* 2016). With these models, we

Table 3. The table displays if the sources are detectable and resolvable by the EHT or the GMVA in an 8 hour integration time.

Source	Resolvable (EHT)	Detectable (EHT)	Resolvable (GMVA)	Detectable (GMVA)
Cen A	Yes	Partly the outer region	Outer regions	Only the outer regions
M84	Partially	No	No	No
NGC 4594	Partially	No	No	No
NGC 3998	Yes	Yes	Partially	Partly the outer region
NGC 4278	Outer regions	No	Outer regions	Outer regions

obtained the best fit to the data with a model consisting of the Jet and an ADAF with thermal plus non-thermal electrons. Model A fits the data better and the accretion rate is similar to the result of a GRMHD simulation by Mościbrodzka et al. 2016. Since both the models provide radial profiles which are within the observable regime of the EHT, future EHT observations may help to distinguish the two scenarios.

• We then use this model to obtain the model parameters for each of the 5 sources in our sample of galaxies by comparing the modeled SED with the observed data. Although we may not be able to resolve the region of maximum emission from the ADAF for Cen A, the flow can still be partially observed at larger radii due to the flux from non-thermal electrons. Table. [3] summarizes these predictions.

• With our model fits, we find that the radial profile of NGC 3998 is expected to be resolved very well with both EHT and GMVA. The ADAFs of M84 and NGC 4594 may be fairly resolved by EHT and not with GMVA, but may not be observable within the current flux limit of the EHT. To observe these, we need better sensitivities of the telescope and longer integration times.

References

Bandyopadhyay, B., Xie, F.-G., Nagar, N. M., Schleicher, D.R.G., Ramakrishnan, V., Arévalo, P., López, E., & Diaz, Y. 2019, *MNRAS*, 490, 4606
Blandford, R. D. & Begelman, M. C. 1999, *MNRAS*, 303, L1
Chael, A., Rowan, M., Narayan, R., Johnson, M., & Sironi, L. 2018, *MNRAS*, 478, 5209
Event Horizon Telescope Collaboration 2019, *ApJ*, 875, L1
Li, Y.-P., Yuan, F., & Xie, F.-G. 2016, *ApJ*, 830, 78
Lister, M. L., Aller, M. F., Aller, H. D., Homan, D. C., Kellermann, K. I., Kovalev, Y. Y., Pushkarev, A. B., Richards, J. L., Ros, E., & Savolainen, T. 2016, *AJ*, 152, 12
Mościbrodzka, M., Falcke, H., & Shiokawa, H. 2016, *A&A* (Letters), 586, A38
Narayan, R. & Yi, I. 1994, *ApJ*, 428, L13
Narayan, R. & Yi, I. 1995, *ApJ*, 444, 231
Nemmen, R. S., Storchi-Bergmann, T., & Eracleous, M. 2014, *MNRAS*, 438, 2804
Prieto, M. A., Fernández-Ontiveros, J. A., Markoff, S., Espada, D., & González-Martín, O. 2016, *MNRAS*, 457, 3801
Spada, M., Ghisellini, G., Lazzati, D., & Celotti A. 2001, *MNRAS* 325, 1559
Xie, F.-G. & Yuan, F. 2012, *MNRAS*, 427, 1580
Yuan, F., Peng, Q., Lu, J., & Wang, J. 1999, *ApJ*, 537, 236

A new tool to derive chemical abundances in type-2 active galactic nuclei

Rubén García-Benito[1], Enrique Pérez-Montero[1], Oli L. Dors[2], José M. Vílchez[1], Monica V. Cardaci[3,4] and Guillermo F. Hägele[3,4]

[1]Instituto de Astrofísica de Andalucía, Apartado de correos 3004,
E-18080 Granada, Spain
emails: rgb@iaa.es, epm@iaa.es

[2]Universidade do Vale do Paraíba, Av. Shishima Hifumi, 2911, Cep 12244-000,
São José dos Campos, SP, Brazil

[3]Instituto de Astrofísica de La Plata (CONICET-UNLP), Argentina

[4]Facultad de Ciencias Astronómicas y Geofísicas, Universidad Nacional de La Plata,
Paseo del Bosque s/n, 1900 La Plata, Argentina

Abstract. We present a new tool for the analysis of the optical emission lines of the gas in the Narrow Line Region (NLR) around Active Galactic Nuclei (AGNs). This new tool can be used in large samples of objects in a consistent way using different sets of optical emission-lines taking into the account possible variations from the O/H - N/O relation. The code compares certain observed emission-line ratios with the predictions from a large grid of photoionization models calculated under the most usual conditions in the NLR of AGNs to calculate the total oxygen abundance, nitrogen-to-oxygen ratio and ionization parameter. We applied our method to a sample of Seyfert 2 galaxies with optical emission-line fluxes from the literature. Our results confirm the high metallicity of the objects of the sample and provide consistent values with the direct method. The usage of models to calculate precise ICFs is mandatory when only optical emission lines are available to derive chemical abundances using the direct method in NLRs of AGN.

Keywords. methods: data analysis, ISM: abundances, galaxies: abundances, galaxies: active, galaxies: Seyfert

1. Introduction

The energetic radiation coming from the central black holes in galaxies is partially re-emitted by the surrounding gas as very bright emission lines which in turn can be used to derived the physical conditions in these extreme regions. Since they can be observed up to very high redshifts, Active galactic Nuclei (AGNs) are thus a powerful source for the study of cosmic evolution of galaxies.

It is widely accepted (Ferland & Netzer 1983) that the main mechanism of the narrow-line region (NLR) in AGNs is photoionization. However, it is also known that the total metallicity derived using the direct method (i.e. the T_e method) gives sub-solar metallicities in AGNs, as compared to the predictions from photoionization models. Using a sample of NLRs of AGNs, Dors et al. (2015) found that the T_e-method using the optical lines underestimated the oxygen abundances by an averaged value of $\sim$0.8 dex as compared to calibrations based on photoionization models.

Models are, therefore, a powerful tool to interpret the observed lines and provide valuable information to study chemical abundances. In this work, we describe a new code based on photoionization models to derive chemical abundances in the NLR in AGNs.

2. The code

In Pérez-Montero et al. (2019) we present a full description of a new code to derive the total oxygen abundance, nitrogen-to-oxygen ratio (N/O), and the ionization parameter (U) from the analysis of optical emission lines in the NLR of type-2 AGNs. The code is based on the well proven HII-CHI-MISTRY† code (hereafter HCM, Pérez-Montero 2014) originally developed for the analysis of star-forming regions. The advantages of the code are: a) it can be applied to a large number of objects in an automatic way; b) all objects are analyzed in a consistent way regardless of the set of input emission lines; c) it provides uncertainties for all the estimated quantities; d) it provides an independent estimation of the N/O ratio; and e) it is consistent with the direct method.

The code uses a grid of 5 865 photoionization models run with the code Ferland et al. (2017) v.17.01. The models cover a wide range of the parameters space with typical NLRs conditions (see Pérez-Montero et al. 2019 for further details). The spectral energy distribution (SED) is composed by two components: the Big Blue Bump at 1 Ryd and a power law with spectral index $\alpha_X = -1$. The continuum between 2KeV and 2500 Å is modeled by a power law with spectral index $\alpha_{OX} = -0.8$. All models were calculated using a spherical geometry with a filling factor of 0.1, a standard dust-to-gas ratio and a constant density of 500 particles per cm^{-3}. In addition we checked the effect of changing in the models the α(ox) down to -1.2 and enhancing the electron density up to $2\,000$ cm^{-3} but no noticeable changes were found in the calculation of the chemical abundances using the method described here. For more details on the results of these comparison see Pérez-Montero et al. (2019). The models cover the range of $12 + \log(O/H)$ from 6.9 to 9.1 in bins of 0.1 dex. The N/O range goes from -2.0 to 0.0 in bins of 0.125 dex and $\log U$ from -4.0 to -0.5 in bins of 0.25 dex.

The code uses as input the reddening-corrected relative-to-Hβ emission line intensities from [[O II]] λ3727 Å, [Ne III] λ3868 Å, [[O III]] λ4363 Å, [[O III]] λ5007 Å, [[N II]] λ6583 Å, and [S II] $\lambda\lambda$6717+6731 Å with their corresponding errors. However, the code is adapted to provide also a solution in case one or several of these lines are not given.

In short, the work-flow of the code is as follows. First, the code constrain the parameter space searching for N/O as a weighted mean over all models, using optical emission lines for similar excitation, such as the ratio [[N II]]λ6583/[[O II]]λ3727 or [[N II]]λ6583/[S II]$\lambda\lambda$6717+6731. These ratios do not show almost any dependence on excitation and therefore N/O can be calculated without any assumption about the ionization parameter. Using the uncertainties of all the input observed lines, the code calculates the error using a Monte Carlo simulation. A set of line ratios such as [[O III]]λ5007/[[O III]]λ4363, [[N II]]λ6583/Hα,

$$R23 = \frac{[OII]\lambda 3727 + [OIII]\lambda\lambda 4959 + 5007}{H\beta},$$

$$O3N2 = \log\left(\frac{[OIII]\lambda 5007}{H\beta} \cdot \frac{H\alpha}{[NII]\lambda 6583}\right),$$

or

$$O2Ne3 = \frac{[OII]\lambda 3727 + [NeIII]\lambda 3868}{H\beta}$$

(depending on the availability of the observed lines) are used in a second iteration to sample a subset of models constrained to the N/O values previously calculated to obtain the oxygen abundance and the ionization parameter.

† Publicly available in the webpage https://www.iaa.csic.es/ epm/HII-CHI-mistry.html.

Table 1. Mean and standard deviation of the residuals between the O/H from Dors et al. (2017) and the values derived by HCM using different input lines as shown in Figure 1. In the table [[O II]] stands for $\lambda 3727$ Å, [[O III]]$_n$ for $\lambda 5007$ Å, [[N II]] for $\lambda 6583$ Å, and [S II] for $\lambda\lambda 6717+6731$ ÅÅ.

Input emission lines	Mean Δ(O/H)	St.dev. Δ(O/H)
All lines	−0.01	0.21
[[O III]]$_n$, [[N II]], [S II]	+0.15	0.26
[[O II]], [[O III]]$_n$	−0.11	0.21
[[O III]]$_n$, [[N II]]	−0.24	0.15
[[N II]]	−0.25	0.16
[[N II]], [S II]	+0.29	0.29

3. Results

3.1. The control sample

No empirical derivation of chemical abundances (i.e. no abundances using the direct method) in the NLR of AGNs using optical emission lines are available in the literature. Therefore, we use as a control sample the abundance estimations by Dors et al. (2017) obtained from detailed tailored photoionization models using the CLOUDY code. They compiled a sample of 47 Seyfert 1.9 and 2 galaxies at a redshift $z \leqslant 0.1$ providing the most prominent optical emission lines, including the auroral line [OIII] 4363 Å. They do not provide an error estimation of the oxygen abundances obtained from their models.

3.2. Comparisons

In Fig. 1 we compare the total oxygen abundance derived for the control sample by Dors et al. (2017) with those obtained by HCM when all or only some of the input lines are used†. The option of restricting the number of input lines simulates common observing conditions when only limited sensitivity or spectral coverage of the detector is available. The left upper panel shows the best case scenario when all possible emission lines are provided. There is a good agreement between both sets with a dispersion of 0.21 dex and a residual of -0.01 dex. The upper right panel displays the relation when lines [[O III]] λ 4363 Å and [[O II]] λ 3727 Å are not included. This is common case when the blue part of the spectrum is not available (e.g. in the Sloan Digital Sky Survey at very low redshifts) and the [[O III]] λ 4363 Å is to faint to be observed. In this case, the dispersion is nearly the same but the residual increases by 0.1 dex. Even when only a couple of lines or only [NII] λ 6583 Å is available the agreement is good, with deviations from the abundances lower than the usual uncertainties. Table 1 shows the mean and standard deviation of the residuals of the comparison cases presented in Fig. 1.

3.3. Consistency with the direct method

There is known discrepancy between the chemical abundances derived using the T_e method in NLRs of type-2 AGNs, leading to very low values if compared to those obtained from some photoionization models (e.g. Dors et al. 2015). The code HCM has proved to be in accordance with the T_e method in star-forming regions (Pérez-Montero 2014). Thus, we can use HCM to investigate the possible origin of the discrepancies in AGNs.

In Fig. 2 we show the total oxygen abundance derived by Dors et al. (2017) for their sample of Sy2 galaxies, compared to the addition of the abundances of the most prominent

† More detailed comparisons including N/O and the ionization parameter can be found in Pérez-Montero et al. (2019).

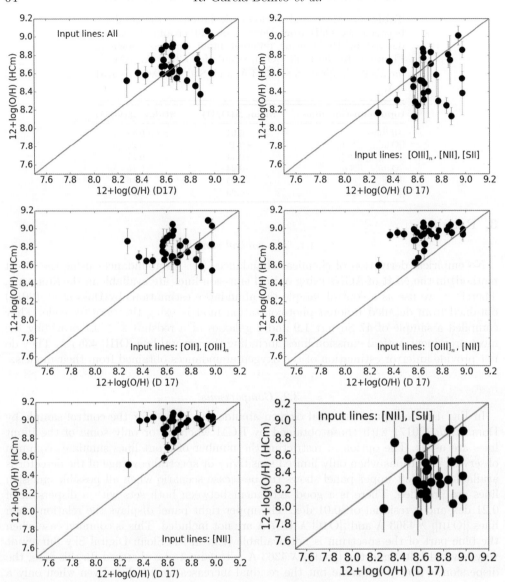

Figure 1. Comparison between total oxygen abundances $12 + \log(O/H)$ derived using the method described in this work (HCM) and those taken from Dors et al. (2017) from tailored photoionization models. *Upper left*: comparison when all the lines are used. *Upper right*: comparison in the absence of [[O III]] λ 4363 Å and [[O II]] λ 3727 Å. *Middle left*: comparison when only [[O II]] λ 3727 Å and [[O III]] λ 5007 Å are included. *Middle right*: comparison when only [[O III]] λ 5007 Å and [[N II]] λ 6583 Å are included. *Bottom left*: comparison when only [[N II]] λ 6583 Å is provided. *Bottom right*: comparison when only [[N II]] λ 6583 Å and [S II] λ 6717+6731 ÅÅ are included. The solid line represents the 1:1 relation. In the legend [[O II]] stands for λ3727 Å, [[O III]]$_n$ for λ5007 Å, [[N II]] for λ6583 Å, and [S II] for $\lambda\lambda$6717+6731 ÅÅ.

oxygen ionic species in the optical part of the spectrum, i.e. O^+ and O^{2+}, calculated using the T_e method. The addition of the relative ionic abundances of the oxygen is 0.7 dex lower that the one derived by the models. Figure 2 shows also predictions from the grid of models for different ionization parameter values. The difference is well explained as an important dependence on the total metallicity and ionization parameter. This result

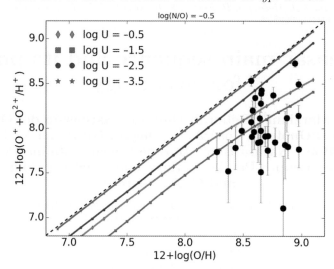

Figure 2. Comparison between total oxygen abundance and the addition of the abundances of the two main oxygen ions O^+ and O^{2+} observed in the optical part of the spectrum. Models are represented using solid lines for different values of U. Black circles represent the data from Dors et al. (2017) whose total abundances where calculated using tailored models, while their ionic abundances were calculated following the T_e method. The dashed black line represents the 1:1 relation.

highlights the importance of using models to derive the total oxygen abundance in NLRs of AGNs when only opital lines are available, as ionization correction factors (ICFs) are far to be negligible, contrary to star-forming regions.

References

Dors, O. L., Cardaci, M. V., Hägele, G. F., Rodrigues, I., Grebel, E. K., Pilyugin, L. S., Freitas-Lemes, P., Krabbe, A. C., et al. 2015, *MNRAS* 453, 4102
Dors, Jr. O. L., Arellano-Córdova, K. Z., Cardaci, M. V., Hägele, G. F., et al. 2017, *MNRAS*, 468, L113
Ferland, G. J. & Netzer, H. 1983, *ApJ*, 264, 105
Ferland et al. 2017, *Rev. Mexicana AyA* 53, 385
Pérez-Montero, E. 2014, *MNRAS* 441, 2663
Pérez-Montero, E., Dors, O. L., Vílchez, J. M., García-Benito, R., Cardaci, M. V., Hägele, G. F., et al. 2019, *MNRAS*, 489, 2652

The quasar main sequence and its potential for cosmology

Paola Marziani[1], Deborah Dultzin[2], Ascensión del Olmo[3]†, Mauro D'Onofrio[4], José A. de Diego[2], Giovanna M. Stirpe[5], Edi Bon[6], Natasa Bon[6], Bożena Czerny[7], Jaime Perea[3], Swayamtrupta Panda[7,8], Mary Loli Martinez-Aldama[3,7] and C. A. Negrete[2]

[1]INAF, Osservatorio Astronomico di Padova, Italy
email: paola.marziani@inaf.it

[2]Instituto de Astronomía, UNAM, Mexico

[3]IAA (CSIC), Granada, Spain

[4]Università di Padova, Italy

[5]INAF, OASS, Bologna, Italy

[6]Belgrade Observatory, Serbia

[7]Center For Theoretical Physics, Polish Academy of Sciences, Warsaw, Poland

[8]Nicolaus Copernicus Astronomical Center, Polish Academy of Sciences, Warsaw, Poland

Abstract. The main sequence offers a method for the systematization of quasar spectral properties. Extreme FeII emitters (or extreme Population A, xA) are believed to be sources accreting matter at very high rates. They are easily identifiable along the quasar main sequence, in large spectroscopic surveys over a broad redshift range. The very high accretion rate makes it possible that massive black holes hosted in xA quasars radiate at a stable, extreme luminosity-to-mass ratio. After reviewing the basic interpretation of the main sequence, we report on the possibility of identifying virial broadening estimators from low-ionization line widths, and provide evidence of the conceptual validity of redshift-independent luminosities based on virial broadening for a known luminosity-to-mass ratio.

Keywords. quasars: general, quasars: emission lines, line: profiles, cosmological parameters, distance scale, dark matter

1. Introduction

The concept of the quasar main sequence originated from a Principal Component Analysis of parameters measured on the optical spectra of ~ 80 PG quasars (Boroson & Green 1992). The first eigenvector 1 (E1) computed by the analysis was found to be mainly associated with two anti-correlations, between strength of FeIIλ4570 and prominence of [OIII]$\lambda\lambda$4959, 5007, and strength of FeIIλ4570 and FWHM of the HI Balmer line Hβ. In the plane FWHM Hβ vs. prominence of the optical FeII emission (where the FeII prominence is measured by the parameter R_{FeII} defined as the intensity ratio of the FeII λ4570 blend and Hβ), the occupation of data points representing low redshift quasars takes the form of an elbow-shaped sequence (Fig. 1). Quasar spectra show a wide range of line widths, profile shapes, R_{FeII}, line shifts, line intensities which imply differences in

† AdO acknowledges financial support from the Spanish grants AYA2016-76682-C3-1-P and the "Center of Excellence Severo Ochoa" award for the IAA (SEV-2017-0709).

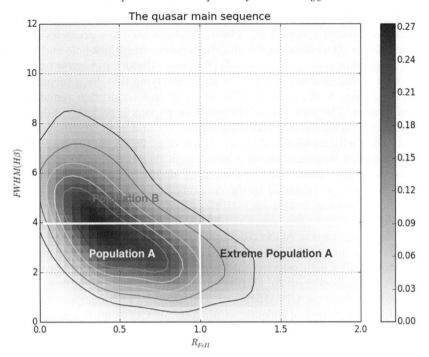

Figure 1. The optical plane of the E1 MS, FWHM of the broad component of Hβ vs $R_{\rm FeII}$. Isodensity curves and shading represent the source occupation from the sample of Zamfir et al. (2010), with ≈ 300 quasars. The labels identify the loci of the main populations: Population A with FWHM $\leqslant 4000$ km s^{-1}; Population B with FWHM > 4000 km s^{-1}; extreme Population A, $R_{\rm FeII} > 1$. See text for more details.

line emitting gas dynamics and ionization levels: the main sequence organizes different properties (Sulentic et al. 2000).

Why are these two parameters – FWHM Hβ and $R_{\rm FeII}$ – so important? FeII emission extends from UV to the IR and can dominate the thermal balance of the low-ionization part of the broad-line region (BLR, Marinello et al. 2016). FeII emission is self-similar (at least to a first approximation) in quasars but FeII intensity with respect to Hβ changes from object to object. The FWHM(Hβ) is explained mainly by Doppler broadening, and is related to projection of the velocity field in the low-ionization BLR along the line of sight. There has been a growing consensus that the low-ionization BLR is predominantly virialized, since the early results of the first major reverberation mapping campaigns (Peterson & Wandel 1999). Therefore $R_{\rm FeII}$ and FWHM Hβ can be considered tracers of the physical and dynamical conditions in the low-ionization BLR. Sulentic et al. (2000) introduced the distinction between two Populations. The FWHM H$\beta \leqslant 4000$ km s^{-1} condition selects narrower sources that are preferentially moderate-to-strong FeII emitters (Population A). Broader sources (FWHM H$\beta > 4000$ km s^{-1}, Population B) tend to have low FeII emission and are believed to be *predominantly* sources radiating at lower Eddington ratio ($L/L_{\rm Edd}$) than the quasars of Population A (Marziani et al. 2018a).

2. The main driver of the quasar main sequence

Several approaches consistently support a relation between Eddington ratio and $R_{\rm FeII}$. For instance, according to the fundamental plane of accreting massive black holes for reverberation-mapped active galactic nuclei (AGNs) – a relation connecting $R_{\rm FeII}$, $L/L_{\rm Edd}$ and a parameter D dependent on line shape (Du et al. 2016) – $R_{\rm FeII} > 1$ implies

$L/L_\mathrm{Edd} > 1$. An independent confirmation is provided by the analysis of the stellar velocity dispersion of the host spheroid, $\sigma_\star$. The $\sigma_\star$ has been used as a proxy for the black hole mass in accordance with established scaling laws connecting black hole and host spheroid mass (e.g., Magorrian et al. 1998). If the AGNs are subdivided in narrow luminosity bins, the $\sigma_\star$ is found to decrease as a function of R_FeII, implying that R_FeII increases as a function of L/L_Edd (Sun & Shen 2015). The origin of this connection remains unclear at the time of writing. The structure of the emitting regions is probably influenced by the balance between gravitation and radiation forces (Ferland et al. 2009), and by a change in accretion mode with increasing L/L_Edd (Wang et al. 2014). So the correlation may be a secondary effect of structural changes induced on the BLR (Panda et al. 2019). Since R_FeII is also dependent on metallicity (Panda et al. 2018), high L/L_Edd might be associated with enriched gas provided to the emitting region by nuclear and circumnuclear star formation (D'Onofrio & Marziani 2018).

Outflows from mildly-ionized gases producing blue shifted lines from ionic species of IP $\lesssim 50$ eV are ubiquitous in AGNs (Richards et al. 2011). The extent and the energetics of the outflows are not yet fully appreciated. For example, it has been possible only in recent year to consider high velocity outflows from very hot gas (the so-called ultra-fast outflows, UFOs, Tombesi et al. 2011). However, the prominence of outflows traced by the mildly-ionized gas emitting CIVλ1549 and [OIII]$\lambda\lambda$4959,5007 increases along the main sequence (MS, see Figure 4 of Sulentic et al. 2000 for CIV) and reaches a maximum in correspondence of the strongest FeII emitters, at extreme values of L/L_Edd.

Several interpretations of the MS (not necessarily in contradiction among themselves) have been proposed. The occupation of data points can be accounted for by a trend between Eddington ratio and R_FeII, convolved with the effect of orientation on the FWHM of Hβ (Marziani et al. 2001; Shen & Ho 2014). This is especially true in the case of a small range of black hole masses. There is a degeneracy between the effect of black hole mass and orientation as both tend to increase the FWHM of Hβ (Marziani et al. 2018a). Panda et al. (2019) showed that, for a fixed black hole mass, there is a limit in FWHM beyond which the orientation effects cannot go. Even if limits to the FWHM range spanned by orientation broadening can be set, the overlap of mass and orientation effects implies that, in the 2D representation of the MS, it is impossible to independently retrieve these parameters (viewing angle, L/L_Edd, black hole mass) for individual quasars. A 3D representation of the MS or additional constraints are necessary. Another important side effect of the degeneracy between mass and viewing angle is that a broad range of black hole mass leads to the MS with a wedge-shaped occupation area (Shen & Ho 2014).

An alternative view considers that Population B quasars are more massive and radiate at lower values of the L/L_Edd than quasars belonging to Population A. This difference between the two populations helps define the shape of the main sequence *as a sequence* in the optical plane, and is most likely a consequence of the down-sizing of nuclear activity at low redshift (Fraix-Burnet et al. 2017). So it is possible to establish an evolutionary connection from the sources of extreme Population A to the ones of Population B, where the cosmic arrow of time is provided by the black hole mass which can only grow: the larger the mass the older the source.

3. Highly-accreting quasars

Extreme Population A (xA) quasars satisfy the condition $R_\mathrm{FeII} > 1$; they are those $\sim 10\%$ of quasars in low-z ($\lesssim 1$), optically selected samples with extreme FeII emission. Their prevalence is steeply decreasing toward higher R_FeII values: in the range $1.0 \leqslant R_\mathrm{FeII} < 1.5$ we find ≈ 7 % of all quasars; in the range $1.5 \leqslant R_\mathrm{FeII} < 2$, ≈ 3 %. Quasars with $R_\mathrm{FeII} > 2$ account for less than 1% of the optically-selected quasar population (Marziani et al. 2013). xA quasars show distinctive features: their UV continuum is usually not reddened, they are often with extremely weak UV emission lines (i.e., weak-lined quasars with W(CIV)λ1549 $\leqslant$ 10Å following Diamond-Stanic et al. 2009). The prominent AlIIIλ1860, and very weak CIII]λ1909 allow for easy UV selection criteria: if (1) $R_{\rm FeII} >$ 1.0 is satisfied, then (2) UV AlIII λ1860/SiIII]λ1892 > 0.5 & SiIII]λ1892/CIII]λ1909 > 1.

Physical parameters are correspondingly extreme. First of all, their $L/L_{\rm Edd}$ is at the high end of the distribution along the main sequence, with small dispersion (Marziani & Sulentic 2014). In other words xA quasars selected according to criteria 1 and 2 are extreme radiators, with maximum radiative output per unit mass close to their Eddington limit. This condition is predicted by accretion disk theory at high (possibly super-Eddington) accretion rates: radiative efficiency should be low, and $L/L_{\rm Edd}$ saturate toward a limiting value (Sadowski et al. 2014, and references therein). The star formation rate (SFR) in the host galaxy as estimated from radio observations can be up to $\sim$ a few $10^3 M_\odot$ yr^{-1} (Ganci et al. 2019, $z \lesssim 1$). The broad emission line intensity ratios in the UV suggest extremely high values for density ($n \gtrsim 10^{12} - 10^{13}$ cm^{-3}), very low ionization (ionization parameter $\sim 10^{-3} - 10^{-2.5}$, Negrete et al. 2012), and high metallicity ($Z \gtrsim 20$ Z$_\odot$, Martínez-Aldama et al. 2018).

Broad emission lines in xA sources are produced by gas apparently enriched by a circumnuclear Starburst. It is tempting to consider that xAs could be the first unobscured stage emerging from obscured stages of AGN evolution. This hypothesis fits an evolutionary sequence (Sanders et al. 1988; Dultzin-Hacyan et al. 2003): merging and strong interaction lead to accumulation of gas in the galaxy central parsecs, and to coeval Starburst and nuclear activity, whereas winds from massive stars and supernovæ provide enriched material. Feedback effects on the host galaxies induced by the extreme outflows of xA quasars can be significant if the AGN luminosity is high, as the outflow thrust and kinetic powers are dependent on the emission line luminosity. This means that xA quasars could likely be a major factor in galactic evolution at moderate-to-high redshift ($z \gtrsim 1$). It is however important to stress that xA sources are not necessarily luminous sources – xAs are the low-mass (relatively rare) high accretors in the local Universe as well as the most distant quasars at $z \gtrsim 6$ (Wang et al. 2019).

Almost symmetric Hβ and AlIIIλ1860 line profiles coexist with a CIVλ1549 profile shifted by several thousand km/s, even at the highest luminosity and when radiation forces predominate over gravity, i.e., when $L/L_{\rm Edd}$ is high (Sulentic et al. 2017; Vietri et al. 2018; Bischetti et al. 2017). A virialized subsystem emitting low ionization lines and a subsystem due to winds or outflows constitute two regions that are, at least in part, appearing as kinematically disjoint in the line profiles.

In addition, we can count on xA quasars' spectral invariance: intensity ratios remain the same; only the line width increases with luminosity. This implies that the radius of the emitting region should rigorously scale as $L^{\frac{1}{2}}$: if not, the ionization parameter should change with luminosity. No significant spectral change from very high luminosity to low luminosity sources has been detected yet. Putting together (1) $L/L_{\rm Edd} = const.$, (2) $r \propto L^{\frac{1}{2}}$ and the virial condition (3) $M_{\rm BH} \propto r{\rm FWHM}^2$, we obtain a relation linking luminosity and line width as $L \propto {\rm FWHM}^4$ (Marziani & Sulentic 2014).

Building the Hubble diagram distance modulus versus redshift (see e.g., Risaliti & Lusso 2015) from the redshift-independent virial luminosity estimates with xA quasars over the redshift range $0.1 \lesssim z \lesssim 3$, we obtain a distribution consistent with concordance ΛCDM (Fig. 2). Constraints on the energy density of matter $\Omega_{\rm M}$ (0.30 ± 0.06) are better than the ones from supernovæ, because of the $z \sim 2$ coverage of the quasar sample. The significant scatter of individual measurements ($\sim 1.1 - 1.3$ mag) may be associated with (a) uncertainty of FWHM, which enters with the 4th power in the luminosity relation; (b) orientation that is likely to be the main source of scatter in the classical scaling

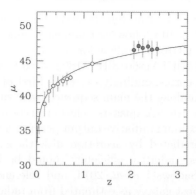

Figure 2. The Hubble diagram distance modulus μ vs z for a sample of quasars (Dultzin *et al.* 2020). Individual measurements have been averaged over redshift bins. Error bars are sample standard deviations for z and μ. Yellow data point refer to the use of the Hβ FWHM as a virial broadening estimator, magenta ones are computed from the FWHM of the AlIIIλ1860 doublet individual components. Error bars show sample standard deviations in μ and z for each bin.

relations (Marziani *et al.* 2019); (c) differences in intrinsic properties of the xA quasars i.e., spectral energy distribution, ionizing photon flux, etc.

4. Conclusion

The MS offer contextualization of quasar observational and physical properties (Marziani *et al.* 2018b). Several MS trends are ultimately associated with Eddington ratio which apparently reaches an extreme value in correspondence of the extreme Population A (xA). xA quasars include the strongest FeII emitters, satisfying the condition $R_{\rm FeII} > 1$. They show a relatively high prevalence (10%) and are easily recognizable thanks to their peculiar and luminosity-invariant spectral properties. In addition, they are sources that tend to show low intrinsic variability (Du *et al.* 2018), and their low ionization lines are apparently emitted in a virialized BLR (see also Swayamtrupta Panda's contributions in this volume). These properties make xA quasars suitable as possible *Eddington* standard candles, where the invariant properties is not intrinsic luminosity, but Eddington ratio.

References

Bischetti, M., Piconcelli, E., Vietri, G., *et al.* 2017, *AAp*, 598, A122
Boroson T. A. & Green R. F. 1992, *ApJS*, 80, 109
Diamond-Stanic, A. M., Fan, X., Brandt, W. N., *et al.* 2009, *ApJ*, 699, 782
D'Onofrio, M. & Marziani, P. 2018, *FrASS*, 5, 31
Du, P., Zhang, Z.-X., Wang, K., *et al.* 2018, *ApJ*, 856, 6
Du, P., Wang, J.-M., Hu, C., *et al.* 2016, *ApJL*, 818, L14
Dultzin, D., Marziani, P., de Diego, J. A., Negrete, C. A., *et al.* 2020, *FrASS*, 6, 80
Dultzin-Hacyan, D., Krongold, Y., & Marziani, P. 2003, *RMexAA CS*, 79
Ferland, G. J., Hu, C., Wang, J.-M., *et al.* 2009, *ApJL*, 707, L82
Fraix-Burnet, D., D'Onofrio, M., & Marziani, P. 2017, *FrASS*, 4, 20
Ganci, V., Marziani, P., D'Onofrio, M., *et al.* 2019, *A&A*, 630, A110
Magorrian, J., Tremaine, S., Richstone, D., *et al.* 1998, *AJ*, 115, 2285
Marinello, M., Rodríguez-Ardila, A., Garcia-Rissmann, A., *et al.* 2016, *ApJ*, 820, 116
Martínez-Aldama, M. L., del Olmo, A., Marziani, P., *et al.* 2018, *AAp*, 618, A179
Marziani, P., Sulentic, J. W., Plauchu-Frayn, I., del Olmo, A., *et al.* 2013, *A&A*, 555, 89
Marziani, P., Sulentic, J. W., Zwitter, T., *et al.* 2001, *ApJ*, 558, 553
Marziani, P., del Olmo, A., D'Onofrio, M., *et al.* 2018a, *PoS*, 328, 2
Marziani, P., Dultzin, D., Sulentic. J, W., Del Olmo, A., *et al.* 2018b, *FrASS*, 5, 6

Marziani, P., del Olmo, A., Martínez-Carballo, M. A., et al. 2019, AAp, 627, A88
Marziani, P. & Sulentic, J. W. 2014, MNRAS, 442, 1211
Negrete, C. A., Dultzin, D., Marziani, P., et al. 2012, ApJ, 757, 62
Negrete, C. A., Dultzin, D., Marziani, P., et al. 2018, AAp, 620, A118
Panda, S., Czerny, B., Adhikari, T. P., et al. 2018, ApJ, 866, 115
Panda, S., Marziani, P., & Czerny, B. 2019, ApJ, 882, 79
Peterson, B. M. and Wandel, A. 1999, ApJL, 521, L95
Richards, G. T., Kruczek, N. E., Gallagher, S. C., et al. 2011, AJ, 141, 167
Risaliti, G. & Lusso, E. 2015, ApJ, 815, 33
Sadowski, A., Narayan, R., McKinney, J. C., et al. 2014, MNRAS, 439, 503
Sanders, D. B., Soifer, B. T., Elias, J. H., et al. 1988, ApJL, 328, L35
Shen, Y. & Ho, L. C. 2014, Nature, 513, 210
Sulentic, J. W., del Olmo, A., Marziani, P., et al. 2017, AAp, 608, A122
Sulentic, J. W., Marziani, P., & Dultzin-Hacyan, D. 2000, ARAAp, 38, 521
Sulentic, J. W., Zwitter, T., Marziani, P., et al. 2000, ApJL, 536, L5
Sun, J. & Shen, Y. 2015, ApJL, 804, L15
Tombesi, F., Cappi, M., Reeves, J. N., et al. 2011, ApJ, 742, 44
Vietri, G., Piconcelli, E., Bischetti, M., et al. 2018, AAp, 617, A81
Wang, F., Yang, J., Fan, X., et al. 2019, ApJ, 884, 30
Wang, J.-M., Qiu, J., Du, P., et al. 2014, ApJ, 797, 65
Zamfir, S., Sulentic, J. W., Marziani, P., et al. 2010, MNRAS, 403, 1759

Spectroscopic properties of radio-loud and radio-quiet quasars

Avinanda Chakraborty[1] and Anirban Bhattacharjee[2]

[1]Presidency University, Kolkata, Pincode 700073, 86/1 College Street, West Bengal, India
email: avinanda.rs@presiuniv.ac.in

[2]Sul Ross State University, Texas, Box C-64, Alpine, TX 79832, Texas, USA
email: anirbanbhattacharjeee@gmail.com

Abstract. Surveys have shown radio-loud (RL) quasars constitute 10%-15% of the total quasar population and rest are radio-quiet (RQ). However, it is unknown if this radio-loud fraction (RLF) remains consistent among different parameter spaces. This study shows that RLF increases for increasing full width half maximum (FWHM) velocity of the Hβ broad emission line ($z < 0.75$). To analyse the reason, we compared bolometric luminosity of RL and RQ quasars sample which have FWHM of Hβ broad emission line greater than 15000km/s (High Broad Line or HBL) with which have FWHM of Hβ emission line less than 2500km/s (Low Broad Line or LBL). From the distributions we can conclude for the HBL, RQ and RL quasars are peaking separately and RL quasars are having higher values whereas for the LBL the peaks are almost indistinguishable. We predicted selection effects could be the possible reason but to conclude anything more analysis is needed. Then we compared our result with Wills & Brotherton (1995) and have shown that some objects from our sample do not follow the pattern of the logR vs FWHM plot where R is the ratio of 5 GHz radio core flux density with the extended radio lobe flux density.

Keywords. Surveys, quasars, redshift, luminosity, emission line, jet

1. Introduction

Quasars are the most luminous active galactic nuclei (AGN) and are powered by accretion of supermassive black holes (SMBHs) (Salpeter 1964; Lynden-Bell 1969). We still don't understand properly why some active galactic nuclei have strong radio sources and others do not (Lynden-Bell 1969). Surveys have shown that other than radio surveys there is no such difference between RLQs and RQQs (Kratzer & Richards 2015). Although Radio-loud quasars (RLQs) were first detected as radio sources. Only 10% of the total quasars are RL (Sandage 1965). The main difference between both RLQs and Radio-quiet quasars (RQQs) is the presence of powerful radio jets (e.g. Bridle *et al.* 1994; Mullin *et al.* 2008). However, there is evidence of weak radio jets in RQQs also (Ulvestad *et al.* 2005; Leipski *et al.* 2006).

Wills & Browne (1995) found a significant correlation between the full width half maximum (FWHM) and of broad Hβ lines and the logR where R is defined as "Ratio of 5 GHz core to extended component flux density by Wills & Brotherton (1995)." The parameter R has been used as a measure of orientation. These authors have shown the distribution of logR to be highly asymmetric and biased toward small Hβ FWHMs with a cut off near 2000km/s.

There is evidence that the broad-line width measurement in quasar is dependant on the source orientation and consistent with the idea of flattened or disc like broad-line regions (Jarvis & McLure 2006). These authors have also presented a significant correlation between radio spectral index and broad-line width of the Hβ and Mg II emission lines ($\gg$99.99%). These authors showed spectral index can be used as a proxy for source orientation.

It has also been shown that normalizing the radio core luminosity by the optical continuum luminosity ($\log R_v$) (K-corrected) is a superior orientation indicator (Van Gorkom et al. 2015). Van Gorkom et al. 2015 compared between $\log R$ and $\log R_v$ and two other indicators of orientation, the ratio of the optical continuum luminosity and emission-line luminosity (Yee & Oke (1978)) and the ratio of the jet power and the luminosity of the narrow-line region Rawlings & Saunders (1981).

2. Overview

Our work is basically focused on investigating different properties of RLQs and RQQs to find the reason behind the high value of RLF at higher FWHM and comparing our results with some other literature and reach to some conclusion. Initial work was done by Bhattacharjee, Gilbert & Brotherton et al. (2018) To check the consistency in different parameter spaces they first looked for the variation of RLF with FWHM of broad Hβ. We then analysed fundamental Hβ line properties of RLQs and RQQs for the HBL region (FWHM > 15000km/s) and compared them with the LBL (FWHM < 2500km/s) region properties to check the reason of high RLF for higher FWHM and lastly we compared our result with other literature. Here are the data samples we have used:

Sloan Digital Sky Survey (SDSS): Our main quasar catalogue comes from the SDSS (York et al. 2000) Data Release 7 (Abazajian et al. 2009) Quasar catalogue (Shen et al. 2011). It consists of 105,783 quasars brighter than $M_i = -22.0$ and are spectroscopically confirmed.

Faint Images of the Radio Sky at Twenty-cm (FIRST): Shen et al. 2011 cross matched the quasar catalogue of SDSS DR7 and FIRST survey of VLA. The quasars having only one FIRST source within 5" are classified as core-dominated radio sources and those having multiple FIRST sources within 30" are classified as lobe dominated. These two categories are together named as RLQs by Shen et al. 2011. And those with only one FIRST match between 5" and 30" are classified as RQQs. We checked optical spectra of the HBL and LBL quasars from SDSS and quasars with some issues with their Hβ line are manually discarded from our sample. So our final Hβ sample contains 298 RLQs and 1,910 RQQs. Among RLQs 56 are HBL (FWHM > 15000 km/s) and 242 are LBL (FWHM < 2500 km/s) and in RQQs 41 are HBL and 1869 are LBL sources.

3. Implications

Radio-loud fraction: Figure 1 shows the variation of RLF across FWHM. From this figure we can see RLF is increasing with FWHM which implies that in the HBL region quasars are more radio loud.

Analysis of Hβ line properties: Figure 2 shows normalised distributions of bolometric luminosity of RLQs and RQQs with Gaussian fits for the HBL and LBL respectively. Now it is clear from luminosity analysis that for the HBL, RLQs and RQQs distributions are different and RLQs are peaking at higher values and for the LBL, RLQ and RQQ distributions are consistent almost.

Orientation of the quasars: Now to compare our result with other literature we looked for the ratio of 5 GHz core to extended component flux density R as a function of FWHM for the broad Hβ line for quasars plot of Wills & Browne (1986) From the plot we are getting high logR value for low FWHM. Wills & Browne (1986) also said that core dominated quasars will have lower FWHM. Then we replot this with our sample but our sample limit is almost beyond their limit. So for the HBL and LBL region, we have

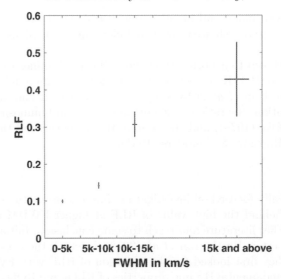

Figure 1. Variation of Radio-Loud Fraction across different FWHM of broad Hβ with FWHM. From this plot we can see that the RLF increases with FWHM.

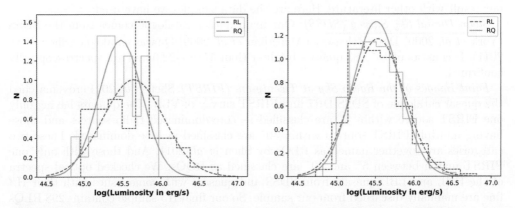

Figure 2. Normalised distribution of bolometric luminosity of RLQs and RQQs for the HBL with Gaussian fits. Here we have taken the luminosity in log scale and its unit is 10^{-7} watt or erg/s and normalised distribution of bolometric luminosity of RLQs and RQQs for the LBL with Gaussian fits. Here we have taken the luminosity in log scale and it's unit is 10^{-7} watt or erg/s respectively.

calculated logR_v. For the HBL (Figure 3) some objects from our sample do not obey the pattern of Wills & Browne (1986) plot. But the LBL region of our sample is consistent with their plot.

4. Discussion

Our main goal is to investigate whether RLF is consistent across different parameter spaces and here we consider only broad Hβ lines. And we have seen it increases with Hβ FWHM so now to find the reason we chose objects with an exceptionally high full width half maximum (FWHM > 15,000km/s) and compared them with widely used low full width half maximum objects (FWHM < 2,500km/s).

We compared their bolometric luminosity distributions, and we can see for the HBL RLQs have higher luminosities. Now detection probability of higher luminous objects should be high so this could be a possible reason for getting high RLF in high line

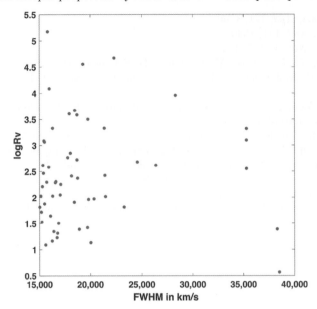

Figure 3. logR_v vs FWHM of Hβ line plot for RLQs with FWHM > 15,000km/s.

widths. We checked for the LBL region also but the distributions for RLQs and RQQs are consistent in that region. More analysis is required to say about the exact reason. We then tried to compare our result with other literature. We took the ratio of 5 GHz core to extended component flux density R as a function of FWHM for the broad Hβ line for quasars plot of Wills & Browne (1986) and compared it with logR_v as a function of FWHM for our HBL and LBL sample because Wills & Brotherton (1995) have suggested a relation between logR and logR_v. From our logR_v vs FWHM plot, we saw that some objects do not fall in the pattern described by Wills & Browne (1986). Further investigation is required for these objects.

Acknowledgement

We would like to thank Dr. Mike Brotherton and Ms. Jaya Maithili from University of Wyoming, Wyoming and Dr. Suchetana Chaterjee from Presidency University, Kolkata and Ms. Miranda Gilbert of Sul Ross State University, Texas for their valuable inputs and DST-SERB for providing financial support through the ECR grant of Dr. Suchetana Chatterjee.

References

Abazajian, K. N., Adelman-McCarthy, J. K., Agüeros, M. A., Allam, S. S., et al. 2009, *ApJS*, 182, 543

Bhattacharjee, A., Gilbert, M., Brotherton, M. S., et al. 2018, *AAS*, 232, 23232203B

Bridle, A. H., Hough, D. H., Lonsdale, C. J., Burns, J. O., & Laing, R. A. 1994, *AJ*, 108, 766

Edge, D. O., Shakeshaft, J. R., McAdam, W. B., Baldwin, J. E., & Archer, S. 1959, *Mem. RAS*, 68, 37

Jarvis, M. J. & McLure, R. J. 2006, *MNRAS*, 369, 182

Kratzer, R. M. & Richards, G. T. 2015, *The Astronomical Journal*, 149, 61

Leipski, C., Falcke, H., Bennert, N., & Hüttemeister, S. 2006, *A&A*, 455, 161

Lynden-Bell, D. 1969, *Nature*, 223, 690

Mullin, L. M., Riley, J. M., & Hardcastle, M. J. 2008, *MNRAS*, 390, 595

Rawlings, S. & Saunders, R. 1991, *Nature*, 349, 138

Salpeter, E. E. 1964, *ApJ*, 141, 1560

Sandage, A. 1965, *ApJ*, 141, 1560

Schmidt, M. 1963, *Nature*, 197, 1040

Shen, Y., Richards, G. T., Strauss, M. A., Hall, P. B., Schneider, D. P., Snedden, S., Bizyaev, D., Brewington, H., Malanushenko, V., Malanushenko, E., Oravetz, D., Pan, K., & Simmons, A. 2011, *ApJ*, 194, 45

Ulvestad, J. S., Wong, D. S., Taylor, G. B., F. Gallimore, J. F., & Mundell, C. G. 2005, *The Astronomical Journal*, 130, 936

Van Gorkom, K. J., Wardle, J. F. C., Rauch, A. P., & Gobeille, D. B. 2015, *MNRAS*, 450, 4240

White, R. L., Helfand, D. J., Becker, R. H., Glikman, E., & de Vries, W. 2007, *ApJ*, 654, 99

Wills, B. J. & Brotherton, M. S. 1995, *The Astrophysical Journal*, 448

Wills, B. J. & Browne, I. W. A. 1986, *ApJ*, 302, 56

Yee, H.K.C. & Oke, J.B. 1978, *ApJ*, 226, 753

York, D. G., Adelman, J., Anderson, J. E., Jr., Anderson, S. F., Annis, J., Bahcall, N. A.,*et al.* 2000, *AJ*, 120, 1579

FeII emission in NLS1s – originating from denser regions with higher abundances?

Swayamtrupta Panda[1,2], Paola Marziani[3] and Bożena Czerny[1]

[1]Center For Theoretical Physics, Polish Academy of Sciences, Al. Lotników 32/46,
02-668 Warsaw, Poland
email: panda@cft.edu.pl

[2]Nicolaus Copernicus Astronomical Center, Polish Academy of Sciences,
ul. Bartycka 18, 00-716 Warsaw, Poland

[3]INAF-Astronomical Observatory of Padova, Vicolo dell'Osservatorio, 5,
35122 Padova PD, Italy

Abstract. The interpretation of the main sequence of quasars has become a frontier subject in the last years. This considers the effect of a highly flattened, axially symmetric geometry for the broad line region (BLR) on the parameters related to the distribution of quasars along their main sequence. We utilize the photoionization code CLOUDY to model the BLR, assuming 'unconstant' virial factor with a strong dependence on the viewing angle. We show the preliminary results of the analysis to highlight the co-dependence of the Eigenvector 1 parameter, R_{FeII} on the broad Hβ FWHM (i.e. the line dispersion) and the inclination angle (θ), assuming fixed values for the Eddington ratio (L_{bol}/L_{Edd}), black hole mass (M_{BH}) and spectral energy distribution (SED) shape. We consider four cases with changing cloud density (n_H) and composition. Understanding the Fe II emitting region is crucial as this knowledge can be extended to the use of quasars as distance indicators for Cosmology.†

Keywords. accretion, accretion disks, radiation mechanisms: thermal, radiative transfer, galaxies: active, (galaxies:) quasars: emission lines, galaxies: Seyfert

1. Introduction

The quasar main sequence contextualizes and eases the interpretation of classes of active galactic nuclei (AGN) whose origin has been debated for decades. An important class is the one of Narrow-Line Seyfert 1 (NLS1) galaxies which constitute a class of Type-1 active galaxies with "narrow" broad profiles. Their supermassive black holes (BH) are believed to have masses lower than the typical broad-line Seyfert galaxies. Black hole masses are estimated assuming that the line broadening is due to Doppler effect associated with the emitting gas motion with respect to the observer. In addition, the motions are believed to be predominantly virial (Peterson & Wandel 1999). If the virial assumption is verified, the M_{BH} can be written as a function of (i) the radius of the broad line emitting region (BLR); and (ii) the FWHM of the emission lines emitted by gas whose motions are assumed virialized. The BLR radius (R_{BLR}) is derived via reverberation mapping (Peterson 1993) i.e., by measuring the light-travel time from the central ionizing source to the line emitting medium. The line FWHM can be reliably measured from high S/N spectroscopy.

† The project was partially supported by NCN grant no. 2017/26/A/ST9/00756 (MAESTRO 9) and MNiSW grant DIR/WK/2018/12. PM acknowledges the INAF PRIN-SKA 2017 program 1.05.01.88.04.

© The Author(s), 2021. Published by Cambridge University Press on behalf of International Astronomical Union

The BLR is a complex region, even if its physics is overwhelmingly driven by the process of photoionization. It cannot be characterized by a single quantity or number. The origin of different ionic species from this region and the advent of the reverberation mapping to probe more emission lines, have shown that the BLR is indeed stratified in terms of its density and structure. Newer observations, such as of the Super-Eddington sources (see Du et al. 2018 and references therein), have opened up a new field in the study of quasars. And one such immediate application is the use of these Super-Eddington sources, which are primarily NLS1s, as "standardizable" Eddington candles furthering the use of quasars in cosmology (see Marziani et al. 2019; Martínez-Aldama et al. 2019 and references therein).

We address this aspect of the geometry of the quasars using photoionisation modelling with CLOUDY in the context of understanding better the main sequence of quasars (see Panda et al. 2019a,b for more details). We focus on modelling the Fe II emission in quasars as a function of the 7 key parameters – (i) black hole mass (M_{BH}); (ii) Eddington ratio (L_{bol}/L_{Edd}); (iii) shape of the broad-band ionizing continuum (SED); (iv) mean cloud density (n_H); (v) cloud metallicity; (vi) micro-turbulence; and (vii) Hβ FWHM distribution. This multi-parameter space is then visualized as a function of the inclination angle (θ) of the source with respect to the observer. Here, we illustrate the results for R_{FeII} estimated from a BLR cloud primarily as a function of the FWHM and θ.

2. Method

We assume a single cloud model where the density (n_H) of the ionized gas cloud is varied from 10^9 cm^{-3} to 10^{13} cm^{-3} with a step-size of 0.25 (in log-scale). We utilize the *GASS10* model Grevesse et al. (2010) to recover the solar-like abundances and vary the metallicity within the gas cloud, going from a sub-solar type (0.1 $Z_\odot$) to super-solar (100 $Z_\odot$) with a step-size of 0.25 (in log-scale). The total luminosity of the ionizing continuum is derived assuming a value of the Eddington ratio (L_{bol}/L_{Edd}) and the respective value for the black hole mass (here, we assume an $L_{bol}/L_{Edd} = 0.25$ and a $M_{BH} = 10^8$ M$_\odot$). These values are appropriate for the part of Population A in spectral types. The shape of the ionizing continuum used here is taken from Korista et al. (1997). The size of the BLR is estimated from the virial relation, assuming a black hole mass, a distribution in the viewing angle [0-90 degrees] and FWHM (for more details see Panda et al. 2019a,b). The cloud column density (N_H) is assumed to be 10^{24} cm^{-2}.

The virial relation can be expressed as

$$R_{BLR} = \frac{GM_{BH}}{f * FWHM^2} = \frac{4GM_{BH}\left[\kappa^2 + \sin^2\theta\right]}{FWHM^2} \quad (2.1)$$

Substituting the values for the M_{BH} and κ (= 0.1 that is consistent with a flat, keplerian-like gas distribution) in the virial relation, we have

$$R_{BLR} \approx 5.31 \times 10^{24} \left[\frac{0.01 + \sin^2\theta}{FWHM^2}\right] \quad \text{(in cm)} \quad (2.2)$$

3. Results and Conclusions

In the left panel of Figure 1, we assume the mean cloud density (n_H) at 10^{10} cm^{-3}. The peak of the R_{FeII} (∼0.8415) is located at ∼60° for FWHM = 1000 km s^{-1}. Within the realms of Type-1 AGNs, i.e., $\theta \lesssim 60°$, $R_{FeII} \propto \theta$. On the other hand, R_{FeII} is inversely related to the FWHM. From the virial relation we have, $R_{BLR} \propto \frac{1}{FWHM^2}$. This implies, $R_{FeII} \propto \sqrt{R_{BLR}}$. In other words, increasing FWHM decreases R_{BLR} which means higher radiation flux on the cloud that leads to depletion in Fe II emission. Hence, R_{FeII} decreases. Increasing the metallicity from solar ($Z_\odot$) to $10Z_\odot$ shifts the peak of the R_{FeII} to ∼81°

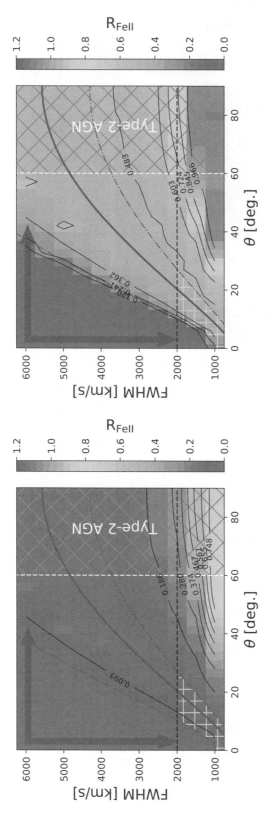

Figure 1. 2D histogram showing the dependence of the parameter R_{FeII} on the Hβ FWHM and the inclination angle (θ). The Hβ FWHM ranges from 1000 km s^{-1} to 6000 km s^{-1} with a step size of 500 km s^{-1}. Similarly, the θ values range from 0°–90° with a step size of 3°. The black hole mass is assumed to be $10^8 M_\odot$, the Eddington ratio, $L_{bol}/L_{Edd} = 0.25$ and a SED consistent with Korista et al. (1997) is used. The value of the $\kappa = 0.1$ consistent with a flat, keplerian-like gas distribution around the central supermassive black hole. The mean cloud density (n_H) is 10^{10} cm^{-3} and the column density is 10^{24} cm^{-2}. The white dashed line marks the upper limit on the θ consistent with Type-1 sources. The hatched region marks the Type-2 AGN zone beyond $\theta = 60°$. The solid green line traces the FWHM-θ for the R_{BLR} estimated from the standard $R_{H\beta} - L_{5100}$ relation, and the dot-dashed green lines correspond to the 1σ scatter in the $R_{H\beta} - L_{5100}$ relation. The yellow patch for FWHM $\lesssim 2000$ km s^{-1} and within the 1σ scatter around the standard $R_{H\beta} - L_{5100}$ relation marks the zone of acceptance for the NLS1s (typical Population A type sources). The solid green arrows point in the direction of increasing R_{BLR} based on the virial relation. **LEFT**: at solar abundance ($Z_\odot$); **RIGHT**: at $10 Z_\odot$.

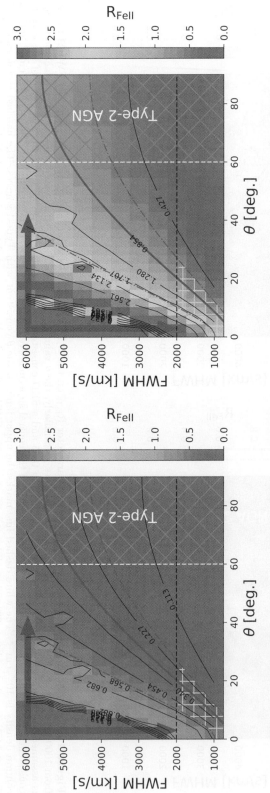

Figure 2. Same as Figure 1. The mean cloud density (n_H) is increased to 10^{12} cm^{-3}. **LEFT**: at solar abundance ($Z_\odot$); **RIGHT**: at $10Z_\odot$.

still for the case with FWHM = 1000 km s^{-1}. Within the limits of $\theta \lesssim 60°$, the maximum value of R$_{\text{FeII}}$ is at $\theta \sim 45°$ (R$_{\text{FeII}} \sim 1.0835$). Trends of R$_{\text{FeII}}$ with respect to θ and FWHM respectively, remain consistent to the previous case (at solar abundance).

In Figure 2, we increase the value of n$_{\text{H}}$ from 10^{10} to 10^{12} cm^{-3}, keeping the other parameters exactly the same as before. The current value of density, i.e. 10^{12} cm^{-3}, is consistent with previous works related to the study of the main sequence of quasars (see Panda et al. 2018 and references therein). This change in the density changes the picture significantly. In the left panel of Figure 2, the peak value moves along one of the contour lines and within 2000 km s^{-1} $\lesssim$ FWHM $\lesssim$ 6000 km s^{-1}, the peak moves from $\sim 3°$ (for ~ 2000 km s^{-1}) to $\sim 18°$ (for ~ 6000 km s^{-1}). These peak values of R$_{\text{FeII}}$ remain at 0.77 ± 0.01. The ionization parameter (U) also changes accordingly. For instance, considering the R_{BLR} from Eq. 2.2 at FWHM = 2000km s^{-1} and at $\theta = 3°$, gives $U \sim 0.23$ which in the low density case (10^{10} cm^{-3}) corresponds to a very high value, i.e. $U \sim 23$!

Hence, the ionisation parameter governs the appearance of the plots. A considerable region in the Figures 1 and 2 falls in "zones of avoidance" where U is either too high or too low to sustain significant FeII emission (dark blue areas in the Figures). In the case with the high density the peak emission is very close to the allowed zone by the $R_{\text{H}\beta} - L_{5100}$ relation (the region within the 1σ scatter of the $R_{\text{H}\beta} - L_{5100}$ is shown with green dashed lines in Figures 1 and 2). This implies that the NLS1s that are high Fe II emitters need to have a high density to boost their Fe II. R$_{\text{FeII}}$ is even more enhanced in case of higher metallicity. However, along the $R_{\text{H}\beta} - L_{5100}$ line R$_{\text{FeII}}$ is constant, as expected since in this case all parameters affecting Fe II intensity are set to a fixed value in our model. These results require further analysis which will be presented in a subsequent paper.

References

Collin, S., Kawaguchi, T., Peterson, B. M., & Vestergaard, M. 2006, A&A, 456, 75
Du, P., Zhang, Z.-X., Wang, K., Huang, Y.-K., Zhang, Y., Lu, K.-X., Hu, C., Li, Y.-R., Bai, J.-M., Bian, W.-H., Yuan, Y.-F., Ho, L. C., Wang, J.-M., & SEAMBH Collaboration 2018, ApJ, 856, 6
Ferland, G. J., Chatzikos, M., Guzmán, F., Lykins, M. L., van Hoof, P. A. M., Williams, R. J. R., Abel, N. P., Badnell, N. R., Keenan, F. P., Porter, R. L., & Stancil, P. C. 2017, RMxAA, 53, 385
Grevesse, N., Asplund, M., Sauval, A. J., & Scott, P. 2010, ApSS, 328, 179
Korista, K., Baldwin, J., Ferland, G., & Verner, D. 1997, ApJS, 108, 401
Martínez-Aldama, M. L., Czerny, B., Panda, S., Kawka, D., Karas, V., Zajaček, M., & Życki, P. T. 2019, ApJ, 883, 2
Marziani, P., Bon, E., Bon, N., del Olmo, A., Martínez-Aldama, M. L., D'Onofrio, M., Dultzin, D., Negrete, C. A., & Stirpe, G. 2019, Atoms, 7, 1
Panda, S., Czerny, B., Adhikari, T. P., Hryniewicz, K., Wildy, C., Kuraszkiewicz, J., & Śniegowska, M. 2018, ApJ, 866, 115
Panda, S., Marziani, P., & Czerny, B. 2019a, ApJ, 882, 2
Panda, S., Marziani, P., & Czerny, B. 2019b, Contributions of the Astronomical Observatory Skalnaté Pleso, in press
Panda, S., Marziani, P., & Czerny, B. 2019c, Proceedings of the International Astronomical Union (IAU), 356, 1
Peterson, B. M. 1993, PASP, 105, 247
Peterson, B. M. & Wandel, A. 1999, ApJL, 521, L95
Sulentic, J. W., Zwitter, T., Marziani, P. & Dultzin-Hacyan, D. 2000, ApJL, 536, L5

The role of failed accretion disk winds in active galactic nuclei

Margherita Giustini[1] and Daniel Proga[2]

[1] Centro de Astrobiología (CSIC-INTA), Departamento de Astrofísica,
Camino Bajo del Castillo s/n, Villanueva de la Cañada, E-28692 Madrid, Spain
email: mgiustini@cab.inta-csic.es

[2] Department of Physics & Astronomy, University of Nevada, Las Vegas,
4505 South Maryland Parkway, NV 89154-4002 Las Vegas, USA
email: dproga@physics.unlv.edu

Abstract. Both observational and theoretical evidence point at outflows originating from accretion disks as fundamental ingredients of active galactic nuclei (AGN). These outflows can have more than one component, for example an unbound supersonic wind and a failed wind (FW). The latter is a prediction of the simulations of radiation-driven disk outflows which show that the former is accompanied by an inner failed component, where the flow struggles to escape from the strong gravitational pull of the supermassive black hole. This FW component could provide a physical framework to interpret various phenomenological components of AGN. Here we briefly discuss a few of them: the broad line region, the X-ray obscurer, and the X-ray corona.

Keywords. accretion, accretion disks; black hole physics; galaxies: active

1. Introduction

The inner structure of luminous active galactic nuclei (AGN) is shaped by the presence of winds launched on accretion disk scales, as indicated by their large terminal velocities ($v_{\rm out} \gg 5000$ km s^{-1} and up to several 0.1c). Such winds cannot be sustained by thermal pressure, but must be driven by either magnetic or radiative forces (e.g., see for a recent review Giustini & Proga 2019). In the case of radiation-driven accretion disk winds (specifically line-driven, LD), a chaotic, dense, struggling inner component is always accompanying the larger scale successful mass outflow: it is the failed wind (FW; Proga, Stone & Kallman 2000; Proga & Kallman 2004; Proga 2005). Here "successful" or "failed" means the ability of inability of the gas in reaching the escape velocity $v_{\rm esc}$. In LD disk wind models, the FW is a fundamental ingredient of the mass flow of luminous AGN as the wind itself. In fact, its formation is more robust than the formation of the wind. Therefore if LD disk wind models hold, observable quantities related to the FW should be identifiable in AGN as well. In the following section, we briefly discuss the possible role of the FW in the inner accretion flow of luminous AGN.

2. Failed line-driven accretion disk winds in the AGN inner structure

Line driving can deposit much more momentum in the gas than pure electron scattering, allowing the launching of material with velocity greater than $v_{\rm esc}$ in sub-Eddington regimes (Castor, Abbott & Klein 1975). For LD to be effective, the presence of spectral transitions is therefore fundamental. The spectral energy distribution of luminous AGN allows for LD winds to be launched from accretion disk scales pushing on the many UV transitions available (e.g., Murray *et al.* 1995; Proga, Stone & Kallman 2000; Proga & Kallman 2004). A large X-ray flux is also characteristic of luminous AGN, and while the X-ray photons will push on relatively few available X-ray lines, they will mostly concur to strip the electrons off the UV-absorbing atoms, thus destroying the many UV spectral transitions available and effectively "overionizing" the wind material (e.g., Dannen *et al.* 2019, and references therein). *The term "overionization" is referred to a level of ionization that is too large to produce the observed UV lines and to sustain LD winds above local $v_{\rm esc}$.* Therefore the ratio between the UV and X-ray radiation flux is crucial for the successful launch and acceleration of LD winds in AGN.

In the first models of LD accretion disk winds in AGN, a layer of dense gas (a *shield*), absorbing the strong ionizing X-ray flux, was assumed to exist between the X-ray continuum source and the UV-absorbing wind. This is the "hitchhiking gas" of Murray *et al.* 1995, which postulated the presence of this gas just in front of the flow that is effectively accelerated out of the system. It was speculated that a gradient in pressure would then cause the hitchhiking gas to accelerate together with the farther out wind (hence the nickname).

Hydrodynamical simulations performed by Proga and collaborators (Proga, Stone & Kallman 2000; Proga & Kallman 2004) showed that for massive (black hole mass $M_{\rm BH} > 10^8\,M_\odot$), luminous AGN (Eddington ratio $\dot{m} > 0.5$), the accretion flow settles in a hot polar flow, a fast equatorial outflow, and an inner transitional zone where the material is lifted up by the strong radiation pressure. The gas in the transition zone is exposed to the strong ionizing continuum, loses most or all of its bond electrons, thus losing line driving force. This material is unable to reach $v_{\rm esc}$ before getting overionized, and it falls back toward the disk. *The failure or success of the wind is measured in terms of overcoming or not the local $v_{\rm esc}$.* The inner FW effectively protects (shields) the material located farther out from the strong ionizing continuum radiation, therefore allowing for the successful launch of the wind at radii larger than where the FW dominates, and where the radiation pressure is large enough to overcome BH gravity (Proga & Kallman 2004; Risaliti & Elvis 2010).

In the scenario proposed by Giustini & Proga (2019), for $M_{\rm BH} \gtrsim 10^8\,M_\odot$ and $\dot{m} \gtrsim 0.01$, the AGN inner structure is dominated by the presence of a LD disk wind and its inner FW component. A LD wind is launched at all radii where the local radiation pressure overcomes gravity; in the inner portions of disk the wind is unable to escape because of overionization, and therefore forms a FW. The AGN appearance will be more or less dominated by the FW (and thus have slower or faster LD winds), depending on how large is the X-ray/UV flux ratio as seen by the gas. Although the Giustini & Proga (2019) scenario is still qualitative, it already makes some predictions. We discuss in the following three phenomenological components of AGN that might be explained by the FW in AGN, in order of decreasing distances from the central supermassive black hole (SMBH): the (high-ionization) broad line region in Section 2.1, the X-ray obscurer in Section 2.2, and the X-ray corona(e) in Section 2.3.

2.1. *The FW as BLR*

The broad line region (BLR) is a fundamental ingredient of luminous AGN: it consists of gas photoionized by the AGN continuum and whose motion responds to the gravitational potential of the central SMBH (e.g., Peterson *et al.* 2004). The BLR is phenomenologically divided into a low-ionization (e.g., Mg II, Hβ) and a high-ionization (e.g., C IV, Lyα) component which show distinct kinematics (e.g., Marziani *et al.* 1996). In particular, a difference in peak position between low-ionization and high-ionization emission lines is indicative of strong radial motions of the gas producing the latter (Gaskell 1982).

The high-ionization emission lines in luminous AGN can in fact be blueshifted by several hundreds (up to thousands) km s^{-1} with respect to the low-ionization emission lines and the host galaxy (e.g., Richards et al. 2011): part of the high-ionization BLR must be a wind. Most of the low-ionization BLR likely originates at large scales, close to the dust sublimation radius, where radiation pressure on dust grains can form an outer wind and a FW (Czerny & Hryniewicz 2011). Radiation pressure on UV lines, that forms a LD accretion disk wind and an inner FW, can explain instead the bulk of the high-ionization emission lines (Proga & Kallman 2004). Part of the low-ionization emission lines can also be produced within dense clumps at the base of the wind, and would then also be blueshifted (Waters et al. 2016).

Blueshifted broad absorption lines in high-ionization UV transitions† are also observed in a large number of luminous AGN (up to 40%, Allen et al. 2011). These are the so-called broad absorption line quasars (BAL QSOs), and display the most direct evidence for the presence along the line of sight of strong winds, which can reach velocities of several 0.1c. Such high velocities must be produced close to the SMBH, on accretion disk scales. The presence of the broad absorption troughs alone indicates that a lot of momentum has been deposited in the gas by radiation. Remarkably, the most recent observations of large samples of AGN have demonstrated that high-ionization BAL QSOs correspond to quasars in general (Rankine et al. 2020), thus supporting accretion disk wind scenarios for luminous AGN in general. If driven by radiation, a disk wind will be accompanied by the inner FW, and this will also contribute to the emission of the BLR. But how?

As summarised in Giustini & Proga (2019), in LD disk winds scenarios the BLR appears in luminous AGN at $\dot{m} \gtrsim 0.01$, when the inner accretion and ejection flow consists of a disk, LD wind, FW, and inner X-ray source; the LD wind + FW then produce the high-ionization BLR. In particular, the production of the symmetric portion of the high-ionization BLR is associated to the FW, while its blueshifted and blue-skewed portion, to the wind itself.

The strongest (fastest, densest) LD disk winds are those produced in AGN with a low X-ray/UV flux ratio (*X-ray weak*), either because of high $\dot{m}$ and/or a large M_{BH}. These have a vast radial zone of the inner flow dominated by winds, and only a small inner region where the wind fails. Thus they produce winds with a large range of velocities, including large terminal velocities $\gg 5,000$ km s^{-1} and up to several 0.1c when launched in the innermost regions of the disk. On the contrary, in the case of AGN with a large X-ray/UV flux ratio (*X-ray bright*), a larger inner region of the accretion flow is dominated by the FW. In these AGN, successful winds are only launched at larger scales, thus reaching lower terminal velocities.

In LD disk winds scenarios, the BLR of X-ray weak AGN is dynamically dominated by the wind: the emission lines of e.g. C IV are strongly blueshifted and blue-skewed. Their equivalent width is lower than the one of the same emission lines produced in X-ray bright AGN: here the dynamics of the BLR is dominated by the FW, that does not reach v_{esc}. The emission lines have a larger equivalent width, a more symmetric profile, and little or no blueshift with respect to the redshift of the host galaxy. In other words, the FW extent regulates the extent of the symmetric, non-shifted BLR at the expense of the skewed, blueshifted BLR produced in the wind. When the disk is observed through the wind, X-ray weak AGN will display deeper, broader, and more blueshifted absorption troughs compared to X-ray bright AGN. Broadly speaking, the first type of AGN would correspond to the population A of quasars along their main sequence, while the second type of AGN to their population B (Sulentic et al. 2000; Sulentic & Marziani 2015).

† Low-ionization broad absorption lines are observed in a small fraction (about 5–10%) of broad absorption line quasars, the low-ionization BAL QSOs; those who do not display them are called high-ionization broad absorption line quasars.

2.2. The FW as the obscurer

The presence of dense, variable layers of X-ray absorbing gas on BLR-scales has been recently inferred by high-quality observations of local Seyfert 1 galaxies (Kaastra et al. 2014; Ebrero et al. 2016; Mehdipour et al. 2015; Kriss et al. 2019). This gas is called "obscurer", as it absorbs the X-ray continuum flux and obscures the view of the strong ionizing X-ray continuum to the material located further out. This further out material, in fact, responds to the changes in X-ray ionizing flux, as strong UV absorption lines are observed emerging in concomitance with the appearance of strong X-ray absorption.

In LD accretion disk wind scenarios, the FW is the material located close to the source of X-rays, that gets all the ionizing continuum, and thus fails reaching $v_{\rm esc}$ and falls back toward the disk. The FW motion is complex: highly dynamical, with locally variable motion made of upward and downward components, and dense filaments and knots embedded in a much hotter medium (Proga 2005). The FW absorbs the X-ray continuum photons, effectively shielding the gas located farther out that can be then accelerated by radiation pressure on UV spectral lines. The FW has therefore all the characteristics to be identified with the "obscurer" of local Seyfert galaxies.

2.3. The FW as the X-ray warm corona(e)

Much closer to the central SMBH than the wind, but maybe partially co-spatial with the inner FW, lies the source of X-ray photons. The X-ray radiation is the clearest signature of accreting BHs, yet its physical origin is still unclear. We know that some compact and hot region must be responsible for the bulk of the intense and variable X-ray emission of AGN, and we call it X-ray "hot corona". The X-ray hot corona has become a synonym for a low-density (optical depth $\tau \ll 1$) medium full of hot (temperature $kT_{\rm e} > 100$ keV) electrons, that by interacting with the much slower UV photons emerging from the thermalized accretion disk, increase their energy through inverse Compton scattering (e.g., Haardt & Maraschi 1993). This hot coronal emission is able to overionize the wind material, therefore concurring in destroying the wind, i.e., creating a FW. In recent years, the presence in the inner regions of luminous AGN of material also able to Compton-upscatter UV photons, but with much lower temperature and much higher optical depth ($kT_{\rm e} \sim 100 - 300$ eV, $\tau \sim 10$), has emerged: this is called the X-ray "warm corona" (Done et al. 2012; Mehdipour et al. 2015; Petrucci et al. 2018). These general physical properties of the X-ray warm corona look similar to those of the FW. An exchange of energy is expected between the source of hard X-ray photons (the hot corona) and the FW, with dense knots within the FW able to emit X-ray bremsstrahlung and thus, in the most extreme cases, switching the main cooling mechanism for the X-ray emitting plasma (Proga 2005). In these cases the warm corona can dominate over the hot corona in terms of density and hence matter cooling/radiation emission, and produce genuinely hard X-ray weak AGN where most of the flux of photons at $E \gtrsim 2$ keV is suppressed: the FW would then become the X-ray corona itself.

3. Conclusions

Accretion disk winds have been recognized as fundamental ingredients of the inner regions of luminous AGN. In the case of LD disk winds, the inner FW component might help interpreting in a physical framework phenomenological features of AGN such as the high-ionization BLR, the obscurer, and the X-ray coronae. The FW solutions of the inner accretion and ejection flow of AGN deserve further attention, in order to assess whether they can change significantly the physical and geometrical structure of the very inner accretion flow around highly accreting SMBHs.

Acknowledgements

MG warmly thanks the IAU 356 symposium organizers for a memorable, transformational meeting. We thank G. Richards, G. Miniutti, E. Lusso, and M. Mehdipour for interesting discussions. MG is supported by the "Programa de Atracción de Talento" of the Comunidad de Madrid, grant number 2018-T1/TIC-11733, and by the Spanish State Research Agency (AEI) Projects number ESP2017-86582-C4-1-R and ESP2015-65597-C4-1-R. This research has been partially funded by the AEI Project number MDM-2017-0737 Unidad de Excelencia "María de Maeztu" – Centro de Astrobiología (INTA-CSIC).

References

Allen, J. T., Hewett, P. C., Maddox, N., Richards, G. T., et al. 2011, *MNRAS*,4105, 860
Castor, J. I., Abbott, D. C., & Klein, R. I. 1975, *ApJ*, 195, 157
Czerny, B. & Hryniewicz, K. 2011, *A&A*, 525, 8
Dannen, R. C., Proga, D., Kallman, T. R., & Waters, T. 2019, *ApJ*, 882, 99
Done, C., Davis, S. W., Jin, C., Blaes, O., & Ward, M. 2012, *MNRAS*, 420, 1848
Ebrero, J., Kriss, G. A., Kaastra, J. S., & Ely, J. C. 2016, *Ap&A*, 586, 72
Gaskell, C. M. 1982, *ApJ*, 263, 79
Giustini, M. & Proga, D. 2019, *A&A*, 630, 94
Haardt, F. & Maraschi, L. 1993, *ApJ*, 413, 507
Kaastra, J. S., Kriss, G. A., Cappi, M., Mehdipour, M., et al. 2014, *Science*, 345, 64
Kriss, G. A., Mehdipour, M., Kaastra, J. S., Rau, A., et al. 2019, *A&A*, 621, 12
Marziani, P., Sulentic, J. W., Dultzin-Hacyan, D., Calvani, M., et al. 1996, *ApJ*, 104, 37
Mehdipour, M., Kaastra, J. S., Kriss, G. A., Cappi, M., et al. 2015, *A&A*, 575, 22
Mehdipour, M., Kaastra, J. S., Kriss, G. A., Arav, N., et al. 2017, *A&A*, 607, 28
Murray, N., Chiang, J., Grossman, S. A., & Voit, G. M. 1995, *ApJ*, 451, 498
Peterson, B. M., Ferrarese, L., Gilbert, K. M., Kaspi, S., et al. 2004, *ApJ*, 613, 682
Petrucci, P.-O., Ursini, F., De Rosa, A., Bianchi, S., et al. 2018, *A&A*, 611, 59
Proga, D., Stone, J. M., & Kallman, T. R. 2000, *ApJ*, 543, 686
Proga, D. & Kallman, T. R. 2004, *ApJ*, 616, 688
Proga, D. 2005, *ApJ*, 630, 9
Rankine, A. L., Hewett, P. C., Banerji, M., & Richards, G. T. 2020, *MNRAS in press*, arXiv:1912.08700
Richards, G. T., Kruczek, N. E., Gallagher, S. C., Hall, P. B., et al. 2011, *AJ*, 141, 167
Risaliti, G. & Elvis, M. 2010, *A&A*, 516, 89
Sulentic, J. W., Zwitter, T., Marziani, P., & Dultzin-Hacyan, D. 2000, *ApJ*, 536, 5
Sulentic, J. W. & Marziani, P. 2015, *FrASS*, 2, 6
Waters, T., Kashi, A., Proga, D., Eracleous, M., et al. 2016, *ApJ*, 827, 53

Optical spectroscopy of nearby type1-LINERs

Sara Cazzoli[1], Isabel Márquez[1], Josefa Masegosa[1], Ascensión del Olmo[1], Mirjana Pović[2,1], Omaira González-Martín[3], Barbara Balmaverde[4], Lorena Hernández-García[5] and Santiago García-Burillo[6]

[1]IAA - Instituto de Astrofísica de Andalucía (CSIC), Apdo. 3004, 18080, Granada, Spain
email: sara@iaa.es

[2]ESSTI/EORC - Ethiopian Space Science and Technology Institute, Entoto Observatory and Research Center, P.O. Box 33679, Addis Ababa, Ethiopia

[3]IRyA - Instituto de Radioastronomía y Astrofísica, 3-72 Xangari, 8701, Morelia, Mexico

[4]INAF - Osservatorio Astronomico di Brera, via Brera 28, I-20121 Milano, Italy

[5]Universidad de Valparaíso, Gran Bretana 1111, Playa Ancha, Valparaíso, Chile

[6]OAN - Observatorio Astronómico Nacional, Alfonso XII, 3, 28014, Madrid, Spain

Abstract. We present the highlights from our recent study of 22 local ($z < 0.025$) type-1 LINERs from the Palomar Survey, on the basis of optical long-slit spectroscopic observations taken with TWIN/CAHA, ALFOSC/NOT and HST/STIS. Our goals were threefold: (a) explore the AGN-nature of these LINERs by studying the broad (BLR-originated) H$\alpha\lambda$6563 component; (b) derive a reliable interpretation for the multiple narrow components of emission lines by studying their kinematics and ionisation mechanism (via standard BPTs); (c) probe the neutral gas in the nuclei of these LINERs for the first time. Hence, kinematics and fluxes of a set of emission lines, from H$\beta\lambda$4861 to [SII]$\lambda\lambda$6716,6731, and the NaD$\lambda\lambda$5890,5896 doublet in absorption have been modelled and measured, after the subtraction of the underlying light from the stellar component.

Keywords. galaxies: active, galaxies: kinematics and dynamics, techniques: spectroscopic

1. Introduction

Low ionisation nuclear emission-line regions (LINERs) are a class of low-luminosity AGNs showing strong low-ionisation and faint high-ionisation emission lines (Heckman 1980). LINERs are interesting objects since they are the most numerous local AGN population bridging the gap between normal and active galaxies (Ho 2008).

Over the past 20 years, the ionising source in LINERs has been studied through a multi-wavelength approach via different tracers (Ho 2008). Nevertheless, a long standing issue is the origin and excitation mechanism of the ionised gas studied via optical emission lines. In addition to the AGN scenario, two more alternatives, such as pAGBs stars and shocks, have been proposed to explain the optical properties of LINERs (e.g. Singh *et al.* 2013).

In LINERs, outflows are common as suggested by their Hα nuclear morphology (Masegosa *et al.* 2011). To open a new window to explore the AGN-nature and the excitation mechanism in LINERs, we propose to infer the role of outflows (identified as relatively broad component) in the broadening of emission lines. This broadening effect may limit the spectroscopic classification, as the contribution of outflows may overcome the determination of an eventually faint and broad Hα component from the BLR.

Ionised gas outflows are observed in starbursts and AGNs via long slit (e.g. Harrison *et al.* 2012) and integral field spectroscopy (e.g. Maiolino *et al.* 2017) of emission lines.

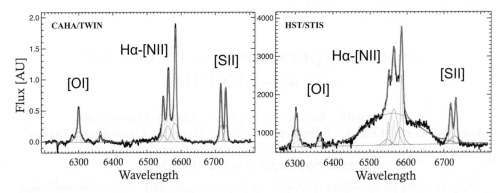

Figure 1. Gaussian fit to the ground- (left) and space- (right) based spectra for NGC4450 in the Hα region. We marked with different colours the components required to model the emission lines. The red curve shows the total contribution from the fit.

The neutral gas in outflows have been studied in detail only in starbursts and luminous and ultra-luminous infrared galaxies (U/LIRGs) via the NaD absorption (e.g. Cazzoli et al. 2016).

In Cazzoli et al. 2018, hereafter C18, our goals were to investigate the AGN nature of type-1 LINERs and to characterize all the components by studying their kinematics and ionisation mechanisms. We also aimed to probe and study the neutral gas properties.

For type-2 LINERs, see the contribution by L. Hermosa-Muñoz in this volume.

2. Sample, Data and Methods

The sample contains nearby ($z < 0.025$) 22 type-1 LINERs (L1) selected from the Palomar Survey (see Ho et al. 1997). Spectroscopic data were gathered with the TWIN Spectrograph mounted on the 3.5m telescope of the Calar Alto Observatory (CAHA) and with ALFOSC attached to the 2.6m North Optical Telescope (NOT). We also analyzed archival spectra (red bandpass) for 12 LINERs (see Balmaverde et al. 2014 for details) obtained with the Space Telescope Imaging Spectrograph (STIS) on board the Hubble Space Telescope (*HST*). The data analysis is organised in three main steps:

Stellar Subtraction

We applied the penalized PiXel fitting (pPXF; Cappellari 2017) and the STARLIGHT methods (Cid Fernandes et al. 2009) for modeling the stellar continuum. The stellar model is then subtracted to the observed to one obtain a interstellar medium spectrum.

Emission Lines

The fit was performed simultaneously for emission lines from [OI]$\lambda\lambda$6300,6363 to [SII], with single or multiple Gaussian kinematic components (up two for forbidden lines and narrow Hα). For the modeling of the Hα-[NII]$\lambda\lambda$6548,6584 blend, we tested three distinct models. Specifically, we considered either [SII] or [OI] (S- and O- models) or both ('mixed' M-model) as reference for tying central wavelengths and line widths (see example in Fig. 1). The latter model takes into account possible stratification density in the narrow line region. Then, a broad Hα component is added if needed to reduce significantly the residual. Finally, the best fitting (i.e. model, components, velocity shift and line widths) has been constraint to be the same for [OIII]$\lambda\lambda$4959,5007 and Hβ lines. Intensity ratios for [NII], [OI] and [OIII] lines were imposed following Osterbrock et al. 2006.

Absorption Lines

The NaD absorption doublet (8 detections) was modelled with one (i.e. two Gaussian profiles) or two components as in Cazzoli et al. 2014. The ratio of the equivalent widths of the two lines was allowed to vary from 1 to 2 (i.e. optically thick/thin limits, Spitzer 1978).

3. Main Results

NGC 4203 represents an extreme case as three line components are not sufficient to reproduce well the Hα profile, therefore, we excluded this L1 from the analysis.

For ground-based data, the S- and O- models reproduce well the line profiles in six of the cases each, while a larger fraction of cases (i.e. 9/21) require M-models for a satisfactory fit. Of the four possible combinations of the three components, as single narrow Gaussian per forbidden line is adequate in 6 out of 21 cases. A broad Hα component is required in the remaining four cases. In most cases (15/21), two Gaussians per forbidden line are required for a satisfactory modelling. Among these 15 cases, only in three cases a broad Hα component is required to reproduce well the observed profiles.

Velocities of the narrow components are close to rest frame varying within ± 110 km s^{-1}. The average velocity dispersion value for the narrow components is $\sigma = 157$ km s^{-1}. For the second components the velocity range is large, from -350 km s^{-1} to 100 km s^{-1}. The velocity dispersion varies between 150 and 800 km s^{-1} being generally broader (on average $\sigma = 429$ km s^{-1}) than for narrow components. A broad Hα component is required only in 7 out of the 21 LINERs, with FWHMs from 1277 km s^{-1} to 3158 km s^{-1}.

For none of the 11 *HST*/STIS spectra, the adopted best fit is obtained using the O-model, finding a slightly large prevalence of best fits with M-models. In four cases, one Gaussian per forbidden line and narrow Hα is adequate. In the remaining cases, two Gaussians are required for a good fit.

Narrow components have velocities between -100 and 200 km s^{-1}; the velocity dispersions vary between 120 and 270 km s^{-1} (176 km s^{-1}, on average). Similarly, the velocities of second component range from -200 to 150 km s^{-1}. These second components are however broader, with velocity dispersion values between 300 and 750 km s^{-1} (433 km s^{-1}, on average). The broad component in is ubiquitous. The FWHM of the broad Hα components in *HST* spectra range from 2152 km s^{-1} to 7359 km s^{-1} (3270 km s^{-1}, on average).

For the NaD absorption, in 7 out of 8 targets, a single kinematic component gives a good fit. Velocities of the neutral gas narrow components vary between -165 and 165 km s^{-1}; velocity dispersions values are in the range 104–335 km s^{-1} (220 km s^{-1}, on average).

4. Discussion

Probing the BLR in L1

The analysis of Palomar spectra by Ho et al. (1997) indicated that all the LINERs in our selected sample show a broad Hα component resulting in their classification as L1. Nevertheless, for ground-based data our detection rate for the broad component is only 33 %, questioning the classification as L1 by Ho et al. (1997). For space-based data the broad component is ubiquitous in agreement with Balmaverde et al. 2014. By comparing our strategy and measurements with those by previous works (Sect. 5.1 in C18), we conclude that the detectability of the BLR-component is sensitive to the starlight decontamination and the choice of the template for the Hα-[NII] blend. Moreover, a single Gaussian fit for the forbidden lines is an oversimplification in many cases.

Kinematic classification of the components

The distribution of the velocity for narrow and second components as a function of their velocity dispersion is presented in Fig. 2. We identified four areas corresponding to different kinematical explanations: rotation, candidate for non-rotational motions and non-rotational-motions (with broad blue/redshifted lines produced by outflows/inflows). For both ground- and space-based data, the kinematics of the narrow component can be explained with rotation in all cases (Fig. 2 left) whereas that of the second components encompass all possibilities (Fig. 2, right). From our ground-based data, we identified

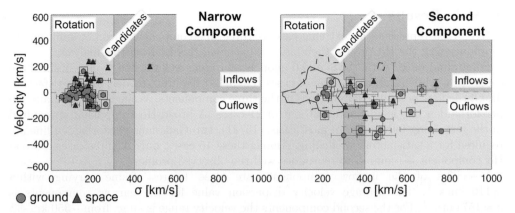

Figure 2. Observed velocity dispersion - velocity plane for narrow (left) and second (right) components. Red circles an purple triangles mark the measurements from ground- and space-based data, respectively. An additional blue box marks those LINERs for which the fitting of the Hα profile is less reliable (see C18). In the right panel, we report the measurements of the narrow component with contours (continuos and dashed lines are for ground- and space-based data, respectively). The coloured areas indicate different classifications (labelled on the panels).

6 out of 15 (40%) cases that may be interpreted as outflows. Outflow-components have velocities varying from $-15\,\mathrm{km\,s^{-1}}$ to $-340\,\mathrm{km\,s^{-1}}$, and velocity dispersions in the range of $450-770\,\mathrm{km\,s^{-1}}$ (on average, $575\,\mathrm{km\,s^{-1}}$). We did not interpreted as outflows any case in *HST*/STIS data. These results partially disagree with studies of the Hα morphology in LINERs which indicate that outflows are frequent in LINERs. A possible explanation is that the extended nature of outflows is not fully captured by the *HST*/STIS spectra.

Ionisation mechanisms from standard 'BPT (Baldwin, Phillips & Terlevich)-diagrams'
The line ratios for the narrow component are generally consistent with those observed in AGNs (either Seyfert or LINERs), excluding the star-formation or pAGBs as the dominant ionization mechanism (see Fig. 9 in C18). For the second component, we reproduced the observed line ratios with the shock-models by Groves *et al.* 2004. We combined the data points and models for the [O I]/Hα BPT diagram (Baldwin, Phillips & Terlevich 1981), the most reliable for studying shocks (Allen*et al.* 2008) shown in Fig. 3, with the kinematical classification shown in Fig. 2. Models at low velocities ($<300\,\mathrm{km\,s^{-1}}$) indicate the presence of mild-shocks associated to perturbations of rotation (Fig. 3 center). At higher velocities ($>400\,\mathrm{km\,s^{-1}}$) shocks are produced by non-rotational motions (Fig. 3 right).

A lack of neutral outflows?
According to the adopted kinematic classification all the neutral gas kinematic components (except one) could be interpreted as rotation. The possible explanation of the lack of neutral gas non-rotational motions, such as outflows, is twofold. First, the neutral component in outflows is possibly less significant in AGNs than in starbursts galaxies and U/LIRGs. Secondly, such a null detection rate might be a consequence of the conservative limits we assumed, as ionised and neutral gas correspond to different phases of the outflows, and hence these may have a different kinematics (e.g. Cazzoli *et al.* 2016).

5. Conclusions

• *The AGN nature of L1.* The detection of the BLR-component is sensitive to the starlight subtraction, the template for the Hα-[NII] blend and the assumption of a single Gaussian fit (often an oversimplification); NLR stratification might be often present in L1.

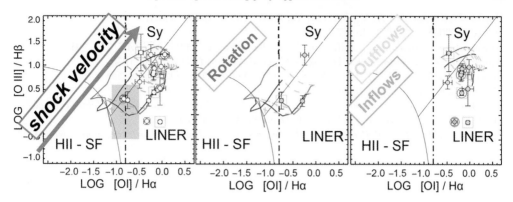

Figure 3. *Left:* optical standard [OI] BPT diagram (Baldwin, Phillips & Terlevich 1981) for the second component obtained from our ground-based spectroscopy (circles). In all the panels an additional circle marks those LINERs for which a broad component is required to reproduce the Hα profile. Light-green square are as in Fig. 2. Black lines represent the dividing curves between HII regions, Seyferts, and LINERs from Kewley *et al.* 2006 and Kauffmann *et al.* 2003, and weak-[OI] and strong-[OI] LINERs from Filippenko & Terlevich 1992. Gray boxes show the predictions of photoionisation models by pAGB stars by Binette *et al.* 1994. The predictions of shock+precursorionisation models from Groves *et al.* 2004 with $n_{el} = 100\,\mathrm{cm}^{-3}$ (blue) and $n_{el} = 1000\,\mathrm{cm}^{-3}$ (red) are overlaid. Iso-velocities are marked with yellow dashed-lines. *Center* and *Right* panels are the same but considering considering different shock-velocities (colour coded according to Fig. 2).

- <u>*Kinematics of emission lines and their classification.*</u> The kinematics of the narrow component can be explained with rotation in all cases whereas that of the second components encompass all possibilities. From our ground-based data, the detection rate of outflows is 40 %. We did not interpreted as outflows any case in *HST*/STIS data.
- <u>*Ionisation mechanisms.*</u> Our results favor the AGN photoionisation as the dominant mechanism of ionisation for the narrow component. Shocks models (at the observed velocities) are able to reproduce the observed line ratios of the second component.
- <u>*Neutral gas in L1.*</u> The neutral gas is found to be in rotation. Neutral gas outflows are possibly less significant in AGNs or the limits we assumed are too conservative.

Acknowledgements

We thank the financial support by the Spanish MCIU and MEC, grants SEV-2017-0709 and AYA 2016-76682-C3. SC thanks the IAU for the travel grant.

References

Allen, M. G., Groves, B. A., Dopita, M. A., *et al.* 2008, *ApJS*, 178, 20
Balmaverde & Capetti 2014, *A&A*, 563, A119
Baldwin, Phillips & Terlevich 1981, *PASP*, 93, 5
Binette, L., Magris, C. G., Stasinska, G., Bruzual, A. G 1994, *A&A*, 292, 13
Cazzoli, S., Marquez, I., Masegosa, J. *et al.* 2018, *MNRAS*, 480, 1106
Cazzoli, S., Arribas, S., Maiolino, R., & Colina, L. 2016, *A&A*, 590, A125
Cazzoli, S., Arribas, S., Colina, L., *et al.* 2014, *A&A*, 569, A14
Cappellari, M. 2017, *MNRAS*, 466, 798
Cid Fernandes, R., Stasinska, G., Schlickmann, M., *et al.* 2009, *Rev. Mex. Astron. Astrofis.*, 35, 127
Filippenko, A. V. & Terlevich R. 1992, *ApJ*, 397, L79
Groves, B. A., Dopita, M. A. & Sutherland, R. S. 2004, *ApJS*, 153, 75
Harrison, C. M., *et al.* 2012, *MNRAS*, 426, 1073
Heckman 1980, *A&A*, 87, 152

Ho, L. C., Filippenko, A. V., Sargent, W. L. W. 1997, *ApJS*, 112, 315
Ho, L. C. 2008, *ARA&A*, 46, 475
Kewley, L. J., Groves, B., Kauffmann, G. & Heckman, T. 2006, *MNRAS*, 372, 961
Kauffmann, G., Heckman, T., Tremonti, C. *et al.* 2003, *MNRAS*, 346, 1055
Maiolino, R., Russel, H., Fabian, A. *et al.* 2017, *Nature*, 544, 202
Masegosa, J., Marquez, I., Ramirez, A. *et al.* 2011, *A&A*, 527, A23
Osterbrock, D. E., *et al.* 2006, *Astrophysics of Gaseous Nebulae and Active Galactic Nuclei*
Singh R. G., van de Ven K., Jahnke, M. *et al.* 2013, *A&A*, 558, A43
Spitzer L. 1978, *Physical Processes in the Interstellar Medium. Wiley-Interscience, New York*

Study of the diversity of AGN dust models

Omaira González-Martín

IRyA - Instituto de Radioastronomía y Astrofísica, 3-72 Xangari, 8701, Morelia, Mexico

Abstract. The dust component of active galactic nuclei (AGN) produces a broad infrared spectral energy distribution (SED), whose power and shape depends on the fraction of the source absorbed, and the geometry of the absorber respectively. This emitting region is expected to be concentrated within the inner ~ 5 pc of the AGN which makes almost impossible to image it with the current instruments. The study the infrared SED by comparison between infrared AGN spectra and predicted models is one of the few way to infer the properties of this dust component. We explore a set of six dusty models of AGN with available SEDs, namely Fritz et al. (2006), Nenkova et al. (2008), Hoenig & Kishimoto (2010), Siebenmorgen et al. (2015), Stalevski et al. (2016), and Hoenig & Kishimoto (2017). They cover a wide range of morphologies, dust distributions and compositions. We explore the discrimination among models and parameter restriction using synthetic spectra (Gonzalez-Martin et al. 2019A, submitted), and perform spectral fitting of a sample of 110 AGN with Spitzer/IRS drawn from the Swift/BAT survey (Gonzalez-Martin et al. 2019B, submitted). Our conclusion is that most of these models can be discriminated using only mid-infrared spectroscopy as long as the host galaxy contribution is less than 50%. The best model describing sample is the clumpy disk-wind model by Hoenig & Kishimoto (2017). However, large residuals are shown irrespective of the model used, indicating that AGN dust is more complex than models. We found that the parameter space covered by models is not completely adequate. This talk will give tips for observers and modelers to actually answer the question: how is the dust arrange in AGN? This question will be one of the main subjects of future research with JWST in the AGN field.

Keywords. galaxies: active, galaxies: dusty models, galaxies: spectral energy distribution

References

Fritz, J., *et al.* 2006, *MNRAS*, 366, 767
Hoenig, S. F. & Kishimoto, M. 2010, *A&A*, 523, 27
Hoenig, S. F. & Kishimoto, M. 2017, *ApJL*, 838, L20
Nenkova, M., *et al.* 2008, *ApJ*, 685, 160
Siebenmorgen, R., *et al.* 2015, *A&A*, 583, 120
Stalevski, M., *et al.* 2016, *MNRAS*, 458, 228

AGN evolution as seen in spectral lines: The case of narrow-line Seyfert 1s

Marco Berton[1,2]

[1] Finnish Centre for Astronomy with ESO (FINCA), University of Turku, Vesilinnantie 5, FI-20014 University of Turku, Finland

[2] Aalto University Metsähovi Radio Observatory, Metsähovintie 114, FI-02540 Kylmälä, Finland

Abstract. Line profiles can provide fundamental information on the physics of active galactic nuclei (AGN). In the case of narrow-line Seyfert 1 galaxies (NLS1s) this is of particular importance since past studies revealed how their permitted line profiles are well reproduced by a Lorentzian function instead of a Gaussian. This has been explained with different properties of the broad-line region (BLR), which may present more pronounced turbulent motions in NLS1s with respect to other AGN. We investigated the line profiles in a recent large NLS1 sample classified using SDSS, and we divided the sources into two subsamples according to their line shapes, Gaussian or Lorentzian. The line profiles seem to separate all the properties of NLS1s. Black hole mass, Eddington ratio, [OIII] luminosity, and Fe II strength are all very different in the Lorentzian and Gaussian samples, as well as their position on the quasar main sequence. We interpret this in terms of evolution within the class of NLS1s. The Lorentzian sources may be the youngest objects, while Gaussian profiles may be typically associated to more evolved objects. Further detailed spectroscopic studies are needed to fully confirm our hypothesis.

Keywords. galaxies: active, galaxies: narrow-line Seyfert 1, galaxies: evolution

The properties of the dusty inner regions of nearby QSOs

Itziar Aretxaga

Instituto Nacional de Astrofísica, Óptica y Electrónica (INAOE), Luis Enrique Erro 1, Sta. Ma. Tonantzintla, Puebla, México

Abstract. We present MIR spectroscopy and photometry obtained with CanariCam on the 10.4 m Gran Telescopio CANARIAS for a sample of 20 nearby, MIR bright and X-ray luminous quasi-stellar objects (QSOs). We find that for the majority of QSOs the MIR emission is unresolved at angular scales ∼0.3 arcsec. We derive the properties of the dusti tori that surround the nucleus based on these observations and find significant differences in the parameters compared with a sample of Seyfert 1 and 2 nuclei. We also find evidence for polycyclic aromatic hydrocarbon (PAH) features in the spectra, indicative of star formation, more centrally peaked (on scales of a few hundred pc) than previously believed.

Keywords. galaxies: active, galaxies: quasars, methods: spectroscopy, methods: photometry

AGN1 vs AGN2 dichotomy as seen from the point of view of ionized outflows

Eleonora Sani

European Southern Observatory (ESO), Santiago de Chile, Chile

Abstract. I present a detailed study of ionized outflows in a large sample of 650 hard X-ray detected AGN. Taking advantage of the legacy value of the BAT AGN Spectroscopic Survey (BASS, DR1), we are able to reveal the faintest wings of the [OIII] emission lines associated with outflows. The sample allows us to derive the incidence of outflows covering a wide range of AGN bolometric luminosity and test how the outflow parameters are related with various AGN power tracers, such as black hole mass, Eddington ratio, luminosity. I'll show how ionized outflows are more frequently found in type 1.9 and type 1 AGN (50% and 40%) with respect to the low fraction in type 2 AGN (20%). Within such a framework, I'll demonstrate how type 2 AGN outflows are almost evenly balanced between blue- and red-shifted winds. This, in strong contrast with type 1 and type 1.9 AGN outflows which are almost exclusively blue-shifted. Finally, I'll prove how the outflow occurrence is driven by the accretion rate, whereas the dependence of outflow properties with respect to the other AGN power tracers happens to be quite mild.

Keywords. galaxies: active, galaxies: outflows, surveys: X-rays

Elusive accretion discs in low luminosity AGN

M. Almudena Prieto[1] and Juan A. Fernandez-Ontiveros[2]

[1]Instituto Astrofisica de Canarias (IAC), Spain
email: aprieto@iac.es

[2]Istituto di Astrofisica e Planetologia Spaziali (INAF-IAPS), Italy
email: j.a.fernandez.ontiveros@gmail.com

Abstract. Low luminosity AGN represent the vast majority of the AGN population in the near universe, and still the least conforming class with the standard AGN scenario. Their low luminosity is at odds with their often very high black hole masses and powerful jets. I will review the challenges that parsec-scale observations across the electromagnetic spectrum of some of the nearest ones are opening on the true nature of their emission, their transition from the most luminous to the feeble ones, and their accretion power. The strict limits imposed by these observations on their accretion power are confronted with the high mechanical energy inferred for their jets. Possible scenarios for these nuclei including the extraction of power form the black hole spin are discussed (Prieto *et al.* 2016; Fernandez-Ontiveros *et al.* 2019).

Keywords. galaxies: active, galaxies: nuclei

References

Fernandez-Ontiveros, J. A., López-Gonzaga, N., Prieto, M. A., *et al.* 2019, *MNRAS*, 485, 5377
Prieto, M. A., Fernandez-Ontiveros, J. A., *et al.* 2016, *MNRAS*, 457, 3081

CHAPTER III. Variability

A status report on AGN variability

Paulina Lira

Departamento de Astronomía, Universidad de Chile, Camino el Observatorio 1515,
Las Condes, Santiago, Chile
email: plira@das.uchile.cl

Abstract. Active Galactic Nuclei (AGN) are ubiquitous variable sources. This trademark property allows the study of many aspects of AGN physics which are not possible by other means. In this review I summarize what has been learnt by the close monitoring of AGN flux variations with special emphasis in studies conducted in optical and near-infrared domain. I also highlight what knowledge is still missing from our picture of AGN phenomena, as well as possible developments expected in this new era of time-domain astronomy.

Keywords. galaxies: active, galaxies: nuclei, surveys, methods: statistical

1. Introduction

The Preface to the Proceedings of the First Texas Symposium on Relativistic Astrophysics, titled "Quasi-Stellar Sources and Gravitational Collapse", includes the very words that Peter Bergman sent around in June 1963 to motivate the symposium, which include: 'The source 3C273B seems to be a superstar, and according to Harlan Smith, has a diameter of about a light-week. It is the brightest known object in the universe, about a million million times brighter than the sun. According to Sandage and Smith, its brightness varies by about 50 percent'.

More than 50 years later, we know that these sources are powered by Super Massive Black Holes (SMBHs) found in the centers of galaxies. When actively accreting matter, gravitational energy is released as radiation and the SMBHs are classified as Active Galactic Nuclei (AGN). As with the case of 3C273B, AGN emission is highly variable at all wavelengths, from the radio to the gamma-rays. As the radiation shines onto the surroundings of the SMBH, this material responds to the variable signal. The study of the primary variable radiation source and the response from nearby structures has been one of the most important tools used by astronomers to study the structure and physical nature of AGN.

In this review I will focus on recent advancements in our understanding of AGN variability. This includes clues on the nature of the primary variable source itself, as well as of those structures that surround SMBHs, such as the accretion disk, the Broad Line Region (BLR) and the dusty torus. I will focus on optical and infrared studies and touch upon X-ray findings, while I redirect the reader to other works that cover high-energy and radio emission, wavelengths which trace the X-ray corona and jet, such as those by Uttley (2015) and the review on radio properties of AGN by Tadhunter (2016) and the results by Mundell *et al.* (2009).

2. Continuum emission variability below 1μm: the primary source and the accretion disk

Most AGN show UV, optical and near-infrared variability on *short time scales*, this is, time scales of the order of days to weeks for SMBHs of masses in the 10^6–10^8 M$_\odot$

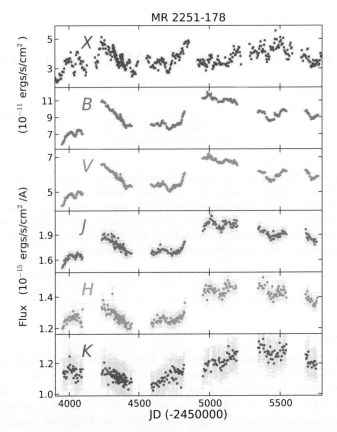

Figure 1. X-ray, optical and near-infrared light curves for the nearby quasar MR2251-178.

mass range or weeks to months for a $10^8 - 10^{10}$ $M_\odot$ mass range. Exception to this rule seem to be a not well quantified fraction of Narrow Line Seyfert 1 nuclei (NLS1s), which tend to show flat light curves despite showing variability in the X-ray and UV domain (e.g., Shemmer & Netzer 2000; Klimek et al. 2004; Yip et al. 2009; Bachev et al. 2009). A particular Spectral Energy Distribution (SED) might be responsible for this behaviour, as I will discuss later. Notice, however, that samples of optically varying NLS1s have also been identified and closely studied (Du et al. 2014).

Continuum variability below $1\mu m$ seems to be highly correlated in all sources monitored with well sampled multiwavelength light curves (e.g., Cackett et al. 2007; Shappee et al. 2014; Edelson et al. 2015; Fausnaugh et al. 2016). An example is shown in Fig. 1, which presents X-ray, optical and near-infrared light curves for the nearby quasar MR2251-178 during ∼5.5 years of monitoring. Partial results from this campaign were already presented in Arévalo et al. (2008) and Lira et al. (2011). Besides the clearly observed correlation, other results are the following : 1) there is a small but significant delay between light curves, with longer wavelengths following shorter wavelengths; 2) short time scale variability decreases with wavelength; 3) the 2–10 keV X-ray emission, while well correlated during the first 3 years of the monitoring, is not longer correlated during the remaining of the campaign.

The results on MR2251-178 are replicated in many sources. Time lag determinations between UV and optical continuum light curves give results consistent with only a few day-lights for typical Seyfert and low-mass quasars. Fitting a power-law to the observed lag versus wavelength correlation results in values close to those predicted by the classical

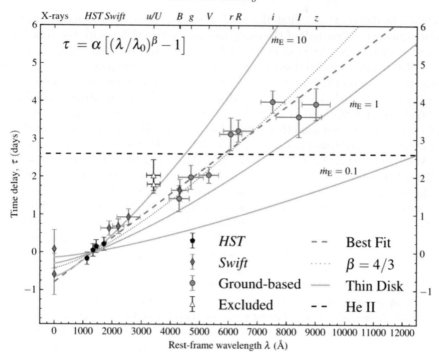

Figure 2. Time delays between X-ray, UV and continuum optical light curves determined for NGC5548. Lines represent model responses to a central illuminating source for accretion disks following the prescription $\tau \propto \lambda^\beta$ (top left corner). Cyan curves represent models with a fixed $\beta = 4/3$, as prescribed by standard accretion disk models, and accretion rates $\dot{m}_{Edd} = 0.1$, 1 and 10. The dotted line gives the best fit for $\beta = 4/3$, but its zero point requires an accretion rate much too high for what is inferred for this AGN. The best fit model, shown with a dashed line, requires $\beta \sim 1$. Taken from Fausnaugh et al. (2016).

accretion disk proposed by Shakura & Sunyaev (1973), i.e., $f_\nu \sim \nu^{1/3}$ although with a larger zero point by about a factor of 3 (see Fig. 2). In other words, the varying signal matches well the time it would take for light to travel from the center of the system to the peripheries of the accretion disk. What we are witnessing is not intrinsic disk variability, but the response from the disk to a centrally illuminating variable source. The compilation of the RMS spectrum for 849 quasars observed by SDSS-IV is astonishingly close to a $\nu^{1/3}$ power-law, supporting again the prescription of Shakura & Sunyaev (Horne et al. in prep). The mismatch in the zero point, which is a function of SMBH mass and accretion rate, is a current matter of debate and could point to the failings of the simplicity of classical accretion disk models, as suggested by numerical simulations (Mishra et al. 2016, 2019; Sadowski 2016; Jiang et al. 2013, 2016, 2019; Gronkiewicz & Różańska 2020 – see further discussion in Section 4).

One very important question has not been answered yet: what is the central variable source? SED modelling can shed some light onto this question. The illuminating source must be found close to the SMBH and should therefore be highly energetic, at frequencies beyond the near-UV. Energetically, it needs to be powerful enough to imprint a variable signal which can be detected *against* the intrinsic emission of the accretion disk. Hence, the ratio between the emission seen (or estimated) in the optical/near-UV (from the disk) over that seen (or estimated) in the far-UV/X-rays (the central source?) can tell us whether short-term variability will be observed as result of the central source illumination the accretion disk. It has been argued that in the case of NLS1s, the dominant disk does not show such variability as it swamps the central varying source (Jin et al. 2017a,b).

MR2251-178 demonstrates that the hard 2–10 keV X-ray emission, although a good candidate during the first half of the campaign and with a delay with the B and V bands consistent with 0 days (Arévalo et al. 2008), cannot be the illuminating source, as it goes completely uncorrelated during the second half of the campaign. Lack of correlation between the X-rays and UV/optical emission is also seen in many other Seyfert galaxies, such as NGC 7469 (Behar et al. 2020), Mrk 817 (Morales et al. 2019), Mrk 335 (Gallo et al. 2018), Ark 120 (Gliozzi et al. 2017), PKS 0558-504 (Gliozzi et al. 2013). But some level of correlation is seen sometimes, as is the case of MR2251-178 (Arévalo et al. 2008, Fig. 1), NGC3783 (Arévalo et al. 2009), NGC4395 (McHardy et al. 2006; Cameron et al. 2012), NGC4593 (Pal & Naik 2018), NGC4051 (Breedt et al. 2010), NGC2617 (Shappee et al. 2014; Fausnaugh et al. 2018), NGC6814 (Troyer et al. 2016), Mrk79 (Breedt et al. 2009). In fact, some sources have shown a good correlation during some campaigns to become uncorrelated at some other time, e.g., NGC5548 (McHardy et al. 2014; Starkey et al. 2017) and F9 (Pal et al. 2017; Lohfink et al. 2014). This complex scenario can be summarized as follows: the highly variable X-ray emission arises from a different structure than the much smoother and well correlated UV/optical emission. This is in good agreement with our general hypothesis that the X-rays originate from a 'corona' of fast electrons that upscatter seed photons provenient from the accretion disk. The accretion disk, at the same time, reprocesses high-energy photons in a well behaved manner, showing variability time scales in agreement with the travel time between regions characterized by different temperatures in the stratified accretion disk. If there is correlation between the X-rays and the longer wavelength light curves, the lags are consistent with ∼1 day, although some longer lags have also been measured (Shappee et al. 2014; Edelson et al. 2017), making things harder for the reprocessing model.

Gardner & Done (2017) analyze the lack of correlation between the X-ray emission and that observed in the UV and optical bands for NGC5548. They propose that a new region has to be added to our paradigm in order to understand these observations. The new region would be found between the central high-energy source and the disk and would reprocess the X-ray central emission into far-UV photons that finally heat the accretion disk (Mehdipour et al. 2011; Middei et al. 2018; Ursini et al. 2019).

3. Continuum emission variability beyond 1μm: the dusty torus

Dust grains will survive to the high temperatures maintained by the central emission from the so-called sublimation radius and beyond. Dust exposed to the AGN radiation will absorb short-wavelength emission and re-emit it with typical temperatures of many hundreds to a thousand Kelvin (i.e., much higher than those experienced in starburst galaxies).

The assumption that this hot dust is arranged into a toroidal shape is based on observational evidence (see Netzer 2015 for a comprehensive review): the axisymmetric nature of the accretion flow suggests that, if the dust is associated to the parsec-scale gas dynamics, then a similar axisymmetric dust structure could be expected; in some objects collimation of the ionizing continuum that reaches to large distances from the central SMBH is thought to be caused by the dusty torus; the presence of polarized broad emission lines in systems that do not show these components in direct light has been considered the strongest evidence for the so-called Unification Model of AGN, where the line of sight of the observer to the central source, either intercepting or not the toroidal structure, determines whether the AGN appears as a 'type-I' or 'type-II' nucleus.

Evidence supporting the Unified Model is vast and shows that even though the situation is probably much more complex than original thought, and that not all AGN follow the model strictly, the presence of something similar to an axisymmetric dusty structure in many AGN is true. Recently, ALMA has been able to detect the presence of dust emission from the center of a few very nearby Seyfert galaxies that might indeed correspond to the torus (Combes et al. 2019).

The cross-over between the emission from the accretion disk and the torus is found at about 1μm, i.e., at the limit between the optical and near-infrared domain. Hence, the K-band corresponds to the near-infrared wavelength where the torus emission dominates, while the J and H bands might still show a strong disk contribution. Cross-correlation between K-band light curves and those obtained in the optical show that the torus follows the variations seen in the accretion disk but with a lag corresponding to the light travel time to the sublimation radius (Clavel et al. 1989; Suganuma et al. 2006; Koshida et al. 2014; Lira et al. 2011, 2015; Pozo et al. 2014, 2015; Mandal et al. 2018). The lags observed in local Seyfert galaxies are in the 40–400 day range, corresponding to distances of fractions of a pc to a few pc. These distances are larger than those inferred for the BLR (see next Section), therefore supporting the most basic requirement of the Unified Model, i.e., that the presence of the dusty torus is able to hide the central region of the AGN from the viewer.

4. Line emission variability: the physics of the BLR

The idea that studying the response of the BLR to the variable central continuum could reveal important clues about the nature of the BLR was first suggested by Bahcall et al. (1973) and Cherepashchuk & Lyutyi (1973). It was put into mathematical formalism by Blandford & McKee (1982), who also coined the term Reverberation Mapping (RM). The first observational campaigns were carried out in the 1980s. The principle is simple: through spectroscopic monitoring of a galactic nuclei it is possible to isolate the continuum and line fluxes, allowing for the construction of separate light curves. The cross-correlation of continuum and line-emission light curves would then allow to determine the time lag, or light travel distance between the central engine and the BLR. Such campaigns, are however, very costly in terms of observing time, and require a very careful analysis to secure the proper flux calibration of the light curves and meaningful error estimations.

The determination of lags using cross-correlation analysis can be done in different ways. The z-transformed discrete correlation function (ZDCF – Alexander 1997) works solely with the observed values of the light curves and it is based on the discrete correlation function (DCF) of Edelson & Krolik (1988). The widely used interpolated cross correlation function (ICCF, Peterson et al. 1998, 2004) interpolates fluxes to a desired cadence assuming that the line and continuum fluxes in gaps between two observed points are properly approximated by a linear interpolation in time between the two. Finally, more recent methods such as JAVELIN (Zu et al. 2011) and CREAM (Starkey et al. 2016) model the light curves as a damped random walk process (see below) to determine a lag and its significance. Their basic assumption is that the emission line light curves are the result of the response to an ionizing continuum which is changing exactly in the same way as the observed continuum used in the lag determination, which however, is not always observed to be the case (Goad et al. 2016; Lira et al. 2018). For well sampled light curves all methods give results which are consistent with each other, as expected.

Early results from RM campaigns immediately yielded significant findings. The BLR was an extended and stratified structure, with different emission lines been produced by gas located at different distances from the central source. This allowed radiative transfer models to reproduce many of the BLR traits, now assuming that different 'cloud' conditions yielded the diverse family of emission lines seen in AGN spectra.

Later findings gave evidence for a 'Virialized' BLR, this is, the BLR is a gravitationally bound structure (Peterson & Wandel 1999; Peterson et al. 2004). This is in fact a corner stone for SMBH mass determinations based on RM of the BLR. The relation

$M_{\rm BH} = fGv^2/R$ shows that we can determine the mass of the SMBH ($M_{\rm BH}$) if we know the location (R) and speed (v) at which an object that is under the gravitational influence of the black hole travels. While the width of the Doppler-shifted broad emission lines gives v, RM analysis yields the much sought after value of R. f, remains a fairly unconstrained factor that encompasses the details of the geometry and kinematics of the BLR and it is expected to vary from source to source (Collin et al. 2006), although recent progress seems to agree about a BLR in the form of a rotating flattened system (Pancoast et al. 2014; Grier et al. 2017; Williams et al. 2018; Mejía-Restrepo et al. 2018; Li et al. 2018; Martínez-Aldama et al. 2019).

One of the most important results that came from RM studies of the BLR is the so-called 'Radius-Luminosity' (RL) relation. This is an observational result that correlates the luminosity of the continuum emission with the distance at which a BLR particular line is produced. The RL relation allows to obtain SMBH masses with a single spectrum! (the so-called 'single-epoch' method), since the continuum luminosity and line width can be determined from one single, flux-calibrated spectroscopic observation. Until now, well determined RL relations based on RM results have only been established for Hβ (Kaspi et al., 2000, 2005; Peterson et al. 2004; Bentz et al. 2006, 2009, 2013; Grier et al. 2017), although cross-calibration with MgII λ2798 shows that no significant biases are present when using this line (McLure & Dunlop 2004; Shen & Liu 2012; Zuo et al. 2015; Mejía-Restrepo et al. 2016), allowing the determination of $M_{\rm BH}$ up to $z \sim 2$ when using optical observations (e.g., Trakhtenbrot & Netzer 2012) and up to $z \sim 5$ when observing MgII λ2798 in the observed-frame near-infrared (e.g., Jiang et al. 2007; Kurk et al. 2007; Willott et al. 2010; Trakhtenbrot et al. 2011; De Rosa et al. 2014; Mazzucchelli et al. 2017; Shen et al. 2019; Onoue et al. 2019).

RM studies are now witnessing a new era, where the boundary is been pushed in terms of redshift, observed lines, and number of studied sources, with recent results (Shen et al. 2016; Lira et al. 2018; Grier et al. 2019; Hoormann et al. 2019) and future projects, such as the SDSSV Black Hole Mapper (Kollmeier et al. 2017) and the 4MOST Time-Domain Extragalactic Survey (TiDES) program (Swann et al. 2019) expanding our knowledge of the BLR. These huge datasets will also bring important challenges, as has already been shown by recent RM results presenting systematic offsets between these new determinations and the 'classical' results of Bentz et al. (2014). Czernic et al. (2019) explores these differences and argues that they can be understood if a range of spin values is allowed when interpreting RL relations.

The very best RM data can afford a more thorough investigation of the BLR using 'tomography', this is, the study of the line response as a function of velocity. As different regions of the broad lines correspond to regions traveling at different velocities, their changes as a function of time allow us to have a close view of the BLR kinematics. In fact, exquisite data might one day allow us to study the presence of spiral arms in the accretion disk of AGN, but so far tomography has been used to put constrains on the presence of bulk BLR motions, such as outflows and inflows (Horne et al. 2004).

The most intense RM campaign is that of the nearby Seyfter galaxy NGC5548 which took place in 2014 using ground and space-based facilities. Continuum lags once again showed that the best-fit model to the lag vs. wavelength relation corresponds to a power-law with index $\sim 1/3$ (Edelson et al. 2015), in good agreement with the prescriptions of classical accretion disks, but with a zero point, which is a function of the $M_{BH}\dot{M}_\odot$ product (where $\dot{M}_\odot$ is the accretion rate – Fausnaugh et al. 2016), too large by about a factor 3, a result found in several other monitoring campaigns (McHardy et al. 2014; Shappee et al. 2014; Lira et al. 2015). However, arguably the most interesting result of the campaign was that of the 'BLR-holiday', corresponding to nearly half of the campaign length, where the line flux light curve did not seem to be responding to the observed continuum, but instead showed an uncorrelated behaviour (Goad et al. 2016 – see Fig. 3).

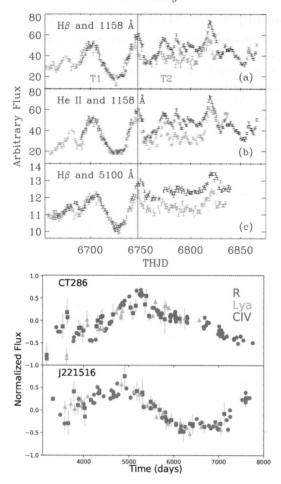

Figure 3. Continuum and emission line light curves for three different AGN. Top panel shows the observed 'BLR-holiday' (from the vertical red line onwards) of the emission lines in NGC5548, which after ∼100 days of monitoring present an uncorrelated behaviour with respect to the UV and optical continuum. Taken from Pei *et al.* (2017). Bottom panel shows the response of Lyα and CIV emission lines to the observed R-band (rest-frame ∼1500 Å) continuum. While emission lines and continuum follow each other almost perfectly in J221516, the response in CT286 is clearly more complex. Adapted from Lira *et al.* (2018).

The anomalous behaviour shown by NGC5548 highlights one of the most important shortcomings during RM analysis: that the *observed continuum* is not the *ionizing continuum* responsible for the formation of emission lines. In particular, recombination Hydrogen lines require photons with energies in excess of 13.6 eV (or wavelengths shorter than 912Å), a range which is inaccessible to us observers. Nature does not allow us to observe any photons of astrophysical origin at the Lyman edge, while Galactic emission in the very soft X-ray domain (0.001–0.01 KeV) severely limits our chances to study extragalactic sources. Hence, the true far-UV emission from AGN can only be studied by indirect methods, which represents a huge challenge and introduces significant problems in our understanding of AGN physics as it corresponds to the peak of the emission coming from the innermost region of the accretion disk. The physics of SMBH spin is largely hidden in this unobserved wavelength range.

5. Variability on longer time-scales: Changing Look AGN

It is expected that while short-time scale variations are dominated by the accretion disk response to the illumination from the central variable source, changes in the accretion flow itself may introduce changes on *longer-time* scales. Depending on the physical mechanism, the accretion flow can show fluctuations with time scales ranging from many months to hundreds of years for a SMBH with $M_{\rm BH} = 10^8$ M$_\odot$, quantities that scale linearly with SMBH mass (e.g., Graham *et al.* 2020).

MR2251-178, with an estimated mass of $\sim 2 \times 10^8$ M$_\odot$ fits well this scenario. As proposed by Arévalo *et al.* (2008), the short-term (days to weeks) variability observed in the optical bands is well explained as illumination of the highly variable X-ray emission (or some other similar central source), while the long-term (months to years) variability shows too large amplitude variations to be explained as reprocessing of the X-ray fluctuations. Hence, changes in the intrinsic accretion flow are invoked.

Changes in the accretion flow have been claimed to be responsible for a much more dramatic behaviour observed in active nuclei: that of Changing Look (CL) AGN. The CL terminology was borrowed from X-ray astronomy (where it refers to changes in the level of obscuration at these wavelengths), to first represent the appearing or disappearing of broad components to the Balmer lines on time scales of months to years, and later to represent also significant continuum variations (e.g., Fig. 4). This redefinition allows to explore large photometric surveys to look for significant flux changes. As an example, in a study of more than 8000 quasars, Rumbaugh *et al.* (2018) found that $\sim 10\%$ of them exhibited more than one magnitude change in flux ($|\Delta g| > 1$) within a time scale of 15 years and claimed that the true fraction could reach 30–50%. Continuum and spectral changes, however, do not necessarily go hand-in-hand, as discussed by Graham *et al.* (2020).

CL AGN has been an exploding area of research in the last decade due to the large number of time domain spectroscopic and photometric measurements now available (e.g., LaMassa *et al.* 2015; McElroy *et al.* 2016; MacLeod *et al.* 2016; Ruan *et al.* 2016; Gezari *et al.* 2017; Ross *et al.* 2018; Rumbaugh *et al.* 2018; Stern *et al.* 2018; Shapovalova *et al.* 2019; Graham *et al.* 2020), although many serendipity detections of CL activity were claimed since the mid 70s.

Occultation of the BLR and accretion disks by intervening material along the line of sight seemed since early on as a possible mechanism responsible for CL systems. Indeed, this process has been clearly identified as the origin for the rapid changes in the X-ray obscuration towards the active nucleus in the type-II Seyfert galaxy NGC1365 (Risaliti *et al.* 2005). In this case, BLR 'clouds' were responsible for producing the transitions in the amount of absorption towards the X-ray emitting region in this nucleus. A similar mechanism for CL AGN, however, usually fails. LaMassa *et al.* (2015) showed that the crossing time across the line of sight of any occulting material towards the BLR was larger than the observed CL time scales.

Using the single-epoch technique, Ruan *et al.* (2016) showed that the inferred SMBH mass estimates for the active nuclei in SDSS J015957.64 + 003310.5 remain consistent with each other when using spectroscopic observations obtained at different epochs in this CL nucleus. The observed *broader when dimmer* effect can be interpreted as the switching-off of distant BLR clouds that produce the broad lines when the continuum becomes dimmer, while new clouds, closer in to the SMBH experiencing a more intense gravitational pull and hence emitting broader lines, switches-on. This results also imply that the BLR remained bound throughout and that the presence of the broad components is not due to other external mechanisms, such as supernova explosions or Tidal Disruption Events (TDEs).

Things can never be that simple, however, as complex behaviour is appearing in some of the best studied CL AGN, as shown in Trakhtenbrot *et al.* (2019) and Ricci *et al.* (2020). They present results from the CL nucleus in 1ES 1927 + 654 which changed from

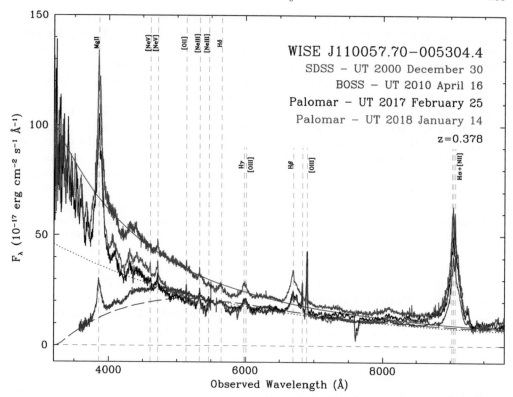

Figure 4. SDSS spectra of a quasar at $z \sim 0.4$ presenting dramatic changes at the blue end within the time scale of 20 years, indicating changes in the inner part of the accretion disk. BLR and near-infrared torus-related emission should follow suit with the expected delays of a few years. Taken from Ross et al. (2018).

a "true" Type-II to a Type-I system in the time scale of months. However, its light curve looks suspiciously similar to that predicted for a TDE, while its X-ray emission showed complex and extremely variable behaviour. This could be freak a case, where the change in the look of this AGN is not due to internal changes, but instead triggered by the infall of a star or a massive cloud onto the central black hole.

The physical mechanism behind CL AGN needs to be explained withing the context of accretion physics. People have looked at what is already known in black hole hosting X-ray Binaries (BHBs), systems that vary in much shorter time scales because of their much smaller sizes. Accreting BHBs have already been united with their radio-loud SMBH counterparts through the "fundamental plane for accreting black holes", a relation between L_{xrays}, L_{radio}, and M_{BH} which covers many orders of magnitudes in all three quantities involved (Merloni et al. 2003; Falcke et al. 2004).

BHBs are known to show two characteristic emission 'states', the Low/Hard and the High/Soft states. This denominations refer to the observed changes in their X-ray spectra, window where disk emission appears in this low-mass objects. These changes have long been understood as changes in the nature of the accretion flow, with the High/Soft state corresponding to high-accretion periods, with a luminous classical accretion disk, while the Low/Hard state is characterized by the inner truncation of the disk, been replaced by a flow that cools inefficiently during a low accretion-rate period (for reviews see Done et al. 2007 and Belloni 2010).

The question that then arises is whether CL AGN correspond to changes of the accretion flow onto SMBHs as seen in BHBs. Some evidence supports this scenario, as both changes are characterized by similar tracks in a Eddington ratio vs spectral-slope plane (Ruan et al. 2019). Some evidence seems to support that the AGN CL episodes would also happen when the accretion rate lowers below 1%, in agreement with what is seen in BHBs. However, a crucial problem remains: the time scales are seriously off when scaled based on black hole mass. State transitions in BHBs occur at time scales of $\sim$ days, which would correspond to $\sim 10^5$ years for SMBHs in AGN (Ruan et al. 2020).

In a very recent work, Graham et al. (2020) analysed the variability properties of a sample of CL AGN candidates selected from the CATALINA survey by requiring a change in the Hβ/OIII ratio larger than 30%, besides other criteria like optical and near-infrared significant variability. They found that 111/717 of nuclei presented the required spectroscopic variability, i.e., $\sim$15%. The variability time scales of the optical continuum in their sample are in reasonable agreement with that predicted for thermal fronts. These were proposed in the early 80s to explain the observed behaviour during outbursts in X-ray binaries and dwarf novae, as partially ionized gas can undergo abrupt changes between hot and cold conditions. The disk structure then cycles between these configurations which also carry sudden changes in mass accretion rate and therefore in the release of gravitational energy. Transition waves or thermal fronts are responsible for this structural changes, either heating or cooling the disk (Meyer 1985; Menou et al. 1999). In AGN, thermal fronts have been proposed to be driven also by changes in the magnetic torques exerted in the innermost part of the disk (Ross et al. 2018), or by instabilities in radiation pressure (Śniegowska & Czerny 2019).

Clearly, the jury is still out concerning CL AGN and it is very likely that this phenomena will in fact group together more than one type of physical mechanisms. Needless to say things will greatly change once the Vera Rubin Telescope (LSST) is operating since biases due to sparse sampling and flux limits will be greatly reduced. Spectroscopic follow-up for the study of the emission lines will require a great effort from the astronomical community but it will no doubt be needed to secure a thorough understanding of these systems.

6. Selecting AGN through variability

Traditionally, AGN have been found by their blue optical colors. In color-color plots (e.g., Richards et al. 2009), AGN have been found to occupy a well defined region that separates them from other sources such as stars and galaxies, which usually look redder. This selection works if the AGN are 1) high luminosity (so that the redder colors of their hosts do not contaminate the AGN emission), 2) unobscured (so that the AGN is not reddened by dust absorption), 3) below $z < 3$ (so that AGN remain well separated from stars). To find AGN that do not fulfill these requirements is a challenge.

AGN are, however, ubiquitously variable. Hence, variable sources located in the centers of galaxies are excellent candidates for accreting SMBH, even if they do not meet one or more of the requirements listed above. With this in mind, many works have focused on the idea of finding AGN using variability as a search criterion (e.g., Butler & Bloom 2011; Myers et al. 2015; Peters et al. 2015; Palanque-Delabrouille et al. 2016; Tie et al. 2017).

More recently Sánchez-Sáez et al. (2019) have shown that using variability and colors not only is more effective than using colors only, but that a new population of redder AGN can be identified. This population appears to be dominated by low-luminosity AGN, with colors some times completely dominated by the host emission. Obscured AGN are however, mostly missed by this technique, as the presence of significant amounts of dust can completely quench the AGN variable continuum.

Variability searches also open a new window for the detection of Intermediate Mass Black Holes (IMBHs), those with $M_{\rm BH} < 10^6$ M$_\odot$ and above the stellar mass range. This population is particularly interesting, as it might help to disentangle between possible scenarios for the formation of SMBHs at high redshift, the so-called 'BH seeds' (Greene, Strader & Ho 2019). The demographics of black holes that have not grown much since their formation will help to constrain which of these scenarios are viable by confronting occupation numbers and mass distributions with theoretical predictions (Bellovary et al. 2019). Also, BH masses in the 10^5–10^6 $M_\odot$ represent the primary targets for the Laser Interferometer Space Antennae (LISA), which is expected to be launched in the next decade. LISA will detect Gravitational Waves from merging black holes from $z \sim 0$ to $z \sim 20$, overlapping with the cosmological times when the *seeding* of black holes occurred and mapping their presence all the way to the present universe.

Finding IMBHs has not been an easy task and the examples known today probably constitute the tip of the iceberg of the underlying population. The most successful method to probe *bona-fide* IMBHs is to look for broad but weak emission Balmer lines, which together with the radius–luminosity scaling relations, has yielded masses for a few hundred sources after combing the whole SDSS (Liu et al. 2018; Chilingarian et al. 2018). The search for faint and compact radio sources ('radio cores') associated to the nuclei of galaxies in the COSMOS field has successfully determine the presence of AGN in 35 dwarf galaxies (Mezcua et al. 2019), although their bolometric luminosities are found beyond that predicted by numerical simulations

More recently, variability has shown to be able to identify many candidate IMBHs. Martínez-Palomera et al. (2020) have selected galaxies which show rapid, low-amplitude variability in their nuclei and shown that this method increases the number density of IMBHs by a factor of 40-50, when compared with spectroscopic searches. About 20% of this candidates have already been confirmed as AGN by the presence of broad Balmer lines, radio cores or by their emission line ratios, typical of AGN.

7. Statistical analysis and description of AGN variability

As has already been noted in this article, many things we have learnt about AGN are the result of what was already known for black holes in stellar systems. The Power Spectrum (or Power-Spectral Density, PSD), this is, the amount of variance as a function of frequency, had already been used to describe the wide range of variability time scales observed in the X-ray light curves of accreting compact stellar systems (e.g., Nolan et al. 1981) before been applied to observations of X-rays from AGN (e.g., Lawrence et al. 1987).

The PSD of AGN is in fact a close relative to those seen in BHBs during their Soft/High state (for a review see Uttley 2006). They are characterized by $P(\nu) \propto \nu^{-2}$, the so called 'red noise', but break to a $P(\nu) \propto \nu^{-1}$, a 'flicker noise' regime, at low frequencies. In fact, this characteristic break frequency has been shown to strongly correlate with $M_{\rm BH}$ (and possibly inversely with accretion rate), a correlation that spans many order of magnitude (McHardy et al. 2016; González-Martín & Vaughan 2012) from BHBs to massive AGN.

In the UV, optical and near-infrared, most of the statistical analysis of light curves has been done in the time domain, instead of frequency. This is because light curves are usually not dense enough and do not have enough data points to apply Fourier Transform techniques. Instead, auto-correlation, cross-correlation and structure function analysis are usually adopted. In particular, the structure function (SF) has been widely used to describe AGN light curves as it is closely related to the PSD (see Fig. 5). For example, a single power law description of the PSD can be recovered by the slope observed in the SF. The problem with techniques that work in the time domain, however, is that the measurements are strongly correlated by design, and therefore can fail to provide a robust description of more complex structure in the variability signal (Emmanoulopoulos et al. 2010). Dense sampling of AGN in the optical have been achieved only recently

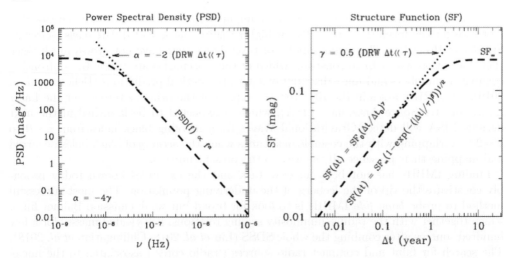

Figure 5. Power Density Spectrum and Structure Function of a DRW signal. Taken from Kozłowski 2016.

using space missions for the search of transiting planets, such Kepler and TESS, allowing for the determination of PSDs for individual sources Smith et al. (2018a) show that 6/21 sources present breaks in their PSDs at low frequency, the remaining sources been consistent with a single power-law. High-frequency exponents are found to be ~ -2.5.

A PSD characterized by $P(\nu) \propto \nu^\alpha$, with $\alpha \leqslant -2$ corresponds to a highly correlated behaviour, such as those displayed by 'random-walk' or 'Brownian motion', or as already mentioned, 'red noise'. Here we understand correlation as the level of independence between *events* (or flux values in a light curve). Random-walk, for example, ties subsequent events, this is, the event that is occurring now has memory of the previous event, and so on. No correlation (characterized by a PSD with $\alpha = 0$), on the other hand, is referred to as 'white-noise'.

The level of correlation or 'memory' in a time sequence can be expressed mathematically in a simple manner when the events are equally spaced in time (fixed Δt). So for example, if the process driving a light curve is that of a random-walk, then every event x_i at time t_i can be expressed as $x_i = \mu + \phi x_{i-1} + \epsilon_i$, where μ is the mean, x_{i-1} corresponds to the event that occurred at t_{i-1}, $\phi = 1$ and ϵ_i is a random term usually drawn from a Gaussian distribution (Moreno et al. 2019). If $\alpha = 2$ the process is called 'damped random-walk' (DRW) and ϕ^{-1} corresponds to the characteristic time scale of the break in the PSD. DRW has been found to be a very good description of AGN light curves (Kelly et al. 2009), although deviations are also found (Kasliwal et al. 2015; Simm et al. 2016; Sánchez et al. 2017). As real data show gaps and unequal time spacing, instead of a using a discrete characterization of the underlying process, a continuous mathematical description is necessary. In its more generic form, this is termed a CARMA process. A CARMA(1,0) process corresponds to a DRW.

Further correlation is found when x_i depends on the last two previous events and the previous perturbation terms (Moreno et al. 2019). In its continuous form this is a CARMA(2,1) process, also known as a 'damped harmonic oscillator' (DHO), characterized by two time scales related to the terms ϕ_1 and ϕ_2. A particular case of a DHO corresponds to a 'quasi-periodic oscillator' (QPO). High and low-frequency QPOs have been observed in a few BHBs, IMBH and AGN (Smith et al. 2018a and references therein) and growing evidence suggests that these are good black hole mass estimators.

With the future of time-domain astronomy and data science looking extremely bright, in the next decade we will be able to learn about the physics of black holes and accretion at an increasing pace, and perhaps unravel some of the best kept secrets of these fantastic systems.

References

Alexander, T. 1997, *Astronomical Time Series*, 163
Arévalo, P., Uttley, P., Kaspi, S., et al. 2008, *MNRAS*, 389, 1479
Arévalo, P., Uttley, P., Lira, P., et al. 2009, *MNRAS*, 397, 2004
Bachev, R., Grupe, D., Boeva, S., et al. 2009, *MNRAS*, 399, 750
Bahcall, J. N., Joss, P. C., Cohen, J. G., et al. 1973, *ApJ*, 184, 57B
Behar, E., Kaspi, S., Paubert, G., et al. 2020, *MNRAS*, 491, 3523
Belloni, T. M. 2010, Lecture Notes in 'Physics, Berlin Springer Verlag, 53
Bellovary, J. M., Cleary, C. E., Munshi, F., et al. 2019, *MNRAS*, 482, 2913
Bentz, M. C., Peterson, B. M., Pogge, R. W., et al. 2006, *ApJ*, 644, 133
Bentz, M. C., Peterson, B. M., Netzer, H., et al. 2009, *ApJ*, 697, 160
Bentz, M. C., Denney, K. D., Grier, C. J., et al. 2013, *ApJ*, 767, 149
Blandford, R. D. & McKee, C. F. 1982, *ApJ*, 255, 419
Breedt, E., Ar'evalo, P., McHardy, I. M., et al. 2009, *MNRAS*, 394, 427
Breedt, E., McHardy, I. M., Arévalo, P., et al. 2010, *MNRAS*, 403, 605
Butler, N. R. & Bloom, J. S. 2011, *AJ*, 141, 93
Cackett, E. M., Horne, K., & Winkler, H. 2007, *MNRAS*, 380, 669
Cameron, D. T., McHardy, I., Dwelly, T., et al. 2012, *MNRAS*, 422, 902
Cherepashchuk, A. M. & Lyutyi, V. M. 1973, *ApL*, 13, 165
Chilingarian, I. V., Katkov, I. Y., Zolotukhin, I. Y., et al. 2018, *ApJ*, 863, 1
Clavel, J., Wamsteker, W., & Glass, I. S. 1989, *ApJ*, 337, 236
Collin, S., Kawaguchi, T., Peterson, B. M., et al. 2006, *A&A*, 456, 75
Combes, F., García-Burillo, S., Audibert, A., et al. 2019, *A&A*, 623, A79
Czerny, B., Wang, J.-M., Du, P., et al. 2019, *ApJ*, 870, 84
De Rosa, G., Venemans, B. P., Decarli, R., et al. 2014, *ApJ*, 790, 145
Done, C., Gierliński, M., & Kubota, A. 2007, *A&A Rev.*, 15, 1
Du, P., Hu, C., Lu, K.-X., et al. 2014, *ApJ*, 782, 45
Edelson, R. A. & Krolik, J. H. 1988, *ApJ*, 333, 646
Edelson, R., Gelbord, J. M., Horne, K., et al. 2015, *ApJ*, 806, 129
Edelson, R., Gelbord, J., Cackett, E., et al. 2017, *ApJ*, 840, 41
Emmanoulopoulos, D., McHardy, I. M., & Uttley, P. 2010, *MNRAS*, 404, 931
Falcke, H., Körding, E., & Markoff, S. 2004, *A&A*, 414, 895
Fausnaugh, M. M., Denney, K. D., Barth, A. J., et al. 2016, *ApJ*, 821, 56
Fausnaugh, M. M., Starkey, D. A., Horne, K., et al. 2018, *ApJ*, 854, 107
Gallo, L. C., Blue, D. M., Grupe, D., et al. 2018, *MNRAS*, 478, 2557
Gardner, E. & Done, C. 2017, *MNRAS*, 470, 3591
Gezari, S., Hung, T., Cenko, S. B., et al. 2017, *ApJ*, 835, 144
Gliozzi, M., Papadakis, I. E., Grupe, D., et al. 2013, *MNRAS*, 433, 1709
Gliozzi, M., Papadakis, I. E., Grupe, D., et al. 2017, *MNRAS*, 464, 3955
Goad, M. R., Korista, K. T., De Rosa, G., et al. 2016, *ApJ*, 824, 11
González-Martín, O. & Vaughan, S. 2012, *A&A*544, A80
Graham, M. J., Ross, N. P., Stern, D., et al. 2020, *MNRAS*, 491, 4925
Greene, J. E., Strader, J., & Ho, L. C. 2019, arXiv e-prints, arXiv:1911.09678
Grier, C. J., Pancoast, A., Barth, A. J., et al. 2017, *ApJ*, 849, 146
Grier, C. J., Trump, J. R., Shen, Y., et al. 2017, *ApJ*, 851, 21
Grier, C. J., Shen, Y., Horne, K., et al. 2019, *ApJ*, 887, 38
Gronkiewicz, D. & Różańska, A., 2020, *A&A*, 633, 35
Hoormann, J. K., Martini, P., Davis, T. M., et al. 2019, *MNRAS*, 487, 3650
Horne, K., Peterson, B. M., Collier, S. J., et al. 2004, *PASP*, 116, 465

Jiang, L., Fan, X., Vestergaard, M., *et al.* 2007, *AJ*, 134, 1150
Jiang, Y.-F., Stone, J. M., & Davis, S. W. 2013, *ApJ*, 778, 65
Jiang, Y.-F., Davis, S. W., & Stone, J. M. 2016, *ApJ*, 827, 10
Jiang, Y.-F., Blaes, O., Stone, J. M., *et al.* 2019, *ApJ*, 885, 144
Jiang, Y.-F., Stone, J. M., & Davis, S. W. 2019, *ApJ*, 880, 67
Jin, C., Done, C., Ward, M., *et al.* 2017b, *MNRAS*, 471, 706
Jin, C., Done, C., & Ward, M. 2017a, *MNRAS*, 468, 3663
Kasliwal, V. P., Vogeley, M. S., & Richards, G. T. 2015, *MNRAS*, 451, 4328
Kaspi, S., Smith, P. S., Netzer, H., *et al.* 2000, *ApJ*, 533, 631
Kaspi, S., Maoz, D., Netzer, H., *et al.* 2005, *ApJ*, 629, 61
Kelly, B. C., Bechtold, J., & Siemiginowska, A. 2009, *ApJ*, 698, 895
Kelly, B. C., Sobolewska, M., & Siemiginowska, A. 2011, *ApJ*, 730, 52
Klimek, E. S., Gaskell, C. M., & Hedrick, C. H. 2004, *ApJ*, 609, 69
Kollmeier, J. A., Zasowski, G., Rix, H.-W., *et al.* 2017, arXiv e-prints, arXiv:1711.03234
Koshida, S., Minezaki, T., Yoshii, Y., *et al.* 2014, *ApJ*, 788, 159
Kozłowski, S. 2016, *ApJ*, 826, 118
Kurk, J. D., Walter, F., Fan, X., *et al.* 2007, *ApJ*, 669, 32
LaMassa, S. M., Cales, S., Moran, E. C., *et al.* 2015, *ApJ*, 800, 144
Lawrence, A., Watson, M. G., Pounds, K. A., *et al.* 1987, *Nature*, 325, 694
Li, Y.-R., Songsheng, Y.-Y., Qiu, J., *et al.* 2018, *ApJ*, 869, 137
Lira, P., Arévalo, P., Uttley, P., *et al.* 2015, *MNRAS*, 454, 368
Lira, P., Arévalo, P., Uttley, P., *et al.* 2011, *MNRAS*, 415, 1290
Lira, P., Kaspi, S., Netzer, H., *et al.* 2018, *ApJ*, 865, 56
Liu, H.-Y., Yuan, W., Dong, X.-B., *et al.* 2018, *ApJS*, 235, 40
Lohfink, A. M., Reynolds, C. S., Vasudevan, R., *et al.* 2014, *ApJ*, 788, 10
MacLeod, C. L., Ross, N. P., Lawrence, A., *et al.* 2016, *MNRAS*, 457, 389
Mandal, A. K., Rakshit, S., Kurian, K. S., *et al.* 2018, *MNRAS*, 475, 5330
Martínez-Palomera, J., Lira, P., Bhalla-Ladd, I., *et al.* 2019, arXiv e-prints, arXiv:1912.02860
Martínez-Aldama, M. L., Czerny, B., Kawka, D., *et al.* 2019, *ApJ*, 883, 170
Martínez-Palomera, J., Lira, P., Bhalla-Ladd, I., *et al.* 2020, *ApJ*, 889, 113
Mazzucchelli, C., Bañados, E., Venemans, B. P., *et al.* 2017, *ApJ*, 849, 91
McElroy, R. E., Husemann, B., Croom, S. M., *et al.* 2016, *A&A*, 593, L8
McHardy, I. M., Koerding, E., Knigge, C., *et al.* 2006, *Nature*, 444, 730
McHardy, I. M., Cameron, D. T., Dwelly, T., *et al.* 2014, *MNRAS*, 444, 1469
McHardy, I. M., Connolly, S. D., Peterson, B. M., *et al.* 2016, *Astronomische Nachrichten*, 337, 500
McLure, R. J. & Dunlop, J. S. 2004, *MNRAS*, 352, 1390
Mehdipour, M., Branduardi-Raymont, G., Kaastra, J. S., *et al.* 2011, *A&A*, 534, A39
Mejía-Restrepo, J. E., Trakhtenbrot, B., Lira, P. *et al.* 2016, *MNRAS*, 460, 187
Mejía-Restrepo, J. E., Lira, P., Netzer, H., *et al.* 2018, *Nature Astronomy*, 2, 63
Menou, K., Hameury, J.-M., & Stehle, R. 1999, *MNRAS*, 305, 79
Merloni, A., Heinz, S., & di Matteo, T. 2003, *MNRAS*, 345, 1057
Mezcua, M., Suh, H., & Civano, F. 2019, *MNRAS*, 488, 685
Meyer, F. 1985, Recent Results on Cataclysmic Variables. The Importance of IUE and Exosat Results on Cataclysmic Variables and Low-mass X-ray Binaries, 83
Middei, R., Bianchi, S., Cappi, M., *et al.* 2018, *A&A*, 615, A163
Mishra, B., Fragile, P. C., Johnson, L. C., *et al.* 2016, *MNRAS*, 463, 3437
Mishra, B., Kluźniak, W., Fragile, P. C., *et al.* 2019, *MNRAS*, 483, 4811
Morales, A. M., Miller, J. M., Cackett, E. M., *et al.* 2019, *ApJ*, 870, 54
Moreno, J., Vogeley, M. S., Richards, G. T., *et al.* 2019, *PASP*, 131, 063001
Mundell, C. G., Ferruit, P., Nagar, N., *et al.* 2009, *ApJ*, 703, 802
Myers, A. D., Palanque-Delabrouille, N., Prakash, A., *et al.* 2015, *ApJS*, 221, 27
Netzer, H. 2015, *ARA&A*, 53, 365
Nolan, P. L., Gruber, D. E., Matteson, J. L., *et al.* 1981, *ApJ*, 246, 494

Onoue, M., Kashikawa, N., Matsuoka, Y., et al. 2019, *ApJ*, 880, 77
Pal, M., Dewangan, G. C., Connolly, S. D., et al. 2017, *MNRAS*, 466, 1777
Pal, M. & Naik, S. 2018, *MNRAS*, 474, 5351
Palanque-Delabrouille, N., Magneville, C., Yèche, C., et al. 2016, *A&A*, 587, A41
Pancoast, A., Brewer, B. J., Treu, T., et al. 2014, *MNRAS*, 445, 3073
Pei, L., Fausnaugh, M. M., Barth, A. J., et al. 2017, *ApJ*, 837, 131
Peters, C. M., Richards, G. T., Myers, A. D., et al. 2015, *ApJ*, 811, 95
Peterson, B. M. & Wandel, A. 1999, *ApJL*, 521, L95
Peterson, B. M., Wanders, I., Horne, K., et al. 1998, *PASP*, 110, 660
Peterson, B. M., Ferrarese, L., Gilbert, K. M., et al. 2004, *ApJ*, 613, 682
Pozo Nuñez, F., Haas, M., Chini, R., et al. 2014, *A&A*, 561, L8
Pozo Nuñez, F., Ramolla, M., Westhues, C., et al. 2015, *A&A*, 576, A73
Ricci, C., Kara, E., Loewenstein, M., et al. 2020, *ApJ*, 898, L1
Richards, G. T., Myers, A. D., Gray, A. G., et al. 2009, *ApJS*, 180, 67
Risaliti, G., Elvis, M., Fabbiano, G., et al. 2005, *ApJL*, 623, L93
Ross, N. P., Ford, K. E. S., Graham, M., et al. 2018, *MNRAS*, 480, 4468
Ruan, J. J., Anderson, S. F., Cales, S. L., et al. 2016, *ApJ*, 826, 188
Ruan, J. J., Anderson, S. F., Eracleous, M., et al. 2019, *ApJ*, 883, 76
Rumbaugh, N., Shen, Y., Morganson, E., et al. 2018, *ApJ*, 854, 160
Sánchez, P., Lira, P., Cartier, R., et al. 2017, *ApJ*, 849, 110
Sánchez-Sáez, P., Lira, P., Cartier, R., et al. 2019, *ApJS*, 242, 10
Sadowski, A. 2016, *MNRAS*, 459, 4397
Shakura, N. I. & Sunyaev, R. A. 1973, *A&A*, 500, 33
Shapovalova, A. I., Popović, L. Č., et al. 2019, *MNRAS*, 485, 4790
Shappee, B. J., Prieto, J. L., Grupe, D., et al. 2014, *ApJ*, 788, 48
Shemmer, O., & Netzer, H. 2000, arXiv e-prints, astro-ph/0005163
Shen, Y. & Liu, X. 2012, *ApJ*, 753, 125
Shen, Y., Horne, K., Grier, C. J., et al. 2016, *ApJ*, 818, 30
Shen, Y., Wu, J., Jiang, L., et al. 2019, *ApJ*, 873, 35
Simm, T., Salvato, M., Saglia, R., et al. 2016, *A&A*, 585, A129
Smith, K. L., Mushotzky, R. F., Boyd, P. T., et al. 2018a, *ApJ*, 857, 141
Smith, K. L., Mushotzky, R. F., Boyd, P. T., et al. 2018b, *ApJL*, 860, L10
Śniegowska, M. & Czerny, B. 2019, arXiv e-prints, arXiv:1904.06767
Starkey, D. A., Horne, K., & Villforth, C. 2016, *MNRAS*, 456, 1960
Starkey, D., Horne, K., Fausnaugh, M. M., et al. 2017, *ApJ*, 835, 65
Stern, D., McKernan, B., Graham, M. J., et al. 2018, *ApJ*, 864, 27
Suganuma, M., Yoshii, Y., Kobayashi, Y., et al. 2006, *ApJ*, 639, 46
Swann, E., Sullivan, M., Carrick, J., et al. 2019, The Messenger, 175, 58
Tadhunter, C. 2016, *A&A Rev.*, 24, 10
Tie, S. S., Martini, P., Mudd, D., et al. 2017, *AJ*, 153, 107
Trakhtenbrot, B. & Netzer, H. 2012, *MNRAS*, 427, 3081
Trakhtenbrot, B., Netzer, H., Lira, P., et al. 2011, *ApJ*, 730, 7
Trakhtenbrot, B., Arcavi, I., MacLeod, C. L., et al. 2019, *ApJ*, 883, 94
Troyer, J., Starkey, D., Cackett, E. M., et al. 2016, *MNRAS*, 456, 4040
Ursini, F., Petrucci, P.-O., Bianchi, S., et al. 2019, arXiv e-prints, arXiv:1912.08720
Uttley, P. 2006, AGN Variability from X-rays to Radio Waves, 101
Uttley, P. 2015, in The Extremes of Black Hole Accretion, Proceedings of the conference held 8–10 June, 2015 in Madrid, Spain
Williams, P. R., Pancoast, A., Treu, T., et al. 2018, *ApJ*, 866, 75
Willott, C. J., Albert, L., Arzoumanian, D., et al. 2010, *AJ*, 140, 546
Yip, C. W., Connolly, A. J., Vanden Berk, D. E., et al. 2009, *AJ*, 137, 5120
Zu, Y., Kochanek, C. S., & Peterson, B. M. 2011, *ApJ*, 735, 80
Zuo, W., Wu, X.-B., Fan, X., et al. 2015, *ApJ*, 799, 189

Recent results of measuring black hole masses via reverberation mapping

Shai Kaspi

School of Physics and Astronomy and Wise Observatory, Raymond and Beverly Sackler
Faculty of Exact Sciences, Tel-Aviv University, Tel-Aviv 6997801, Israel
email: shai@wise.tau.ac.il

Abstract. Over the past three decades more than 100 Active Galactic Nuclei (AGNs) were measured using the reverberation mapping technique. This technique uses the response of the line emission in the Broad Line Region (BLR) to continuum emission variation and yields a measure for the distance of the BLR from the central Black Hole (BH). This in turn is used to measure the BH's mass. Almost all of these measurements are of low-luminosity AGNs while for quasars with luminosities higher than 10^{46} erg s^{-1} there are hardly any attempts of reverberation mapping. This contribution reports on recent results from a two-decades campaigns to measure the BH mass in high-luminosity quasars using the reverberation mapping technique. BLR distance from the BH, BH mass, and AGN UV luminosity relations over eight orders of magnitude in luminosity are presented, pushing the luminosity limit to the highest point so far.

Keywords. galaxies: active, quasars: emission lines, galaxies: Seyfert

1. Introduction

Reverberation Mapping (RM) is a well developed method which is being used over the past three decades to measure the distance of the Broad Line Region (BLR) from the central black hole (BH) in Active Galactic Nuclei (AGNs), e.g., Bahcall et al. (1972); Blandford & McKee (1982); Peterson (1993, 2006); Netzer & Peterson (1997). This method uses the time lag between continuum variation of the AGN and the response Broad Line Region (BLR) emission line variation to estimate that distance. Once this distance is determined the AGN's central BH mass can be computed using the width of the emission line, and assumptions about the geometry of the BLR. Thus far, More than a hundred AGNs have sufficient data to significantly estimate the BLR size and the BH mass (e.g., Kaspi et al. 2000, 2005; Du et al. 2016; Bentz & Katz 2015; Grier et al. 2017). These studies resulted with a firm relation between the BLR's size and the AGN's Luminosity, and a relation between the BH's mass and the AGN's Luminosity. Both relations are roughly consistent with that the size and the mass are scaling with the square root of the luminosity. These relations are widely used to estimate the BLR size and BH mass in thousands of AGNs using a single epoch spectrum, in order to study cosmological structure and scales, the mass function of AGNs, accretion history, etc.

Although these size–mass–luminosity relations are used for AGNs at all luminosities, RM studies have been limited until very recently to intermediate luminosity AGNs in the range $10^{44} < \lambda L_\lambda(5100\text{Å}) < 10^{46}$ erg s^{-1} and at low redshift, mostly $z < 0.5$. Most RM studies were carried out using the Hβ broad emission line which is convivially located in the center of the optical wavelength region and is accessible for most optical spectrograph. In order to study if the size–mass–luminosity relations hold for higher-luminosity and higher-redshift AGNs there is a need to carry RM campaigns on such objects. Though, due to the high redshift the Hβ line is redshifted to the IR and the C IVλ 1550 and Lyα broad emission lines are redshifted to the optical region and can be used for RM with optical telescopes.

However, such studies are difficult to carry out due to a number of reasons: RM studies, specially toward high-luminosity and high-redshift AGNs, require a lot of telescope resources. Such objects have much larger BLR and much longer variability timescales, and thus need long monitoring periods - of order of a decade. Time Allocation Committees do not want to commit telescopes for such long periods. The light curves are smeared by the cosmic time dilution, and the continuum variations are smeared by the large BLR size, thus it is hard to detect the line variations in response to the continuum variations.

Because of these reasons only a few attempts were carried out so far to study RM of high-luminosity and high-redshift AGNs (e.g., Welsh *et al.* 2000; Trevese *et al.* 2007, 2014). This contribution presents two RM studies carried out on these type of AGNs. Together with results from past and recently published studies, the size–mass–luminosity relations will be presented for the the C IVλ 1550 emission line over the whole luminosity range of AGNs.

2. Las Campanas Observatory, Chile, Campaign

We† monitored a sample of 17 quasars from the Calan-Tololo and the SDSS catalogs. The luminosities of the quasars in this sample are $\lambda L_\lambda(1350\text{Å}) > 10^{46.5}$ erg s^{-1} and their redshift range is $2.3 < z < 3.4$. Photometric observations were carried out using the 0.9m at the CTIO every about two months since 2005. Spectroscopic observations were carried out using the 2.5m DuPont telescope at LCO. The results from that decade long campaign are published in Lira *et al.* (2018). An example of a light curve and cross corralations to determine the time lags from the C IVλ 1550 and Lyα broad emission lines are shown in Figure 1. In this study we significantly and reliably measure time lags for Lyα in 8 objects, for C IV in 8 objects, for Si IV in 3 objects, and for C III] in 1 object. Altogether significant time lags were measured for 10 distinct objects. We present an updated C IV radius–luminosity relation in which the radius scales like the UV luminosity to the power of 0.46 ± 0.08, and show for the first time a radius–luminosity relation for the other three lines. We conclude that the regions responsible for the emission of Lyα, Si IV, C IV, and C III] are commonly interior to that producing Hβ, but there is no clear stratification among them. Three out of the 17 sources in Lira *et al.* (2018) show an unexpected behavior in some emission lines in the sense that the line light curves do not appear to follow the observed UV continuum variations. This is an interesting behavior which is also seen in low luminosity AGNs (e.g., Goad *et al.* 2016).

3. HET—Wise Campaign

We‡ monitored a sample of 6 quasars in the luminosity range $10^{46} < \lambda L_\lambda(1350\text{Å}) < 10^{48}$ erg s^{-1} and redshift range of $2.3 < z < 3.4$. These objects are among the most luminous quasars and lie within the primary epoch of BH growth for luminous quasars. Photometric observations were carried out since 1994 at the 1 m telescope of the Wise Observatory (Israel) for 18 years and spectroscopic observations were carried out since 1999 at the 9 m Hobby-Eberly Telescope (HET) for 13 years. Preliminary results from that study were presented in Kaspi *et al.* (2007) where we show a preliminary C IV time lag for S5 0836 + 71 at z = 2.172 of 188^{+27}_{-37} days in the rest frame.

† In collaboration with Paulina Lira, Ismael Botti, Hagai Netzer, Nidia Morrell, Julián Mejía-Restrepo, Paula Sanchéz, Jorge Martínez, and Paula López.
‡ In collaboration with Niel Brandt, Dan Maoz, Hagai Netzer, Donald Schneider, Ohad Shemmer, and Kate Grier.

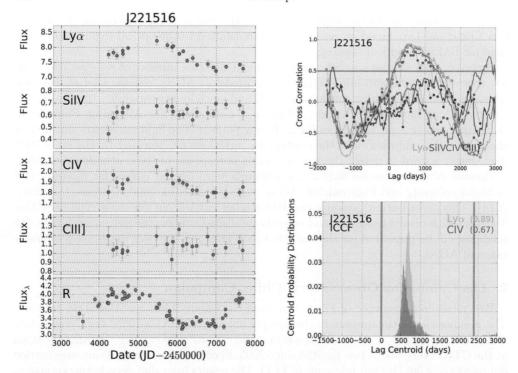

Figure 1. Continuum and emission lines light curves for the quasar 2QZ J221516 (left) together with cross correlation functions between the different emission lines and the continuum light curves (top right), and cross correlation centroid distribution for the two lines which show significant time lags (bottom right). For details see Lira *et al.* (2018).

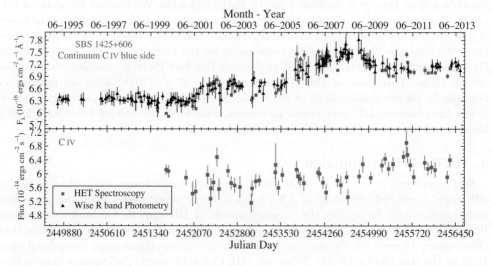

Figure 2. Continuum (top) and C IV (bottom) light curves of the quasar SBS 1425+606.

The results from the full period of this campaign will be presented in Kaspi *et al.* (2020, in preparation). An example of a light curve of one object is shown in Figure 2. All objects in the sample show continuum variability of order 10%–70%. C IV time lags are detected for 3 objects and C III] time lag was detected in one object. The C IV time

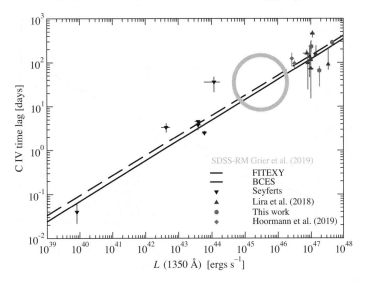

Figure 3. C IV time lag as a function of the UV luminosity.

lags are of order 100 to 250 light days. In the following we use these results together with results from previous studies to construct the relations between the BLR distance from the BH, BH mass, and AGN UV luminosity.

4. C IV Size–Mass–Luminosity Relations

C IV RM studies were carried out during the past three decades toward several low luminosity AGNs (Seyfert Galaxies), e.g., Peterson *et al.* (2004, 2005), De Rosa *et al.* (2015), Metzroth *et al.* (2006). Hoormann *et al.* (2019) report C IV RM time lags for two objects from the Dark Energy Survey Supernova Program (DES-SN) and the Australian Dark Energy Survey (OzDES) at redshifts of 1.9 and 2.6. In that study the photometric monitoring covers five years while the spectroscopic monitoring was 3–4 years. Out of the 393 objects with C IV in their sample, 23 were variable and had high cadence data but only two had significant time-lag measurements. These objects have 1350 Å luminosity of a few 10^{46} erg s^{-1}. Recently, Grier *et al.* (2019) report on C IV time lags in 52 AGNs from the Sloan Digital Sky Survey RM Project, with an estimated false-positive detection rate of 10%. 18 of these AGNs are defined as their "gold sample" with the highest-confidence lag measurements. These objects have redshift range of $1.4 < z < 2.8$ and luminosity range of $10^{44.5} < \lambda L_\lambda(1350\text{Å}) < 10^{45.6}$ erg s^{-1}. Adding all these results to the ones presented here, in the previous two sections, a size—luminosity relation can be determined between the C IV time lags and the 1350Å luminosity. That relation is presented in Figure 3 and the fitted relation has a slope of 0.46 ± 0.03, and an intercept of 0.17 ± 0.04. This result is consistent with that measured by Lira *et al.* (2018).

The central masses of AGNs can be estimated using $M_{BH} = fG^{-1}V^2r$, where V is an estimate of the velocity of the BLR around the central mass, and r is an estimate of the typical distance between the BLR and the central BH. f is a scaling factor which depends on the BLR geometry and velocity field. Although it was found that using the C IV line to estimate the BH mass in AGNs has some drawbacks and limitations (e.g., Baskin & Laor 2005; Trakhtenbrot & Netzer 2012), we do present here BH mass—UV luminosity relation for this whole sample, since for the high-luminosity AGN we only have RM measurements of C IV. This relation is shown in Figure 4 and the fitted relation has a slope of 0.49 ± 0.06, and an intercept of $0.88^{+0.57}_{-0.34}$.

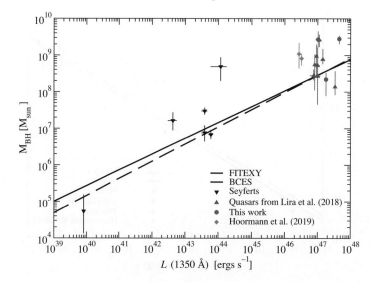

Figure 4. BH mass as a function of the UV luminosity for the sample of AGNs with measured C IV time lag.

The results presented here about RM of the C IV broad emission line, together with some recent similar studies, have populated the high-luminosity end of the size—mass—luminosity relations in AGNs (Kaspi *et al.* 2020, in preparation). We find that both the BH mass and the distance of the C IV emitting region from the BH, are scaling as the square root of the UV luminosity, to within the uncertainties. However, the low luminosity end at $\lambda L_\lambda(1350\text{Å}) < 10^{43}$ erg s^{-1} is poor with data points and further RM of the C IV line in low-luminosity AGNs are needed to better populate that region.

References

Bahcall, J. N., Kozlovsky, B.-Z., & Salpeter, E. E. 1972, *ApJ*, 171, 467
Baskin, A. & Laor, A. 2005, *MNRAS*, 356, 1029
Bentz, M. C. & Katz, S. 2015, *PASP*, 127, 67
Blandford, R. D. & McKee, C. F. 1982, *ApJ*, 255, 419
De Rosa, G., Peterson, B. M., Ely, J., *et al.* 2015, *ApJ*, 806, 128
Du, P., Lu, K.-X., Zhang, Z.-X., *et al.* 2016, *ApJ*, 825, 126
Goad, M. R., Korista, K. T., De Rosa, G., *et al.* 2016, *ApJ*, 824, 11
Grier, C. J., Trump, J. R., Shen, Y., *et al.* 2017, *ApJ*, 851, 21 (erratum 868, 76 [2018])
Grier, C. J., Shen, Y., Horne, K., *et al.* 2019, *ApJ*, 887, 38
Hoormann, J. K., Martini, P., Davis, T. M., *et al.* 2019, *MNRAS*, 487, 3650
Kaspi, S., Smith, P. S., Netzer, H., Maoz, D., Jannuzi, B. T., & Giveon, U. 2000, *ApJ*, 533, 631
Kaspi, S., Maoz, D., Netzer, H., Peterson, B. M., Vestergaard, M., & Jannuzi, B. T. 2005, *ApJ*, 629, 61
Kaspi, S., Brandt, W. N., Maoz, D., Netzer, H., Schneider, D. P., & Shemmer, O. 2007, *ApJ*, 659, 997
Lira, P., Kaspi, S., Netzer, H., *et al.* 2018, *ApJ*, 865, 56
Metzroth, K. G., Onken, C. A., & Peterson, B. M. 2006, *ApJ*, 647, 901
Netzer, H. & Peterson, B. M. 1997, in D. Maoz, A. Sternberg and E. Leibowitz (eds.), *Astronomical Time Series*, (Dordrecht: Kluwer Academic Publishers), p. 85
Peterson, B. M. 1993, *PASP*, 105, 247
Peterson, B. M., Ferrarese, L., Gilbert, K. M., *et al.* 2004, *ApJ*, 613, 682

Peterson, B. M., Bentz, M. C., Desroches, L.-B., *et al.* 2005, *ApJ*, 632, 799 (erratum 641, 638 [2006])
Peterson, B. M. 2006, in C. M. Gaskell, I. M. McHardy, B. M. Peterson and S. G. Sergeev (eds.) *AGN Variability from X-Rays to Radio Waves*, (Astronomical Society of the Pacific Conference Series) 360, 191
Saturni, F. G., Trevese, D., Vagnetti, F., Perna, M., & Dadina, M. 2016, *A&A*, 587, A43
Trakhtenbrot, B. & Netzer, H. 2012, *MNRAS*, 427, 3081
Trevese, D., Paris, D., Stirpe, G. M., Vagnetti, F., & Zitelli, V. 2007, *A&A*, 470, 491
Trevese, D., Perna, M., Vagnetti, F., Saturni, F. G., & Dadina, M. 2014, *ApJ*, 795, 164
Welsh, W., Robinson, E. L., Hill, G., *et al.* 2000, *BAAS*, 32, 1458

Revisiting old (AGN) friends - what's changed in their spectral looks

Hartmut Winkler

Department of Physics, University of Johannesburg, PO Box 524,
2006 Auckland Park, Johannesburg, South Africa
email: hwinkler@uj.ac.za

Abstract. Active Galactic Nuclei (AGN) have long been known to be variable, but the amplitude, timescale and nature of these changes can often differ dramatically from object to object. The richest source of information about the properties of AGN and the physical processes driving these remains the optical spectrum. While this spectrum has remained remarkably steady over decades for some AGN, other objects, referred to as Changing Look AGN, have experienced a comprehensive spectral transformation. Developments in the detection technology have enabled detailed probing in other wavebands, highlighting for example often quite different variability patterns for high energy emission. This paper explores the current characteristics of some long-known (and almost forgotten) Seyfert galaxies. It compares their present optical spectral properties, determined from recent observations at the South African Astronomical Observatory, with those from much earlier epochs. It furthermore considers the implication of the changes that have taken place, alternatively the endurance of specific spectral features, on our understanding of the mechanisms of the observed targets in particular, and on AGN models in general.

Keywords. galaxies: Seyfert, line: profiles, galaxies: individual (Fairall 9, IC 4329A, Mrk 926)

1. Introduction

Seyfert galaxies were identified as a class in the 1940's (Seyfert (1943)), and have elicited astrophysical interest ever since. While initially restricted to a few, mostly nearby objects, the list of galaxies belonging to the class grew substantially in the 1970's, due to systematic surveys such as the one of Markarian (Markarian *et al.* (1981)), and as a result of the first x-ray source identification programmes. The AGN explored in this paper, Fairall 9, IC 4329A and Mkn 926, were discovered to be Seyferts during such investigations, and their names have been amongst the most recognisable in the discipline ever since. All three have at one stage been suggested as candidates for the title of 'nearest quasar'.

These AGN have now been known to exhibit broad-line Seyfert galaxy spectra for over 40 years. Over that period they have been frequently (though not regularly) re-observed, and in addition considerable data sets have also been collated in other wave bands, such as in x-rays and in the infrared. They have become suitable probes to investigate the medium-term (∼20–50 year period) behaviour of luminous broad-line AGN.

Inspecting the nature of variations over the medium-term also enables one to determine the commonality of the relatively recently identified 'changing look' phenomenon. This refers to a relatively rapid, dramatic change in the optical spectrum. Typically, initially strong broad lines of a changing look AGN disappear almost entirely over a few months, meaning that what was once a Seyfert 1 or 1.5 is transformed into a Seyfert 1.8 or 1.9 over a time period considered too short for major nuclear obscuration changes. Changing look AGN therefore challenge the unified AGN model that considers Seyfert classes to

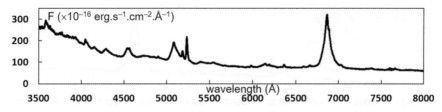

Figure 1. Optical spectrum of Fairall 9 on 9 July 2019.

merely be products of different inclination angles of a dusty torus constituting the outer parts of the nucleus (see, e.g., the discussion in Oknyansky et al. (2019)).

2. Observations

This paper discusses optical spectroscopic observations carried out on the night of 9 Jul 2019 with the SpUpNIC spectrograph (Crause et al. (2019)) mounted on the 1.9 m telescope at the South African Astronomical Observatory in Sutherland. Two 1200 s integrations were made for each object, bracketed by 10 s Ar wavelength comparison spectra immediately before and after each AGN spectrum. The grating employed was one giving a low resolution and resulting in a useful spectral wavelength range 3250–8500 Å, and the slit width was set to 2.7 arcsec on the sky. The flux calibration was achieved by means of a spectrum of the standard star LTT 3864. The processing of the frames included standard bias subtractions, flatfield corrections and the cleaning of pixels affected by cosmic rays.

3. Fairall 9

Fairall 9 was identified as a quasar-like emission-line object during a spectral survey of galaxies with compact nuclei (Fairall (1977)). Ward et al. (1978) independently identified the object as the optical counterpart of the x-ray source 2A 0120–591, and published the first detailed spectrum displaying strong broad lines. Confirming the spectrum and presenting photometric measurements indicating $V \sim 13.2$, $B - V \sim 0.2$ and $U - B \sim -1.0$, West et al. (1978) classified the object as a quasar.

From about 1981, Fairall 9 experienced a remarkable drop in luminosity. By 1984, the Balmer lines had weakened to about 20% of their 1981 values relative to the forbidden lines (Kollatschny & Fricke (1985); Wamsteker et al. (1985)). The luminosity decline, and the subsequent slow partial rebrightening, was also closely monitored in the ultraviolet and infrared (Clavel et al. (1989)). Only minimal spectral changes were detected in 1987–1988 (Winkler (1992)). A 1993 spectrum displays a rather weak broad component, with a red shoulder now quite clearly distinguishable (Marziani et al. (2003)). In 1994, a large reverberation mapping campaign determined a broad-line radius of 23 days (Santos-Lleò et al. (1997)).

Thereafter, spectral data was secured a lot less frequently as researchers moved to study other AGN. While a lot of data has still been collected for Fairall 9 post-1994, very little of it covers the optical regime. The optical spectral behaviour over this lengthy period can therefore only be inferred from what was measured in other wave bands. There is no evidence of brightening comparable to the levels witnessed in the late 1970's, neither does anything suggest that the AGN 'switched itself off' at any stage.

The July 2019 spectrum of Fairall 9 is illustrated in Fig. 1. It looks remarkably similar to the spectrum from the late 1980's shown in Winkler (1992), when the luminosity of Fairall 9 had strengthened to intermediate from the 1985 minimum. There is no more obvious sign of the 'red shoulder', and the FeII emission bands also resemble the same lines in the corresponding earlier spectra.

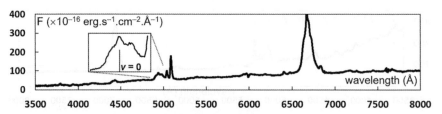

Figure 2. Optical spectrum of IC 4329A on 9 July 2019, including an enlarged display of the Hβ profile.

4. IC 4329A

The AGN nature of IC 4329A, an edge-on galaxy with a striking dust lane (see, e.g. Malkan et al. (1998)), was established in 1972 (Disney (1973)). The discovery paper highlights the strong broad-line spectrum, with Balmer line zero intensity widths of up to 13000 km.s^{-1}. It furthermore records an Hα-to-Hβ ratio of over 8, a clear sign of significant nuclear reddening and extinction. The high Balmer-line flux ratios were confirmed in many later studies (e.g., Wilson & Penston (1979)), although it is now also clear that the intrinsic ratio may be somewhat greater than was initially assumed, meaning that the intrinsic nuclear luminosity is not high enough to warrant classification as a quasar.

The discovery spectrum also revealed a secondary peak (red shoulder) of the broad Hβ, corresponding to a rest wavelength of 4900 Å, and a recession velocity of roughly 2300 km.s^{-1} relative to the primary Hβ peak. Significantly, this secondary peak also appears in varying degrees in many later spectra (e.g., Winkler (1992); Winge et al. (1996)).

Figure 2 displays the spectrum most recently obtained for IC 4329A. It does not differ markedly from the other spectra published over the past almost 50 years. Notably, the redshifted second peak of the Hβ emission line is still clearly distinguishable, implying that its presence is associated with a stable nuclear configuration that the AGN returns to frequently. It is pointed out that there are spectra where the red peak appears absent (e.g. Morris & Ward (1988)).

5. Mkn 926

Mkn 926, also frequently referred to as MCG–2-58-22, was identified as the optical counterpart of the bright x-ray source 2A 2302–088 by Ward et al. (1978). Morris & Ward (1988) show a spectrum recorded in 1984, when the object was in a luminous state. The broad-line component weakened substantially from about 1987 (Winkler (1992)), and a further significant broad-line luminosity decline ensued in around 1993 (Kollatschny et al. (2006)). Mkn 926 was also observed spectroscopically during the Sloan Digital Sky Survey (Ahn et al. (2014)), and throughout the early years of the present century the Balmer broad lines remained weak (consistent with a Seyfert 1.8 classification), though with a complex profile that included a distinct red saddle at $\sim$10000 km.s^{-1}.

The evolution of Mkn 926 since its discovery is illustrated in Fig. 3 in the form of an approximate V-magnitude light curve. Three distinct phases can be distinguished: i) a bright phase lasting to late 1986, when the object was one of the more luminous AGN in the not-too-distant ($z < 0.05$) universe, ii) and intermediate phase lasting to about 1993 where Mkn 926 was much less luminous than before, but still exhibited typical broad-line Seyfert characteristics and medium-term luminosity fluctuations, and iii) a faint phase from about 1994, where the spectrum is dominated by narrow lines (including comparatively strong lines associated with low-ionisation), though a weak broad Balmer line spectrum always remains present.

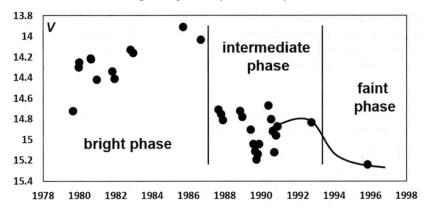

Figure 3. V light curve of Mkn 926 from 1978–1996. Magnitudes from Doroshenko & Terebizh (1981); Mallama (1983); de Ruiter & Lub (1986) (converted to V-magnitudes using the conversion equations given in that paper), Hamuy & Maza (1987), Winkler et al. (1992), Winkler (1997), and converted to a 10 arcsec aperture diameter. The curved line estimates the magnitudes for 1990–1995 from spectral continuum decline determined by Kollatschny et al. (2006).

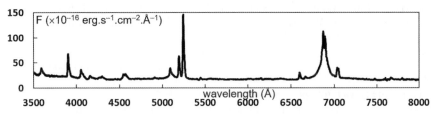

Figure 4. Optical spectrum of Mkn 926 on 9 July 2019.

The latest spectrum of Mkn 926 is shown in Fig. 4. The broad component is now even weaker than it was around the time of the SDSS observations, and the 'red saddle' evident in some earlier spectra is no longer visible. The broad component of Hβ is only detectable because of the high S/N, and a spectrum of lower quality might have led to Mkn 926 being described as a Seyfert 1.9.

While the Balmer and helium emission in the current spectrum is dramatically lower than 40 years ago, indeed so different that Mkn 926 can be classified as a changing look AGN, the object's peculiar narrow line spectrum has not undergone any obvious change. In particular, the comparatively strong [O II] and [S II] lines are more reminiscent of a LINER, and this may in some way relate to this AGN's peculiar spectral evolution.

6. Discussion and Conclusion

Three well-known, comparatively nearby Seyfert 1 galaxies, Fairall 9, IC 4329A and Mkn 926, all of which were studied extensively spectroscopically in the last quarter of the 20th century, have been reobserved. The spectra of Fairall 9 and IC 4329A were found to closely match spectra from 30 years ago. Fairall 9 has been significantly brighter (and fainter) in some earlier epochs, which suggests that comparatively stable states do exist that the AGN returns to after its bright phases. While the nuclear luminosity of IC 4329A appears comparatively stable, some line profile changes have been evident. In particular, an emission peak 2300 km.s^{-1} redward of the Hβ peak, suggestive of Keplerian motion in the nuclear regions and seen in spectra from ~ 25 years ago, remains visible.

In contrast to the other two AGN investigated here, Mkn 926 has shown no signs of re-entering the brighter states it was in from the mid-1977s to the early 1990s. If there are cycles of activity, the timeframe of this cycle is at least 50 years long. Its semi-quiescent state has now endured for about 25 years. In all cases, the medium-term spectral evolution is determined by physical conditions inside the nucleus, and as such constrains models to explain the physical mechanism driving the AGN.

References

Ahn, C. P., Alexandroff, R., Allende Prieto, C., *et al.* 2014, *ApJS*, 211, A17
Clavel, J., Wamsteker, W., Glass, I. S. 1989, *ApJ*, 337, 236
Crause, L. A., Gilbank, G., van Gend, C., *et al.* 2019, *J. Astr. Tel. Instr. Sys.*, 5(2), 024007
de Ruiter, H. R. & Lub, J. 1987, *A&AS*, 63, 59
Disney, M. J. 1973, *ApJ* (Letters), 181, L55
Doroshenko, V. T. & Terebizh, V. Y. 1981, *Astrophysics*, 17, 358
Fairall, A. P. 1977, *MNRAS*, 180, 391
Glass, I. S. 2004, *MNRAS*, 350, 1049
Hamuy, M. & Maza, H. 1987, *A&AS*, 68, 383
Kollatschny, W. & Fricke, K. J. 1985, *A&A*, 146, L11
Kollatschny, W., Zetzl, M., Dietrich, M., *et al.* 2006, *A&A*, 454, 459
Mallama, A. D. 1983, *J. Ass. Var. Star Obs.*, 12, 69
Malkan, M. A., Gorjian, V., Tam, R., *et al.* 1998, *ApJS*, 117, 25
Markarian, B. E., Lipotevskii, V. A., Stepanian, D. A., *et al.* 1981, *Af*, 17, 619
Marziani, P., Sulentic, J. W., Zamanov, R., Calvani, M., Dultzin-Hacyan, D., Bachev, R., Zwitter, T., *et al.* 1997, *ApJS*, 145, 199
Morris, S. L. & Ward, M. J. 1979, *MNRAS*, 230, 639
Oknyansky, V. L., Winkler, H., Tsygankov, S. S., *et al.* 2019, *MNRAS*, 483, 558
Santos-Lleó, M., Chatzichristou, E., Mendes de Oliveira, C., *et al.* 1997, *ApJS*, 112, 271
Seyfert, C. K. 1943, *ApJ*, 97, 28
Wamsteker, W., Alloin, D., Pelat, D., Gilmozzi, R., *et al.* 1985, *ApJ* (Letters), 295, L33
Ward, M. J., Wilson, A. S., Penston, M. V., *et al.* 1997, *ApJ*, 223, 788
West, R. M., Danks, A. C., Alcaino, G., *et al.* 1978, *A&A*, 62, L13
Wilson, A. S. & Penston, M. V. 1979, *ApJ*, 232, 389
Winge, C., Peterson, B. M., Paltoriza, M. G., Storchi-Bergmann, T., *et al.* 1996, *ApJ*, 469, 648
Winkler, H. 1992, *MNRAS*, 257, 677
Winkler, H. 1997, *MNRAS*, 292, 273
Winkler, H., Glass, I. S., van Wyk, F., *et al.* 1992, *MNRAS*, 257, 659

Discovery of new changing look in NGC 1566

Victor L. Oknyansky[1], Sergey S. Tsygankov[2,3], Vladimir M. Lipunov[1], Evgeny S. Gorbovskoy[1] and Nataly V. Tyurina[1]

[1]M. V. Lomonosov Moscow State University, Sternberg Astronomical Institute,
119234, Moscow, Universitetsky pr-t, 13, Russian Federation
e-mail: oknyan@mail.ru

[2]Department of Physics and Astronomy, FI-20014 University of Turku, Finland

[3]Space Research Institute of the Russian Academy of Sciences,
Profsoyuznaya Str. 84/32, Moscow 117997, Russia

Abstract. We present continuation of the multi-wavelength (from X-ray to optical) monitoring of the nearby changing look (CL) active galactic nucleus in the galaxy NGC 1566 performed with the *Neil Gehrels Swift Observatory*,the MASTER Global Robotic Network over the period 2007–2019. We also present continuation of optical spectroscopy using the South African Astronomical Observatory 1.9-m telescope between Aug. 2018 and Mar. 2019. We investigate remarkable re-brightenings in of the light curve following the decline from the bright phase observed at Dec. 2018 and at the end of May 2019. For the last optical spectra (31 Nov. 2018–28 Mar. 2019) we see dramatic changes compared to 2 Aug. 2018, accompanied by the fading of broad emission lines and high-ionization [FeX]6374Å line. Effectively, one more CL was observed for this object: changing from Sy1.2 to the low state as Sy 1.8–Sy1.9 type. Some possible explanations of the observed CL are discussed.

Keywords. AGN, Spectra, photometry, X-ray, variability, NGC 1566

1. Introduction

The NGC 1566 was discovered at 1826 by James Dunlop and has long history of investigations (see e.g. references at da Silva *et al.* (2017); Oknyansky *et al.* (2019); Parker *et al.* (2019)). The broad nuclear emission lines characteristic of the Seyfert phenomenon in this object were discovered in 1956 by de Vaucouleurs & de Vaucouleurs (1961); de Vaucouleurs (1973) and confirmed in 1962 by Shobbrook (1966). The NGC 1566 nucleus was later classified as type 1. It is one of the brightest ($V \approx 10^m.0$ but $V_{AGN} \approx 13^m.0$ in 5 arcsec radius aperture) and nearest galaxies with AGN in the South Hemisphere. This object is also nearest Changing Look (CL) AGN (Oknyansky *et al.* (2019)). The nucleus of NGC 1566 has a low brightness relative to the host galaxy ($V \approx 13^m.0$) and CL events there were probably discovered just because it is so close. The Changing Look active galactic nuclei (CL AGNs) are objects which undergo dramatic variability of the emission line profiles and classification type, which can change from type 1 (showing both broad and narrow lines) to type 1.9 (where the broad lines almost disappear) or vice versa within a short time interval (typically a few months). The dramatical spectral variations in NGC 1566 had been reported firstly by Pastoriza & Gerola (1970) after comparison of the spectrum obtained in 1969 with some of the earliest spectroscopic investigations, in 1956 (de Vaucouleurs & de Vaucouleurs (1961)), in 1962 (Shobbrook (1966)) where broad Hβ line was much more intensive. That was done soon after discovery of variability of AGNs in continuum (Fitch *et al.* (1967)) and variability of emission lines was not too unexpected. The object had several dramatic changes of its spectrum during the past tens of years (da Silva et al. (2017)) but was not identified as a CL AGN until the 2018 event Oknyansky et al. (2018, 2019)) since this designation only came into common use in the past decade.

NGC 1566 is a galaxy with a very well-studied variable active nucleus, but most intensive multiwave photometric observations were done during past year after discovery of a new reawakening (Kuin et al. (2018); Ferrigno et al. (2018); Grupe et al. (2018a); Parker et al. (2019); Cutri (2018)) and new CL phase (see Oknyansky et al. (2018, 2019) and references there). A substantial increase of X-ray flux by 1.5 orders of magnitude was observed following the brightening in the UV and optical bands during the last year. After a maximum was reached at the beginning of July 2018 the fluxes in all bands decreased with some fluctuations. The amplitude of the flux variability is strongest in the X-ray band and decreases with increasing wavelength.

2. Observational data and results

We summarize a study of optical, UV and X-ray light curves of the nearby changing look active galactic nucleus in the galaxy NGC 1566 obtained with the Neil Gehrels Swift Observatory and the MASTER Global Robotic Network over the period 2007–2019. The light curves for 2007-2018 are presented by (Oknyansky et al. (2019)) and those for the years 2018–2019 are shown in Fig. 1. It can be seen there that all variations in the optical, UV and X-ray are well correlated. We also report on optical spectroscopy using the South African Astronomical Observatory 1.9-m telescope between Aug 2018 (Oknyansky et al. (2018, 2019)) and Mar 2019. A substantial increase in X-ray flux by 1.5 orders of magnitude was observed following the brightening in the UV and optical bands during the first half of 2018 (Ducci et al. (2018); Kuin et al. (2018); Ferrigno et al. (2018); Grupe et al. (2018a); Oknyansky et al. (2018, 2019); Parker et al. (2019)). The amplitude of the flux variability is strongest in the X-ray band and decreases with increasing wavelength. After a maximum was reached at the beginning of July 2018, the fluxes in all bands decreased with some fluctuations. The most remarkable re-brightening in the light curve following the decline from the bright phase was observed at MJD range 58440-58494 (Event 1) and 58603-58654 (Event 2). Event 1 and Event 2 are indicated in Fig. 1 (see also Grupe et al. (2018b) and Grupe et al. (2019)). The amplitudes of the re-brightening in UV and optical bands are significantly higher for Event 1 than for Event 2. That is different from the X-ray variations for which fluxes in the maxima were about the same. This difference is well seen at Fig. 1. If we take into account the host galaxy contamination in the aperture used then the relative decreases from the maximum in July 2018 to minimum in June 2019 in the different UV/Opt bands were about the same (∼9 times).

Low-resolution spectra (2 Aug 2018) (see Fig. 2) reveal a dramatic strengthening of the broad emission as well as high-ionization [FeX]6374Å lines. These lines were not detected so strongly in the past published spectra. The change in the type of the optical spectrum (Oknyansky et al. (2018, 2019)) was accompanied by a significant change in the X-ray spectrum. For the last spectra (30 Nov 2018–28 Mar 2019) we see dramatic changes compared to Aug 2018, accompanied by the fading of the broad emission lines. Effectively, two changing look were observed by us for this object during the past year.

3. Conclusion

NGC 1566 is one of the typical examples of a CL AGN, since it demonstrates dramatical variability of broad emission lines, UV continuum, high ionisation lines like [FeVII] and [FeX], and also recurrent brightening and dimming events. NGC 1566 is one of the

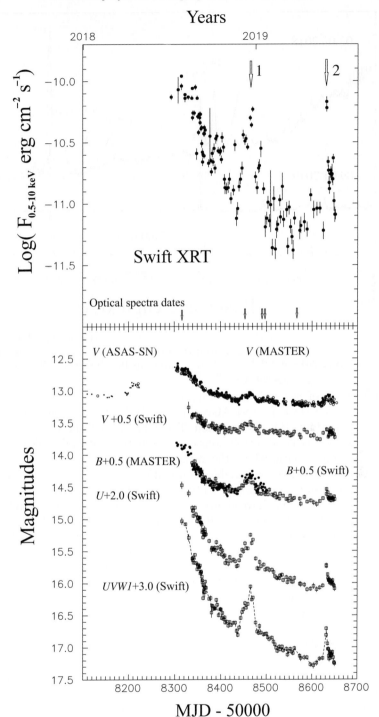

Figure 1. Multi-wavelength observations of NGC 1566 shown for 2018-2019 only. *Top Panel:* The *Swift*/XRT 0.5–10 keV X-ray flux (in erg cm^{-2} s^{-1}) – (filled circles). The big arrows show locations of the Event 1 and the Event 2 (see in the text). *Bottom Panel:* The large open circles are MASTER unfiltered optical photometry of NGC 1566 reduced to the V system while the points are V ASAS-SN (nightly means) reduced to the *Swift* V system. The filed circles are MASTER BV photometry. The open boxes are $UVW1$ and UBV data obtained by *Swift*.

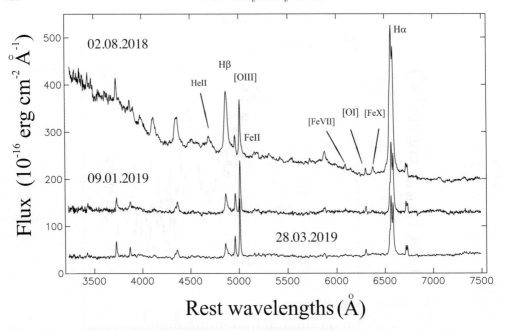

Figure 2. The isolated nuclear (low resolution) nonstellar spectra in NGC 1566 obtained by subtraction of the host galaxy spectrum from the original spectrum. (See details in the text). The spectra for 2 Aug 2018 and 9 Jan 2019 are shifted up $10^{-14} erg\ cm^{-2} A^{-1}$ for good seeing

clearest illustrations of Seyfert spectra ranging from type 1.2 to type 1.9 AGNs at different epochs. The object is nearest AGN and CL AGN and so it offers one of the best opportunities for studying this phenomenon. The light curves from X-ray to optical bands presented here show very good correlations over a long time interval. This result is mostly in agreement with a scenario where the variability across several wavebands (spanning X-rays–UV/Optical) is driven by variable illumination of the accretion disc by soft X-rays (see the same conclusions for another CL object, NGC2617 Shappee et al. (2014) and Oknyansky et al. (2017)) . We have shown, using spectroscopy (1.9 m SAAO) and multi-wavelength photometry (MASTER, *Swift* Ultraviolet/Optical and XRT Telescopes), that NGC 1566 recently experienced a dramatic outburst in all wavelengths, including a considerable strengthening of broad permitted and high ionisation [FeX]6374Å lines, as well as substantial changes in the shape of the optical and X-ray continua Oknyansky et al. (2019). After a maximum was reached at the beginning of July 2018, the fluxes in all bands dramatically decreased with some fluctuations. The strength of the broad permitted, high ionisation [FeX]6374Å lines and UV continuum dramatically decreased (end of March 2019) and the object can then be classified as Sy1.8–Sy1.9. So we witnessed a second CL in NGC1566.

Despite the successes of the simple orientation-based AGN unification scheme, there are significant problems that cannot be explained solely by different orientations. A major challenge to the simple model is the existence of CL AGNs. Orientation obviously cannot change on the time-scale of the observed type changes, and hence some other explanation is needed. What must happen to make such a dramatic changes possible? An alternative explanation is that transitions from type 1 to type Sy2 AGNs or vice versa are connected with some dramatic variability of the ionizing radiation, such as temporarily switching on or switching off their engine Lyutyj et al. (1984); Penston & Perez (1984); Runnoe et al. (2016); Katebi et al. (2019). More references on the possible explanations for the CL phenomenon can be found in discussions by MacLeod et al. (2019).

Acknowledgements

We thank the South African Astronomical Observatory for the generous allocation of telescope time. We thank to H. Winkler and F. van Wyk for the spectral observations which also resulted in the spectra presented in this paper. We also express our thanks to the *Swift* ToO team for organizing and executing the observations. This work was supported in part by the Russian Foundation for Basic Research through grant 17-52-80139 BRICS-a and by the BRICS Multilateral Joint Science and Technology Research Collaboration grant 110480. MASTER work was supported by Lomonosov Moscow State University Development Programme and RSF grant 16-12-00085. DB is supported by the National Research Foundation of South Africa. We are grateful to K. Malanchev and P. Ivanov for useful discussions. VO thanks to the IAU for the travel grant.

References

Cutri, R. M., Mainzer, A. K., Dyk, S. D. V., & Jiang, N. 2018, *The Astronomer's Telegram*, 11913
da Silva, P., Steiner, J. E., & Menezes, R. B. 2017, *MNRAS*, 470, 3850
de Vaucouleurs, G. & de Vaucouleurs, A. 1961, *MemRAS*, 68, 69
de Vaucouleurs, G. 1973, *ApJ*, 181 31
Ducci, L., Siegert, T., Diehl, R., et al. 2018, *The Astronomer's Telegram*, 11754
Ferrigno, C., Siegert, T., Sanchez-Fernandez, C., et al. 2018, *Astronomer's Telegram*, 11783
Fitch, W. S., Pacholczyk, A. G., & Weymann, R. J. 1967, *ApJ*, 150, L67.
Grupe, D., Komossa, S., & Schartel, N. 2018a, *The Astronomer's Telegram*, 11903
Grupe, D., Mikula, R., Komossa, S., et al. 2018b, *The Astronomer's Telegram*, 12314
Grupe, D., Mikula, R., Komossa, S., et al. 2019, *The Astronomer's Telegram*, 12826
Katebi, R., Chornock, R., Berger, E., et al. 2019 *MNRAS*, 487, 4057
Kuin, P., Bozzo, E., Ferrigno, C., et al. 2018, *Astronomer's Telegram*, 11786
Lyutyj, V. M., Oknyanskij, V. L., & Chuvaev, K. K. 1984, *Soviet Astronomy Letters*, 10, 335
MacLeod, C. L., Green, P. J., Anderson, S. F., et al. 2019, *ApJ*, 874, 8
Oknyansky, V. L., Gaskell, C. M., Huseynov, N. A., et al. 2017, *MNRAS*, 467,1496
Oknyansky, V. L., Lipunov, V. M., Gorbovskoy, E. S., et al. 2018, *The Astronomer's Telegram*, 11915
Oknyansky, V. L., Winkler, H., Tsygankov, S. S., et al. 2019, *MNRAS*, 483, 558
Parker, M. L., Schartel, N., Grupe, D., et al. 2019, *MNRAS*, 483, L88
Pastoriza, M. & Gerola, H. 1970, *Ap.Lett*, 6, 155.
Penston, M. V. & Perez, E. 1984, *MNRAS*, 211, 33P
Runnoe, J. C., Cales, S., Ruan, J. J., et al. 2016, *MNRAS*, 455, 1691
Shappee, B. J., Prieto, J. L., Grupe, D., et al. 2014, *ApJ* 788, 48
Shobbrook, R. R. 1966, *MNRAS*, 131,365

The study of variability of 8 blazar candidates among the *Fermi*-LAT unidentified gamma-ray sources

Pheneas Nkundabakura[1], Jean D'amour Kamanzi[2], Jean D. Mbarubucyeye[1,3] and Tom Mutabazi[2]

[1]University of Rwanda, College of Education, P.O. Box 5039, Kigali, Rwanda
[2]Mbarara University of Science and Technology, Department of Physics, Mbarara, Uganda
[3]Deutsches Elektronen-Synchrotron (DESY), Platanenallee 6, 15738 Zeuthen, Germany

Abstract. We discuss the time-series behavior of 8 extragalactic 3FGL sources away from the Galactic plane (i.e., $|b| \geqslant 10°$) whose uncertainty ellipse contains a single X-ray and one radio source. The analysis was done using the standard Fermi *ScienceTools*, package of version v10r0p5. The results show that sources in the study sample display a slight indication of flux variability in γ-ray on monthly timescale. Furthermore, based on the object location on the variability index versus spectral index diagram, the positions of 4 objects in the sample were found to fall in the region of the already known BL Lac positions.

Keywords. galaxies: active - galaxies: jets - gamma rays: galaxies - BL Lacertae objects: general - radiation mechanism: non-thermal

1. Introduction

The study of variability is particularly important in γ-ray astronomy primarily due to different advantages such as assisting in the identification of the correct radio/optical/X-ray source within the γ-ray position box, with the observations at other wavelengths (De Cicco *et al.* 2015; Ferrara *et al.* 2015). For unidentified sources, variability characteristics can also support the recognition of the correct source class (Nolan *et al.* 2003).

Fortunately, the Large Area Telescope (LAT) aboard the *Fermi Gamma-ray Space Telescope* has revolutionised the field of γ-ray astronomy by detecting a wealth of new γ-ray sources and allowing the study of previously known sources with unprecedented details (Zechlin & Horns 2015). Previous studies show that most of the sources detected by the *Fermi*-LAT are blazars (Ackermann *et al.* 2015). The 3FGL (Acero *et al.* 2015) and the 4FGL (The Fermi-LAT Collaboration 2019) catalogs reported a significantly large fraction of sources compared to the previous ones.

However, the majority of 3FGL and 4FGL sources remain unassociated with low-energy counterparts, hence understanding their nature is an open question in high-energy astrophysics. In addition, it seems plausible that most of the unassociated high-latitude γ-ray sources are expected to be faint AGN, which may include blazar sub-class (Mirabal *et al.* 2012; Massaro *et al.* 2012; Ackermann *et al.* 2012). These unidentified γ-ray sources represent a discovery area for the new source classes or new members of existing source classes which may include different types of AGN.

For instance, previous studies show a combined effort to isolate potential blazar candidates among this large population (e.g. Massaro *et al.* 2012; Zechlin & Horns 2015; Paiano *et al.* 2017). Some studies used the analysis of the multiwavelength Spectral Energy Distribution (SED) through detecting a double peaked spectrum. This indicated that the radiation among the selected sample originates mainly from synchrotron and the inverse-Compton emission in the so-called synchrotron-Compton blazars (Mbarubucyeye J. D., Krauß F., & Nkundabakura P. 2019, in prep...), though the SED alone is not enough to fully characterise the blazar nature based on their broad band properties. Since blazars display intrinsic variability and more significantly in the γ-ray energy band (Ulrich *et al.* 1997), it is needed to use this property to characterise individual synchrotron-Compton blazar candidates that may be present in the γ-ray unidentified and unassociated population. In this paper, we discuss the time-series behavior of 8 extragalactic 3FGL sources away from the Galactic plane (i.e., $|b| \geqslant 10°$) which were carefully selected among the Unidentified *Fermi*-LAT sources with the purpose to detect any sign of variability which can be linked to the blazar nature of these sources.

2. Sample selection

The following selection criteria were used to obtain a study sample:
(*a*) Being unidentified sources at high Galactic latitudes, $|b| \geqslant 10°$,
(*b*) Being unidentified sources which have a single X-ray and one radio source in its uncertainty region,
(*c*) Being unidentified sources that were reported in the 4FGL catalog.

Applying all cuts to the population of unidentified sources listed in 3FGL, a sample of 8 unidentified sources thought to be potential blazar candidates was isolated.

3. Data analysis

The astrophysical data analysis of LAT begins with a list of counts detected. This list results from processing made by the LAT instrument team, which reconstructs events for the signals from different parts of LAT. Two principal types of analysis were applied in this study, they were performed in a systematic way such that the results from the first analysis became the input of the next one. The types of analysis performed are:
- Global analysis which was performed using *Fermi ScienceTools* v10r0p5. This provided the fluxes and spectral parameters of all objects in our study sample. The photon counts within a region of interest of 25 degree radius were taken into account. We selected events within the energy range 100 MeV–300 GeV, a maximum zenith angle of 90 degrees and event type 3.
- Time-series analysis (light-curve analysis & variability analysis). This provided the γ-ray light-curves for the period of 9 years and the variability indices of target sources, together with the significance of the observed variability in light-curves.

The observed variability was obtained using the following equation as in Nolan *et al.* 2003:

$$TS_{var} = 2 \sum_i \frac{\Delta F_i^2}{\Delta F_i^2 + f^2 F_{const}^2} \ln\left(\frac{\mathcal{L}_i(F_i)}{\mathcal{L}_i(F_{const})}\right), \quad (3.1)$$

where $f = 0.02$, i.e., a 2%, is a systematic correction factor, F_i and ΔF_i are the flux and error in flux in the ith bin, respectively. $\mathcal{L}_i(F_{const})$ is the value of the likelihood in the ith bin under the null hypothesis where the source flux is constant across the full period and F_{const} is the constant flux for this hypothesis. $\mathcal{L}_i(F_i)$ is the value of the likelihood in the ith bin under the alternate hypothesis where the flux in the ith bin is optimised.

4. Results and discussion

4.1. Average results

Before performing the light-curve analysis, a fit of the entire 108 months LAT data using a power-law model was performed through binned likelihood analysis for each

Table 1. Gamma-Ray average fluxes and spectral characteristics of the sample sources.

No	3FGL Name (1)	F_γ (2)	Γ_γ (3)	σ (4)
1	J0049.0+4224	0.93 ± 0.43	1.81 ± 0.14	7.02
2	J1119.8−2647	2.99 ± 0.81	1.94 ± 0.11	10.03
3	J1132.0−4736	3.72 ± 1.21	2.00 ± 0.09	10.11
4	J1220.0−2502	7.62 ± 1.26	2.16 ± 0.20	7.46
5	J1220.1−3715	3.44 ± 0.81	1.96 ± 0.09	10.24
6	J1619.1+7538	0.85 ± 0.23	1.78 ± 0.10	10.01
7	J1923.2−7452	8.61 ± 0.68	2.04 ± 0.10	14.05
8	J2015.3−1431	4.63 ± 1.52	2.23 ± 0.19	5.03

Note: Column 1, 2, 3 and 4 show the source 3FGL name, the average γ-ray flux for 108 months LAT data in scale of 10^{-9} ph cm^{-2} s^{-1}, the γ-ray spectral index corresponding to column (2), the significance in sigma units corresponding to column (2), respectively.

target source. This provided the sample average results fluxes and spectral properties in a period of 9 years. Sources in the sample were found to be faint γ-ray emitters with γ-ray spectral index, $\Gamma \sim 2$ (see Table 1), which is consistent with the previous studies (e.g., Ackermann et al. (2015)).

4.2. Monthly γ-ray light-curves

To determine the trends of flux change and variability of sources for a period of 9 years, light-curve analysis was performed. This was done through extracting the monthly fluxes along this period, and plotting light-curves.

Generally, we found that sources in our sample do not show significant signal in their light-curves, which is an indication that they are relatively faint in γ-ray. This was also suggested by Acero et al. (2015). Therefore, the light-curves indicate that signals are not significantly detectable in many monthly bins (represented as an upper limit). This implies that their fluxes are close to zero, hence summing them over the full period (9 years) tends to lower the source average flux as shown in Figure 1. The sample light-curves display behaviours commonly shown by blazars such as: non periodic flux change characterised by undefined and no specific trends, associated with unpredictable and sudden flux rise seen across the whole period of 9 years (see Figure 1). However, the large error bars on the data points does not allow to firmly establish such sharp flux rises.

4.3. Variability indices

The light-curves presented in this study show many upper limits that correspond to the time when the signal in monthly bins was not significant enough to characterise a source. It is also clear that the error bars corresponding to the significant flux points are relatively large. Therefore, we used the '*variability index*' defined in Equation 3.1 to quantify the observed variability in light-curves, in which the information of the upper limits is properly considered. To compare the already known classification in 3FGL with our sources that are lacking classifications, variability-spectral index diagram for all 3FGL sources including the study sample was plotted. Ackermann et al. (2015) observed that blazars are located in different zones on the variability-spectral index diagram, according to their subtypes (FSRQs and BL Lacs), though there is also a large recovery zone (see Figure 2).

The significance of the observed variability from sample light-curves was estimated by using a χ^2 distribution. This provided the variability index (TS_{var}) threshold at which

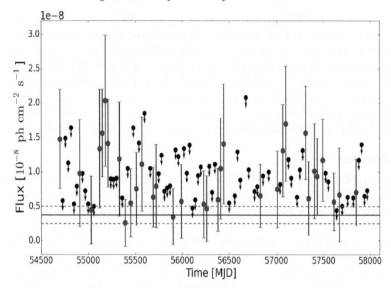

Figure 1. The 100 MeV–300 GeV monthly light-curves for 3FGL J1132.0−4736. The horizontal solid line along with two dashed lines present the 9-years average flux and its 1σ error range derived in the global analysis, respectively. The blue points represent flux with its 1σ error bar, while the downward arrows together with black points represent the 95% upper limits.

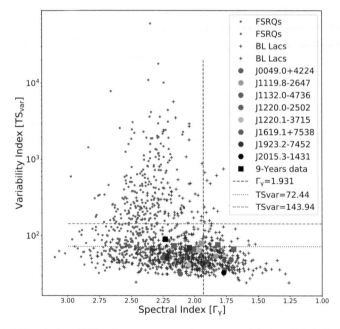

Figure 2. Variability index (TS_{var}) versus power law spectral index (PL index) diagram of all 3FGL extragalactic sources. Red points: Flat Spectrum Radio Quasars, blue crosses: BL Lacs, and sources in the study sample. Circles in different colours indicate data of sources in the selected sample, while solid squares in same colours present results of the same sources obtained in this study. The black horizontal dotted line indicates the 3FGL variability index threshold (72.44), while the green horizontal dashed line shows the 9-years variability index threshold (143.94) obtained in this study. The black vertical dashed line shows the spectral index (1.931) below which the region is populated by BL Lacs.

we assigned the source a 99% probability of being variable (on a timescale of $\gtrsim 1$ month). For 9-years data, we found that variability is considered significant with 99% confidence level if the variability index is greater than 143.94. However, sources in our study sample have TS_{var} values much lower than the threshold (i.e., $TS_{var} < 143.94$). This implies that we can not conclude at 99% confidence level that our target sources are variable due to lack of statistics. The variability significance of all sources in the study sample was found to be in the range of 0.5% to 12%. The variability significance of 3FGL J2015.3-1431 was found to be the highest compared to other sources in the study sample.

5. Conclusions

Although probing the γ-ray variability of blazar candidate sources is of definite interest in the study of AGN properties towards a better classification of the sources, definite classification is expected to be properly achieved by multiwavelength studies of their spectral energy distribution together with their optical spectra. Indeed, variability is well understood when it is studied across the electromagnetic spectrum (Radio, Optical/UV and X-rays), and on different timescales. This contributes to checking the variability correlation in different energy bands and testing whether variability exists for all timescales. Therefore, future studies are expected to consider multi-waveband variability and on different timescales. The Variability can be applied to estimate physical parameters of AGN such as the size of the emitting region, timescale of variability, magnetic field in the jets, mass of the central engine (blackhole), etc. However, the estimation of all these parameters requires primarily to know the object's redshift, which can be obtained through spectroscopic studies. Therefore, future studies through the analysis of optical spectra of sources listed in our studied sample should be considered. Such observations from ground-based optical telescopes (such as a 10-meter class telescope) would be the ideal program to determine the nature of blazar candidates.

Acknowledgements

We acknowledge the useful contribution of Richard J.G. Britto, University of the Free State – South Africa. Financial support from the Swedish International Development Cooperation Agency (SIDA) through the International Science Programme (ISP) is also gratefully acknowledged.

References

Acero, F., Ackermann, M., Ajello, M., et al. 2015, ApJS, 218, 23
Ackermann, M., Ajello, M., Allafort, A., et al. 2012 ApJ, 753, 83
Ackermann, M., Ajello, M., Atwood, W. B., et al. 2015, ApJ, 810, 14
De Cicco, D., Paolillo, M., Covone et al. 1995, A&A, 574, A112
Ferrara, E. C., Mirabal, N. R., & Fermi-LAT Collaboration 2015, in American Astronomical Society Meeting Abstracts # 225. p. 336.02
Mbarubucyeye, J. D., Krauß, F., & Nkundabakura, P. 2019, in prep..., MNRAS
Massaro, F., D'Abrusco, R., Tosti, G., et al. 2012 ApJ, 752, 61
Mirabal, N., Frías-Martinez, V., Hassan, T., et al. 2012 MNRAS, 424, L64
Nolan, P. L., Tompkins, W. F., Grenier et al. 2003, ApJ, 597,615
Paiano, S., Falomo, R., Franceschini, A., et al. 2017 ApJ, 851, 135
The Fermi-LAT collaboration. 2019, arXiv e-prints, arXiv:1902.10045
Zechlin, H.-S. & Horns, D. 2015, J. Cosmology Astropart. Phys., 2, E01
Ulrich, M. H., Maraschi, L., & Urry, C. W. 1997. ARA&A, 35, 445–502
Urry, C. M. & Padovani, P. 1995, PASP, 107, 803

High-resolution radio astronomy: An outlook for Africa

Leonid I. Gurvits[1,2,3], **Robert Beswick**[4], **Melvin Hoare**[5], **Ann Njeri**[6], **Jay Blanchard**[7,1], **Carla Sharpe**[8], **Adrian Tiplady**[8] **and Aletha de Witt**[8]

[1]Joint Institute for VLBI ERIC, Oude Hoogeveensedijk 4,
7991 PD Dwingeloo, The Netherlands
email: lgurvits@jive.eu

[2]Dept. of Astrodynaics & Space Missions, Delft University of Technology,
Kluyverweg 1, 2629 HS Delft, The Netherlands

[3]CSIRO Astronomy and Space Science, PO Box 76, Epping, NSW 1710, Australia

[4]Jodrell Bank Centre for Astrophysics, The Alan Turing Building, Department of Physics and Astronomy, Oxford Road, The University of Manchester, M13 9PL, UK

[5]School of Physics & Astronomy, University of Leeds, Leeds, LS2 9JT, UK

[6]Department of Physics and Astronomy, Oxford Road,
The University of Manchester, M13 9PL, UK

[7]National Radio Astronomy Observatory, PO Box O, 1003 Lopezville Rd.,
Socorro, NM 87801, USA

[8]South African Radio Astronomy Observatory, 2 Fir Street, Black River Park,
Cape Town, 7925, South Africa

Abstract. Very Long Baseline Interferometry (VLBI) offers unrivalled resolution in studies of celestial radio sources. The subjects of interest of the current IAU Symposium, the Active Galactic Nuclei (AGN) of all types, constitute the major observing sample of modern VLBI networks. At present, the largest in the world in terms of the number of telescopes and geographical coverage is the European VLBI Network (EVN), which operates under the "open sky" policy via peer-reviewed observing proposals. Recent EVN observations cover a broad range of science themes from high-sensitivity monitoring of structural changes in inner AGN areas to observations of tidal eruptions in AGN cores and investigation of redshift-dependent properties of parsec-scale radio structures of AGN. All the topics above should be considered as potentially rewarding scientific activities of the prospective African VLBI Network (AVN), a natural "scientific ally" of EVN. This contribution briefly describe the status and near-term strategy for the AVN development as a southern extension of the EVN-AVN alliance and as an eventual bridge to the Square Kilometre Array (SKA) with its mid-frequency core in South Africa.

Keywords. Radio astronomy, VLBI

1. Introduction

High angular resolution studies of Active Galactic Nuclei (AGN) constitute a sizeable fraction of the science topics of the current IAU Symposium No. 356. In radio domain, these studies are conducted by using Very Long Baseline Interferometry (VLBI) systems spread over the entire surface of Earth and extended to Space. The 'imaging angular resolution of VLBI reaches record high values of tens of microarcseconds as demonstrated recently in the ground-based observations of the Event Horizon Telescope (EHT) at 1.5 mm (EHT 2019) and in the Space VLBI observations by the RadioAstron mission at

1.3 cm (Bruni et al. 2019, and references therein). The geometry of a VLBI system is one of its key characteristics: in addition to the size (which does matter in interferometry), a distribution of interferometric elements defines the quality of images obtained by such a system.

For almost the entire history of VLBI studies which began in 1967 (Moran 1998), the African continent was present in the VLBI world only by the Hartebeesthoek Observatory in South Africa (Gaylard & Nicolson 2007). This single VLBI station on the entire continent provided an important baseline extension for the European and global VLBI Networks (EVN)†, the Australian Long Baseline Array (LBA)‡ and other networks requiring a VLBI station in the Southern hemisphere. However, the large "telescope-free" area between Hartebeesthoek and Eurasia was always seen as a potentially attractive region for placing VLBI in the interests of various science tasks.

A new momentum for VLBI in Africa came on the wave of the Square Kilometre Array (SKA) developments. The SKA, with its mid-frequency core in South Africa, has VLBI as one of its science-driven operational modes (Paragi et al. 2015). A natural synergy between developing the SKA as such and its partner observatories throughout the continent was realised at the first African SKA Partnership meeting at the Hartebeesthoek Observatory in 2003. Over the following years, three major components of the pan-African cooperation were formulated: development of the African VLBI Network (AVN), human capital development in areas related to radio astronomy, and the SKA governance. These components got a strong political support in the Pretoria Resolutions of 2014 and Memorandum of Understanding and Joint Development Plan of 2017.

2. The African VLBI Network

A typical 30-m class VLBI radio telescope has a great deal of common technical features with a generic space communication antenna. Not surprisingly, many modern VLBI telescopes were designed or even built originally as Earth-Space communication facilities. The already mentioned first VLBI station in Africa, the 26 m radio telescope at the Hartebeesthoek Observatory in South Africa has began its life in 1961 as a NASA tracking station (Gaylard & Nicolson 2007). More recently, a former Soviet military 32 m antenna in Irbene, Latvia, has become a radio telescope involved, among other science applications, in VLBI observations as a member of the EVN (Upnere et al. 2013). A 32 m satellite communication antenna at Warkworth in New Zealand has also become a radio telescope which, in particular, conducts LBA and global VLBI observations (Woodburn et al. 2015). The latter example is of particular interest for the African VLBI Network as the most relevant example.

In the first decade of the 21st century, the global communication infrastructure has shifted its main data transport to fibre-optical cables. This shift released from active duty many satellite communication antennas. A census of such facilities defined about a dozen of decommissioned or nearly-decommissioned 30-m-class antennas distributed over the African continent. Fig. 1 presents the expected geographical distribution of the prospective AVN telescopes. The deployment of AVN is seen as a two-stage process, with the first wave of telescopes, led by the radio astronomy cluster in South Africa and 32-m antennas in Ghana, Kenya, Madagascar, Mauritius, Namibia and Zambia. These telescopes will be followed by antennas in Egypt, Ethiopia, Morocco, Nigeria and Senegal.

The geometry of an interferometer is one of its main characteristics. Ideally, the geometry should be optimised using complex criteria describing the quality of reconstructed

† https://www.evlbi.org/home, accessed 2019.11.17
‡ https://www.atnf.csiro.au/vlbi/overview/index.html, accessed 2019.11.17

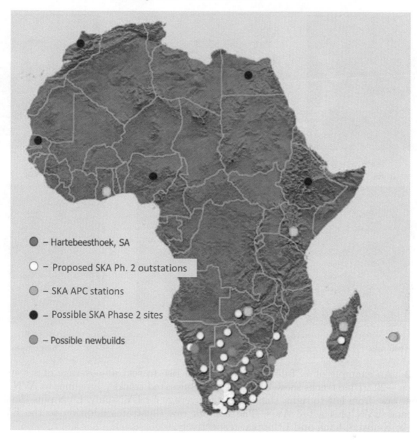

Figure 1. Geography of prospective African VLBI Network radio telescopes. The red spot indicates the existing Hartebeesthoek VLBI telescopes, yellow – the SKA core, white – SKA outstations. Green dots indicate prospective AVN stations in SKA Africa Partner Countries (APC), and black and blue dots – AVN stations under further considerations, possibly as SKA Phase 2 sites.

images under various limitations and boundary conditions. Such the optimisation was exercised, e.g., for the dedicated VLBI network – the Very Long Baseline Array (VLBA, Napier et al. 1994) and the SKA (Lal et al. 2010, and references therein). Due to the ad hoc approach to the formation of the AVN, such the optimisation is obviously impossible – antennas' locations should be taken as a given. Nevertheless, an analysis of the AVN configuration is a necessary component of the overall project as it allows us to evaluate the observational capability of the network. For an interferometric array, the most straightforward representation of its observational potential is given by the 2D distribution of sampling of the spectra of spatial frequencies, the so-called uv-coverage. Fig. 2 illustrates the value of new AVN stations in combination with the existing EVN stations. Even single additional antennas (illustrated in the bottom raw) provide a sensible "filling-in" improvement in the uv-gap at the north-south intermediate baselines. The configuration consisting from full EVN and AVN arrays offers a nearly perfect uv-coverage.

3. Kuntunse radio telescope, Ghana

The Ghana Radio Astronomy Observatory in Kuntunse (also spelled sometimes as Kutunse or Nkuntunse), located about 25 km north-west of the capital city of Accra,

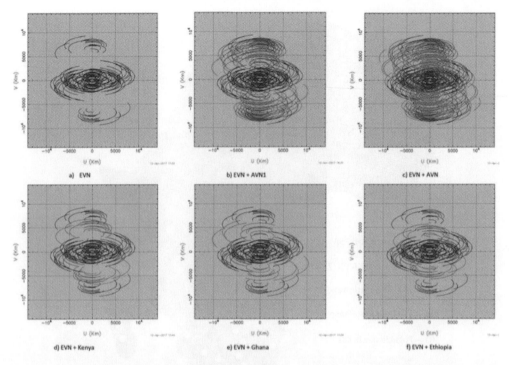

Figure 2. An example of a "full-track" (i.e., from rise to set) uv-coverage of a source at the declination $+20°$. Blue tracks show EVN-only baselines, red tracks – baselines to AVN antennas. The upper raw, from left to right, shows the uv-coverage for EVN-only, EVN plus the first wave of AVN, and EVN plus a full AVN. The bottom raw illustrates addition to the EVN single antennas in Kenya, Ghana and Ethiopia, respectively.

is the first AVN observatory undergoing extensive refurbishment of its instrumentation (Asabere et al. 2015). Its 32-m antenna (Fig. 3, left panel) operated as a telecommunication facility between 1981 and 2008. In 2011, the antenna's ownership was transferred to the newly established Ghana Space Science and Technology Institute (GSSTI). In collaboration with the SKA–SA, now known as the South African Radio Astronomy Observatory (SARAO) and EVN institutes, the GSSTI staff prepared the facility for the first test VLBI observations in 2017. The test observation was conducted with the C-band (3.8 – 6.4 GHz) communication receiver. Fig. 3 (right panel) presents the clear VLBI fringes obtained at JIVE on baselines between Kuntunse and several telescopes in Europe and South Africa.

4. Educational potential of AVN

The radio astronomy infrastructure in Africa and in particular AVN-related activities serve as an efficient setting for educational and public outreach activities on the continent. A number of joint schools (e.g., a school in Kuntunse, May 2018, Fig. 4) workshops and exchange visits conducted in the past several years assist in building up the professional radio astronomy community around the prospective AVN observatories. Hopefully, at the successive symposium in Addis Ababa of 2024, young researchers from African institutes will presets studies conducted with the African VLBI Network.

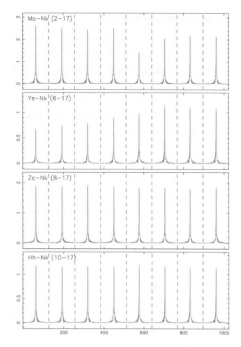

Figure 3. *Left*: The Kuntunse 32 m satellite communication antenna, Ghana. *Right*: The first VLBI fringes to the Kuntunse telescope on baselines to the telescopes (top–down) Medicina (Italy), Yebes (Spain), Zelenchukskaya (Russia) and Hartebeesthoek (South Africa).

Figure 4. Participants of the radio astronomy school at Kuntunse, organised jointly by the JUMPIG JIVE and DARA projects, May 2018.

Acknowledgements

Collaborative activities with the prospective African VLBI Network are supported by the European Commission Horizon 2020 Research and Innovation Programme under grant agreement No. 730884 (JUMPING JIVE).The DARA project is funded by the UK's Newton Fund via grant ST/R001103/1 from the Science and Technology Facilities Council and by South Africa's National Research Foundation. LIG expresses gratitude to the Leids Kerkhoven-Bosscha Fonds for partial support of participation in the IAU Symposium No. 356.

References

Asabere, B. D., Gaylard, M. J., Horellou, C., et al. 2015, arXiv:1503.08850
Bruni, G., Savolainen, T., Gómez, J. L., et al. 2020, Adv. Sp. Res., 65(2), 712
Event Horizon Telescope Collaboration 2019, ApJL, 875, L1
Gaylard, M. J. & Nicolson, G. D. 2007, African Sky, 11, 49
Lal, D. V., Lobanov, A. P., & Jiménez-Monferrer, S. 2010, arXiv:1001.1477
Moran, J. M. 1998, in J. A. Zensus, G. B. Taylor & J. M. Wrobel (eds.) *Radio Emission from Galactic and Extragalactic Compact Sources*, IAU Coll. 164, ASP Conf. Series, 144, 1
Napier, P. J., Bagri, D. S., Clark, B. G., et al. 1994, Proc. IEEE, 82, No. 5, 658
Paragi., Z., Godfrey, L., Reynolds, C., et al. 2015, in *Advancing Astrophysics with the Square Kilometre Array*, PoS(AASKA14)143
Upnere, S., Jekabsons, N., & Joffe, R. 2013, J. of Theoretical and Applied Mechanics, 43, 39
Woodburn, L., Natusch, T., Weston, et al. 2015, PASA, 32, e017

X-ray properties of reverberation-mapped AGNs with super-Eddington accreting massive black holes

Jaya Maithil[1], Michael S. Brotherton[1], Bin Luo[2], Ohad Shemmer[3], Sarah C. Gallagher[4], Du Pu[5], Hu Chen[5], Jian-Min Wang[5] and Yan-Rong Li[5]

[1]Department of Physics & Astronomy, University of Wyoming, Laramie, WY 82071, USA

[2]School of Astronomy and Space Science, Nanjing University, Nanjing, Jiangsu 210093, China

[3]Department of Physics, University of North Texas, Denton, TX 76203, USA

[4]Department of Physics & Astronomy, University of Western Ontario, London, ON N6C 1T7, Canada

[5]Key Laboratory for Particle Astrophysics, Institute of High Energy Physics, Chinese Academy of Sciences, Beijing 100049, China

Abstract. Active Galactic Nuclei (AGN) exhibit multi-wavelength properties that are representative of the underlying physical processes taking place in the vicinity of the accreting supermassive black hole. The black hole mass and the accretion rate are fundamental for understanding the growth of black holes, their evolution, and the impact on the host galaxies. Recent results on reverberation-mapped AGNs show that the highest accretion rate objects have systematic shorter time-lags. These super-Eddington accreting massive black holes (SEAMBHs) show BLR size 3-8 times smaller than predicted by the Radius-Luminosity (R-L) relationship. Hence, the single-epoch virial black hole mass estimates of highly accreting AGNs have an overestimation of a factor of 3-8 times. SEAMBHs likely have a slim accretion disk rather than a thin disk that is diagnostic in X-ray. I will present the extreme X-ray properties of a sample of dozen of SEAMBHs. They indeed have a steep hard X-ray photon index, Γ, and demonstrate a steeper power-law slope, α_{ox}.

Keywords. galaxies: active, galaxies: X-rays

The BLR physics from the long-term optical monitoring of type-1 AGN

Dragana Ilić

Department of Astronomy, Faculty of Mathematics, University of Belgrade,
Studentski trg 16, 11000 Belgrade, Serbia

Abstract. The variation of optical continuum and broad emission lines is observed in all type 1 active galactic nuclei (AGN). In some cases even extreme variability is detected when broad-line profiles completely disappear as is the case in the co-called changing-look AGN, which raise new question on the theoretical model of AGN. This variability is an important tool to study the physics and geometry of the broad line region (BLR), e.g. it can be used to estimate its size through the reverberation mapping technique. Especially, long-term campaigns give new insights, like the detection of the periodic signals or discoveries of changing-look AGN. Here we will present the results of our long-term monitoring campaign of several well-known AGN, as e.g. NGC 3516 for which we confirm that it is the changing-look AGN, putting special attention of the applications for future large time-domain spectroscopic surveys, like the MaunaKea Spectroscopic Explorer project.

Keywords. galaxies: active, galaxies: variability

CHAPTER IV. Properties of AGN host galaxies

Properties of X-ray detected far-IR AGN in the green valley

Antoine Mahoro[1,2], Mirjana Pović[3,4], Petri Väisänen[1,5], Pheneas Nkundabakura[6], Beatrice Nyiransengiyumva[6] and Kurt van der Heyden[2]

[1]South African Astronomical Observatory, P.O. Box 9 Observatory, Cape Town, South Africa

[2]Department of Astronomy, University of Cape Town, Private Bag X3, Rondebosch 7701, South Africa

[3]Ethiopian Space Science and Technology Institute (ESSTI), Entoto Observatory and Research Center (EORC),
Astronomy and Astrophysics Research Division, P.O. Box 33679, Addis Ababa, Ethiopia

[4]Instituto de Astrofísica de Andalucía (IAA-CSIC), Glorieta de la Astronomía s/n, 18008 Granada, Spain

[5]Southern African Large Telescope, P.O. Box 9 Observatory, Cape Town, South Africa

[6]MSPE Department, School of Education, College of Education, University of Rwanda, P.O. Box 5039, Kigali, Rwanda

Abstract. In this study, we analysed active galactic nuclei in the "green valley" by comparing active and non-active galaxies using data from the COSMOS field. We found that most of our X-ray detected active galactic nuclei with far-infrared emission have star formation rates higher than the ones of normal galaxies of the same stellar mass range.

Keywords. galaxies: active, galaxies: evolution, galaxies: high redshift, galaxies: star formation, galaxies: structure, infrared: galaxies

---

1. Introduction

In optical colour-magnitude diagrams (CMDs) galaxies predominantly lie along either the "red sequence," characterised with red colours and elliptical or spheroidal morphologies, or in the "blue cloud," characterised by blue colours and disk or irregular morphologies. The differences between these two groups create a bimodality in the color-magnitude distribution of galaxies. The two populations have a nontrivial region of overlap between them termed the "green valley" (Wyder et al. 2007; Salim et al. 2009), through which past attempts have been made to place a quantitative divider that splits the two populations on the basis of colour.

Previous works that studied green valley galaxies suggested that they are a transitional phase between the blue cloud and red sequence in terms of star formation, colours, stellar mass, luminosity, and different morphological parameters (Pović et al. 2012; Schawinski et al. 2014; Salim 2014; Lee et al. 2015; Phillipps et al. 2019).

Interestingly, the rate of active galactic nuclei (AGN) detection is high in green valley galaxies, whether AGN are selected by deep X-ray surveys (Nandra et al. 2007; Coil et al. 2009; Pović et al. 2012) or by optical line-ratio diagnostics (Salim et al. 2007).

Previous works proposed that AGN negative feedback plays a key role in the galaxies' process of quenching star formation (SF) and galaxy transformation. In this work, we went one step further by looking at the transition processes from the blue cloud to the red sequence on one side, and at AGN triggering mechanisms and their connection with normal galaxies on the other, by using AGN and normal galaxies in the green valley.

2. Data

In this work, we used the data from COSMOS survey, aimed at studying the formation and evolution of galaxies as a function of both, cosmic time and the local galaxy environment (Scoville et al. 2007).

Our main catalogue was taken from the Tasca et al. (2009). In order to select AGN we used the ratio between the X-ray flux (F_x) in the hard 210 keV band and the optical i band flux (F_o) (Bundy et al. 2007; Trump et al. 2009). Non-AGN galaxies were selected by removing sources selected as AGN, which is explained in Mahoro et al. (2017).

There have been many ways of defining the green valley galaxies, all being based on the bimodal distribution of galaxies when using different colours. In this work, we used the U-B rest-frame colour and criteria $0.8 \leqslant U - B \leqslant 1.2$ (e.g. Nandra et al. 2007). To obtain U-B rest-frame colours, we first run KCORRECT code (Blanton & Roweis 2007) to apply the k-correction on both CFHT u and Subaru B bands. The far-infrared data was obtained by cross-matching the previously obtained sample with Herschel/PACS data as explained in Mahoro et al. (2017).

3. Star formation rates (SFR)

We determine the SFRs of the FIR selected green valley AGN and non-AGN galaxies using the spectral energy distribution (SED)-fitting method the LEPHARE code (Ilbert et al. 2006) and by assuming that all the FIR luminosity is due to star formation for non-activate galaxies. To measure SFRs we use Kennicutt et al. (1998) relation, as explained in Mahoro et al. (2017). To fit AGN we use Kirkpatrick et al. (2015) templates with known AGN contribution to the IR luminosity. Non-AGN galaxies were fitted using Chary & Elbaz (2001) libraries.

In Fig. 5 we present the relation between SFR and stellar mass (e.g. Netzer et al. 2016; Pović et al. 2016), and study the location of our AGN and non-AGN sources in relation to the main sequence of star-forming galaxies taken from Elbaz et al. (2011).

Our sample of FIR green valley AGN do not show signs of SF quenching (as suggested in the majority of previous studies of X-ray detected AGN) since 68% and 14% of of sources are located on and above the MS of SF, respectively. On the other hand we have 70% of non-AGN FIR green valley galaxies on the MS, with 9% being above it.

4. Morphological classification and analysis

We also carried out the morphological analysis of selected FIR AGN and FIR non-AGN green valley galaxies to understand better their location with respect to the main sequence of star formation obtained in Mahoro et al. (2017). By using the HST/ACS F814W images, we did visual morphological classification of all FIR AGN and non-AGN galaxies samples. Classification was done by three independent classifiers, separating all galaxies into:
- class 1: elliptical, S0 or S0/S0a,
- class 2: spiral,
- class 3: irregular,
- class 4: peculiar and
- class 5: unclassified.

Fig. 2 shows the normalised distribution for final visual classification. We can clearly see a difference in class 4, with 38% for FIR AGN and 19% of non-AGN galaxies being peculiar, with clear signs of interactions and mergers. On the other hand, we also obtained

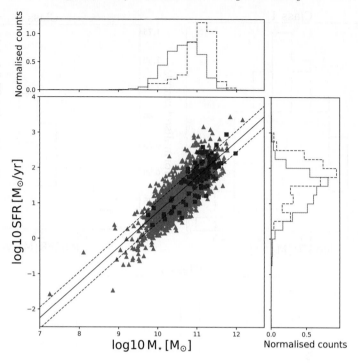

Figure 1. SFR vs. stellar mass. AGN are represented by blue squares, while non-AGN are represented by red triangles. The solid black line shows the Elbaz et al. (2011) fit for the MS, while the dashed lines represent the MS width of ±0.3 dex. Top and right histograms show the normalised distributions of stellar mass and SFR respectively, and comparison between the AGN (blue dash lines) and non-AGN (red solid lines) samples.

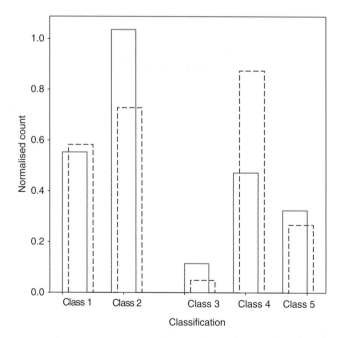

Figure 2. Normalised distributions of visually classified morphological types of FIR green valley AGN (blue dashed lines) and non-AGN (red solid lines).

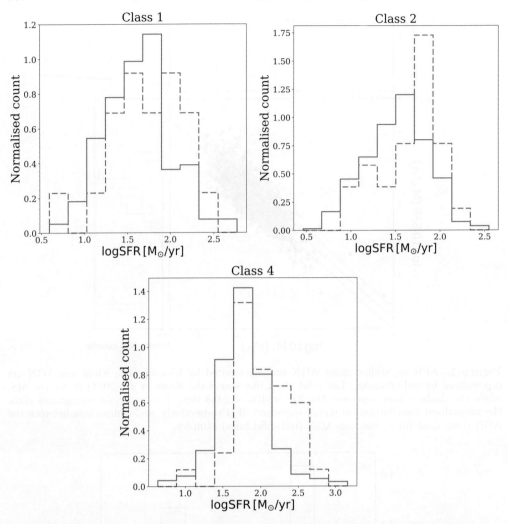

Figure 3. Normalised distributions of SFR of FIR AGN (blue dashed histograms) and non-AGN (red solid histograms) for a fixed stellar mass range of $\log M^* = 10.6 M_\odot - 11.6 M_\odot$ in relation to morphology.

an important difference in class 2, finding higher fractions of FIR non-AGN (46%) in comparison to AGN (26%). We compared our visual classifications with available non-parametric classifications in COSMOS, and we did more analysis of the distributions of different morphological parameters by comparing FIR AGN and FIR non-AGN green valley galaxies. We found the standard behaviour of morphological parameters, as explained in Mahoro et al. (2019).

5. Discussion

We found that FIR AGN have higher SFRs in comparison to FIR non-AGN green valley galaxies. Fig. 3 shows the comparison of SFRs of AGN and non-AGN samples for the same stellar mass ranges taking into account morphology. In all cases the AGN sample shows higher SFRs in comparison to non-AGN galaxies, independently of morphology (Mahoro et al. 2019). In case of class 4, interactions and mergers could explain higher values of SFRs observed in AGN. However, taking into account higher SFRs in the rest of AGN green valley sample (e.g., in 26% and 25% of class 1 and class 2 galaxies, respectively), we suggested that interactions and mergers alone cannot explain the results of Mahoro et al. (2019). Therefore, if there is an impact of AGN feedback on star formation in case of selected FIR emitters it looks to be rather positive than negative.

Acknowledgments

This work was supported by the National Research Foundation of South Africa (Grant Numbers 110816). AM acknowledges financial support from the Swedish International Development Cooperation Agency (SIDA) through the International Science Programme (ISP) - Uppsala University to University of Rwanda through the Rwanda Astrophysics, Space and Climate Science Research Group (RASCSRG), East African Astro-physics Research Network (EAARN) are gratefully acknowledged. MP acknowledges financial supports from the Ethiopian Space Science and Technology Institute (ESSTI) under the Ethiopian Ministry of Innovation and Technology (MInT), and from the Spanish Ministry of Science, Innovation and Universities (MICIU) through projects AYA2016-76682C3-1-P and the State Agency for Research of the Spanish MCIU through the Center of Excellence Severo Ochoa award to the IAA-CSIC (SEV-2017-0709). PV acknowledges support from the National Research Foundation of South Africa.

References

Blanton, M. R. & Roweis, S. 2007, *AJ*, 133, 734
Bundy, K., Treu, T., & Ellis, R. S. 2007, *ApJL*, 665, L5
Chary, R. & Elbaz, D. 2001, *ApJ*, 556, 562
Coil, A. L., Georgakakis, A., Newman, J. A., et al. 2009, *ApJ*, 701, 1484
Elbaz, D., Dickinson, M., Hwang, H. S., et al. 2011, *A&A*, 533, A119
Ilbert, O., Arnouts, S., McCracken, H. J., et al. 2006, *A&A*, 457, 841
Kennicutt, Robert, C. J., Stetson, P. B., Saha, A., et al. 1998, *ApJ*, 498, 181
Kirkpatrick, A., Pope, A., Sajina, A., et al. 2015, *ApJ*, 814, 9
Lee, G.-H., Hwang, H. S., Lee, M. G., et al. 2015, *ApJ*, 800, 80
Mahoro, A., Povi¢, M., & Nkundabakura, P. 2017, *MNRAS*, 471, 3226
Mahoro, A., Povi¢, M., Nkundabakura, P., Nyiransengiyumva, B., & Väisänen, P. 2019, *MNRAS*, 485, 452
Nandra, K., Georgakakis, A., Willmer, C. N. A., et al. 2007, *ApJL*, 660, L11
Netzer, H., Lani, C., Nordon, R., et al. 2016, *ApJ*, 819, 123
Phillipps, S., Bremer, M. N., Hopkins, A. M., et al. 2019, *MNRAS*, 485, 5559
Povi¢, M., Márquez, I., Netzer, H., et al. 2016, *MNRAS*, 462, 2878
Povi¢, M., Sánchez-Portal, M., Pérez García, A. M., et al. 2012, *A&A*, 541, A118
Salim, S. 2014, *Serbian Astronomical Journal*, 189, 1
Salim, S., Dickinson, M., Michael Rich, R., et al. 2009, *ApJ*, 700, 161
Salim, S., Rich, R. M., Charlot, S., et al. 2007, *ApJS*, 173, 267
Schawinski, K., Urry, C. M., Simmons, B. D., et al. 2014, *MNRAS*, 440, 889
Scoville, N., Aussel, H., Brusa, M., et al. 2007, *ApJS*, 172, 1
Tasca, L. A. M., Kneib, J. P., Iovino, A., et al. 2009, *A&A*, 503, 379
Trump, J. R., Impey, C. D., Elvis, M., et al. 2009, *ApJ*, 696, 1195
Wyder, T. K., Martin, D. C., Schiminovich, D., et al. 2007, *ApJS*, 173, 293

Properties of green valley galaxies in relation to their selection criteria

Beatrice Nyiransengiyumva[1,2], Mirjana Pović[3,4], Pheneas Nkundabakura[2] and Antoine Mahoro[5,6]

[1]Mbarara University of Science and Technology (MUST), P.O. Box 1410, Uganda

[2]University of Rwanda, College of Education, P.O. Box 5039, Kigali, Rwanda

[3]Ethiopian Space Science and Technology Institute (ESSTI), Entoto Observatory and Research Center (EORC), Astronomy and Astrophysics Research and Development Division, P.O. Box 33679, Addis Ababa, Ethiopia

[4]Instituto de Astrofísica de Andalucía (IAA-CSIC), Glorieta de la Astronomía s/n, 18008 Granada, Spain

[5]South African Astronomical Observatory, P.O. BOX: 9 Observatory, Cape Town, South Africa

[6]Department of Astronomy, University of Cape Town, Private Bag X3, Rondebosch 7701, South Africa

Abstract. The distribution of galaxies has been studied to show the difference between the blue cloud and red sequence and to define the green valley region. However, there are still many open questions regarding the importance of the green valley for understanding the morphological transformation and evolution of galaxies, how galaxies change from late-type to early-type and the role of AGN in galaxy formation and evolution scenario. The work focused on studying in more details the properties of green valley galaxies by testing the six most used selection criteria, differences between them, and how they may affect the main results and conclusions. The main findings are that, by selecting the green valley galaxies using different criteria, we are selecting different types of galaxies in terms of their stellar masses, sSFR, SFR, spectroscopic classification and morphological properties, where the difference was more significant for colour criteria than for sSFR and SFR vs. M_* criteria.

Keywords. galaxies: active, galaxies: star formation, galaxies: fundamental parameters, galaxies: evolution, ultraviolet: galaxies, methods: statistical

1. Introduction

The bi-modality in the distribution of galaxies usually obtained in colour-mass, colour-magnitude (CMD) or colour-star formation rate diagrams has been studied to show the difference between the blue cloud and red sequence galaxies and to define the intermediate green valley region. In general, blue cloud galaxies are activity star-forming sources that are rich in gas, while red sequence galaxies are mainly abundant quiescent sources and a small fraction of dusty star-forming galaxies and edge-on systems (Blanton & Moustakas 2009). Between the red sequence and the blue cloud there is a sparsely populated green valley region that has been viewed as the crossroads of galaxy evolution where the galaxies in it are thought to represent the transition population between the blue cloud of star-forming galaxies and the red sequence of quenched and passively evolving galaxies (Schiminovich et al. 2007; Pović et al. 2012; Salim 2014; Lin et al. 2017; Coenda et al. 2018; Bryukhareva & Moiseev 2019; Phillipps et al. 2019).

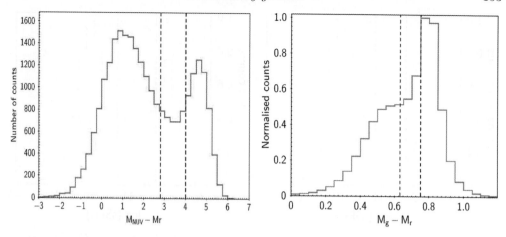

Figure 1. Distributions of $M_{NUV} - M_r$ colour for UV sample (left panel) and $M_{NUV} - M_r$ colour for optical sample (right panel). Green valley region is represented in between the two black dashed lines.

Several selection criteria have been used to define green valley in order to study the properties of the green galaxy population, using colours such as U – V (Brammer et al. 2009), U – B (Mendez et al. 2011; Mahoro et al. 2017, 2019), NUV – r (Wyder et al. 2007; Lee et al. 2015; Coenda et al. 2018), g – r (Trayford et al. 2015, 2017; Walker et al. 2013, Eales et al. 2018), u – r (Bremer et al. 2018; Eales et al. 2018; Kelvin et al. 2018; Ge et al. 2019; Phillipps et al. 2019), specific SFR (sSFR) (Schiminovich et al. 2007; Salim et al. 2009; Salim 2014; Phillipps et al. 2019; Starkenburg et al. 2019), and the SFR-stellar mass (M_*) diagram (Noeske et al. 2007; Chang et al. 2015). Therefore, the study of green valley galaxies provides us crucial clues to connect the red sequence and blue cloud galaxies in terms of their star formation quenching and evolution.

2. Data and sample selection

We used optical and ultraviolet (UV) data in this study. The optical data are from the Sloan Digital Sky Survey (SDSS) Data Release 7 (DR7) with photometric system of five filters named u, g, r, i and z with limiting magnitudes of 22.0, 22.2, 22.2, 21.3 and 20.3 respectively (York et al. 2003). For SFR, M_* and sSFR we used MPA-JHU catalogue (Kauffmann et al. 2003; Brinchmann et al. 2004; Tremonti et al. 2004). The ultraviolet data are from the GALEX-MIS (GR5) survey with limiting magnitudes of 22.6 and 22.7 in far-UV (FUV) and near-UV (NUV) respectively (Martin et al. 2005).

Six criteria commonly used in previous studies were tested to select the green valley samples.

(*a*) The NUV absolute magnitude minus the r absolute magnitude ($M_{NUV} - M_r$) and the g absolute magnitude minus r absolute magnitude ($M_g - M_r$). The range of green valley galaxies is defined in between $2.8 < M_{NUV} - M_r < 4$ and $0.63 < M_g - M_r < 0.75$, as shown in Figure 1 which is in line with previous studies (Belfioree et al. 2017; Taylor et al. 2017; Eales et al. 2018).

(*b*) sSFR in UV and in optical. The green valley sample is defined in the range of $-11.6 <$ sSFR < -10.8 for both criteria as shown in Figure 2 which is in line with Salim et al. 2009 and Salim 2014.

(*c*) SFR vs. M_* in UV and in optical. The green valley sample is defined as shown in Figure 3 (in line with Noeske et al. 2007; Chang et al. 2015).

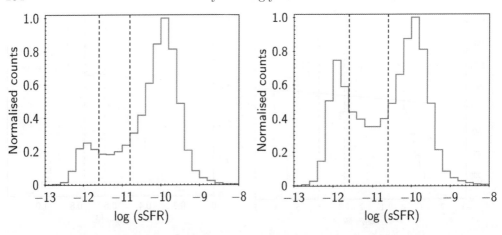

Figure 2. Distributions of sSFR for UV (left panel) and optical (right panel) samples. Green valley region is represented in between the two black dashed lines.

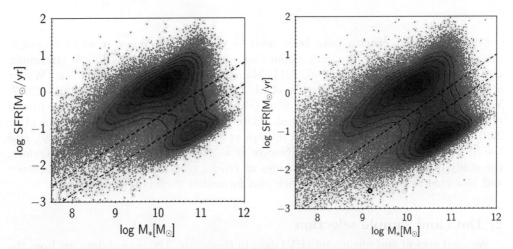

Figure 3. SFR versus stellar mass for UV sample (left panel) and optical sample (right panel). Green valley region is represented in between the two black dashed lines.

3. Analysis and results

We analysed the distributions of stellar mass, SFR and sSFR in different green valley samples. Figure 4 represents distributions of these parameters, respectively, in optical (top panels) and UV (bottom panels). In each panel, three samples are compared using: SFR-SM criteria (red solid line), colour criteria (blue dashed line), and sSFR criteria (green dotted line). It can be seen that when using different criteria we are selecting different galaxies in terms of their stellar mass, where more massive galaxies are selected when using UV data. When observing the total samples, in general SFR and sSFR are higher in optically selected green valley galaxies than in UV, however this difference is more significant when using colour and/or SFR vs. stellar mass criteria.

Using BPT-NII diagram (Baldwin et al. 1981), we classified green valley samples into star forming, composites, Seyfert 2 and LINERs galaxies. Most of the galaxies in the green valley are classified as star forming and Composites where the main difference is when using colour criteria. We found higher fraction of star forming and composite

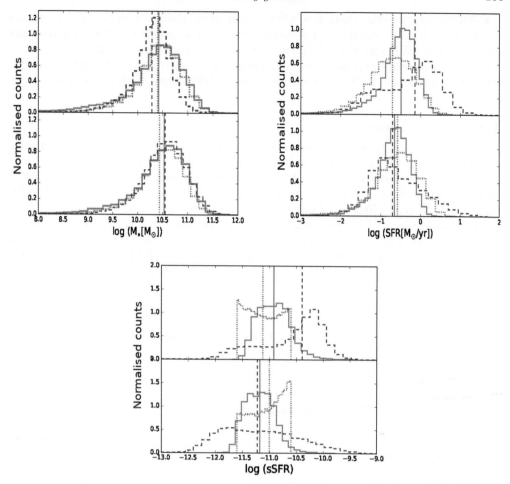

Figure 4. Comparison of stellar mass, SFR, and sSFR in optical (top panels) and UV (bottom panels). The blue dashed, green dotted, and red solid histograms in all figures represent the green valley sample selected using colours (M_{NUV}-M_r and M_g-M_r), sSFR, SFR vs. stellar mass criterion. The vertical dashed lines in all figures represent the median values for each histogram, where each median has the same colour as the one of the corresponding histogram.

galaxies being selected in optical and UV colour criterion, respectively, while in general higher fraction of Seyfert 2 and LINER galaxies have been selected in UV.

We used the visual morphological classification from the Galaxy Zoo survey (Lintott et al. 2011), where galaxies have been classified as elliptical, spiral, and uncertain. When comparing the fraction of elliptical and spiral galaxies selected by different green valley criteria, the only significant difference was found between the two colour criteria where we are detecting more spiral and less elliptical galaxies in optical in comparison to UV (41% vs. 24% for spirals, and 3% vs. 14% for ellipticals, respectively).

4. Conclusions

This work conducted a study on properties of galaxies in different green valley samples selected based on different criteria and analysed the differences and similarities from one criterion to another. Our main findings are:

• By selecting the green valley galaxies based on UV and optical data, we are selecting different types of galaxies in terms of their stellar masses, SFRs, sSFRs, morphological classification, spectroscopic types etc.

- With colour criteria, there is a difference in terms that with optical colour we are selecting much more spirals than ellipticals but for UV colour criteria, we have more ellipticals. For sSFR and SFR-M, the difference is not more pronounced.
- With spectroscopic types, more star forming galaxies were selected in optical sample while composites and AGNs were selected more in UV sample. This is very pronounced for colour criteria.

Acknowledgements

Financial support from the Swedish International Development Cooperation Agency (SIDA) through the International Science Programme (ISP) to the East African Astrophysics Research Network (EAARN) is gratefully acknowledged. MP acknowledges financial support from the Ethiopian Space Science and Technology Institute (ESSTI) under the Ethiopian Ministry of Innovation and Technology (MInT), and support from the Spanish Ministry of Science, Innovation and Universities (MICIU) through project AYA2016-76682C3-1-P and the State Agency for Research of the Spanish MCIU through the Center of Excellence Severo Ochoa award to the IAA-CSIC (SEV-2017-0709). AM acknowledges support from the National Research Foundation of South Africa (Grant Numbers 110816)

References

Baldry, I. K., Glazebrook, K., Brinkmann, J., Ivezic̀, Ž., Lupton, R. H., Nichol, R. C., Szalay, A. S., et al. 2004, *ApJ*, 600, 681
Baldwin, J. A., Phillips, M. M., Terlevich, R., et al. 1981, *PASP*, 93, 5
Belfiore, F. et al. 2017, preprint, (arXiv:1710.05034)
Blanton, M. R. & Moustakas, J. 2009, *ARA&A*, 47, 159
Brammer, G. B. et al. 2009, *ApJ*, 706, L173
Bremer, M. N. et al. 2018, *MNRAS*, 476, 12
Brinchmann, J., Charlot, S., White, S. D. M., Tremonti, C., Kauffmann, G., Heckman, T., Brinkmann, J., et al. 2004, *MNRAS*, 351, 1151
Bryukhareva, T. S. & Moiseev, A. V. 2019, *MNRAS*, 489, 3174
Chang, Y.-Y., van der Wel, A., da Cunha, E., Rix, H.-W., et al. 2015, *ApJS*, 219, 8
Coenda, V., Martìnez, H. J., Muriel, H., et al. 2018, *PASP*, 473, 5617
Eales, S. A. et al. 2018, *MNRAS*, 481, 1183
Ge, X., Gu, Q.-S., Chen, Y.-Y., Ding, N., et al. 2019, *Research in Astronomy and Astrophysics*, 19, 027
Kauffmann, G. et al 2003, *MNRAS*, 341, 33
Kelvin, L. S. et al. 2018, *MNRAS*, 477, 4116
Kewley, L. J., Dopita, M. A., Sutherland, R. S., Heisler, C. A., Trevena, J., et al. 2001, *ApJ*, 556, 121
Lee, G.-H., Hwang, H. S., Lee, M. G., Ko, J., Sohn, J., Shim, H., Diaferio, A., et al. 2015, *ApJ*, 800, 80
Lin, L. et al. 2017, *ApJ*, 851, 18
Lintott, C. et al. 2011, *MNRAS*, 410, 166
Mahoro, A., Povic̀, M., Nkundabakura, P., et al. 2017, *MNRAS*, 471, 3226
Mahoro, A., Povic̀, M., Nkundabakura, P., Nyiransengiyumva, B., Väisänen, P., et al. 2019, *MNRAS*, 485, 452
Martin, D. C. et al. 2005, *ApJ*, 619, L1
Mendez, A. J., Coil, A. L., Lotz, J., Salim, S., Moustakas, J., Simard, L., et al. 2011, *ApJ*, 736, 110
Noeske, K. G. et al. 2007, *ApJ*, 660, L43
Phillipps, S. et al. 2019, *MNRAS*, 485, 5559
Povic̀, M. et al. 2012, *A&A*, 541, A118
Salim, S. 2014, 1995, *Serbian Astronomical Journal*, 181, 1
Salim, S. et al. 2009, *ApJ*, 700, 161

Schawinski, K., Thomas, D., Sarzi, M., Maraston, C., Kaviraj, S., Joo, S.-J., Yi, S. K., Silk, J., et al. 2007, *MNRAS*, 382, 1415
Schiminovich, D. et al. 2007, *ApJS*, 173, 315
Starkenburg, T. K., Tonnesen, S., Kopenhafer, C., et al. 2019, *ApJ*, 874, L17
Taylor, M. 2017, arXiv:1707.02160
Trayford, J. W. et al. 2017, preprint (arXiv:1705.02331)
Trayford, J. W. et al. 2015, *MNRAS*, 452, 2879
Tremonti, C. A. et al. 2004, *ApJ*, 613, 898
Walker, L. M. et al. 2013, *ApJ*, 775, 129
Wyder, T. K. et al. 2007, *ApJS*, 173, 293
York, D. G. et al. 2000, *AJ*, 120, 1579

Study of AGN contribution on morphological parameters of their host galaxies

Tilahun Getachew-Woreta[1,2,3], Mirjana Población[1,4], Josefa Masegosa[4], Jaime Perea[4], Zeleke Beyoro-Amado[1,2,5] and Isabel Márquez[4]

[1]Ethiopian Space Science and Technology Institute (ESSTI), Entoto Observatory and Research Center (EORC), Astronomy and Astrophysics Research and Development Division, P.O.Box 33679, Addis Ababa, Ethiopia

[2]Addis Ababa University (AAU), P.O.Box 1176, Addis Ababa, Ethiopia

[3]Bule Hora University (BHU), P.O.Box 144, Bule Hora, Ethiopia

[4]Instituto de Astrofísica de Andalucía (IAA-CSIC), Glorieta de de la Astronomía, s/n, 18008, Granada, Spain

[5]Kotebe Metropolitan University (KMU), P.O.Box 31248, Addis Ababa, Ethiopia

Abstract. We tested how the AGN contribution (5%–75% of the total flux) may affect different morphological parameters commonly used in galaxy classification. We carried out all analysis at $z \sim 0$ and at higher redshifts that correspond to the COSMOS field. Using a local training sample of >2000 visually classified galaxies, we carried out all measurements with and without the central source, and quantified how the contribution of a bright nuclear point source could affect different morphological parameters, such as: Abraham and Concelice-Bershady indices, Gini, Asymmetry, $M20$ moment of light, and Smoothness. We found that concentration indexes are less sensitive to both redshift and brightness in comparison to the other parameters. We also found that all parameters change significantly with AGN contribution. At $z \sim 0$, up to a 10% of AGN contribution the morphological classification will not be significantly affect, but for $\geqslant 25\%$ of AGN contribution late-type spirals follow the range of parameters of elliptical galaxies and can therefore be misclassified early types.

Keywords. AGN, morphology, galSVM

1. Introduction

Morphology is a key element used to study the properties of AGN host galaxies, their connection with AGN, and their evolution (Población et al. 2012). Earlier morphological studies of local active galaxies suggested that most of AGN reside in spiral galaxies (e.g. Adams 1977; Heckman 1978; Filippenko 1995; Ho 2008). Later on Kauffmann et al. (2003) analysed thousands of active galaxies from the Sloan Digital Sky Survey (SDSS) at low redshifts ($z \leqslant 0.4$). They found that most AGN reside in massive galaxies, whose distribution of sizes, stellar surfaces, mass densities and concentration all resemble those of early-spiral SDSS galaxies. This goes in line with the results by Población et al. (2012) where in the SXDS survey (at median $z \sim 1.0$) most of X-ray detected active galaxies were found to reside in spheroidal or bulge-dominated sources.

The methods for morphological classification of galaxies can be separated into three categories: 1) Visual methods, used for classifying nearby and well-resolved sources. They can provide detailed information about galaxy structure, but also can be subjective and are time-consuming, especially when dealing with large data sets (e.g., Nair & Abraham 2010; Willett et al. 2013). 2) Parametric methods, based on galaxy physical and mathematical parameters, where an analytic model for fitting the galaxy is assumed.

They fail to provide again a correct description if the galaxies are not well resolved (e.g., at higher redshifts) and are time-consuming (e.g., Peng et al. 2002, 2010). 3) Non-parametric methods, that do not assume any particular analytic model, and are based on measuring different galaxy quantities that correlate with morphological types, i.e. colour, spectral properties or light distribution, galaxy shape, etc. They are less time-consuming in comparison with other methods and can provide an easy and fast separation between early- and late-type galaxies up to intermediate redshifts ($z \sim 1.5$) or higher if dealing with space-based instead of ground-based data (Pović et al. 2012, 2013, 2015).

In this work, we went a step further in understanding how the presence of an AGN affects different morphological parameters commonly used in morphological classification at $z \sim 0$ and at higher redshifts. We assumed a concordance cosmology with $\Omega_\Lambda = 0.7$, $\Omega_M = 0.3$, and $H_0 = 70$ km s^{-1} Mpc^{-1}. All magnitudes given in this paper are in the AB system.

2. Data

2.1. Local sample

To study the AGN contribution on morphological parameters of active galaxies, we used an initial sample of 8000 local galaxies in the range $0.01 \leqslant z \leqslant 0.1$ (with a mean redshift of 0.04), observed in the SDSS (York et al. 2003) Data Release 4 (DR4) down to an apparent extinction-corrected magnitude of $g < 16$, and visually classified in the g-band (Nair & Abraham 2010).

The galaxies were selected randomly out of $\sim 14,000$ sources contained in the catalogue, making sure that the selected sub-sample is a fair representation of the whole sample. The detailed description of the training sample can be found in Beyoro-Amado et al. (2019) and Pović et al. (2013).

We used the SDSS DR7 spectroscopic data (Abazajian et al. 2009) from the MPA-JHU emission line catalogue to select the non-AGN sample. By locating the 8000 local galaxies in the BPT diagram (Kewley et al. 2006) we obtained that 2744 (35%) can be classified as non-AGN, 1918 (24%) as composite galaxies, 594 (7%) as Seyfert2, and 2684 (34%) as LINERs.

After selecting the 2744 non-AGN galaxies, we further removed galaxies with foreground stars superposed onto them and those with evident signs of interactions and mergers. Finally we ended up with a sample of 2301 non-active galaxies.

2.2. Non-local sample (COSMOS)

To study morphology at higher redshift, we chose as an example the Cosmic Evolution Survey (COSMOS; Scoville et al. 2007). This survey is the largest project ever undertaken by the Hubble Space Telescope (HST) covering an area of 2deg^2 to a depth of $I_{F814W} = 27.8$ mag (5σ, AB). The observations from the HST-ACS survey consist on a sample of more than 1.2×10^6 sources down to a magnitude limit of 26.5 in F814W filter. This field samples scale from $30 - 180$Mpc with $z = 0.2 - 4$ including a million galaxies in overall volume of 10^7Mpc^3. In this work we used the photometric images taken by the HST-ACS in the I_{F814W} filter.

3. Methodology and analysis

We used the galSVM code (Huertas-Company et al. 2008) for measuring the morphological parameters. It is a freely available code written as an IDL library that, combined with the similarly freely available library libSVM (Chang & Lin 2011) enables a morphological classification of galaxies in an automated way. This code has been especially useful when dealing with low-resolution and high redshift data (Pović et al. 2015).

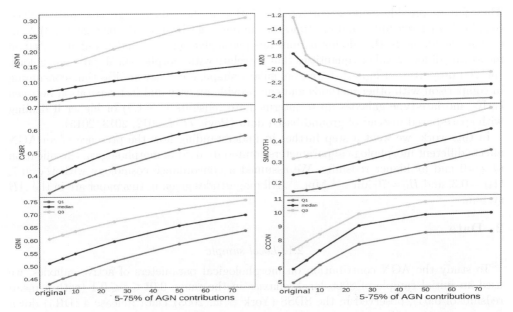

Figure 1. Effect of AGN contribution (5-75%) on six morphological parameters at $z \sim 0$. The red line indicates quartile 1 (Q1), blue for median and lime for quartile 3 (Q3)

The input information that the code needs, related to the source position, size, ellipticity parameters, and background, were obtained with SExtractor (Bertin & Arnouts 1996. The six morphological parameters used by large commonly to distinguish between early- (elliptical and lenticular) and late-type (spiral and irregular) galaxies analysed in this work are: Abraham concentration index (hereafter CABR; Abraham et al. 1996), Asymmetry index (A; Abraham et al. 1994), Gini coefficient (GINI; Abraham et al. 2003; Lotz et al. 2004), Conselice-Bershady concentration index (CCON; Bershady et al. 2000), M_{20} moment of light (M20; Lotz et al. 2004) and Smoothness (or clumpiness) index (Conselice et al. 2000).

In this work we went through the following: a) We measured the morphological parameters of the local sample of 2301 visually classified galaxies. b) We obtained simulated galaxy images by adding to the original images an AGN contribution from 5-75% of the total flux by using PSF images and a Moffat (1969) law representing the central profile. We then measured morphological parameters on all simulated images. In this way we can estimate differences on all parameters depending on the AGN contribution to the total flux. c) We simulated the images of local sample by moving them to the rest-frame magnitude (from 21 to 25) and redshift distribution of the COSMOS field, and we measured the parameters following the procedures described in (Pović et al. 2015). With this step we can test how morphological parameters change with magnitude and redshift. d) Finally, we repeated the previous step, but now taking into account the simulated images with different AGN contributions obtained in step b). The aim of this final step is to study the effect on morphology when having both, AGN contribution and the effect of brightness and high redshift. In this paper we will principally focus on the results obtained in steps a), b) and c).

4. Results

In this paper, we show the results obtained on the morphological parameters. Figure 1 shows the variations of the morphological parameters depending on the AGN contribution

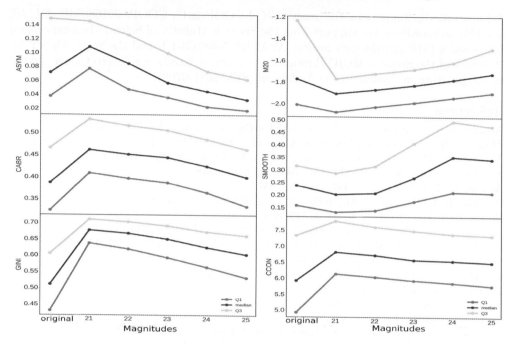

Figure 2. Comparison of six morphological parameters measured on the original sample (at $z \sim 0$) and once moved to the redshift distribution in the COSMOS field and to fainter magnitudes (21–25).

to the total flux. It can be seen that all parameters change significantly when the AGN contribution is larger than 25% of the total flux. Figure 2 shows the same kind of analysis when moving to higher redshifts and fainter magnitudes. The results are similar to those obtained by Pović et al. 2015, although in that work no separation was done regarding the effect that the AGN can have on the morphological classification of active galaxies. The three concentration indices, CABR, CCON, and GINI show small variation when going to fainter magnitudes (typically no larger than 20%), but ASYM and SMOOTH show to be more sensitive to both redshift and brightness.

5. Conclusions

The main objective of this work has been to study the effects of the AGN on the morphological parameters of their host galaxies and to answer some of the fundamental questions still open related to the effect of AGN on them. For the total sample, all parameters change significantly with the AGN contribution (with ASYM being less affect than others). At $z = 0$ for most of the parameters up to a 10% AGN contribution morphological classification will not be significantly affected, but for 25% and higher late-type spirals follow the range of parameters typical of elliptical galaxies (Pović et al. 2015). When moving galaxies to higher z and fainter mag, we conclude that the concentration indices CABR and CCON are less sensitive to redshift and magnitude than GINI, ASYM, M20, and SMOOTH.

Acknowledgements

TGW acknowledges the support from Bule Hora University. TGW, MP, and ZBA acknowledge financial support from the Ethiopian Space Science and Technology Institute (ESSTI) under the Ethiopian Ministry of Innovation and Technology (MInT). TGW and

JM acknowledge support by the grant CSIC I-COOP 2017, COOPA20168. MP, JM, JP, and IMP acknowledge the support from the Spanish Ministry of Science, Innovation and Universities (MICIU) through project AYA2016-76682C3-1-P and the State Agency for Research of the Spanish MCIU through the Center of Excellence Severo Ochoa" award to the Instituto de Astrofísica de Andalucía (SEV-2017-0709).

References

Abazajian, K. N., Jennifer, K., Adelman-McCarthy, et al. 2009, ApJS, 182, 543A
Abraham, R. G., Valdes, F., Yee, H. K. C., et al. 1994,ApJ, 432, 75
Abraham, R. G., van den Bergh, S., Glazebrook, K., et al. 1996,ApJS, 107, 1
Abraham, R. G., van den Bergh, et al. 2003, ApJ, 588, 218
Adams, M. T. & Rudnick, L. 1977, AJ, 82, 857A
Baldwin, J. A., Phillips, M. M., Terlevich, et al. 1981, PASP, 93, 5B
Bertin, E. & Arnouts, S. 1996, A&A, 117, 393
Bershady M. A., Jangren A., Conselice C. J., et al. 2000, AJ, 119, 2645
Beyoro-Amado, Z., Pović, M., Sánchez-Portal, M., et al. 2019, MNRAS, 485, 1528A
Chang, C. C. & Lin, C. J., 2011, ACM-T IST, 2, 27
Conselice, C. J., Bershady, M. A., Dickinson, M., et al. 2000, ApJ, 529, 886
Filippenko, A. 1995, hst..prop, 5792F
Heckman, T. M., Balick, B., & Sullivan, W. T., III. 1978, ApJ, 224, 745H
Ho Luis C., 2008, ARAA, 46, 475-539
Huertas-Company, M., Rouan, D., Tasca, L., et al. 2008, A&A, 478, 971
Kauffmann, G., Heckman, T. M., et al. 2003, MNRAS, 346, 1055
Kewley, Lisa J. Groves, Brent, et al. 2006, MNRAS, 372, 961K
Lotz, J. M., Primack, J., Madau, et al. 2004, AJ, 128, 163, MNRAS, 485, 452M
Moffat, J.W. 1969, PhRv, 177, 2456M
Nair P. B. & Abraham R. 2010, ApJS, 186, 427
Peng C. J. et al. 2002, AAS, 201, 1303P
Peng C. J. et al. 2010, AAS, 139, 2097
Pierce, C. M., Lotz, J. M. et al. 2010, MNRAS, 405, 718
Pović M., Sánchez-Portal M., et al. 2012, A&A, 541, A118
Pović, M., Huertas-Company, M., et al. 2013, MNRAS, 435, 3444P
Pović M., Márquez I., et al. 2015, MNRAS, 453, 1644
Scoville, N., Aussel, H., Brusa, M., et al. 2007, ApJS, 172, 1
Willett, K. W. et al. 2013, MNRAS, 435, 2835
York, D. G., Adelman, J., et al. 2000, AJ, 120, 1579

Galaxy evolution studies in clusters: the case of Cl 0024 + 1652 cluster galaxies at $z \sim 0.4$

Zeleke Beyoro-Amado[1,2,3], Mirjana Pović[1,4], Miguel Sánchez-Portal[5], Solomon Belay Tessema[1], Tilahun Getachew-Woreta[1,2,6] and the GLACE team

[1]Ethiopian Space Science and Technology Institute (ESSTI), Entoto Observatory and Research Centre (EORC), Astronomy and Astrophysics Research and Development Division, P.O.Box 33679, Addis Ababa, Ethiopia

[2]Addis Ababa University (AAU), P.O.Box 1176, Addis Ababa, Ethiopia

[3]Kotebe Metropolitan University, College of Natural and Computational Sciences, Department of Physics, P.O.Box 31248, Addis Ababa, Ethiopia

[4]Instituto de Astrofísica de Andalucía (IAA-CSIC), Glorieta de la Astronomía s/n, 18008, Granada, Spain

[5]Instituto de Radioastronomía Milimétrica, Av. Divina Pastora 7, Núcleo Central, E-18012 Granada, Spain

[6]Bule Hora University, P.O.Box 144, Bule Hora, Ethiopia

Abstract. Studying the transformation of cluster galaxies contributes a lot to have a clear picture of evolution of the universe. Towards that we are studying different properties (morphology, star formation, AGN contribution and metallicity) of galaxies in clusters up to $z \sim 1.0$ taking three different clusters: ZwCl 0024 + 1652 at $z \sim 0.4$, RXJ1257 + 4738 at $z \sim 0.9$ and Virgo at $z \sim 0.0038$. For ZwCl 0024 + 1652 and RXJ1257 + 4738 clusters we used tunable filters data from GLACE survey taken with GTC 10.4 m telescope and other public data, while for Virgo we used public data. We did the morphological classification of 180 galaxies in ZwCl 0024 + 1652 using galSVM, where 54% and 46% of galaxies were classified as early-type (ET) and late-type (LT) respectively. We did a comparison between the three clusters within the clustercentric distance of 1 Mpc and found that ET proportion (decreasing with redshift) dominates over the LT (increasing with redshift) throughout. We finalized the data reduction for ZwCl 0024 + 1652 cluster and identified 46 [OIII] and 73 Hβ emission lines. For this cluster we have classified 22 emission line galaxies (ELGs) using BPT-NII diagnostic diagram resulting with 14 composite, 1 AGN and 7 star forming (SF) galaxies. We are using these results, together with the public data, for further analysis of the variations of properties in relation to redshift within $z < 1.0$.

Keywords. ZwCl 0024 + 1652, galaxy cluster, early type, late type, ELGs, AGN, Tunable Filter data

1. Introduction

The way galaxies form and evolve in both field and clusters still remains one of the open questions in modern cosmology. Understanding how galaxies transform inside the clusters presents one of the main steps in disentangling the picture of galaxy formation and evolution, and the formation of the universe at large. In this regard significant differences have been observed between the field and cluster galaxies (e.g., Koopman & Kenny 1998).

Research results show that significant evolution in properties of galaxies within clusters are observed as a function of redshift. According to previous studies, the cores of low redshift clusters are dominated by red ET galaxies while the blue LT galaxy population dominates the higher redshift (z) clusters (e.g., Butcher & Oemler 1984; Beyoro-Amado et al. 2019). Concerning the star formation activity, an increase of the obscured SF in mid-infrared (MIR) and far-infrared (FIR) surveys of distant clusters has been indicated (e.g., Haines et al. 2009). In addition to these, higher population of AGN have been observed at higher redshifts (e.g., Bufanda et al. 2017).

Kodama & Bower (2001) and Oh et al. (2018) described that properties of galaxies in clusters also transform significantly with environment. Different physical processes were suggested for affecting the galaxy evolution in clusters. According to Treu et al. (2003) tidal halo stripping, tidal triggering star formation, and ram pressure stripping affect the galaxy evolution in the cluster core ranging to about 2 Mpc from the cluster center while the effects of starvation, harassment and merging remain significant up to a clustercentric distance of about 5 Mpc.

Morphological classification of galaxies could be performed either by traditional (visual inspection; e.g., Lintott et al. 2011) or modern techniques. The modern techniques can be either parametric, performed by fitting some parameters assuming a predefined parametric model (e.g., Tarsitano et al. 2018), or non parametric, where no specific analytic model is assumed while performed by measuring a set of well-chosen observables (e.g., Pović et al. 2015; Huertas-Company et al. 2015; Pintos-Castro et al. 2016; Beyoro-Amado et al. 2019). Previous studies show that local density has effects on galaxy morphology (e.g., Nantais et al. 2013); as well as SF activities (e.g., Woo et al. 2013). Generally speaking, galaxy properties like morphology, nuclear activity, SF activity, metallicities, and color vary in relation to redshift and environment (e.g., Laganá & Ulmer 2018).

Several controversial results exist in relation to AGN activities in clusters. For instance there are studies describing that AGN activities do not depend on local densities (e.g., Miller et al. 2003) in one hand while there are others indicating that high density regions avoid luminous AGNs (e.g., Powell et al. 2018). AGN fraction versus environment, metallicities, population of ELGs in relation to radial distance, and morphological transformations with radial distance and redshift for galaxy clusters still remain some open issues. This motivated us to carry out a PhD research (for the first author here) on galaxy evolution and transformation in clusters by studying different properties of three galaxy clusters.

Our work is generally aimed at studying the properties of galaxies in clusters up to $z \sim 1.0$ using three galaxy clusters: ZwCl0024+1652 at $z \sim 0.4$, RXJ1257+4738 at $z \sim 0.9$ and Virgo at $z \sim 0.0038$. Hence performing the detailed morphological study, and analysing the AGN versus SF contributions plus the metallicity variations; could give us important clues about galaxy evolution and transformations with cosmic time.

2. Data

For ZwCl0024+1652 cluster, we used the tunable filter (TF) raw observational data from GaLAxy Cluster Evolution (GLACE) survey with targets to [OIII], Hβ, Hα and [NII] emission lines (Sánchez-Portal et al. 2015). The GLACE data were taken with Gran Telescopio Canarias (GTC) 10.4m telescope located at La Palma. For this cluster, we also used HST/ACS data, WFP2 master catalogue (Moran et al. 2005), and visual morphology catalogue (Moran et al. 2007). For the cluster at $z \sim 0.9$, we used the GLACE TF data targeting Hβ and [OII] lines, the FIR data (Pintos-Castro et al. 2013) and the catalogue of morphological properties (Pintos-Castro et al. 2016). We used public data in case of Virgo cluster: the Virgo cluster catalogue (VCC, Binggeli et al. 1985), HST/ACS images (Peng et al. 2013), Herschel reference survey data (HRS, Hughes et al. 2013), and extended Virgo cluster catalogue data (EVCC, Kim et al. 2014).

Table 1. Variation of morphological classifications of cluster galaxies with redshift

Cluster	Redshift (z)	ET	LT
ZwCl0024 + 1652	∼0.4	97 (∼54%)	83 (∼46%)
RXJ1257 + 4738	∼0.9	31 (∼89%)	4 (∼11%)
Virgo	∼0.0038	273 (∼68%)	131 (∼32%)

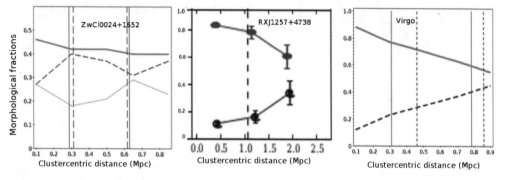

Figure 1. Variations of morphological fractions with clustercentric distance for the three clusters. The red lines stand for ET while the blue lines represent the LT galaxies. The middle plot is recovered from Pintos-Castro et al. (2016).

3. Results of morphological study

We have performed the morphological study for all the three clusters and did a comparison among them. For the morphological classification of ZwCl0024+1652, we used a non parametric public code called galaxy Support Vector Machine (galSVM) operating in IDL environment. The code simultaneously measures six morphological parameters: Abraham concentration (CABR), moment of light (M20), Bershady-Conselice concentration (CCON), Gini coefficient (GINI), asymmetry (ASYM) and smoothness. Using these parameters together with ellipticity from SExtractor, galSVM gives a probability (P_{ET}) for the galaxy to be an ET where the probability for being LT is $1-P_{ET}$. We measured morphological parameters for 231 member galaxies within the clustercentric distance of 1 Mpc, where 180 galaxies have been classified with 97 (∼54%) and 83 (∼46%) galaxies have been classified as ET and LT respectively. The classification is in ∼80% agreement with previous works (Moran et al. 2007) and 121 new galaxies have been classified. With this work we obtained one of the most complete morphological catalogue of ZwCl0024 + 1652 cluster within the clustercentric radius of 1 Mpc (Beyoro-Amado et al. 2019).

To track the trends of evolution of cluster galaxies, we compared the results for the three clusters. The morphological class comparison is presented in Table 1. From the table we can see that in a redshift range to $z \sim 1$ within a clustercentric distance of 1 Mpc, the proportion of ET galaxies dominate over the LT galaxies. Here because of the small number of sources especially for RXJ1257 + 4738 cluster, we couldn't generalize, but from results of ZwCl0024 + 1652 and Virgo clusters, the usual evolution of morphologies of cluster galaxies holds in such a way that the proportion of LT increases while ET proportion decreases as redshift increases. We also present the comparison of the results in morphological fractions for the three clusters at different redshifts as a function of clustercentric distance in Fig. 1. The plots show that for clusters at $z < 1$, the fractions of ET galaxies decrease with an increase in LT galaxies as a function of clustercentric distance out to 1 Mpc. It looks that at lower redshifts the morphological fractions change faster (decrease for ET, increase for LT) than in the case of higher redshifts.

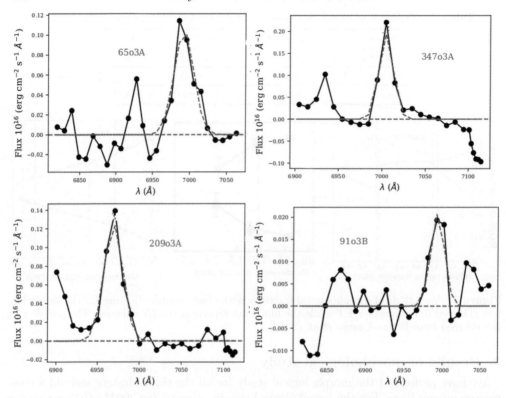

Figure 2. Example of pseudospectra indicating some of the [OIII] emission lines identified. The red label in each plot indicates the unique id of the galaxy as in our results.

4. The GLACE TF data reduction for ZwCl0024 + 1652 and RXJ1257 + 4738 clusters

We had a TF raw data of ZwCl0024 + 1652 and RXJ1257 + 4738 clusters observed under GLACE project targeting [OIII] and Hβ lines. For data reduction, the basic standard steps have been implemented and we used different scripts (in IDL, IRAF and python) with TFREd package suited for TF data. We used GAIA astrometry for mapping the cluster before doing the wavelength and flux calibration. Using Hα results (Sánchez-Portal et al. 2015) of Zwcl0024 + 1652 cluster, we matched our results, measured the fluxes and plotted the pseudospectra of all sources. Pseudospectra of some of the [OIII] lines are presented in Fig. 2 as an example. By relying on better signal to noise ratio and visual inspection of the pseudospectra, a total of 46 [OIII] and 73 Hβ emission lines have been identified (both increasing with clustercentric distance). Combining ours with the published results (Sánchez-Portal et al. 2015) we have identified 22 galaxies with all four emission lines; ([OIII], Hβ, Hα and [NII]). For these member galaxies we plotted the BPT-NII diagnostic diagram (Kewley et al. 2006) to classify them into SF (7), composite (14), and LINER-AGN (1), as shown in Fig. 3.

5. Conclusions and the work in progress

From morphological analysis, we suggested that cores of galaxy clusters at $z < 1.0$ are dominated by ET galaxies whose fraction decreases as a function of clustercentric distance up to 1 Mpc. The LT proportions increase with redshift and separately with the clustercentric distance. We also suggested that ELGs do not populate cluster cores

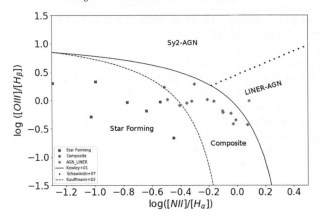

Figure 3. Our results of classifying the ELGs using the BPT-NII diagram

and increase in numbers outwards with clustercentric distance. The majority of ELGs in ZwCl0024 + 1652 cluster are SF with small number of AGNs comparing the two classes; while more than half of the ELGs are composite. Currently, we are working on the SF/AGN, SFR and metallicity analysis of ZwCl0024 + 1652 cluster. Then having the results, comparison with RXJ1257 + 4738 and Virgo will be performed to draw final conclusions on the evolution and transformation of galaxies in clusters up to $z \sim 1.0$.

Acknowledgements

We thank the Ethiopian Space Science and Technology Institute (ESSTI) under the Ethiopian Ministry of Innovation and Technology (MInT) for all the financial support. ZBA especially acknowledges Instituto de Radioastronomía Milimétrica (IRAM) for the financial and logistics supports, and Instituto de Astrofísica de Andalucía (IAA-CSIC) for providing the working space during the research visits for part of this work. MP acknowledges the support from the ESSTI and Ethiopian MInT, Spanish MICIU through project AYA2016-76682C3-1-P and the State Agency for Research of the Spanish MCIU through the Center of Excellence Severo Ochoa award to the IAA-CSIC (SEV-2017-0709).

References

Beyoro-Amado, Z., Pović, M., Sánchez-Portal, M., et al. 2019, *MNRAS*, 485, 1528
Binggeli, B. et al. 1985, *AJ*, 90, 1681
Bufanda, E., Hollowood, D., et al. 2017, *MNRAS*, 465, 2531
Butcher, H. & Oemler, A. 1984, *ApJ*, 285, 426
Haines, C. P. et al. 2009, *ApJ*, 704, 126
Huertas-Company, M. et al. 2015, *ApJS*, 221:8
Hughes, T. M. et al. 2013, *A&A*, 550, A115
Kewley, L. J. et al. 2006, *MNRAS*, 372, 961
Kim, S. et al. 2014, *ApJS*, 215:22
Kodama, T. & Bower, R., 2001, *MNRAS*, 321, 18
Koopmann, R. A. & Kenney, J. D. P. 1998, *ApJ*, 497, L75
Laganá, T. F. & Ulmer, M. P., 2018, *MNRAS*, 475, 523
Lintott, C. J. et al. 2011, *MNRAS*, 410, 166
Martini, P., Miller, E. D., Brodwin, M., et al. 2013, *ApJ*, 768, 1
Miller, C. J. et al. 2003, *ApJ*, 597, 142
Moran, S. M., Ellis, R. S., Teu, T., et al. 2005, *ApJ*, 634, 977
Moran, S. M., Ellis, R. S., Teu, T., et al. 2007, *ApJ*, 671, 1503

Nantais, J. B., Flores, H., Demarco, R., *et al.* 2013, *A&A*, 555, A5
Oh S., Kim K., *et al.* 2018, *AJSS*, 237, 14
Peng, E. W. *et al.* 2008, *ApJ*, 681, 197
Pintos-Castro, I., Porić, M., *et al.* 2016, *A&A*, 592, A108
Pintos-Castro, I., Sánchez-Portal M., *et al.* 2013, *A&A*, 558, A100
Pović M., Márquez I., *et al.* 2015, *MNRAS*, 453, 1644
Powell, M. C. *et al.* 2018, *ApJ*, 858, 110
Sánchez-Portal M., Pintos-Castro I., *et al.* 2015, *A&A*, 578, A30
Tarsitano, F. *et al.* 2018, *MNRAS*, 481, 2018
Treu, T., Ellis, R. S., Kneib, J., *et al.* 2003, *APJ*, 591, 53
Woo J. *et al.* 2013, *MNRAS*, 428, 3306

AGN, host galaxies, and starbursts

Petri Väisänen[1,2]

[1]South African Astronomical Observatory, P.O. Box 9 Observatory, Cape Town, South Africa
[2]Southern African Large Telescope, P.O. Box 9 Observatory, Cape Town, South Africa

Abstract. AGN by definition live in galaxies. Despite a long history of studies, there is still much ongoing research into the interplay of the nucleus and the host galaxy, how do they affect each other, how is their evolution intertwined. This review will briefly go over the historical developments behind the current understanding of AGN host galaxies, their types and characteristics. It will discuss the starburst and AGN connection in particular, and how these phenomena may be connected or influence each other by means of e.g. gas flows. Finally, some examples of AGN/starburst evolution studies from SALT and other large telescopes will be presented.

Keywords. galaxies: active, galaxies: starbursts

Quasar host galaxies and environments in the GAMA survey

Jari Kotilainen

Finnish Centre for Astronomy with ESO (FINCA), University of Turku, Finland

Abstract. We present first results from our study of the host galaxies and environments of quasars in Galaxy And Mass Assembly (GAMA), a multiwavelength photometric and spectroscopic survey for ∼300,000 galaxies over ∼300 deg^2, to a limiting magnitude of r ∼ 20 mag. We use a GAIA-selected sample of ∼350 quasars at z < 0.3 in GAMA. For all the quasars, we determine all surrounding GAMA galaxies and measure their star formation (SF) rate and SF history, and the host galaxy morphology and group membership of the quasars. As a comparison sample of inactive galaxies, we use 1000 subsets of galaxies in GAMA, matched in redshift and galaxy stellar mass to the quasars. We find that quasar activity does not depend on the large-scale environment (cluster/group/void), although quasars tend to prefer satellite location in their environment. Compared to inactive galaxies, quasars are preferentially hosted in bulge-dominated galaxies and have higher SF rates, both overall and averaged over the last 10 and 100 Myr. Quasars also have shorter median SF timescales, shorter median time since the last SF burst, and higher metallicity than inactive galaxies. We discuss these results in terms of triggering mechanisms of the quasar activity and the role of quasars in galaxy evolution.

Keywords. galaxies: active, galaxies: quasars, galaxies: star formation, galaxies: environments

Role of environment on AGN activity

Amirnezam Amiri

Physics Department, Kharazmi University, Tehran, Iran

Abstract. Motivated by the apparently conflicting results reported in the literature on the effect of environment on nuclear activity, we have carried out a new analysis by comparing the fraction of galaxies hosting active galactic nuclei (AGNs) in the most overdense regions (rich galaxy clusters) and the most underdense ones (voids) in the local universe. Exploiting the classical BPT diagnostics, we have extracted volume limited samples of star forming and AGN galaxies. We find that, at variance with star-forming galaxies, AGN galaxies have similar distributions of specific star formation rates and of galactic ages (as indicated by the Dn4000 parameter) both in clusters and in voids. In both environments galaxies hosting AGNs are generally old, with low star formation activity. The AGN fraction increases faster with stellar mass in clusters than in voids, especially above $10^{10.2}\,M_\odot$. Our results indicate that, in the local universe, the nuclear activity correlates with stellar mass and galaxy morphology and is weakly, if at all, affected by the local galaxy density.

Keywords. galaxies: active, galaxies: environments, galaxies: clusters

Hosts of jetted narrow-line Seyfert 1 galaxies in near-infrared

Emilia Järvelä

European Space Agency, European Space Astronomy Centre, C/Bajo el Castillo s/n, 28692 Villanueva de la Cañada, Madrid, Spain

Abstract. Host galaxy morphology studies of jetted narrow-line Seyfert 1 galaxies (NLS1) are scarce. Although it seems that they are mostly hosted by late-type galaxies the results remain inconclusive, mostly due to the small sample size. Increasing the number of studied sources is crucial to achieve statistically significant results and to establish a preferred host type for jetted NLS1s. To this end we observed the host galaxies of nine NLS1s in near-infrared using NOTCam at the Nordic Optical Telescope. Seven of these sources are jetted based on the 37 GHz observations at Metsähovi Radio Observatory, Finland. To determine the morphological types of the hosts we performed photometric decomposition of the near-infrared images using GALFIT. Here we present the results of the host galaxy modelling, discuss the importance of this study to our understanding of the nature of the diverse NLS1 population, as well as its significance and implications for active galactic nuclei research in general.

Keywords. galaxies: active, galaxies: morphology, galaxies: properties

Dirt-cheap gas scaling relations: Using dust attenuation and galaxy radius to predict gas masses for large samples of AGNs

Hassen Yesuf

Kavli Institute for Astronomy and Astrophysics, Peking University, Beijing 100871, China

Abstract. We analyze the molecular and atomic gas data from the GALEX Arecibo SDSS Survey (xGASS) and the extended CO Legacy Database (xCOLD GASS) IRAM survey using novel survival analysis techniques to identify a small number of stellar properties that best correlate with the gas mass. We find that the dust absorption, A_V, and the stellar half-light radius, R50, are likely the two best secondary parameters than improve the Kennicutt - Schmidt type relation between the gas mass and the star formation rate, SFR. We fit multiple regression, taking into account gas mass upper limits, to summarize the median, mean, and the 0.15/0.85 quantile multivariate relationships between the gas mass (atomic or molecular hydrogen), SFR, AV and/or R50. In particular, we find that the A_V of both the stellar continuum and nebular gas emission shows a significant partial correlation with the molecular hydrogen after controlling for the effect of SFR. The partial correlation between the A_V and the atomic gas, however, is weak and their zero-order correlation may be explained by SFR. This is expected since in poorly dust-shielded regions molecular hydrogen is dissociated by the far ultraviolet photons and HI is the dominant phase. Similarly, R50 shows significant partial correlations with both atomic and molecular gas masses. This hints at the importance of environment (e.g., galacto-centric distance) on the gas contents galaxies and on the interplay between gas and star formation rate. We apply the gas scaling relations we found to a large sample of type 2 and type 1 AGNs and infer that the gas mass correlates with AGN luminosity. This correlation is inconsistent with the prediction of AGN feedback models that strong AGNs remove or heat cold gas in their host galaxies.

Keywords. galaxies: active, galaxies: properties, galaxies: star formation

CHAPTER V. Triggering, feedback, and shutting off of AGN activity

AGN fueling and feedback

Francoise Combes

Observatoire de Paris, LERMA, Collège de France, CNRS, PSL University,
Sorbonne University, UPMC, Paris
email: francoise.combes@obspm.fr

Abstract. Dynamical mechanisms are essential to exchange angular momentum in galaxies, drive the gas to the center, and fuel the central super-massive black holes. While at 100pc scale, the gas is sometimes stalled in nuclear rings, recent observations reaching ∼10pc scale have revealed, within the sphere of influence of the black hole, smoking gun evidence of fueling. Observations of AGN feedback are described, together with the suspected responsible mechanisms. Molecular outflows are frequently detected in active galaxies with ALMA and NOEMA, with loading factors between 1 and 5. When driven by AGN with escape velocity, these outflows are therefore a clear way to moderate or suppress star formation. Molecular disks, or tori, are detected at 10pc-scale, kinematically decoupled from their host disk, with random orientation. They can be used to measure the black hole mass.

Keywords. galaxies: active, galaxies: general, galaxies: nuclei, galaxies: Seyfert, galaxies: spiral

1. Introduction: the main paradigm

For a long time, the main unification paradigm to explain the large variety of AGN, and in particular the type 1 with broad lines (BLR), and the type 2, with only narrow lines (NLR), has been the obscuration of the accretion disk by a dusty torus, hidding the BLR and showing only the NLR to the observer (Urry & Padovani 1995). This idea is still valid for a certain number of AGN (for instance type 2 Seyfert which reveal their BLR in polarized light), but is known not to be the only parameter distinguishing the various AGN, since there exist intrinsic differences between Sy1 and Sy2, and also a number of changing look AGNs have been discovered, while their transformation from Sy1 to Sy2 an vice-versa in time-scales of dozens of years has nothing to do with obscuration (Denney et al. 2014; McElroy et al. 2016).

In the last decade, high spatial resolution observations in the mid-infrared with the VLT Interferometer (VLTI) showed that the dust on parsec scales is not mainly in a thick torus, but instead in a polar structure, forming like a hollow cone, perpendicular to a thin disk (e.g., Asmus et al. 2016; Hönig 2019) and references therein). A large majority of objects (∼90%) reveal that most (>50%) of the mid-infrared emission of the dust comes from a polar structure, leaving little room for a torus contribution (Asmus 2019). The new view which is sketched now for the cold gas is a two component medium, one inflowing in a thin disk, where millimeter lines have been found, with also H_2O masers, or the rotoibrational lines of warm H_2, and an outflowing component, in the perpendicular direction, responsible for the polar dust emission. Molecular outflows will be driven along the borders of this hollow cone. The inner boundary of the dusty thin disk would correspond to the sublimation of the dust by the AGN radiation. The circum-nuclear molecular disk could be observed to extend down to smaller distances from the center.

In the following, I review recent observations at high-angular resolution of the molecular gas, showing how gravity torques can help feeding the central black hole, through

© The Author(s), 2021. Published by Cambridge University Press on behalf of International Astronomical Union

exchange of angular momentum. These are the consequences of nuclear dynamical features, such as nuclear bars & spirals. The same observations might reveal outflows, occurring simultaneously with inflow, albeit in different directions. These outflows are due to the feedback effect of the AGN, added to the star formation feedback. Frequently in nearby active galaxies, molecular circum-nuclear disks are observed; with decoupled kinematics from the larger-scale galactic disks. These parsec-scale structures may be identified to molecular tori, able to obscure the central accretion disks. Being within the sphere of influence of the black holes, they help to determine their mass.

2. Nuclear trailing spirals

Observed nuclear spirals bring clear evidence of fueling. They were detected first in NGC 1566 (Combes et al. 2014), during the first ALMA observations, with 0.5 arcsecond ∼25pc resolution. In this barred spiral, there is an r = 400pc ring, corresponding to the inner Lindblad resonance (ILR) of the bar. The stellar periodic orbits, which attract all regular orbits, change by 90° at each resonance (Contopoulos & Papayannopoulos 1980). They are parallel to the bar inside corotation, and become perpendicular in-between the two ILRs.

The gas behaviour may be derived from these orbits; the gas elliptical streamlines tend to follow them, but gas clouds are subject to collisions. The ellipses are gradually tilted by 90° and wind up in spiral structures. The precession rate of these elliptical orbits in the epicyclic approximation is equal to $\Omega - \kappa/2$, with Ω the rotation frequency = V/r, and κ being the epicyclic frequency. When the dissipative gas is driven to the center, the precession rate first increases, since $\Omega - \kappa/2$ increases, and the spiral is trailing. But inside the ILR, the precession rate reaches a maximum and declines, and this changes the winding sign of the gas, which is expected to be in a leading spiral. The reversal of the winding sense has also the consequence of reversing the sign of the gravity torques exerted by the stellar bar on the gas (Buta & Combes 1996). The torque is negative from corotation to ILR, but then is positive inside the ILR, this confines the gas in the ILR ring, where it forms stars actively.

However, in the presence of the central black hole, all frequencies Ω and $\Omega - \kappa/2$ increase again towards the center, as shown in Fig. 1. This is able to reverse the winding sense of the gas spiral, if a sufficient amount of gas is within the influence of the black hole. In that case, the gravity torques become negative again, and the gas is driven to the center, to fuel the AGN. The very presence of a nuclear trailing spiral is smoking gun evidence of the AGN fueling.

Such trailing nuclear spirals have been found also in NGC 613 (Audibert et al. 2019) and in NGC 1808 (Audibert et al. 2020, in prep.), as can be seen in Fig. 1. The nuclear spiral is conspicuous in the CO(3-2) emission line, but also in the dense gas tracers like HCN, HCO^+ or CS. The rotation curves of these galaxies, derived both from the stellar potential (traced by near-infrared images) and the observed kinematics of the gas, confirm that the nuclear gas is falling within the sphere of influence of the black holes.

The gravity torques have been measured on the deprojected images of the molecular gas, with the method described in Garcia-Burillo & Combes (2012), and the torques are indeed negative (Audibert et al. 2019). The computation allows to quantify the strength of the torques, and shows that the gas loses most of its angular momentum in one rotation, i.e. 10 Myr.

3. Molecular outflows

In some cases, molecular outflows are observed simultaneously with the evidence of fueling. This is the case of NGC 613, where a very short (23pc) and small velocity

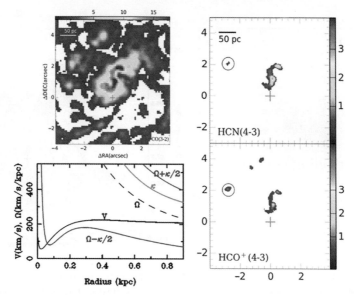

Figure 1. ALMA observations of the Seyfert-2 galaxy NGC 1808. Left is a zoomed 4" × 4" region of the CO(3 − 2) intensity map, showing the nuclear trailing spiral. Right are the intensity maps of the dense gas tracers HCN(4 − 3) and HCO$^+$(4 − 3). The size is given in parsec by the 50pc bar. At the bottom left, is the model rotation curve, taken from both the near-infrared (potential from old stars) and the gas observed velocity (Salak et al. 2016). From Audibert et al. (2020) in prep., and Combes et al. (2019).

(300km/s) outflow is detected on the minor axis, in the same direction of the radio jet, mapped at cm wavelength with the VLA (Audibert et al. 2019). This was possible thanks to the ALMA resolution of 60 mas (5pc). The coincidence with the radio jet strongly suggests that the outflow is AGN feedback in the radio mode.

In other cases (e.g. NGC 1566) no molecular outflow is detected. In NGC 1808, an outflow has been known for a long time, at large scale, from ionized gas and dust ejected perpendicular to the plane, creating an extra-planar medium (Busch et al. 2017). However, at parsec scale, the CO emission from ALMA does not reveal any outflow. In that case, it can be concluded that the outflow is driven by supernovae feedback, but not from the AGN.

In that NGC 1808 galaxy, the starburst and AGN contributions can be distinguished by the diagnostics of line ratios between HCN and HCO$^+$ or CS. Close to the AGN, the HCN line is considerably enhanced (Usero et al. 2004; Krips et al. 2008), due to the X-rays from the AGN.

In these nearby low-luminosity AGN, the main mechanism to drive molecular outflows is entrainment by the radio jets. There are two main modes identified for AGN feedback:

• the quasar mode, or radiative mode with winds. This occurs when the AGN luminosity is close to the Eddington luminosity, for young quasars at high redshift essentially. Then the radiation pressure exerted on the ionized gas (with the Thomson cross-section) can drive an ionized wind. A similar effect can occur with the radiation pressure on dust, with a higher cross-section.

• the radio mode, or kinetic mode, due to radio jets. This occurs when the AGN luminosity is lower than 1/100 th of the Eddington luminosity, typically at low redshift. Massive galaxies, and early-type galaxies are frequently the host of radio-loud AGN. These low-luminosity AGN are radiatively inefficient (ADAF). In the low redshift universe, strong AGN feedback in the radio mode is observed frequently in cooling flows of galaxy clusters.

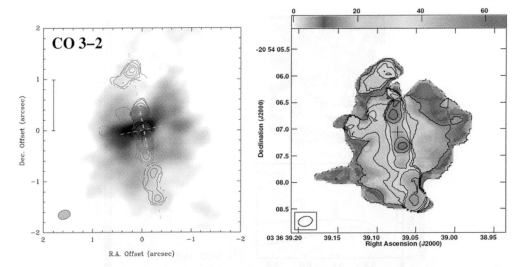

Figure 2. ALMA CO(3-2) map of the galaxy NGC 1377: Left, the grey-scale corresponds to the emission close to systemic velocity, while he high velocity (80 to 150km/s from V_{sys}) is indicated as red and blue for red-shifted and blue-shifted emission respectively. The vertical bar indicates a scale of 100 pc. The dashed line indicates the jet axis (and nuclear disk) orientations. Right: the velocity dispersion (moment 2) map, from 0 to 66 km/s. The cross indicates the position of the continuum peak at 345 GHz. From Aalto et al. (2016).

There is a radio-mode molecular outflow in the proto-typical Seyfert 2 galaxy NGC 1068. The nucleus is off-centered, and the radio jet is not perpendicular to the plane; therefore it is sweeping out some gas in the disk (Garcia-Burillo et al. 2014). The molecular outflow is estimated at $63 M_\odot$/yr, or 10 times the star formation rate in the central region.

In NGC 1068, the high resolution of ALMA has allowed the detection of a molecular torus, with both the CO(6 − 5) emission line, and with the dust emission at 432μm. The radius of the torus depends on the tracer, it is with CO(6 − 5) $5 - 6$ pc, but larger in low-J lines, like CO(2–1). The various tracers yield different aspects of the cold medium in the center. The CO disk is warped, and appears more inclined than the H_2O maser disk (Garcia-Burillo et al. 2016). It is possible that the circum-nuclear disk is unstable, through the Papaloizou-Pringle instability (Papaloizou & Pringle 1984).

The molecular torus is located just at the base of a polar dusty cone. The cone has been mapped in the near-infrared, with SPHERE on the VLT, and in particular the polarisation is revealing beautifully the conical struture (Gratadour et al. 2015).

The radio mode might also be at play in the lenticular galaxy NGC 1377, although no radio jet has been detected, and the galaxy is the most radio-quiet found, in terms of the radio-far-infrared correlation. A weak radio emission is detected at the center, but much weaker than expected from the well known correlation (Helou et al. 1985).

As shown in Fig. 2, ALMA has detected in this galaxy a very narrow molecular outflow (Aalto et al. 2016). The molecular outflow changes sign along the flow, on each side of the galaxy. This very surprising behaviour is very rare, and is interpreted as the precession of the jet, which happens to be very little inclined on the sky plane. Therefore a precession of 10° only is sufficient to reverse the sign of the flow velocity towards the observer. Such precession is observed in micro-quasars jets in the Milky Way, for instance SS433 (Mioduszewski et al. 2005).

The molecular gas in the cone is derived to be 10^8 $M_\odot$, and there is 10^7 $M_\odot$ in the outflow (Aalto et al. 2016). A model of a precessing molecular outflow has been

AGN fueling and feedback, from pc to kpc scale

Table 1. Radii, masses, and inclinations of the molecular tori

Galaxy	Radius (pc)	M(H$_2$) 10^7 M$_\odot$	inc(°) torus	PA(°) torus	inc(°) gal
NGC 613	14±3	3.9±1.4	46±7	0±8	36
NGC 1326	21±5	0.95±0.1	60±5	90±10	53
NGC 1365	26±3	0.74±0.2	27±10	70±10	63
NGC 1433	–	–	–	–	67
NGC 1566	24±5	0.88±0.1	12±12	30±10	48
NGC 1672	27±7	2.5±0.3	66±5	0±10	28
NGC 1808	6±2	0.94±0.1	64±7	65±8	84
NGC 1068	7±3	0.04±0.01	80±7	120±8	24

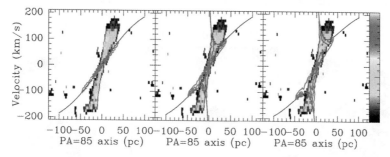

Figure 3. ALMA CO(3-2) observations of the galaxy NGC 1672. The 3 panels are the position-velocity diagrams in color, along the major axis of the molecular disk (or torus), with superposed in red, three models with different values of the central black hole mass (0, 2.5 and 5.0 10^7 M$_\odot$. The predicted circular velocity is reproduced in blue lines. From Combes et al. (2019).

found compatible with the data. The velocity of the flow, according to the model, ranges between 250 and 600km/s. The flow is launched at a distance from the center lower than 10pc. It cannot be driven by supernovae feedback, since there is no starburst in the galaxy, and the flow would not be so collimated. A radio jet must exist at a low level, or has existed in a recent past.

4. Molecular tori

When the high-angular resolution is available with ALMA, the frequency of detection of molecular tori in nearby Seyfert galaxies is quite high, 7 out of 8, as shown in the Table 1.

We call "molecular torus" the circum-nuclear molecular disk, of parsec scale, which is kinematically and spatially decoupled from the rest of the disk. It might not appear as a torus, except in favorable cases, like in NGC 1365. The molecular tori are located within the sphere of influence of the black holes, and can serve to measure their mass. The example of the Sy2 NGC 1672, where the torus is seen almost edge-on, is displayed in Fig. 3. The ALMA resolution is 3pc. The position-velocity diagram reveals a strong velocity gradient of about 180 km/s in 30pc.

A 3D modelisation of the dynamics of the nuclear disk has led to a black hole mass of 5 10^7 M$_\odot$ (Combes et al. 2019). The black hole mass determination from the gas kinematics is a precious method, for these low-luminosity AGN, which are late-type spiral galaxies frequently with a pseudo-bulge. For them the scaling relation between the central velocity dispersion and the mass of their black hole is quite scattered (e.g., Graham et al. 2011). This is not the case of more massive early-type galaxies, where the gas method has also been used with success (Davis et al. 2018).

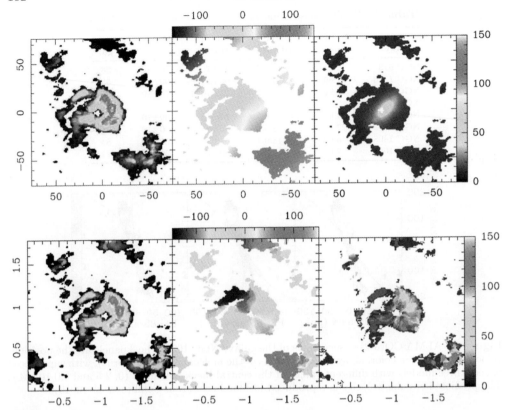

Figure 4. ALMA CO(3-2) observations of the galaxy NGC 1365. In each row, the three panels display the three first moments (intensity, velocity and dispersion). The velocity color scale is in km/s. The model corresponds to a black hole mass of $4\ 10^6$ M$_\odot$. Top is the model, and bottom the observations. The scale is spatial offset in parsec for the model, and in arcsec for the observations.. From Combes et al. (2019).

The 3D modelisation of the molecular torus dynamics for the Sy 1.8 galaxy NGC 1365 is displayed in Fig. 4. The black hole mass is $4\ 10^6$ M$_\odot$ (Combes et al. 2019).

One common feature to the detected molecular tori, is their random orientation and decoupling with the large-scale disk of their host, cf Table 1. This feature is not unexpected, given the very different time-scales at parsec scale and kpc-scales, and the almost spherical potential of the central galaxy at small scale. The material near the black hole will eventually inflow and fuel the AGN, and will be replaced by accreted gas, coming with different orientations and angular momenta. Numerical simulations show examples of central starbursts, where gas is ejected in a fountain through supernovae feedback, and may cool and fall back with random orientation, sometimes in a polar ring (Renaud et al. 2015; Emsellem et al. 2015).

Examples of nuclear disks decoupled from their host disks are frequently seen, as in the Milky-Way, where a circum-nuclear ring of 2-3pc in radius surrounds an almost face-on mini-spiral, or in NGC 4258, where anomalous arms (in fact the radio jet) are winding in the plane of the large-scale disks, with the normal spiral arms.

In the edge-on disc galaxy HE1353-1917, Husemann et al. (2019) have found a radio jet impinging the molecular gas of the host, and producing an outflow; in addition, [OIII] emission in a cone oriented at only $\sim 10°$ from the edge-on plane, reveals gas illuminated by the AGN, collimated by a highly inclined torus.

5. Summary

Recent high spatial resolution observations with ALMA have revealed the role played by dynamical features like bars to drive the gas at parsec scales, and fuel the AGN. The process is occurring in several steps, first from corotation, the gas is driven inwards, and piles up in a ring at the inner Lindblad resonance, at 100 pc-scale. Then, either through a nuclear bar, or through dissipation, the gas may be driven further in; there, within the sphere of influence of the black hole, it rapidly loses its angular momentum, settles in a 10pc-scale disk or torus and fuels the AGN.

Simultaneously, the AGN feedback through radiative or radio mode, according to the Eddington ratio, may drive molecular outflows, in a perpendicular direction. In some cases, the feedback is from the supernovae of a central starburst, in association (or not) to the AGN feedback. The AGN feedback can have a strong coupling with the gas in the host disk, due to the mis-alignment of the nuclear and large-scale disks.

Circum-nuclear disk, or molecular tori are frequently detected in nearby Seyferts, with a random orientation, kinematically decoupled from their host disk. This mis-alignment between small scales and large scales is due to random gas accretion, and different dynamical time-scales.

References

Aalto, S., Costagliola, F., Muller, S., et al. 2016, *A&A*, 590, A73
Asmus, D. 2019, *MNRAS*, 489, 2177
Asmus, D., Hönig, S. F., Gandhi, P., et al. 2016, *Ap. J.*, 822, 109
Audibert, A., Combes, F., Garcia-Burillo, S., et al. 2019, *A&A*, in press., arXiv:1905.01979
Busch, G., Eckart, A., Valencia-S, M., et al. 2017, *A&A*, 598, A55
Buta, R. & Combes, F. 1996, *Fund. Cosmic Phys.*, 17, 95
Combes, F., Garcia-Burillo, S., Casasola, V., et al. 2014, *A&A*, 565, A97
Combes, F., Garcia-Burillo, S., Audibert, A., et al. 2019, *A&A*, 623, A79
Contopoulos, G. & Papayannopoulos, T. 1980, *A&A*, 92, 33
Davis, T. A., Bureau, M., Onishi, K., et al. 2018, *MNRAS*, 473, 3818
Denney, K. D., De Rosa, G., Croxall, K., et al. 2014, *Ap. J.*, 796, 134
Emsellem, E., Renaud, F., Bournaud, F., et al. 2015, *MNRAS*, 446, 2468
Garcia-Burillo, S. & Combes, F. 2012, *JPhCS*, 372, a2050
Garcia-Burillo, S., Combes, F., Usero, A., et al. 2014, *A&A*, 567, A125
Garcia-Burillo, S., Combes, F., Ramos Almeida, C., et al. 2016, *Ap. J.*, 823, L12
Graham, A. W., Onken, C. A., Athanassoula, E., & Combes, F. 2011, *MNRAS*, 412, 2211
Gratadour, D., Rouan, D., Grosset, L., et al. 2015, *A&A*, 581, L8
Helou, G., Soifer, B. T., Rowan-Robinson, M., et al. 1985, *Ap. J.*, 298, L7
Hönig, S. F., 2019, *Ap. J.*, 884, 171
Husemann, B., Scharwächter, J., Davis, T. A., et al. 2019, *A&A*, 627, A53
Krips, M., Neri, R., Garcıa-Burillo, S., et al. 2008, *Ap. J.*, 677, 26
McElroy, R. E., Husemann, B., Croom, S. M., et al. 2016, *A&A*, 593, L8
Mioduszewski A. J., Dhawan, V., Rupen, M. P., et al. 2005, *ASPC*, 340, 281
Papaloizou, J. C. B. & Pringle, J. E. 1984, *MNRAS*, 298, 721
Renaud, F., Bournaud, F., Emsellem, E., et al. 2015, *MNRAS*, 454, 3299
Salak, D, Nakai, N., Hatakeyama, T., Miyamoto, Y., et al. 2016, *Ap. J.*, 823, 68
Urry, C. M. & Padovani, P. 1995, *PASP*, 107, 803
Usero, A., Garcıa-Burillo, S., Fuente, A., et al. 2004, *A&A*, 419, 897

Feedback from supermassive and intermediate-mass black holes at galaxy centers using cosmological hydrodynamical simulations

Paramita Barai

Instituto de Astronomia, Geofísica e Ciências Atmosféricas – Universidade de São Paulo (IAG-USP), Rua do Matão 1226, São Paulo, 05508-090, Brasil
email: paramita.barai@iag.usp.br

Abstract. Accretion of matter onto central Black Holes (BHs) in galaxies liberates enormous amounts of feedback energy, which affects the environment from pc to Mpc scales. These BHs are usually Supermassive BHs (SMBHs: mass $\geqslant 10^6 M_\odot$) existing at the centers of active galactic nuclei (AGN), which are widely observed through their multi-wavelength emission at all cosmic epochs. Relatively recently, Intermediate-Mass BHs (IMBHs: mass $= 100 - 10^6 M_\odot$) have started to be observed hosted in Dwarf Galaxy (DG) centers. Some of the central IMBHs in DGs show signatures of activity in the form of low-luminosity AGN. We have performed Cosmological Hydrodynamical Simulations to probe SMBHs in high-z quasars (Barai *et al.* 2018), and IMBHs in DGs (Barai & de Gouveia Dal Pino 2019). Our simulations employ the 3D TreePM SPH code GADGET-3, and include metal cooling, star formation, chemical enrichment, stellar evolution, supernova feedback, AGN accretion and feedback. Analyzing the simulation output in post-processing, we investigate the growth of the first IMBHs and the first SMBHs, as well as their impact on star-formation.

1. Introduction

Active galactic nuclei (AGN) emit enormous amounts of energy powered by the accretion of gas onto their central supermassive black holes (SMBHs) (e.g., Rees 1984). Feedback from AGN are believed to strongly influence the formation and evolution of galaxies (e.g., Richstone *et al.* 1998). A strong manifestation of AGN feedback are AGN outflows observed in a wide variety of forms (e.g., Crenshaw, Kraemer & George 2003).

Quasars are very powerful AGN existing more commonly at high-z than in the local Universe (e.g., Fan 2006). In the host galaxy of the quasar SDSS J1148 + 5251 at $z = 6.4$, Maiolino *et al.* (2012) detected broad wings of the [CII] line tracing a massive outflow with velocities up to ± 1300 km/s. The physical mechanisms by which quasar outflows affect their host galaxies remain as open questions. SMBHs of mass $\geqslant 10^9 M_\odot$ are observed to be in place in luminous quasars by $z \sim 6$, when the Universe was less than 1 Gyr old (e.g., Wu *et al.* 2015). It is difficult to understand how these early SMBHs formed over such short time-scales, and there are open issues with various plausible scenarios (e.g., Matsumoto *et al.* 2015).

Black holes are usually observed to belong to two populations: stellar-mass ($M_{\rm BH} \leqslant 10 - 100 M_\odot$) BHs, and supermassive ($M_{\rm BH} \geqslant 10^6 M_\odot$) BHs. By natural extension, there should be a population of Intermediate-Mass Black Holes (IMBHs: with mass between $100 - 10^6 M_\odot$) in the Universe. Analogous to SMBHs producing AGN feedback,

Table 1. Zoomed-In Cosmological Hydrodynamical Simulations (for SMBHs)

Run name	AGN feedback algorithm	Reposition of BH to potential-minimum	Geometry of region where feedback is distributed	Half opening angle of effective cone
noAGN	No BH	–	–	–
AGNoffset	Kinetic	No	Bi-Cone	45°
AGNcone	Kinetic	Yes	Bi-Cone	45°
AGNsphere	Kinetic	Yes	Sphere	90°

the IMBHs should also have feedback. AGN feedback mechanism has recently started to be observed in low-mass galaxies (e.g., Marleau et al. 2017; Penny et al. 2017). The concordance ΛCDM cosmological scenario of galaxy formation presents multiple challenges in the dwarf galaxy mass range: e.g. core versus cusp density profile, number of DGs. Recently Silk (2017) made an exciting claim that the presence of IMBHs at the centers of Dwarf Galaxies (DGs) can potentially solve the problems. Early feedback from these IMBHs output energy and affect the gas-rich DGs at $z = 5-8$, can quench star-formation and reduce the number of DGs.

In this work we present results of the growth and feedback of SMBHs in AGN and IMBHs in DGs. We focus on negative BH feedback effects where star-formation is quenched (e.g., Scannapieco, Silk & Bouwens 2005; Schawinski et al. 2006).

2. Numerical Method and Simulations

The initial conditions at $z = 100$ are generated using the MUSIC† software (Hahn & Abel 2011). We use a modified version of the TreePM (particle mesh) – SPH (smoothed particle hydrodynamics) code GADGET-3 (Springel 2005) to perform our cosmological hydrodynamical simulations.

BHs are collisionless sink particles (of mass M_{BH}) in our simulations. A BH (of initial mass M_{BHseed}) is seeded at the center of each galaxy more massive than a total mass $M_{HaloMin}$, which does not contain a BH already. We test different values of minimum halo mass and seed BH mass in the range: $M_{HaloMin} = (10^6 - 10^7) M_\odot$, and $M_{BHseed} = (10^2 - 10^3) M_\odot$. The sub-resolution prescriptions for gas accretion onto BHs and *kinetic* feedback are adopted from (Barai et al. 2014, 2016).

We execute 4 Zoomed-In cosmological hydrodynamical simulations, with characteristics listed in Table 1, all the runs incorporating metal cooling, chemical enrichment, SF and SN feedback. The first run has no AGN included, while the latter three explore different AGN feedback models. We perform cosmological hydrodynamical simulations of small-sized boxes with periodic boundary conditions, to probe dwarf galaxies at high redshifts. We execute 10 simulations, with characteristics listed in Table 2.

3. Results and Discussion

3.1. *Black Hole Accretion and Growth*

The redshift evolution of the most-massive SMBH mass in the three AGN runs of the zoomed-in cosmological simulations is plotted in *Fig. 1 - left panel*. Each SMBH starts as a seed of $M_{BH} = 10^5 M_\odot$, at $z \sim 14$ in the runs *AGNcone* and *AGNsphere* ($z \sim 10$ in *AGNoffset*). The subsequent growth is due to merger with other SMBHs and gas accretion. The dominant mode of SMBH growth occurs over the redshifts $z = 9 - 6$ in runs *AGNcone* and *AGNsphere*, corresponding to Eddington-limited gas accretion where Eddington ratio = 1. The $\dot{M}_{BH}$ has a power-law increase, and the SMBH mass increases

† MUSIC – Multi-scale Initial Conditions for Cosmological Simulations: https://bitbucket.org/ohahn/music

Table 2. Periodic-Box Cosmological Hydrodynamical Simulations (for IMBHs)

Run name	BH present	Min. Halo Mass for BH Seeding, $M_{\rm HaloMin}[M_\odot]$	Seed BH Mass, $M_{\rm BHseed}[M_\odot]$	BH kinetic feedback kick velocity v_w (km/s)
SN	No	–	–	–
BHs2h1e6	Yes	$h^{-1} \times 10^6$	10^2	2000
BHs2h7e7	Yes	$5h^{-1} \times 10^7$	10^2	2000
BHs3h1e7	Yes	1×10^7	10^3	2000
BHs3h2e7	Yes	2×10^7	10^3	2000
BHs3h3e7	Yes	3×10^7	10^3	2000
BHs3h4e7	Yes	4×10^7	10^3	2000
BHs3h4e7v5	Yes	4×10^7	10^3	5000
BHs3h5e7	Yes	5×10^7	10^3	2000
BHs4h4e7	Yes	4×10^7	10^4	2000

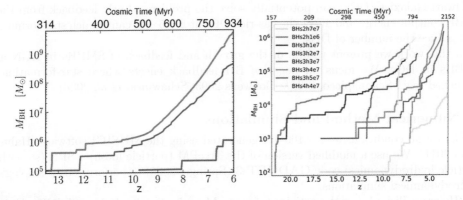

Figure 1. BH mass growth with redshift of the most-massive BH in each run. Left panel: Zoomed-In cosmological simulations showing growth of SMBHs. The different colours discriminate the various runs: *AGNoffset* – violet, *AGNcone* – red, *AGNsphere* – green. Right panel: Periodic-Box cosmological simulations showing growth of IMBHs. The colours indicate the runs: *BHs2h7e7* – cyan, *BHs2h1e6* – red, *BHs3h1e7* – indigo, *BHs3h2e7* – green, *BHs3h3e7* – magenta, *BHs3h4e7* – brown, *BHs3h5e7* – blue, *BHs4h4e7* – yellow.

by a factor $\sim 10^3$. The final properties reached at $z = 6$ depends on the simulation; e.g. $M_{\rm BH} = 4 \times 10^9 M_\odot$ and $\dot{M}_{\rm BH} = 100 M_\odot$/yr in run *AGNcone* (red curve). There is variability of the $\dot{M}_{\rm BH}$, whereby it fluctuates by a factor of up to 100. The SMBH grows 10 times more massive at $z = 6$ in the *AGNcone* case than in the *AGNsphere* run. This is because more gas can inflow along the perpendicular direction to the bi-cone, and accrete onto the SMBH.

We find that first IMBHs are seeded at different cosmic times depending on the value of minimum halo mass for BH seeding, $M_{\rm HaloMin}$. The seeding epoch varies between $z \sim 22$ to $z \sim 16$ in our periodic-box cosmological simulations, when the first halos reach $M_{\rm halo} = h^{-1} \times 10^6 M_\odot$ to $M_{\rm halo} = 5 \times 10^7 M_\odot$. The redshift evolution of the most-massive IMBH mass in these periodic-box simulation runs is plotted in *Fig. 1 – right panel*. Each IMBH starts from an initial seed of $M_{\rm BH} = 10^2 M_\odot$ in the runs named *BHs2**, $10^3 M_\odot$ in the runs named *BHs3**, and $10^4 M_\odot$ in the runs named *BHs4**. The subsequent mass growth is due to merger with other IMBHs (revealed as vertical rises in $M_{\rm BH}$), and gas accretion (visualized as the positive-sloped regions of the $M_{\rm BH}$ versus z). The most-massive BH, considering all the runs, has grown to $M_{\rm BH} = 2 \times 10^6 M_\odot$ at $z = 5$ in run *BHs3h4e7* (brown curve in Fig. 1 – right panel).

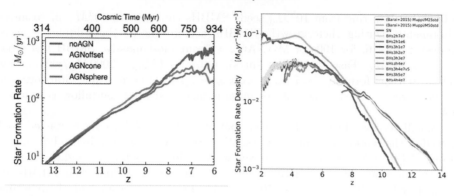

Figure 2. Left panel: Sum total star formation rate (in $M_\odot yr^{-1}$) as a function of redshift, in the Zoomed-In cosmological simulations. Right panel: Total star formation rate density (in $M_\odot yr^{-1} Mpc^{-3}$, total SFR integrated over simulation volume) as a function of redshift, in the Periodic-Box cosmological simulations.

3.2. Star Formation

Stars form in the simulation volume from cold dense gas. The star formation rate (total SFR in the simulation box) versus redshift of the four zoomed-in cosmological simulations is displayed in *Fig. 2 – left panel*. The SFR rises with time in all the runs initially, and continues to increase in the *noAGN* case without a SMBH. The SFR in run *AGNoffset* is almost similar to that in the run *noAGN*, because the SMBHs are too small there to generate enough feedback. A similar outcome happens in the runs *AGNcone* and *AGNsphere* at $z \geqslant 8$, when the SMBHs are too small.

The models suppress SF substantially from $z \sim 8$ onwards, when the SMBHs have grown massive and generate larger feedback energy. Thus, we find that SMBHs need to grow to $M_{BH} > 10^7 M_\odot$, in order to suppress star-formation, even in massive galaxies (of $M_\star = 4 \times 10^{10} M_\odot$, and specific-SFR $= 5 \times 10^{-9}$ yr^{-1}).

The Star Formation Rate Density (SFRD in units of $M_\odot yr^{-1} Mpc^{-3}$, counting stars forming in the whole simulation box) versus redshift of the periodic-box cosmological simulation runs is displayed in *Fig. 2 – right panel*. The SFRD rises with time in the *SN* run (blue curve) initially from $z \sim 15$, reaches a peak at $z \sim 4$ (the peak epoch of star-formation activity in the Universe), and decreases subsequently over $z \sim 4 - 2$. The presence of a IMBH quenches star formation by accreting some gas in, ejecting some gas out of the halo as outflows, and/or heating the gas. The models suppress SF substantially from $z \sim 8$ onwards, when the IMBHs have grown massive. We find that IMBHs need to grow to $M_{BH} > 10^5 M_\odot$, in order to suppress star-formation.

The red curve (run *BHs2h1e6*) already quenches SF as early as $z \sim 8$. This is because the IMBH has already grown to $M_{BH} \sim 5 \times 10^5 M_\odot$ at that epoch, more massive than all the other runs. As another example, the brown (run *BHs3h4e7*) and royal-blue (run *BHs3h5e7*) curves quench SF from $z \sim 4.5$ to $z \sim 3.5$. This is the epoch when the IMBH masses in these runs increase from $M_{BH} = 10^5 M_\odot$ to $M_{BH} = 10^6 M_\odot$ (as can be seen from Fig. 1 – right panel).

4. Conclusions

Gas accretion onto central supermassive black holes of active galaxies and resulting energy feedback, often manifested as AGN outflows, is an important component of galaxy evolution. We investigate outflows in quasar-host galaxies at $z \geqslant 6$ by performing cosmological hydrodynamical simulations. We simulate the $2R_{200}$ region around a $2 \times 10^{12} M_\odot$ halo at $z = 6$, inside a $(500 \text{ Mpc})^3$ comoving volume, using the zoomed-in technique.

We find that, starting from $10^5 M_\odot$ seeds SMBHs can grow to $10^9 M_\odot$ in cosmological environments. During their growth, SMBHs accrete gas at the Eddington accretion rate over $z = 9-6$, for 100s of Myr. At $z = 6$, our most-massive SMBH has grown to $M_{\rm BH} = 4 \times 10^9 M_\odot$. Fast ($v_r > 1000$ km/s), powerful ($\dot{M}_{\rm out} \sim 2000 M_\odot/{\rm yr}$) outflows of shock-heated gas form at $z \sim 7$, and propagate up to hundreds kpc. Star-formation is quenched over $z = 8-6$. The outflow mass is increased (and the inflow is reduced) by $\sim 20\%$.

Intermediate-mass black holes (mass between $100 - 10^6 M_\odot$) have started to been observed at the centers of dwarf galaxies. We perform cosmological hydrodynamical simulations of $(2h^{-1}$ Mpc$)^3$ comoving boxes with periodic boundary conditions, to probe dwarf galaxies and central IMBHs at high redshifts. We conclude that IMBHs at DG centers grow from $10^2 - 10^3 M_\odot$ to $10^5 - 10^6 M_\odot$ by $z \sim 4$ in cosmological environments. Star formation is quenched when the IMBHs have grown to $M_{\rm BH} > 10^5 M_\odot$. Our conclusions, based on numerical simulation results, support the phenomenological ideas made by Silk (2017). IMBHs at the centers of dwarf galaxies can be a strong source of feedback to quench star-formation and generate outflows. At the same time, these IMBHs form the missing link between stellar-mass and SMBHs.

Acknowledgements

This work is supported by the Brazilian Funding Agency FAPESP (grants 2016/01355-5 and 2016/22183-8).

References

Barai, P., Viel, M., Murante, G., Gaspari, M., & Borgani, S. 2014, *MNRAS*, 437, 1456
Barai, P., Murante, G., Borgani, S., Gaspari, M., Granato, G. L., Monaco, P., & Ragone-Figueroa, C. 2016, *MNRAS*, 461, 1548
Barai, P., Gallerani, S., Pallottini, A., Ferrara, A., Marconi, A., Cicone, C., Maiolino, R., & Carniani, S. 2018, *MNRAS*, 473, 4003
Barai, P. & de Gouveia Dal Pino, E. M. 2019, *MNRAS*, 487, 5549
Crenshaw, D. M., Kraemer, S. B., & George, I. M. 2003, *ARA&A*, 41, 117
Fan, X. 2006, *NewAR*, 50, 665
Hahn, O. & Abel, T. 2011, *MNRAS*, 415, 2101
Maiolino, R. *et al.* 2012, *MNRAS*, 425, L66
Marleau, F. R., Clancy, D., Habas, R., & Bianconi, M. 2017, *A&A*, 602, A28
Matsumoto, T., Nakauchi, D., Ioka, K., Heger, A., & Nakamura, T. 2015, *ApJ*, 810, 64
Penny, S. J. *et al.* 2017, submitted to *MNRAS*, eprint arXiv:1710.07568
Rees, M. J. 1984, *ARA&A*, 22, 471
Richstone, D. *et al.* 1998, *Nature*, 395, A14
Scannapieco, E., Silk, J., & Bouwens, R. 2005, *ApJ*, 635, L13
Schawinski, K. *et al.* 2006, *Nature*, 442, 888
Silk, J. 2017, *ApJ*, 839, L13
Springel, V. 2005, *MNRAS*, 364, 1105
Wu, X.-B. *et al.* 2015, *Nature*, 518, 512

Radiative feedback of low-$L_{\rm bol}/L_{\rm Edd}$ AGNs

Fu-Guo Xie

Key Laboratory for Research in Galaxies and Cosmology, Shanghai Astronomical Observatory,
80 Nandan Road, Shanghai 200030, China
email: fgxie@shao.ac.cn

Abstract. AGN feedback, through either radiation or kinematics by expelled medium, plays a crucial role in the coevolution of supermassive black hole (SMBH) and its host galaxy. The nuclei spend most of their time as low-luminosity AGNs (LLAGNs), whose spectra are distinctive to bright AGNs, and the feedback is the hot mode (also named kinetic mode). We thus investigate the radiative heating in the hot mode. We calculate the value of "Compton temperature" $T_{\rm C}$, which defines the heating capability of the radiation at given flux, and find that $T_{\rm C} \sim (5-15) \times 10^7$ K, depending on the spectrum of individual LLAGNs. This work provides a cheap way to include the radiative heating of LLAGNs in the study of AGN feedback.

Keywords. galaxies: Seyfert — galaxies: evolution — accretion, accretion disks

1. Introduction

The co-evolution of the supermassive black hole and its host galaxy is now widely believed to be due to the feedback by the active galactic nuclei (AGNs) (for reviews, Fabian 2012; Kormendy & Ho 2013; Heckman & Best 2014). Accretion onto the supermassive black hole in the galactic center will possibly produce three ingredients, i.e. jet, wind and radiation. These outputs will interact with the interstellar medium (ISM) in the host galaxy, by transferring their momentum and energy to the ISM. The gas will then be heated up or pushed away from the black hole. On one hand, the changes in the temperature and density of the gas will affect the star formation and galaxy evolution. On the other hand, they will also affect the fueling of the black hole by changing the accretion rate, thus the radiation and matter output of accretion and the growth of the black hole mass.

Two feedback modes have been identified, which correspond to two accretion modes (Fabian 2012; Kormendy & Ho 2013; Yuan et al. 2018). One is called the cold mode (or more commonly the radiative mode), which operates when the black hole accretes at a significant fraction of the Eddington rate. In this case, the accretion flow is a cold geometrically thin disk and the corresponding AGNs are luminous. The other mode is called the hot mode (or more commonly the kinetic mode), when the black hole accretes at a low accretion rate. In this case, the accretion flow is described by a hot accretion flow (Yuan & Narayan 2014). The corresponding AGNs are called low-luminosity AGNs (LLAGNs). This is directly analogy with the soft and hard states of black hole X-ray binaries (BHBs), where the boundary between the two modes is $L_{\rm bol} \sim (1-2)\% L_{\rm Edd}$, where $L_{\rm bol}$ is the bolometric luminosity and $L_{\rm Edd}$ is the Eddington luminosity.

In the hot mode of feedback, all three types of output exist. Among them, jet is perhaps most widely considered in the study, mainly because observationally jets are most evident (e.g., Ho 2008). The wind is also considered (e.g., Ciotti et al. 2010), which helps to rapid reddening of moderately massive galaxies without expelling too many baryons (Weinberger et al. 2017).

The radiative heating in hot feedback mode of LLAGNs is ignored by most previous work (but see, e.g., Ciotti & Ostriker 2001; Ciotti et al. 2010; Ostriker et al. 2010; Gan et al. 2014, 2019). However, we argue it is an oversimplification based on the following reasons. First, the luminosity of a hot accretion flow covers a very wide range depending on the accretion rate and can be moderately high. For example, the radiative efficiency of hot accretion flow can be fairly high (Xie & Yuan 2012), with the highest luminosity of hot accretion flow can be $L_{\rm bol} \sim (2-10)\%$ $L_{\rm Edd}$ (e.g., Done et al. 2007 for the case of BHBs in hard state). Second, the spectrum of LLAGNs is distinctive to that of luminous AGNs (Ho 2008), i.e. LLAGNs lack the big blue bump and LLAGNs are more prominent in X-rays. LLAGNs are thus expected to be more effective in radiative heating than bright AGNs (assuming the same bolometric luminosity), see Equation 1.1 below.

Radiative heating mainly includes two processes, one is through photoionization and the other (the focus of this work) is through Compton scattering. The Compton heating rate per unit volume can be evaluated as,

$$q_{\rm Comp} = n^2 \frac{n_e}{n} \frac{k\sigma_T}{\pi m_e c^2} \frac{L_{\rm bol}}{nR^2} (T_{\rm C} - T_e). \tag{1.1}$$

Here all the symbols are of their normal meanings, $L_{\rm bol}/4\pi R^2$ is the radiative intensity at distance R. The "Compton temperature" $T_{\rm C}$ is determined by the energy-weighted average energy of the emitted photons from LLAGNs (Sazonov et al. 2004). The Compton temperature of typical luminous AGNs is $T_{\rm C} \approx 2 \times 10^7$ K (Sazonov et al. 2004).

We calculate the value of $T_{\rm C}$ of LLAGNs. For this aim, we in §2 combine the data from the literature to obtain the broadband spectral energy distribution (SED) of LLAGNs. Special attention will be paid to the hard X-ray spectrum since this is the most important part in the spectrum for heating. The Compton temperature and its applications are given in §3. The final section is devoted to discussions and a short summary.

2. Broadband SED of LLAGNs

It is quite challenging to obtain the "average" broadband SED of LLAGNs. Various sample selection and normalization methods have been developed in literature (e.g. Ho 2008; Winter et al. 2009; Eracleous et al. 2010). As shown in Fig. 1, we adopt the composite SED of LLAGNs from Ho (2008), which has a relatively broad coverage in photon energy, i.e. from radio to soft X-rays ($E \lesssim 10$ keV). We include three sets of SEDs with different ranges of Eddington ratio $\lambda \equiv L_{\rm bol}/L_{\rm Edd}$, i.e. $\lambda < 10^{-3}$, $10^{-3} < \lambda < 10^{-1}$, and $10^{-1} < \lambda < 1$. For comparison, the composite SED averaged over Type 1 and Type 2 AGNs compiled by Sazonov et al. (2004) is also shown here by the black solid curve.

The spectrum from the hard X-ray to soft γ-ray regime ($E = h\nu > 10$ keV, where ν is the frequency), crucial for the evaluation of $T_{\rm C}$, is absent in these composite SED data. We thus complete the SED of this energy range through a power law with an exponential cutoff (named the "cutoff PL"), i.e. $F_E \propto E^{1-\Gamma} \exp(-E/E_c)$, where Γ is the photon index and E_c is the cutoff energy. Γ is better constrained. Both hot accretion theory and observations suggest that the X-ray emission becomes softer as it becomes fainter, i.e. LLAGNs follow a negative $\Gamma - L_X/L_{\rm Edd}$ correlation, see Emmanoulopoulos et al. 2012; Soldi et al. 2014; Yang et al. (2015); Connolly et al. (2016).

The cutoff energy E_c, on the other hand, is poorly constrained. This is because the sensitivity of most existing hard X-ray telescopes/instruments are not high enough to probe the low-luminosity AGNs. There are only limited sources with such measurements, where a negative $E_c - L_X/L_{\rm Edd}$ relationship is reported (for BHBs in hard state, see e.g., Miyakawa et al. 2008; for LLAGNs, see e.g., Ursini et al. 2016; Xie et al. 2017; Ricci et al. 2018). The game changer is *NuSTAR*, whose spectral resolution is also sufficiently high. Recent measurement on E_c by *NuSTAR* includes Pahari et al. (2017); Zoghbi et al. (2017);

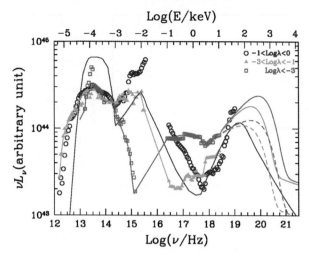

Figure 1. Composite SED of LLAGNs. We include three sets of SEDs with different ranges of Eddington ratio, i.e. $\lambda < 10^{-3}$ (red), $10^{-3} < \lambda < 10^{-1}$ (sky blue), and $10^{-1} < \lambda < 1$ (blue). For comparison, the composite SED averaged over Type 1 and Type 2 AGNs compiled by Sazonov et al. (2004) is also shown by the black solid curve. Above 10 keV the SED is completed by a power-law with exponential cutoff. Taken from Xie et al. (2017).

Rani et al. (2019); Younes et al. (2019). With existing data, we may crudely suggest that LLAGNs have E_c 300 – 500 keV. Note that this value is somewhat smaller than that adopted in Xie et al. (2017). We complete the SED of this energy range through the cutoff PL (normalized at $E = 10$ keV), based on our knowledge of Γ and E_c. Additional weak tail due to jet emission above $2E_c$ is also taken into account. The final SED used in this work is shown in Figure 1.

One additional key uncertainty in the composite SED is the origin of infrared (IR) emission, i.e. it may come from the dusty torus, the circum-nuclear star formation, the central AGN (including the accretion disk, the jet, and sometimes the narrow-line emission clouds), or their combination. It is obviously difficult to discriminate the possible contaminations. Observations with high spatial resolution and sufficient sensitivity are crucial. Over past decades, extensive efforts have been made through infrared interferometric techniques (e.g., Asmus et al. 2011, 2014; González-Martí et al. 2015, 2017), but the contaminations are still difficult to constrain (e.g. Asmus et al. 2011, 2014).

One key progress in IR observations is that, the nuclear IR flux derived from arcsecond-scale resolution observations (e.g. typical resolution in mid-IR of *Spitzer* is $\sim 4''$) may be accurate within a factor of $\lesssim 2 - 8$, and the fraction of nuclei IR emission decreases with decreasing Eddington ratio λ (Asmus et al. 2011, 2014; González-Martí et al. 2017). This result is fairly robust.

3. Compton temperature of LLAGNs

In general, the photon energy from AGNs and/or the electrons to be scattered can be comparable to or even larger than $m_e c^2$. In this case, the Compton scattering process becomes complicated, and simplified Comptonization model may result in estimations that are inaccurate by orders-of-magnitude. Moreover, because of the strong coupling between electrons and photons, there is no exact definition of Compton temperature. We thus follow Guilbert (1986) to derive the accurate Compton heating rate and then use Equation 1.1 to evaluate the "effective" T_C.

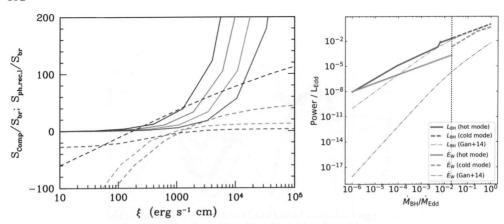

Figure 2. *Left panel:* Radiative heating/cooling versus ionization parameter ξ (solid for $S_{\rm comp}/S_{\rm br}$; dashed for $S_{\rm ph,rec,l}/S_{\rm br}$), from Xie et al. (2017). We set $T_{\rm C} = 1 \times 10^8$ K. The color of each curve represents the temperature of electrons, i.e. $\log(T_{\rm e}) = 4.5$ (black), 5.0 (red), 5.5 (green), and 6.0 (blue). *Right panel:* Power of feedback of different components, taken from Yuan et al. (2018).

We first test our calculations for the $T_{\rm C}$ of the composite SED of bright AGNs by Sazanov et al. (2004), where good consistency is observed. We then apply our numerical calculations to other composite SEDs by Ho (2008), and find that $T_{\rm C} \approx (5-9) \times 10^7$ K. Obviously, harder spectrum will higher $E_{\rm c}$ will result in higher $T_{\rm C}$. The dependence on electron temperature is weak, as long as these electrons are non-relativistic.

If we further consider the case of reducing IR emission by a factor of 10, then we have $T_{\rm C} \approx (1-2) \times 10^8$ K, i.e. approximately a factor of ~ 2 higher than that of normal IR case. Note that this value should be applied for the gas heating within the torus (i.e. at a distance $\lesssim 1$ kpc to the SMBH).

4. Summary and applications

The Compton scattering in principle plays a heating or cooling role, depending on the comparison between the photon energy and electron temperature of the gas. We adopt the composite SED of LLAGNs with different luminosities and calculate the "Compton temperature" $T_{\rm C}$. We find that $T_{\rm C}$ of LLAGNs is $\approx (5-9) \times 10^7$ K outside of torus and $\approx (1-2) \times 10^8$ K within it. This value is about 3–10 times that of bright AGNs.

In general there are three radiation sources. Among them, the Compton scattering ($S_{\rm Comp}$) usually plays a heating role, and the bremsstrahlung ($S_{\rm br}$) is always a cooling term. The photo-ionization, recombination and lines ($S_{\rm ph,rec,l}$), on the other hand, can be either a heating or a cooling term, depending on both the temperature and the ionization parameter ξ ($\xi = \frac{L_{\rm ion}}{nR^2}$, where the ionization luminosity is for photon energy between 13.6 eV and 13.6 keV.). In a given galaxy, ξ decreases with R (e.g., Tombesi et al. 2013). The left panel of Figure 2 illustrates these heating/cooling terms as a function of ξ, where we find that, depending on the electron temperature $T_{\rm e}$, the Compton heating dominates for $\xi \gtrsim 3 \times 10^2$ erg s^{-1} cm.

The new measurement of $T_{\rm C}$ for the Compton heating has been applied in numerical simulations of AGN feedback in isolated elliptical galaxies over cosmic time (e.g., Yuan et al. 2018; Gan et al. 2019; Yoon et al. 2018), or in the investigation of AGN heating on small scales (e.g., Bu & Yang 2019a,b). In order to understand its relative importance, we show in the right panel of Figure 2 the power of different mechanisms as a function of accretion rate. The radiative heating ($L_{\rm BH}$ in the plot) dominates over the kinematic power ($\dot{E}_{\rm w}$ in the plot) in the hot accretion mode (Yuan et al. 2018; Gan et al. 2019; Yoon et al. 2019).

References

Asmus, D., Gandhi, P., Smette, A., Hönig, S. F., Duschl, W. J., et al. 2011, *A&A*, 536, 36
Asmus, D., Hönig, S. F., Gandhi, P., Smette, A., Duschl, W. J., et al. 2014, *MNRAS*, 439, 1648
Bu, D. & Yang, X. 2019a, *MNRAS*, 484, 1724
Bu, D. & Yang, X. 2019b, *ApJ*, 871, 138
Ciotti, L. & Ostriker, J. P. 2001, *ApJ*, 551, 131
Ciotti, L., Ostriker, J. P., & Proga, D., et al. 2010, *ApJ*, 717, 708
Connolly, S. D., McHardy, I. M., Skipper, C. J., Emmanoulopoulos, D., et al. 2016, *MNRAS*, 459, 3963
Done, C., Gierliński, M., & Kubota, A. 2007, *A&Ap. Rev.*, 15, 1
Emmanoulopoulos, D., Papadakis, I. E., McHardy, I. M., Arévalo, P., Calvelo, D. E., Uttley, P., et al. 2012, *MNRAS*, 424, 1327
Eracleous, M., Hwang, J. A., Flohic, H. M. L., et al. 2010, *ApJS*, 187, 135
Fabian, A. C. 2012, *ARA&A*, 50, 455
Gan, Z., Yuan, F., Ostriker, J. P., Ciotti, L., & Novak, G. S. 2014, *ApJ*, 789, 150
Gan, Z., Ciotti, L., Ostriker, J. P., Yuan, F., et al. 2019, *ApJ*, 872, 167
Gaspari, M., Ruszkowski, M., & Oh, S. P. 2013, *MNRAS*, 432, 3401
González-Martín, O., Masegosa, J., Hernán-Caballero, A., et al. 2017, *ApJ*, 841, 37
González-Martín, O., Masegosa, J., Márquez, I., et al. 2015, *A&A*, 578, 74
Guilbert, P. W. 1986, *MNRAS*, 218, 171
Heckman, T. M. & Best, P. N. 2014, *ARA&A*, 52, 589
Ho, L. C. 2008, *ARA&A*, 46, 475
Kormendy, J. & Ho, L. C. 2013, *ARA&A*, 51, 511
Miyakawa, T., Yamaoka, K., Homan, J., et al. 2008, *PASJ*, 60, 637
Ostriker, J. P., Choi, E., Ciotti, L., et al. 2010, *ApJ*, 722, 642
Pahari, M., McHardy, I. M., Mallick, L., Dewangan, G. C., Misra, R. et al. 2017, *MNRAS*, 470, 3239
Rani, P., Stalin, C. S., & Goswami, K. D. 2019, *MNRAS*, 484, 5113
Ricci, C., Ho, L. C., Fabian, A. C., et al. 2018, *MNRAS*, 480, 1819
Sazonov, S. Y., Ostriker, J. P., Sunyaev, R. A., et al. 2004, *MNRAS*, 347, 144
Soldi, S., Beckmann, V., Baumgartner, W. H., et al. 2014, *A&A*, 563, 57
Tombesi, F., Cappi, M., Reeves, J. N., et al. 2013, *MNRAS*, 430, 1102
Ursini, F., Petrucci, P.-O., Matt, G., et al. 2016, *MNRAS*, 463, 382
Weinberger, R., Springel, V., Herquist, L., et al. 2017, *MNRAS*, 465, 3291
Winter, L. M., Mushotzky, R. F., Reynolds, C. S., Tueller, J., et al. 2009, *ApJ*, 690, 1322
Xie, F. & Yuan, F. 2012, *MNRAS*, 427, 1580
Xie, F., Yuan, F., & Ho, L. C., et al. 2017, *ApJ*, 844, 42
Yang, Q. X., Xie, F. G., Yuan, F., et al. 2015, *MNRAS*, 447, 1692
Yoon, D., Yuan, F., Gan, Z., et al. 2018, *ApJ*, 864, 6
Yoon, D., Yuan, F., Ostriker, J. P., et al. 2019, *ApJ*, 885, 16
Younes, G., Ptak, A., Ho, L. C., et al. 2019, *ApJ*, 870, 73
Yuan, F. & Narayan, R. 2014, *ARA&A*, 52, 529
Yuan, F., Yoon, D., Li, Y., et al. 2018, *ApJ*, 857, 121
Zoghbi, A., Matt, G., Miller, J. M., et al. 2017, *ApJ*, 836, 2

The impact of AGN on the life of their host galaxies at z ∼ 2

Chiara Circosta[1,2]

[1] European Southern Observatory, Karl-Schwarzschild-Str. 2,
85748 Garching bei München, Germany

[2] Dept. of Physics & Astronomy, University College London, Gower Street,
London WC1E 6BT, United Kingdom
email: c.circosta@ucl.ac.uk

Abstract. Feedback from active galactic nuclei (AGN) is thought to be key in shaping the life cycle of host galaxies by regulating star formation. Therefore, measuring the molecular gas reservoir out of which stars form is essential to understand the impact of AGN on star formation. In this talk I present an ongoing analysis to study the CO($J = 3-2$) emission in a sample of 25 AGN at $z \sim 2$ using ALMA observations. The CO properties of our AGN have been compared to normal (non-AGN) star-forming galaxies. The comparison between the two samples reveals that, on average, the CO luminosities of AGN at high stellar masses ($\log(M_*/M_\odot) > 11$) are 0.5 dex lower than normal galaxies. We ascribe this difference to the AGN activity, which could be able to change the conditions of the gas through, e.g., excitation, heating or removal of CO.

Keywords. galaxies: active, galaxies: evolution, quasars: general, surveys, ISM: jets and outflows

1. Introduction

Active galactic nuclei (AGN) are powered by accreting supermassive black holes which release a huge amount of energy, injected into the surrounding interstellar medium (ISM). This energy, if efficiently coupled, could remove, heat and/or dissociate the molecular gas, that represents the fuel out of which stars form. The mechanisms by which the energy produced by the AGN is coupled to the ISM is called AGN feedback (Fabian 2012; King & Pounds 2015; Harrison 2017) and is thought to regulate the growth of the host galaxy. Despite being a necessary ingredient in models of galaxy evolution, proving the role of AGN feedback observationally remains a challenge.

A promising approach to move forward requires direct observations of the cold gas reservoir of galaxies through, for example, carbon monoxide (CO) rotational emission lines. The molecular gas provides an instantaneous measure of the raw fuel to form stars and a more direct tracer of potential feedback effects since it is less affected by the timescale issue, unlike the star formation rate (SFR). In particular, characterizing the cold gas phase of AGN host galaxies is key to understand whether this is different from inactive galaxies and quantify effects of AGN feedback on the host galaxy ISM.

Studies of local sources currently report no clear evidence for AGN to affect the ISM component of the host, by tracing the molecular phase (Husemann et al. 2017; Saintonge et al. 2017; Rosario et al. 2018), the atomic one (Ellison et al. 2019), and the dust mass as a proxy of the gas mass (Shangguan & Ho 2019). Interestingly, studies at redshift $z > 1$, near the peak of AGN and star-formation activity in the Universe ($z = 1 - 3$), present contrasting results. Reduced molecular gas fractions ($f_{gas} = M_{mol}/M_*$) are found in AGN compared to the parent population of normal galaxies (Carilli & Walter 2013; Fiore *et al.* 2017; Kakkad *et al.* 2017; Brusa *et al.* 2018; Perna *et al.* 2018). This has been interpreted as an evidence for highly efficient gas consumption possibly related to AGN feedback affecting the gas reservoir of the host galaxies.

Nevertheless, these studies are plagued by several assumptions and limitations. CO measurements at $z > 1$ are usually performed using high-J transitions, and correction factors need to be assumed to estimate the luminosity of the ground-state transition as well as the gas mass. When dealing with SFRs of AGN hosts, an additional complication is the difficulty to properly account for the AGN contribution to the far-infrared luminosity. Different methods to estimate the AGN contribution can lead to different results and place the AGN population on the same star-formation law of normal galaxies (Kirkpatrick *et al.* 2019). Finally, AGN samples at high redshift are usually small and likely biased toward brighter objects (e.g. Brusa *et al.* 2018) or are heterogeneous when assembled from literature data (e.g. Fiore *et al.* 2017; Perna *et al.* 2018; Kirkpatrick *et al.* 2019).

Prompted by the need for a systematic and uniform study of the molecular gas content of AGN at $z \sim 2$, we are conducting an ALMA survey of 25 AGN to infer whether their activity affects the ISM of the host galaxy. Here we present the results of an ongoing analysis exploiting the CO($J = 3-2$) emission line as a tracer, which is the lowest-J transition accessible with ALMA at $z \sim 2$.

2. Sample selection

The sample of 25 AGN was drawn from the *COSMOS-Legacy* survey (Civano *et al.* 2016; Marchesi *et al.* 2016) and the wide-area *XMM-Newton* XXL survey North (Menzel *et al.* 2016; Pierre *et al.* 2016), by selecting targets with an absorption-corrected X-ray luminosity $L_X \geqslant 10^{42}$ erg s^{-1} and secure spectroscopic redshift in the range $z = 2.0 - 2.5$. We followed Circosta *et al.* (2018) to collect the multi-wavelength data spanning from the X-rays to the radio regime and derive the properties of the targets through X-ray spectral analysis and broad-band spectral energy distribution fitting (e.g., stellar mass, SFR, bolometric and X-ray luminosities, obscuring column densities).

We built a comparison sample of star-forming galaxies which do not host an AGN (normal galaxies) by selecting sources from the PHIBSS sample (Tacconi *et al.* 2018), matched in redshift, stellar mass, SFR and CO transition with our AGN sample. By requiring that also the comparison sample is observed in CO($J = 3-2$), we are free from assumptions on the excitation correction, which can be crucial when comparing CO luminosities and bias the results (Kirkpatrick *et al.* 2019). The final sample of normal galaxies used in this work is composed of 47 objects.

In Fig. 1 we show the distribution of our AGN in the SFR-M_* plane, as well as the control sample. Of our AGN (normal galaxies), 44% (60%) sit on the MS, 40% (27%) above and 16% (13%) at the lower boundary of the MS. The coverage in AGN bolometric luminosity is around two orders of magnitude.

3. ALMA data

The dataset is made of ALMA observations carried out in Band 3 during Cycle 4 and 5 (Project codes: 2016.1.00798.S and 2017.1.00893.S; PI: V. Mainieri). The requested angular resolution was 1" in order to probe the total gas reservoir of our targets. ALMA visibilities were calibrated using the CASA software version 4.7.0 for Cycle 4 data and 5.1.1 for Cycle 5, as originally used for the reduction with the pipeline. The final datacubes were generated with the CASA task CLEAN in velocity mode using "natural" weighting, cellsize of 0.2" and velocity bin width of 24 km/s. Final products, generated

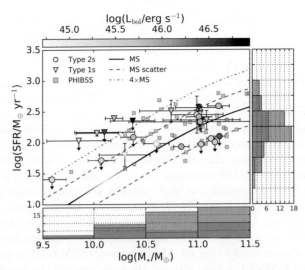

Figure 1. Distribution of host galaxy properties in the SFR-M_* plane for the 25 AGN (type 1s marked by triangles and type 2s marked by circles) in our sample and the comparison sample of normal galaxies (gray squares). The two data points with green edges represent the targets with SFR derived through modeling of the stellar emission with spectral energy distribution fitting. The color coding indicates the AGN bolometric luminosity for each object. The black solid line reproduces the main sequence of star-forming galaxies from Schreiber et al. (2015) at the average redshift of our target sample (i.e., ~ 2.3). The dashed lines mark the scatter of the main sequence (equal to 0.3 dex). The histograms show the projected distribution of the two quantities along each axis, in blue for our sample and gray for the comparison one.

from the continuum-subtracted uv datasets, have angular resolutions of $0.6-1.4''$. The sensitivity reached is 0.1–0.5 mJy per 100 km/s velocity bin.

Lines with S/N $\gtrsim 3$ are considered as detections. Moment 0 maps (i.e., the frequency-integrated flux maps) were constructed by using the task IMMOMENTS and collapsing along the channels with CO detection. The spectra were extracted from a region of the ALMA cubes with 2σ significance. We fit the line using a Gaussian model (Python package LMFIT) in order to retrieve widths and velocity-integrated fluxes. $CO(J = 3-2)$ is detected in 10 out of 25 targets. We measured FWHM in the range 80–700 km/s and $CO(J = 3-2)$ luminosities $\log(L'_{CO}/K$ km/s pc$^2) = 9.15-10.73$. For non detections, we derived the rms from the moment 0 maps by collapsing over a velocity width of 360 km/s (the average over the detections) and we then used 3σ upper limits for our analysis.

4. Comparison with normal galaxies

In Fig. 2 (left) we compare the distribution of L'_{CO} with the stellar masses of AGN and normal galaxies, color-coded by the offset in SFR with respect to the main sequence of star-forming galaxies. We use the $CO(J = 3-2)$ luminosity as a proxy of the molecular gas mass. To quantify the differences between the two samples, in Fig. 2 (right) we divided the targets in bins of stellar mass (width of 0.5 dex) and for each we computed the average L'_{CO}. In order to take into account upper limits, we performed a survival analysis by using the function KMESTM within the ASURV package, which gives the Kaplan-Meier estimator for the distribution function of a sample with upper limits. As for the comparison sample, the quantities shown in Fig. 2 (right) are the mean values for each stellar mass bin and the error bars represent the standard deviation of the distribution. The difference is not statistically significant in the lower mass bins ($\log(M_*/M_\odot) < 10.5$). However, there is a difference in L'_{CO} at the high mass end, which is particularly clear for

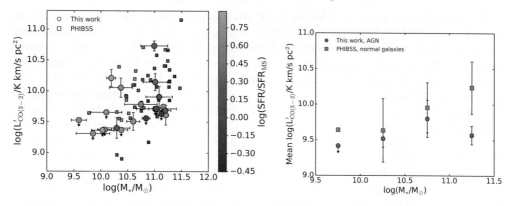

Figure 2. *Left:* L'_{CO} versus stellar masses for our sample of AGN (circles) and normal galaxies (squares; Tacconi *et al.* 2018). The color coding shows the deviation of the targets with respect to the MS (Schreiber *et al.* 2015). The targets with an upper limit in SFR are colored in gray. *Right:* mean L'_{CO} in bins of stellar mass (0.5 dex width) for AGN (circles) and normal galaxies (squares). Normal galaxies appear to have CO luminosities higher than AGN, with a difference of more than 0.5 dex in the high-mass bin.

$\log(M_*/M_\odot) > 11$. Normal galaxies appear to have CO luminosities higher than AGN, with a difference of more than 0.5 dex.

Previous work (Förster-Schreiber *et al.* 2019), studying outflows in galaxies through the Hα emission line, found that incidence, strength, and velocity of AGN-driven winds is strongly correlated with stellar mass. In particular, high-velocity ($\sim 1000-2000$ km s^{-1}) AGN-driven outflows are commonly detected at masses above $\log(M_*/M_\odot) = 10.7$. This is the same stellar-mass threshold above which the difference in CO luminosity between AGN and normal galaxies in our analysis is significant. A deficit of CO emission in the CO($J = 3-2$) transition has been recently found by Kirkpatrick *et al.* (2019). They performed a study of CO transitions for a literature sample spanning from normal galaxies to powerful AGN and found lower CO($J = 3-2$) and CO($J = 1-0$) fluxes for AGN than normal galaxies. Such result is in agreement with what we found for our samples and would translate in lower gas masses in high-redshift AGN. Similarly, Fiore *et al.* (2017) compared molecular gas and stellar masses for an heterogeneous AGN sample collected from the literature and found lower gas masses in AGN than in normal galaxies.

The difference in CO luminosity between the population of AGN and normal galaxies could be ascribed to the presence of the AGN. The central engine may have a role in heating, exciting, dissociating and/or ejecting the gas. However, understanding the mechanism producing such difference requires further data and larger samples, tracing different gas phases and several CO transitions.

5. Interpretation of our results

Possible scenarios to interpret our results are as follows.

The gas is excited by the AGN to higher-J transitions. According to previous work (Weiß *et al.* 2007; Carilli & Walter 2013), the molecular gas of AGN could therefore be more excited than in normal galaxies and, as a consequence, we observe less CO emission in the lower-J transitions. At high redshift the sampling of CO transitions is usually sparser than in the Local Universe, but variations in the line fluxes connected with the presence of an AGN were reported (Carilli & Walter 2013; Brusa *et al.* 2018; Carniani *et al.* 2019). Studies of several CO transitions over large AGN samples, especially out to high-J, are needed to understand the role of AGN ionization in shaping the properties of the molecular gas reservoir.

The gas is heated/dissociated by the AGN. An interesting example of this scenario was recently presented by Rosario et al. (2019). Spatially-resolved observations of different gas phases (both molecular and ionized) for a nearby Seyfert 2 galaxy revealed a central region (< 200 pc) which is weak in CO($J = 2-1$) emission but filled with ionized and warm molecular gas (H_2 MIR rotational lines). The energy liberated by the AGN may influence the molecular gas properties and suppress the CO($J = 2-1$) emission by heating and dissociating the molecular gas. At high redshift, when the overall AGN population is more active, the central engine could have an impact over larger spatial scales.

AGN-driven outflows affect the gas reservoir. Another possible effect of the AGN activity on the molecular gas is through outflows. This possibility is supported by observations of individual objects. For example, Brusa et al. (2018) found low gas fraction in a powerful AGN at $z \sim 1.6$ hosting also a high-velocity molecular and ionized outflow. AGN feedback in action in this target could be depleting the molecular gas reservoir. As for our sample, we are performing a systematic investigation of the ionized gas phase with SINFONI at the ESO's Very Large Telescope as part of the SINFONI Survey for Unveiling the Physics and Effect of Radiative feedback (SUPER; Circosta et al. 2018). These data, available for 19 out of 25 targets, will reveal the presence of outflows.

References

Brusa, M., Cresci, G., Daddi, E., et al. 2018, *A&A*, 612, 29
Carilli, C. L. & Walter, F. 2013, *ARAA*, 51, 105
Carniani, S., Gallerani, S., Vallini, L., et al. 2019, *MNRAS*, 489, 3939
Circosta, C., Mainieri, V., Padovani, P., et al. 2018, *A&A*, 620, 82
Civano, F., Marchesi, S., Comastri, A., et al. 2016, *ApJ*, 819, 62
Ellison, S. L., Brown, T., Catinella, B., & Cortese, L. 2019, *MNRAS*, 482, 5694
Fabian, A. C. 2012, *ARAA*, 50, 455
Fiore, F., Feruglio, C., Shankar, F., et al. 2017, *A&A*, 601, 143
Förster Schreiber, N. M., Übler, H., Davies, R. L., et al. 2019, *ApJ*, 875, 21
Harrison, C. M. 2017, *Nature Astronomy*, 1, 165
Husemann, B., Davis, T. A., Jahnke, K., et al. 2017, *MNRAS*, 470, 1570
Kakkad, D., Mainieri, V., Brusa, M., et al. 2017, *MNRAS*, 468, 4205
King, A. & Pounds, K. 2015, *ARAA*, 53, 115
Kirkpatrick, A., Sharon, C., Keller, Pope, A., et al. 2019, *ApJ*, 879, 41
Marchesi, S., Civano, F., Elvis, M., et al. 2016, *ApJ*, 817, 34
Menzel, M.-L., Merloni, A., Georgakakis, A., et al. 2016, *MNRAS*, 457, 110
Noeske, K. G., Weiner, B. J., Faber, S. M., et al. 2007, *ApJL*, 660, 43
Perna, M., Sargent, M. T., Brusa, M., et al. 2018, *A&A*, 619, 90
Pierre, M., Pacaud, F., Adami, C., et al. 2016, *A&A*, 592, 1
Rosario, D. J., Burtscher, L., Davies, R. I., et al. 2018, *MNRAS*, 473, 5658
Rosario, D. J. Togi, A., Burtscher, L., et al. 2019, *ApJ*, 875, 8
Saintonge, A. Catinella, B., Tacconi, L. J., et al. 2017, *ApJS*, 233, 22
Schreiber, C., Pannella, M., Elbaz, D., et al. 2015, *A&A*, 575, 74
Shangguan, J. & Ho, L. C. 2019, *ApJ*, 873, 90
Tacconi, L. J., Genzel, R., Saintonge, A., et al. 2018, *ApJ*, 853, 179
Weiß, A., Downes, D., Neri, R., et al. 2007, *A&A*, 467, 955

Do AGN really suppress star formation?

Chris M. Harrison[1], David M. Alexander[2], Dalton J. Rosario[2], Jan Scholtz[2,3] and Flora Stanley[3]

[1]School of Mathematics, Statistics and Physics, Newcastle University,
Newcastle Upon Tyne, NE1 7RU, United Kindgom
email: christopher.harrison@newcastle.ac.uk

[2]Centre for Extragalactic Astronomy, Durham University, Department of Physics,
South Road, Durham, DH1 3LE, United Kindgom

[3]Department of Space, Earth and Environment, Chalmers University of Technology,
Onsala Space Observatory, SE-43992 Onsala, Sweden

Abstract. Active galactic nuclei (AGN) are believed to regulate star formation inside their host galaxies through "AGN feedback". We summarise our on-going study of luminous AGN ($z \sim 0.2$–3; $L_{\rm AGN,bol} \gtrsim 10^{43}$ erg s^{-1}), which is designed to search for observational signatures of feedback by combining observed star-formation rate (SFR) measurements from statistical samples with cosmological model predictions. Using the EAGLE hydrodynamical cosmological simulations, in combination with our *Herschel* + ALMA surveys, we show that – even in the presence of AGN feedback – we do not necessarily expect to see any relationships between average galaxy-wide SFRs and instantaneous AGN luminosities. We caution that the correlation with stellar mass for both SFR and AGN luminosity can contribute to apparent observed positive trends between these two quantities. On the other hand, the EAGLE simulations, which reproduce our observations, predict that a signature of AGN feedback can be seen in the wide specific SFR distributions of *all* massive galaxies (not just AGN hosts). Overall, whilst we can not rule out that AGN have an immediate small-scale impact on in-situ star-formation, all of our results are consistent with a feedback model where galaxy-wide in-situ star formation is not rapidly suppressed by AGN, but where the feedback likely acts over a longer timescale than a single AGN episode.

Keywords. galaxies: active, galaxies: evolution

1. Introduction

A fundamental component of galaxy formation models is that the central growing supermassive black holes (i.e., active galactic nuclei; AGN) regulate star formation inside their host galaxies. There have been considerable attempts to search for observational evidence of this "AGN feedback" throughout the last one-to-two decades (e.g., see review in Harrison 2017). One common approach has been to simultaneously measure AGN luminosities and the in-situ star-formation rates (SFRs) of AGN host galaxies. Here we refer to the in-situ SFRs as the measured *galaxy-wide* SFRs of on-going star-formation inside galaxies that host an observable AGN. In this article we summarise our own work following this approach, stressing the importance of testing specific model predictions. In a series of papers we have explored the SFRs and specific SFRs (sSFRs; SFR/stellar mass) of AGN-host galaxies (Stanley *et al.* 2015, 2017, 2018; Harrison 2017; Scholtz *et al.* 2018). In these works we infer SFRs from the observed far-infrared luminosities after subtracting the AGN contribution using careful decomposition of the spectral energy distributions (following Stanley *et al.* 2015). Removing the AGN contribution is important, particularly for the most powerful AGN, where the contribution to the total far-infrared luminosity can become significant (e.g., Fig. 7 of Stanley *et al.* 2017).

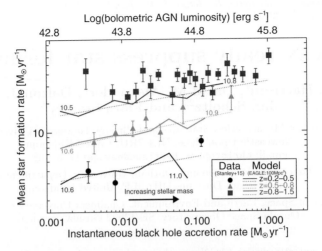

Figure 1. Mean star formation rate versus instantaneous black hole accretion rate for the reference EAGLE simulation and versus AGN luminosity (converted from X-ray luminosity) from observations. The solid curves show the running average (mean) simulation values and the dotted lines are a linear fit to these values. The logarithm of the average stellar masses (in solar mass units) of the first and last values from the simulation are labelled. The slight increase in mean SFR with AGN luminosity is attributed to the increasing stellar masses. Effective star-formation suppression by AGN feedback does not infer that galaxies with a currently visible AGN should have reduced average *in-situ* galaxy-wide SFRs. Figure from Harrison (2017).

2. AGN feedback does not reduce galaxy-wide in-situ SFRs

In Stanley *et al.* (2015) we combined *Herschel* and *Spitzer* photometry for ∼2000 $z=0.2$–2.5 X-ray AGN ($L_{2-10\text{keV}} = 10^{42}$–$10^{45.5}$ erg s^{-1}) and calculated average (mean) SFRs in bins of X-ray luminosity (data points in Fig. 1). Tracking the evolution of the overall galaxy population, a strong evolution of average SFR with redshift is observed. However, we find that the relationship between average SFR and AGN luminosity is only weakly correlated across all AGN luminosities investigated. In Stanley *et al.* (2017) we repeated a similar experiment, over the same redshift range, but on powerful Type 1 quasars ($L_{\text{AGN}} > 10^{45}$ erg s^{-1}) and using *WISE*+*Herschel* data. We found a stronger correlation between SFR and AGN luminosity (Fig. 2); however, we showed that this can be explained by both quantities being correlated with stellar masses (assumed from black hole masses; see details in Stanley *et al.* 2017). This mass effect is comparatively weak in the X-ray AGN sample in Fig. 1, which can be explained due to the different selection effects of X-ray AGN and optical Type 1 quasars, where the latter selects a narrower range of (higher) Eddington-ratios (see discussion in Rosario *et al.* 2013).

Across our work we found no strong dependence of average SFR on AGN luminosity when mass and redshift trends are accounted for. Furthermore, the *mean* SFRs of AGN-host galaxies are broadly consistent with star-forming "main sequence" galaxies (Fig. 2).

In Harrison (2017), to aid the interpretation of our observational results, we compared the observations of Stanley *et al.* (2015) to the reference model of the EAGLE hydrodynamical cosmological simulations (Schaye *et al.* 2015). The simulation models astrophysical processes (including prescriptions for star formation and AGN feedback) inside a 100 Mpc^3 volume of the Universe and contains 1000s of massive galaxies, allowing us to track SFRs and AGN luminosities (inferred from instantaneous black hole accretion rates) across cosmic time. Consequently, we selected simulated galaxies following the same criteria as used for our observations (for details see McAlpine *et al.* 2017; Harrison 2017). EAGLE successfully reproduces the observed average SFRs on AGN (Fig. 1). This leads

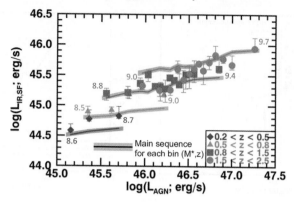

Figure 2. Mean infrared star formation luminosity (a proxy for SFR) versus bolometric AGN luminosity for Type 1 quasars. The solid curves show the expected value for star-forming galaxies at the average redshifts and stellar masses (extrapolated from the black hole masses) in each bin of quasars (Schreiber et al. 2015). The logarithm of average black masses (solar mass units) of the first and last bins are labelled. Increasing mean SFR with increasing AGN luminosity is attributed to the increasing masses. Quasar host galaxies have *mean* SFRs that are consistent with mass- and redshift-matched star-forming galaxies; however, the underlying distributions may still be different (see Section 3 and Fig. 3). Figure adapted from Stanley et al. (2017).

to an important conclusion: *no evidence for "suppressed" galaxy-wide in-situ SFRs in AGN host galaxies is still consistent with a cosmological model including AGN feedback.* This, perhaps counter intuitive result, can be explained if the timescale of an AGN episode is shorter than the timescale of star formation suppression by these (possibly multiple) AGN episode(s) (Harrison 2017; Scholtz et al. 2018 and references there-in).

3. AGN feedback is imprinted on the overall galaxy population

To move beyond simple average SFRs towards characterising distributions we obtained deep ALMA observations of ~110 X-ray AGN, enabling us to achieve 2-10× more sensitive SFR constraints than previously possible (Stanley et al. 2018). Using these results, in Scholtz et al. (2018), we investigated $z=1.5$–3.2 X-ray AGN ($L_{2-10\mathrm{keV}} = 10^{43}$–$10^{45}$ erg s^{-1}), and measured the mode and width of the (specific) SFR distributions of AGN host galaxies (following Mullaney et al. 2015). We found that, for AGN in this luminosity range, whilst the *mean* SFRs of AGN are typically consistent with the main sequence of star-forming galaxies, the *median/mode* SFRs of AGN host galaxies are typically found to be *lower* than star-forming galaxies (Fig. 3). This can be explained because of different underlying distributions, where AGN host galaxies have a broader distribution, extending to lower SFRs compared to the main sequence (Mullaney et al. 2015).†
Nonetheless, for feedback studies we suggest that it is more meaningful to investigate the galaxy population *as a whole* instead of a comparison to the main sequence (which is still not well defined for the highest masses; see dotted/dashed curves in Fig. 3).

In Scholtz et al. (2018), we investigated galaxies in the EAGLE simulations with the AGN turned off (i.e., with no AGN feedback). This enabled us to show that, within the main reference model, the impact of AGN is to decrease the mode, by a factor of ~2–3, and to increase the width, by a factor of ~2–3, of the sSFR distributions of massive galaxies (Fig. 4). This is simply explained by the fact that AGN feedback produces the quiescent galaxy population, spreading the sSFR to lower values (Scholtz et al. 2018).

† We note, with increasing AGN luminosity there may be an increased likelihood for AGN-host galaxies to be star-forming main sequence galaxies (Bernhard et al. 2019; Schultz et al. 2019).

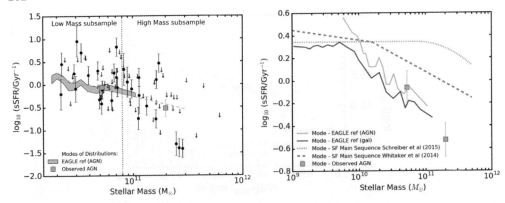

Figure 3. *Left:* sSFR versus stellar mass for X-ray AGN (black data points). Squares indicate the distribution mode for low and high mass subsamples (see vertical dashed line). The shaded region indicates the sSFR distribution mode for redshift-matched AGN host galaxies from EAGLE; the dashed line is an extrapolation to higher masses. *Right:* sSFR distribution modes versus stellar mass for X-ray AGN (data points) and for AGN hosts and all galaxies in the EAGLE reference model (blue and yellow curves, respectively). EAGLE matches the observations for the sSFRs for AGN hosts. However, AGN of these luminosities, have a lower *mode* of sSFRs than galaxies on the main sequence of *star-forming* galaxies (definitions shown from Schreiber et al. 2015 [dotted curve] and Whitaker et al. 2014 [dashed curve]). Figures from Scholtz et al. (2018).

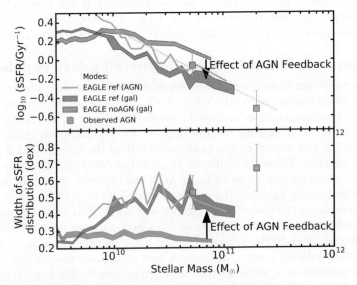

Figure 4. Mode (top panel) and width (bottom panel) of sSFR distributions versus stellar mass for observed AGN hosts (data points) compared to *all* galaxies in both the EAGLE reference simulation and the EAGLE simulation without AGN (blue and red shaded regions, respectively). The orange solid curve is for AGN host galaxies in the reference model (extrapolated to higher masses with the dashed line). In the EAGLE reference simulation (which matches the trends of the data), AGN feedback reduces the mode and increases the width of sSFR distributions. This is seen in the distributions of *all* massive galaxies, even if the AGN happen to be "off" at the time that they are observed. Figure from Scholtz et al. (2018).

At the limit of the spatial resolution of our integral field spectroscopic observations (~few kpc), in combination with high-resolution far-infrared imaging, we have also found no evidence that AGN-driven ionised outflows (traced via high-velocity [O III] emission) have an instantaneous positive or negative significant impact on the in-situ star-formation in a representative sample of 8 $z\sim$1.4–2.6 AGN (Scholtz et al. 2020). This is in qualitative agreement with at least some models that suggest that AGN outflows have no impact upon the *in-situ* star formation (Gabor & Bournaud 2014). Nonetheless, we can not rule out a smaller-scale impact or that outflows have an impact on longer timescales; for example by removing gas which is later prevented from re-accreting onto the galaxy.

We have shown that if AGN feedback is the mechanism to suppress galaxy-wide star formation in massive galaxies it does not necessarily follow that AGN host galaxies have "suppressed" (specific) SFRs compared to, mass- and redshift-matched, galaxies without a visible AGN (Fig. 1). Instead, the signature of AGN feedback is likely to be imprinted on the properties of the entire massive galaxy population (Fig. 4). More work is now required to test specific observable model predictions that use different prescriptions for AGN feedback. For example, assessing if the observed molecular gas content of AGN host galaxies (e.g., Shangguan et al. 2018) is consistent with model predictions.

4. Final remarks: the need for testing specific model predictions

We have investigated the SFRs of large samples of AGN host galaxies. We have emphasised the importance of controlling for redshift and mass when investigating the SFRs of AGN host galaxies - without doing so could result in artificial correlations between SFRs and AGN luminosities. By comparing our observations with the EAGLE simulations we have concluded that: (1) AGN feedback does not reduce galaxy-wide in-situ SFRs of AGN host galaxies; (2) the signature of AGN feedback is imprinted on the overall massive galaxy population. Importantly, any observation which finds that galaxy-wide SFRs of AGN host galaxies are not "suppressed" does not rule out all models of AGN feedback. Instead this observation may only rule out models where the complete suppression of star formation is more rapid than the AGN episode itself. *When looking for observational signatures of AGN feedback, we strongly advocate testing specific model predictions rather than just expecting AGN host galaxies to be special.* Further rigorous model–observation comparisons are required to make progress in understanding how AGN impact upon the star formation in their host galaxies.

References

Bernhard, E., Grimmett, L. P. Mullaney, J. R., et al. 2019, *MNRAS*, 483L, 52
Gabor, J. M. & Bournaud, F. 2014, *MNRAS*, 441, 1615
Harrison, C. M. 2017, *Nature Astronomy*, 1, 165
McAlpine, S., Bower, R. G., Harrison, C. M., et al. 2017, *MNRAS*, 468, 3395
Mullaney, J. R., Alexander, D. M., Aird, J., et al. 2015, *MNRAS*, 453, L83
Rosario, D. J., Trakhtenbrot, B., Lutz, D., et al. 2013, *A&A*, 560, 72
Schaye, J., Crain, R., Bower, R. G., et al. 2015, *MNRAS*, 446, 521
Scholtz, J., Alexander, D. M., Harrison, C. M., et al. 2018, *MNRAS*, 475, 1288
Scholtz, J., Harrison, C. M., Rosario, D. J., et al. 2020, *MNRAS*, 492, 3194
Schreiber, C., Pannella, M., Elbaz, D., et al. 2015, *A&A*, 575, A74
Schulze, A., Silverman, J. D., Daddi, E., et al. 2019, *MNRAS*, 488, 1180
Shangguan, J., Ho, L. C., Xie, Y. 2018, *ApJ*, 854, 158
Stanley, F., Harrison, C. M., Alexander, D. M., et al. 2015, *MNRAS*, 453, 591
Stanley, F., Alexander, D. M., Harrison, C. M., et al. 2017, *MNRAS*, 472, 2221
Stanley, F., Harrison, C. M., Alexander, D. M., et al. 2018, *MNRAS*, 478, 3721
Whitaker, K. E., Franx, M., Leja, J., et al. 2014, *ApJ*, 795, 104

Accretion and star formation in 'radio-quiet' quasars

Sarah V. White[1,2,3], Matt J. Jarvis[3,4], Eleni Kalfountzou[5,6], Martin J. Hardcastle[6], Aprajita Verma[3], José M. Cao Orjales[6] and Jason Stevens[6]

[1] South African Radio Astronomy Observatory (SARAO), 2 Fir Street, Observatory, Cape Town, 7925, South Africa

[2] Department of Physics and Electronics, Rhodes University, PO Box 94, Grahamstown, 6140, South Africa

[3] Astrophysics, University of Oxford, Denys Wilkinson Building, Keble Road, Oxford, OX1 3RH, UK
email: sarahwhite.astro@gmail.com

[4] Department of Physics, University of the Western Cape, Bellville 7535, South Africa

[5] European Space Astronomy Centre (ESAC), Villanueva de la Cañada, E-28692 Madrid, Spain

[6] Centre for Astrophysics Research, School of Physics, Astronomy and Mathematics, University of Hertfordshire, Hatfield, Herts, AL10 9AB, UK

Abstract. Radio observations allow us to identify a wide range of active galactic nuclei (AGN), which play a significant role in the evolution of galaxies. Amongst AGN at low radio-luminosities is the 'radio-quiet' quasar (RQQ) population, but how they contribute to the total radio emission is under debate, with previous studies arguing that it is predominantly through star formation. In this talk, SVW summarised the results of recent papers on RQQs, including the use of far-infrared data to disentangle the radio emission from the AGN and that from star formation. This provides evidence that black-hole accretion, instead, dominates the radio emission in RQQs. In addition, we find that this accretion-related emission is correlated with the optical luminosity of the quasar, whilst a weaker luminosity-dependence is evident for the radio emission connected with star formation. What remains unclear is the process by which this accretion-related emission is produced. Understanding this for RQQs will then allow us to investigate how this type of AGN influences its surroundings. Such studies have important implications for modelling AGN feedback, and for determining the accretion and star-formation histories of the Universe.

Keywords. galaxies: active – galaxies: statistics – galaxies: evolution – galaxies: high-redshift – quasars: general – radio continuum: galaxies

1. Radio emission from 'radio-quiet' quasars

Star formation and black-hole accretion are the key processes that govern how galaxies evolve. Since both processes produce radio emission, deep radio surveys enable complete selection of galaxies out to high redshift, unaffected by dust obscuration (unlike optical surveys) or Compton-thick absorption (unlike X-ray surveys). However, as is the case at other wavelengths, the contributions of star formation and the active galactic nucleus (AGN) towards the total emission need to be disentangled. The techniques we use to do this will prove particularly important for science with the Square Kilometre Array (SKA) and its precursor/pathfinder telescopes (e.g. McAlpine et al. 2015; Jarvis et al. 2015). As we probe to lower radio flux-densities (< 1mJy), a flattening of the source counts implies that we are going from an AGN dominated regime to a star-formation

Table 1. The accretion-related contribution to the radio luminosity, across our sample of RQQs. An object with $L_{1.5\,\text{GHz, acc}}/L_{1.5\,\text{GHz}} > 0.5$ is described as being 'AGN-dominated'. 'Summed radio luminosity' refers to the summation of the total radio luminosity for each object ($\Sigma L_{1.5\,\text{GHz}}$), and the 'fraction that is accretion-related' is given by $\Sigma L_{1.5\,\text{GHz, acc}}/\Sigma L_{1.5\,\text{GHz}}$. 'Upper' and 'lower' limits refer to the fraction of the radio emission that is related to accretion, taking into account cases where the object is undetected (i.e. $<2\sigma$) in the radio and/or the FIR.

Description of objects used (and the $L_{1.5\,\text{GHz, acc}}$ values considered)	No. of objects	Fraction that are AGN-dominated	Summed radio luminosity (W Hz^{-1})	Fraction of summed luminosity that is accretion-related
Whole sample	70	0.80	3.82×10^{25}	0.74
Whole sample (lower limits)	70	$\geqslant 0.47$	$\leqslant 5.28 \times 10^{25}$	$\geqslant 0.60$
Whole sample (upper limits)	70	$\leqslant 0.89$	$\leqslant 5.28 \times 10^{25}$	$\leqslant 0.83$
Radio-detected, FIR-detected	26	0.92	3.07×10^{25}	0.80

dominated regime (Condon et al. 2012). However, it is not only star-forming galaxies that are detected below 1 mJy. We also uncover the population of radio-quiet quasars [RQQs; see predictions by Wilman et al. (2008)], which are a subset of AGN that produce relatively low levels of radio emission when compared to their 'radio-loud' counterparts.

The current assumption is that star formation is responsible for the radio emission from RQQs, with Kimball et al. (2011) and Condon et al. (2013) favouring this interpretation. However, these studies use only radio and optical data for their analyses. For our work, we take a step further by also using far-infrared (FIR) data from the *Herschel Space Observatory*, reduced and presented by Kalfountzou et al. (2017). This allows us to *quantify* the level of star formation within our sample of 70 RQQs, by combining the derived far-infrared luminosities with the far-infrared to radio correlation (FIRC). This very tight relation is observed for star-forming galaxies, and arises due to star formation being associated with both dust (producing the FIR emission) and supernova remnants (producing radio emission).

2. Disentangling black-hole accretion from star formation

Like the quasars used by Kimball et al. (2011) and Condon et al. (2013), our sample is selected from the Sloan Digital Sky Survey (SDSS), but we restrict the redshift range to $0.9 < z < 1.1$ in order to minimise any possible evolutionary effects. The radio data we use is in the form of targeted observations using the Karl G. Jansky Very Large Array, enabling us to reach a sensitivity of $\sim 30\,\mu$Jy in 25 minutes of integration time per source [PI: Jarvis; White et al. (2017)]. When analysed in conjunction with FIR data, we find that for the majority of our RQQs, there is a significant amount of radio emission that cannot be explained by star formation alone. This is illustrated by the RQQs lying to the right of the FIRC in Fig. 1, having 'excess' radio emission. The excess emission must be due to another process – that being black-hole accretion – and so we use the offsets from the FIRC to calculate the accretion-related radio luminosity. As shown by the numbers in Table 1, this accretion component is dominating the total radio emission from our RQQs, in support of previous work by White et al. (2015) and going against the assumption that star formation dominates. Our fractions may even be underestimates, given a study by Wong et al. (2016) that shows a sample of hard X-ray selected AGN lying on the FIRC, mimicking the properties of star-forming galaxies.

Another aspect of our sample is that it has been downselected in order to enforce a uniform distribution in absolute *i*-band magnitude, spanning a factor of ~ 100 in optical luminosity. This allows us to explore whether there is any trend in the star-formation-related radio luminosity, or the accretion-related radio luminosity, with the

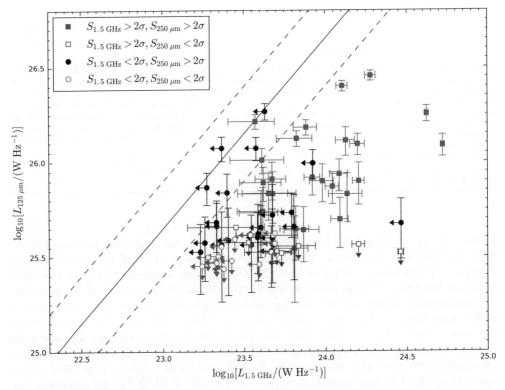

Figure 1. The FIR luminosity at rest-frame 125 μm, $L_{125\,\mu m}$, versus the radio luminosity, $L_{1.5\,\mathrm{GHz}}$. Squares correspond to objects detected in the radio above 2σ, and circles are those below this detection threshold. Note that objects with $\log_{10}[L_{125\,\mu m}] \lesssim 25.6$ are below 2σ at 250 μm (unfilled symbols). Arrows represent 2-σ upper limits in $L_{1.5\,\mathrm{GHz}}$ or $L_{125\,\mu m}$, for quasars undetected at the 2-σ level in the radio (horizontal arrows) or the 250 μm photometry (vertical arrows), respectively. The dashed lines are the lower and upper bounds on q_{125} (2.4 and 2.9, respectively), where $q_{125} = \log_{10}[L_{125\,\mu m}/L_{1.5\,\mathrm{GHz}}]$. The solid line corresponds to the midpoint value, $q_{125} = 2.65$, and represents the FIRC (where star-forming galaxies are expected to lie). *Figure reproduced from White et al. (2017).*

optical luminosity of the quasar. White et al. (2017) find a stronger statistical correlation for the latter combination, which is as expected since the accretion disc of a quasar produces thermal emission that dominates the i band. We therefore have further evidence that this component of the radio emission truly is connected with the AGN. However, a lot of scatter is still seen between the accretion-related radio luminosity and the absolute i-band magnitude (Fig. 2), which could be due to variations in magnetic-field strength, environmental density, or the differing timescales over which the radio emission and optical emission are produced.

3. What accretion-related mechanism is present?

We have concluded that the AGN is dominating the total radio emission from RQQs, but the exact mechanism involved requires further investigation. It is possible that radio jets (typically associated with 'radio-loud' quasars) are being launched (e.g. Hartley et al. 2019) but they are too small to be resolved in our radio images. Supporting this, earlier in the Symposium we were shown images of small jets in 'radio-silent' Seyfert galaxies (Lähteenmäki et al. 2018), which are believed to be low-redshift analogues of quasars. Another suggestion, regarding the origin of the AGN-related radio emission, is that it is

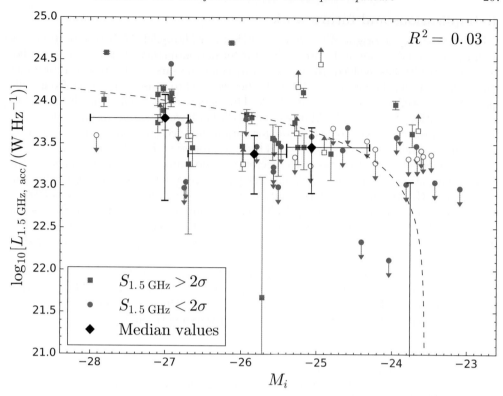

Figure 2. The accretion luminosity, $L_{1.5\,\text{GHz, acc}}$ against the absolute i-band magnitude, M_i. Blue squares correspond to objects detected in the radio above $2\,\sigma$, and red circles are those below this detection threshold. Unfilled symbols correspond to the FIR data being below $2\,\sigma$. Arrows indicate whether the value of $L_{1.5\,\text{GHz, acc}}$ is either an upper or lower limit (at $2\,\sigma$), dependent on whether the object is undetected in both the radio and the FIR, or undetected in the FIR alone. The line of best-fit is given by $L_{1.5\,\text{GHz, acc}} = (-2.99 \pm 0.84) \times 10^{23} M_i - (7.05 \pm 2.11) \times 10^{24}$, and the associated coefficient of determination is shown in the top right-hand corner. (Uncertainties in $L_{1.5\,\text{GHz, SF}}$ and $L_{1.5\,\text{GHz, acc}}$ are used for this fit.) The dashed line is the result of converting this best-fit line into log–linear space. Overplotted are the median luminosities (black diamonds), derived using all objects, binned in M_i. The horizontal error-bars indicate the ranges of the M_i bins, and uncertainties on the median radio-luminosities are given by the vertical error-bars. Note that the values of the luminosities, even if negative, are used for the linear regression analysis and the calculation of the median luminosities, rather than the limit values. *Figure reproduced from White et al. (2017).*

associated with an X-ray corona close to the accretion disc (Laor & Behar 2008), with similar magnetic connection events as seen for coronally-active stars.

Alternatively, the winds associated with the accretion disc may be impacting upon the surrounding medium and creating shock fronts, which in turn accelerate electrons and generate radio emission. Hwang *et al.* (2018) provide evidence for this explanation, by looking at the radio luminosity of quasars against the equivalent width of the [O III] emission line. This quantity acts as a proxy for the quasar's outflow velocity, and they find that an 'extreme' sample of high-redshift quasars appears to lie on the same correlation observed for their previous sample of low-redshift quasars. However, they cannot completely rule out the possibility of compact jets being present, as these could give rise to the same kinematics as quasar winds. Indeed, later during the Symposium, Jarvis *et al.* (2019) presented results for *jetted* RQQs lying on the relation by Hwang *et al.* (2018).

4. Implications and future work

Understanding the origin of the radio emission in radio-quiet AGN is important if we wish to: (i) use faint sources to probe the star-formation history of the Universe, or how accretion activity has evolved; (ii) uncover why some quasars exhibit powerful radio-jets whilst the majority do not; and (iii) see whether we need to consider feedback from radio-quiet AGN as well as from radio-loud AGN. This is particularly important for modelling galaxy evolution. For example, Mancuso et al. (2017) have used our 'disentangled' radio luminosities to test their models of star formation and accretion in the radio band, for which they consider in-situ co-evolution of the two processes. They find that our RQQs lie on the average relationship between the accretion radio-luminosity and the star-formation radio-luminosity, averaged across *all* galaxies (since radio data offer complete selection).

Most recently, Malefahlo et al. (submitted) present radio luminosity functions (RLFs), over multiple redshift ranges, for optically-selected quasars identified in SDSS. For this they use a Bayesian stacking analysis to probe below the 1-mJy flux-limit of the FIRST survey (Becker et al. 1995). White et al. will test the radio-quiet distribution of these RLFs using optical spectroscopy from the Southern African Large Telescope (PI: White) and deep radio data that has been obtained as part of the MIGHTEE survey (Jarvis et al. 2016). Following this is the need to disentangle the star-formation and AGN contributions to the total radio emission, for which very long baseline interferometry would offer excellent spatial information to supplement multi-wavelength analysis.

References

Becker, R. H., White, R. L., Helfand, D. J., 1995, *ApJ*, 450, 559
Condon, J. J., Cotton, W. D., Fomalont, E. B., Kellermann, K. I., Miller, N., Perley, R. A., Scott, D., Vernstrom, T., & Wall, J. V. 2012, *ApJ*, 758, 23
Condon, J. J., Kellermann, K. I., Kimball, A. E., Ivezić, Ž., Perley, R. A. 2013, *ApJ*, 768, 37
Hartley, P., Jackson, N., Sluse, D., Stacey, H. R., & Vives-Arias, H. 2019, *MNRAS*, 485, 3009
Hwang, Hsiang-Chih, Zakamska, N. L., Alexandroff, R. M., Hamann, F., Greene, J. E., Perrotta, S., Richards, & G. T. 2018, *MNRAS*, 477, 830
Jarvis, M. J. et al. 2015, *PoS*, arXiv:1412.5753
Jarvis, M. J. et al. 2016, *PoS*, arXiv:1709.01901
Jarvis, M. E., Harrison, C. M., Thomson, A. P., Circosta, C., Mainieri, V., Alexander, D. M., Edge, A. C., Lansbury, G. B., Molyneux, S.J. & Mullaney, J. R. 2019, *MNRAS*, 485, 2710
Kalfountzou, E., Stevens, J. A., Jarvis, M. J., Hardcastle, M. J., Wilner, D., Elvis, M., Page, M. J., Trichas, M., & Smith, D. J. B. 2017, *MNRAS*, 471, 28
Kimball, A. E., Kellermann, K. I., Condon, J. J., Ivezić, Ž., & Perley, R. A. 2011, *ApJ* (Letters), 739, L29
Lähteenmäki, A., Järvelä, E., Ramakrishnan, V., Tornikoski, M., Tammi, J., Vera, R. J. C., & Chamani, W. 2018, *A&A* (Letters), 614, L1
Laor, A. & Behar, E. 2008, *MNRAS*, 390, 847
Malefahlo, E., Santos, M. G., Jarvis, M. J., White, S. V., & Zwart, J. T. L. 2019, *submitted to MNRAS*, arXiv:1908.05316
Mancuso, C., Lapi, A., Prandoni, I., Obi, I., Gonzalez-Nuevo, J., Perrotta, F., Bressan, A., Celotti, A., & Danese, L. 2017, *ApJ*, 842, 95
McAlpine, K. et al. 2015, *PoS*, arXiv:1412.5771
White, S. V., Jarvis, M. J., Häußler, B., & Maddox, N. 2015, *MNRAS*, 448, 2665
White, S. V., Jarvis, M. J., Kalfountzou, E., Hardcastle, M. J., Verma, A., Cao Orjales, J. M., & Stevens, J. 2017, *MNRAS*, 468, 217
Wilman, R. J., Miller, L., Jarvis, M. J., Mauch, T., Levrier, F., Abdalla, F. B., Rawlings, S., Klöckner, H.-R., Obreschkow, D., Olteanu, D., & Young, S. 2008, *MNRAS*, 388, 1335
Wong, O. I., Koss, M. J., Schawinski, K., Kapińska, A. D., Lamperti, I., Oh, K., Ricci, C., Berney, S., Trakhtenbrot, B., et al. 2016, *MNRAS*, 460, 1588

MUSE-adaptive optics view of the starburst-AGN connection: NGC 7130

Johan H. Knapen[1,2], Sébastien Comerón[3] and Marja K. Seidel[4]

[1]Instituto de Astrofísica de Canarias, E-38205 La Laguna, Tenerife, Spain
[2]Departamento de Astrofísica, Universidad de La Laguna, E-38205 La Laguna, Tenerife, Spain
[3]University of Oulu, Astronomy Research Unit, P.O. Box 3000, FI-90014 Oulu, Finland
[4] Caltech-IPAC, MC 314-6, 1200 E California Blvd, Pasadena, CA 91125, USA

Abstract. We combine ALMA and MUSE-NFM (narrow field mode, with full four-laser adaptive optics correction) data at 0.15 arcsec spatial resolution of the archetypical AGN-starburst "composite" galaxy NGC 7130. We present the discovery of a small $0.''2$ (60 pc) radius kinematically decoupled core or small bi-polar outflow, as well as a larger-scale outflow. We confirm the existence of star-forming knots arranged in an $0.''58$ (185 pc) radius ring around the Seyfert 1.9 nucleus, previously observed from UV and optical *Hubble Space Telescope* and CO(6-5) ALMA imaging. An extinction map derived from the MUSE data highlights the regions of enhanced CO emission as clearly seen in the ALMA data. We determine the position of the nucleus as the location of a peak in gas velocity dispersion. A plume of material extends towards the NE from the nucleus until at least the edge of our field of view at $2''$ (640 pc) radius which we interpret as an outflow originating in the AGN. The plume is not visible morphologically, but is clearly characterised in our data by emission lines ratios characteristic of AGN emission, enhanced gas velocity dispersion, and distinct non-circular gas velocities. Its orientation is roughly perpendicular to the line of nodes of the rotating host galaxy disk. An $0.''2$-radius circumnuclear area of positive and negative velocities indicates a tiny inner disk or a small bipolar outflow, only observable when combining the integral field spectroscopic capabilities of MUSE with full adaptive optics.

Keywords. Galaxies: active – Galaxies: individual: NGC 7130 – Galaxies: ISM – Galaxies: outflows – Galaxies: nuclei – Galaxies: Seyfert

1. Introduction

Accretion onto a supermassive black hole, in combination with increased cold gas availability due to inflow, may simultaneously increase the star formation (SF) and the active galactic nucleus (AGN) activity in the centre of a galaxy.

Inflowing gas must lose by far most of its angular momentum before it reaches the very central regions (e.g., Begelman *et al.* 1984; Shlosman *et al.* 1989), but can do so under the influence of non-axisymmetries in the host galaxy induced by bars or past or present interactions. The resulting availability of gaseous fuel can lead to both starburst and AGN activity, which can even occur simultaneously and which is then referred to as "composite". Studying the gas physics, stellar properties, and kinematics of composite AGN/starburst galaxies is important as it provides direct links to the physics of the gas transport and the activity, star-forming and/or AGN, that may result.

We study one of the best-known examples of such composite AGN-starburst galaxies: the luminous infrared galaxy NGC 7130 (also known as IC 5135), a peculiar Sa (de Vaucouleurs *et al.* 1991) galaxy at a distance of 65.5 Mpc (so $1''$ corresponds to 318 pc). NGC 7130 hosts a Seyfert 1.9 AGN nucleus (Véron-Cetty & Véron 2006) as well as a powerful compact circumnuclear starburst (Phillips *et al.* 1983). We study this

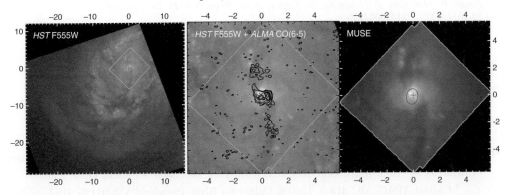

Figure 1. *Left*: *HST* image of NGC 7130 in the F555W band. The red + sign indicates the centre of the galaxy and the green contour indicates the MUSE field of view. *Middle*: enlarged version of the image on the *left*. The black contours indicate the CO(6-5) intensity as observed by *ALMA*. *Right*: white-light image obtained by collapsing the MUSE datacube along its wavelength axis. The red ellipse indicates the outline of the circumnuclear ring. The angular resolution achieved by MUSE+AO is similar to that of the *HST*. Axis labels in arcsec, N is up and E to the right.

galaxy with new integral field spectroscopy obtained with MUSE-NFM (narrow field mode) on the ESO-VLT, which uses the adaptive optics (AO) system GALACSI with four laser guide stars to feed MUSE. Further details are given in Knapen *et al.* (2019, KCS19 hereinafter) and in Comerón, Knapen & Seidel (2020, in prep).

2. Observations

We used the MUSE integral field spectrograph on the VLT in its NFM in conjunction with the AO system GALACSI during September 2018. In addition to GALACSI's four laser guide stars, we used the Seyfert nucleus of NGC 7130 as a natural guide star. The resulting field of view (FOV) is $7.''59 \times 7.''59$ with a sampling of $0.''0253$ per pixel, and the resulting spatial resolution achieved was $0.''15 - 0.''18$, obtained on Sept 18. The resulting data were reduced using v. 2.5.2 of the MUSE pipeline (Weilbacher *et al.* 2012) and additional software; details have been described in KCS19. We fitted the stellar component in the spectra using pPXF (Cappellari, & Emsellem 2004; Cappellari 2017). In addition, we used an archival WFPC3 F555W *HST* image, 600s deep (six different exposures) and with $\sim 0.''1$ angular resolution, and the ALMA CO(6-5) data with a resolution of $0.''20 \times 0.''14$ from Zhao *et al.* (2016).

3. Results

Figure 1 shows how the optical image reconstructed from the MUSE data cube has excellent spatial resolution—comparable to that of the *HST* image. We also show the CO emission as observed with ALMA, outlining spiral-shaped regions of enhanced gas and thus dust obscuration.

The MUSE spectral data allow us to fit many different absorption and emission lines. We concentrate here on gas emission lines, which we showed in KCS19 to offer evidence for an outflow region to the NW of the (AGN) nucleus, clearly offset from the rest of the inner region by AGN-like emission line ratios, enhanced velocity dispersion, and deviant gas velocities.

We now take the next step in our analysis, which is to fit two components to the gas emission (Hβ, [OIII] λ4959, [OIII] λ5007, [NI] λ5198, [NI] λ5200, [OI] λ6300, [OI] λ6364, [NII] λ6548, Hα, [NII] λ6583, [SII] λ6716, and [SII] λ6731 lines). The resulting velocity and

MUSE-NFM view of the starburst-AGN connection: NGC 7130   211

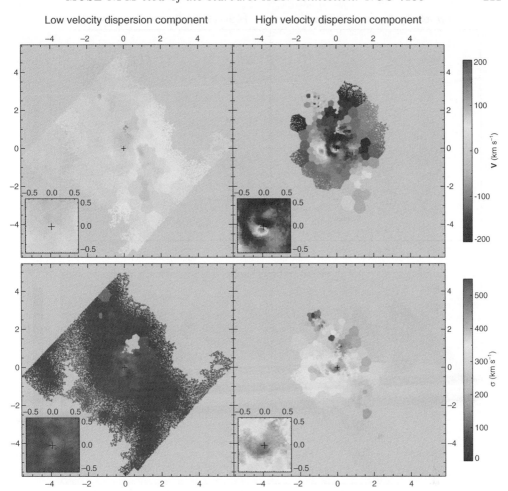

Figure 2. Velocity (V; *upper* panels) and velocity dispersion (σ; *lower* panels) maps for the low-velocity dispersion (cold, *left* column) and the high-velocity dispersion (hot, *right* column) components of the ionised gas in NGC 7130. An enlarged view of the central arcsecond is presented in the insets. The cold component shows rotation and can be associated with the disc of the galaxy. The hot component has typical velocities of up to $200\,\mathrm{km\,s^{-1}}$ with respect to the galaxy disc and can be associated with an outflow. The innermost region of the high-velocity dispersion component shows a tiny structure, $0.''2$ in radius, that can be interpreted as either a nuclear disc or a bipolar outflow. Axis labels in arcsec, N is up and E to the right.

velocity dispersion fields are shown in Fig. 2, separately for the low- (left panels) and high-velocity dispersion (right) components of the gas. We see that the low-velocity component cleanly separates and outlines a rotating disk. The high-velocity dispersion component, in contrast, traces non-rotating structure in the velocity field, whereas the dispersions correlate with the outline of the AGN-related outflow identified in KCS19. KCS19 also identified a small kinematically decoupled core, of around $0.''2$ or 60 pc in radius. From Fig. 2 we now see that this core is only visible in the high-dispersion maps, which is strong evidence in favour of it being connected to, or even part of, the AGN-driven gas outflow.

Figure 3 highlights the different emission line ratios in the low- and high-velocity dispersions as defined above. We show the so-called BPT (after Baldwin, Phillips &

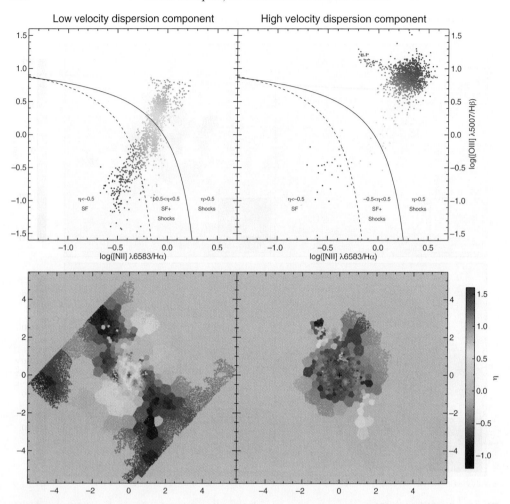

Figure 3. For the low-velocity dispersion component (*left*) and the high-velocity dispersion component (*right*), we show in the *upper* panels: the [O III]/Hβ vs [N II]/Hα diagnostic diagram used to build the BPT maps, and in the *lower* panels: resolved BPT diagrams. Each of the data points corresponds to one of the bins in the *upper* row. The points and the bins are colour-coded according to their η, a parameter that describes the distance between a given point in the diagnostic diagram and the bisector line between the continuous and the dashed lines in the plot (Erroz-Ferrer *et al.* 2019). Blue points indicate regions ionised by star formation whereas red points indicate shock ionisation. The disc of the galaxy (low-velocity dispersion component) is mostly ionised by star formation. The outflow is clearly shock-ionised. Axis labels in arcsec, N is up and E to the right.

Terlevich 1981) line ratio diagnostic diagrams for these two components separately, the low-dispersion component again on the left (upper panels). The points in the BPT diagrams are colour-coded by how far from the central lines they are, and are then mapped, with these colours, back into the spatial plane (lower panels). We see that the disk of the galaxy is mostly blue, which indicates star formation, whereas the high-velocity dispersion component returns alsmost exclusively red colours, which is shocked gas. This confirms that the high-dispersion component is driven by shocks, most likely by the AGN. The outflow zone is indeed seen to be related to both high dispersion and a shocked origin of the emission line ratios, confirming what we described in KCS19.

4. Conclusions

We present new adaptive optics MUSE integral-field observations, combined with archival *HST* and ALMA imaging, of the archetypical AGN-starburst composite galaxy and Seyfert 2 host NGC 7130. We reach *HST*-like spatial resolution, but with the full complement of absorption and emission lines of MUSE at our disposal. We find an outflow region of at least ∼600 pc, presumably driven by the AGN, which is set apart from the rest of the inner region by enhanced gas velocity dispersion, non-circular velocities, and emission line ratios indicative of shocks (see KCS19 for details).

We separate two components of the gas emission, with low and high velocity dispersion, and plot the velocity and dispersion fields as well as BPT diagnostic diagrams separately for these two components. The low-dispersion gas traces the star-forming and rotating inner disk of the galaxy. The high-dispersion component, in contrast, highlights the outflow region and shows the extent to which part of the gas across the inner kpc region is shocked, presumably by the AGN. A small kinematically decoupled core, see in KCS19, manifests itself exclusively in the high-dispersion gas, which is evidence for it being connected to the AGN and/or the outflow, rather than it being a traditional small disk rotating in a different plane.

Acknowledgements

J.H.K. acknowledges financial support from the European Union's Horizon 2020 research and innovation programme under Marie Skłodowska-Curie grant agreement No 721463 to the SUNDIAL ITN network, from the State Research Agency (AEI) of the Spanish Ministry of Science, Innovation and Universities (MCIU) and the European Regional Development Fund (FEDER) under the grant with reference AYA2016-76219-P, from IAC project P/300724, financed by the Ministry of Science, Innovation and Universities, through the State Budget and by the Canary Islands Department of Economy, Knowledge and Employment, through the Regional Budget of the Autonomous Community, and from the Fundación BBVA under its 2017 programme of assistance to scientific research groups, for the project "Using machine-learning techniques to drag galaxies from the noise in deep imaging".

References

Baldwin, J. A., Phillips, M. M., & Terlevich, R. 1981, *PASP*, 93, 5
Begelman, M. C., Blandford, R. D., & Rees, M. J. 1984, *Rev. Mod. Phys.*, 56, 255
Cappellari, M. 2017, *MNRAS*, 466, 798
Cappellari, M. & Emsellem, E. 2004, *PASP*, 116, 138
de Vaucouleurs, G., de Vaucouleurs, A., Corwin, Jr., H. G., *et al.* 1991, Third Reference Catalogue of Bright Galaxies
Erroz-Ferrer, S., Carollo, M. C., den Brok, M., *et al.* 2019, *MNRAS*, 484, 5009
Knapen, J. H., Comerón, S., & Seidel, M. K. 2019, *A&A*, 621, L5 (KCS19)
Phillips, M. M., Charles, P. A., & Baldwin, J. A. 1983, *ApJ*, 266, 485
Shlosman, I., Frank, J., & Begelman, M. C. 1989, *Nature*, 338, 451111
Véron-Cetty, M. P. & Véron, P. 2006, *A&A*, 455, 773
Weilbacher, P. M., Streicher, O., Urrutia, T., *et al.* 2012, *SPIE*, 8451, 84510B
Zhao, Y., Lu, N., Xu, C. K., *et al.* 2016, *ApJ*, 820, 118

The relationship between black hole accretion rate and gas properties at the Bondi radius

De-Fu Bu

Key Laboratory for Research in Galaxies and Cosmology, Shanghai Astronomical Observatory, Chinese Academy of Sciences, 80 Nandan Road, Shanghai 200030, China
email: dfbu@shao.ac.cn

Abstract. The mass accretion rate determines the black hole accretion mode and the corresponding efficiency of active galactic nuclei (AGNs) feedback. In large-scale simulations studying galaxy formation and evolution, the Bondi radius can be at most marginally resolved. In these simulations, the Bondi accretion formula is always used to estimate the black hole accretion rate. The Bondi solution can not represent the real accretion process. We perform 77 simulations with varying density and temperature at Bondi radius. We find a formula to calculate the black hole accretion rate based on gas density and temperature at Bondi radius. We find that the formula can accurately predict the luminosity of observed low-luminosity AGNs. This formula can be used in sub-grid models in large-scale simulations with AGNs feedback.

Keywords. accretion, accretion discs, black hole physics, hydrodynamics, galaxies:active, galaxies:nuclei

1. Introduction

Mass accretion rate is a key parameter in black hole accretion physics. It determines the accretion mode and the spectrum of a black hole system. AGN feedback plays an important role in the evolution of their host galaxy (Fabian 2012). AGNs outputs including wind (Li & Begelman 2014), radiation (Giotti & Ostriker 2001) and jet (Guo 2016) depend on accretion mode of the black hole. Different accretion modes can have very different AGNs feedbacks.

In large-scale simulations studying galaxy formation and evolution, the Bondi radius can be at most marginally resolved (e.g. Springel *et al.* 2005; Yuan *et al.* 2018). In these simulations, the Bondi formula is usually used to estimate the black hole accretion rate. The black hole growth and accretion mode are determined by the Bondi solution. The Bondi accretion model is simple but has some problems (Waters *et al.* 2019). For example, in the Bondi solution, it is assumed that the accretion rate is constant with radius in the region between black hole horizon and Bondi radius. In reality, due to the presence of wind, the accretion rate decreases from the Bondi radius towards the black hole. Some previous works (e.g. Igumenshchev & Narayan 2002; Moscibrodzka 2006; Gaspari *et al.* 2013) have shown that the Bondi formula is not a good estimate when calculating the black hole accretion rate.

We aim to find an analytical formula to calculate the real accretion rate onto the black hole as a function of density and temperature at the Bondi radius (see Bu & Yang 2019). We focus on low luminosity active galactic nuclei (LLAGNs), which has a luminosity lower than 2% Eddington luminosity. The accretion mode for LLAGNs is hot accretion flow (see Yuan & Narayan 2014 for reviews). Although radiation and wind of a LLAGN

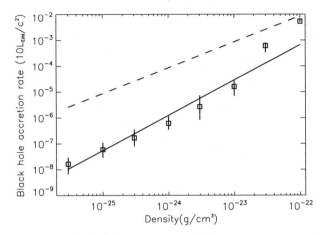

Figure 1. Black hole accretion rate in unit of Eddington rate versus gas density at Boundi radius, with gas temperature at Bondi radius fixed to be 10^7K. The solid line is from Equation (3.1). Squares correspond to time-averaged values of each simulation. The error bars over-plotted represent the change range of simulations owing to fluctuations. The dashed line is calculated by using Bondi formula, assuming adiabatic index $\gamma = 5/3$.

are not as strong as those for luminous AGNs, the AGNs spend most of its time in the LLAGN phase. Therefore, the cumulative effects of LLAGN (or hot accretion mode) feedback should be important (Yuan et al. 2018).

2. Numerical method

We set the black hole mass to be 10^8 solar mass. We consider slowly rotating accretion flows in the region $500r_s$–$10^6 r_s$ (r_s is the Schwarzschild radius). The accretion flow is irradiated by the photons from the region very close to the black hole. Because we focus on LLAGN, we assume that the photons from the region very close to the black hole are all X-ray photons. The gas can be Compton heated/cooled by the X-ray photons. The Compton temperature of the X-ray photons is set to be 10^8K (Xie et al. 2017). We can calculate the mass accretion rate at the inner radial boundary of the simulations. According to the simulation results of hot accretion flow (Yuan et al. 2012), we can know how the accretion rate changes with radius inside our inner radial boundary ($500r_s$). Then the black hole accretion rate can be calculated. The luminosity of the accretion flow can also be calculated by combining accretion rate and radiative efficiency (Xie & Yuan 2012).

3. Results

We perform 77 simulations with different gas densities and temperatures at the Bondi radius. We use power-law function of density and temperature at Bondi radius to fit the simulation results. We find that the black hole accretion rate can be described as follows,

$$\frac{\dot{M}_{BH}}{\dot{M}_{Edd}} = 10^{-3.11} \left(\frac{\rho_0}{10^{-22} \text{g} \cdot \text{cm}^{-3}}\right)^{1.36} \left(\frac{T_0}{10^7 \text{K}}\right)^{-1.9} \quad (3.1)$$

where $\dot{M}_{BH}$ and $\dot{M}_{Edd}$ are black hole mass accretion rate and Eddington accretion rate, respectively. ρ_0 and T_0 are gas density and temperature at Bondi radius.

We compare the results with the Bondi solution. Figure 1 shows the black hole accretion rate as a function of gas density at the Bondi radius. In order to plot Figure 1, we set $T_0 = 10^7$K. It is clear that the Bondi formula can over estimate the accretion rate by almost two orders of magnitude. We find that the reason for the significantly lower

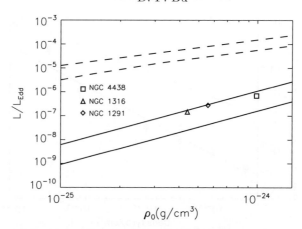

Figure 2. Luminosity of the accretion flow in unit of Eddington luminosity versus gas density at Boundi radius. The two solid lines are calculated using the formula Equation (3.1). The upper and lower solid lines correspond to $T_0 = 3.9 \times 10^6$ K and 7.2×10^6 K, respectively. The upper and lower dashed lines are calculated according to the adiabatic Bondi formula, assuming $T_0 = 3.9 \times 10^6$ K and 7.2×10^6 K, respectively. The square, triangle and diamond represent NGC 4438, 1316 and 1291, respectively.

accretion rate given by Equation (3.1) is that strong wind is present. Winds take away gas, which makes the black hole accretion rate significantly lower than that predicted by the Bondi formula. The mechanisms for driving wind depend on the accretion rate. We find that if $\dot{M}_{BH}/\dot{M}_{Edd} < 10^{-3}$, both radiative cooling and Compton heating/cooling are not important, wind is driven by gas pressure gradient force. If $\dot{M}_{BH}/\dot{M}_{Edd} > 10^{-3}$, wind is episodically driven by Compton heating. In this case, X-ray photons from the central region can heat gas, which makes gas temperature to be higher than virial temperature. Consequently, winds are launched. In AGNs, the winds driven by Compton heating have also been studied by Giotti & Ostriker (2001). Recently, it is found that Compton heating can also drive winds from a thin disk in a black hole X-ray binary system (Higginbottom et al. 2019).

We also do comparison with observations. In order to make the comparison, we need to look for observations of LLAGNs. The LLAGNs must satisfy three conditions. First, the black hole mass is $\sim 10^8$ solar mass; second, the gas density and temperature around the Bondi radius of these LLAGNs are known; finally, the luminosity of these LLAGNs is given by observations. There are three LLAGNs (Pellegrini 2005), namely NGC 4438, NGC 1316 and NGC 1291. The temperature of gas around the Bondi radius of these three LLAGNs is in the range $3.9-7.2 \times 10^6$ K. The results are shown in Figure 2. From this figure, we can see that the Bondi formula overestimates the luminosity of these three LLAGNs by two orders of magnitude. The fitting formula (Equation (3.1)) accurately predicts the luminosity of these three LLAGNs.

4. Implications

Two-dimensional simulations were performed to investigate slowly rotating accretion flows irradiated by a LLAGN at parsec and sub-parsec scales. We obtained a formula to calculate the black hole accretion rate based on the density and temperature of gas at Bondi radius. The formula can well predict the luminosity of LLAGNs. This formula can be used in subgrid models in large-scale simulations studying galaxy formation and evolution with AGNs feedback.

References

Bu, D. & Yang, X. 2019, *MNRAS*, 484, 1724
Ciotti, L. & Ostriker, J. P. 2001, *ApJ*, 551, 131
Fabian, A. C. 2012, *ARA&A*, 50, 455
Gaspari, M., Ruszkowski, M., & Oh, S. P. 2013, *MNRAS*, 432, 3401
Guo, F. 2016, *ApJ*, 826, 17
Higginbottom, N., Knigge, C., Long, K. S. *et al.* 2019, *MNRAS*, 484, 4635
Igumenshchev, I. V. & Narayan, R. 2002, *ApJ*, 566, 137
Li, S. & Begelman, M. C. 2014, *ApJ*, 786, 6
Moscibrodzka, M. 2006, *A&A*, 450, 93
Pellegrini, S. 2005, *ApJ*, 624, 155
Springel, V., Di Matteo, T., & Hernquist, L. 2005, *MNRAS*, 361, 776
Waters, T., Aykutalp, A., Proga, D., et al. 2019, arXiv:1910.01106
Xie, F. & Yuan, F. 2012, *MNRAS*, 427, 1580
Xie, F., Yuan, F., & Ho, L., C. 2017, *ApJ*, 844, 42
Yuan, F., Bu, D., & Wu, M. 2012, *ApJ*, 761, 130
Yuan, F. & Narayan, R. 2014, *ARA&A*, 52, 529
Yuan, F., Yoon, D., Li, Y., *et al.* 2018, *ApJ*, 857, 121

Tracing AGN feedback, from the SMBH horizon up to cluster scales

Francesco Tombesi[1,2,3,4]

[1]Department of Physics, University of Rome 'Tor Vergata', Via della Ricerca Scientifica 1, I-00133 Rome, Italy
[2]INAF Astronomical Observatory of Rome, Via Frascati 33, 00078 Monte Porzio Catone, Italy
[3]Department of Astronomy, University of Maryland, College Park, MD 20742, USA
[4]NASA/Goddard Space Flight Center, Code 662, Greenbelt, MD 20771, USA

Abstract. Observations performed in the last decades have shown that supermassive black holes (SMBHs) and cosmic structures are not separate elements of the Universe. While galaxies extend on spatial scales about ten orders of magnitude larger than the horizon of SMBHs, black holes would not exist without matter feeding them, and cosmic structures would not be the same without feedback from SMBHs. Powerful winds/jets in active galactic nuclei (AGN) may be the basis of this co-evolution. Synergistic observations in the X-rays and other wavebands have been proven to be fundamental to map AGN winds from the event horizon up to galaxy scales, providing a promising avenue to study the multi-phase SMBH feeding and feedback processes. Moreover, a spatially resolved, spectroscopic analysis of AGN in clusters will allow us to probe the multiphase medium ranging from galactic up to cluster scales. Revolutionary advances are expected in the upcoming decade with new multi-wavelength observatories, ranging from radio to X-rays.

Keywords. Active galactic nuclei, supermassive black holes, AGN feedback, X-ray astronomy, AGN winds

1. Introduction

Several observational and theoretical inferences require the existence of fundamental connections between the activity of supermassive black holes (SMBHs) and their host galaxies. For instance, the $M - \sigma$ relation indicates that the larger the SMBH mass, the larger is the velocity dispersion of stars in the bulges of galaxies (e.g., Kormendy & Ho 2013; Shankar et al. 2019). Moreover, active galactic nuclei (AGN) feedback is theoretically required in order to quench star formation in the most massive galaxies and to adequately model the observed distribution (e.g. Bower et al. 2012). How SMBHs, which gravitationally dominate on a linear size more than a billion times smaller than a galaxy and with a mass only about 0.1% of the stellar bulge mass (e.g., Magorrian et al. 1998; Häring & Rix 2004), can affect the galaxy scale environment is still debated.

Nevertheless, it is important to note that the gravitational energy of SMBHs far exceeds the binding energy of an entire galaxy bugle, thus even if a small fraction of such energy is converted and deposited in the galaxy environment, it could have a significant influence (e.g., Hopkins & Elvis 2010). Indeed, AGN radiation, jets or winds can release a significant amount of mass and energy in the host galaxy environment (e.g., Fabian 2012). However, it still remains to quantify the intensity the SMBH feedback on galaxy evolution and to establish the dominant mechanisms.

Accretion disk winds, and in particular ultra-fast outflows (UFOs), may play an important role in AGN feedback. Such UFOs are primarily observed as mildly-relativistic

($v_{out} \simeq 0.1$c) and highly ionized absorbers in the X-ray spectra of $>30\%$ of local radio-quiet and radio-loud AGN (e.g. Tombesi et al. 2010, 2011, 2014; Gofford et al. 2013). Although with relatively large uncertainties, their mass outflow rates and energetics are found to be comparable to the AGN accretion rate and higher than a few percent of the AGN luminosity, respectively (e.g., Tombesi et al. 2012, 2013; Gofford et al. 2015).

2. Overview

2.1. From nuclear to galaxy scale feedback

In order to effectively impact the host galaxies, AGN winds need to propagate from the accretion disk and to entrain/disrupt the host interstellar medium up to galaxy scales. In our Tombesi et al. (2015) work we reported, for the first time observationally, a link between the nuclear, mildly relativistic UFO observed in the X-rays with *Suzaku* and the galaxy-scale molecular outflow in the IR detected with *Herschel* in the ultra-luminous infrared galaxy (ULIRG) IRAS F11119+3257. The OH region of the IR spectrum of the galaxy shows a prominent P-Cygni line profile indicating a molecular outflow with a maximum velocity of $\simeq 1,000$ km s^{-1}, implying a mass outflow rate of ~ 800 $M_\odot$/yr and a distance of > 300 pc from the center Veilleux et al. (2013). Such massive molecular outflows, potentially affecting star formation in the host, were detected in several other sources, but their connection with the AGN was unclear (e.g., Sturm et al. 2011). The X-ray detection of a highly ionized and massive UFO with velocity of $v_{out} \simeq 0.25$c provided the evidence relating the phenomenon to AGN feedback (Tombesi et al. 2015).

The presence and characteristics of the X-ray UFO and the molecular outflow were subsequently independently confirmed using data from *NuSTAR* and *ALMA* in the mm-band, respectively (Tombesi et al. 2017; Veilleux et al. 2017). The energetics of the X-ray UFO and the molecular outflows in IRAS F11119+3257 support theoretical models describing the origin of galaxy-scale molecular outflows as due to a energy-conserving hot bubble inflated by the inner AGN wind, with a coupling efficiency of $\sim 20\%$ (e.g., Faucher-Giguère & Quataert 2012; Zubovas & King 2014).

Recent results on other sources support such picture, but the coupling between the inner UFO and the neutral/molecular outflow is found between the two extreme regimes of momentum- and energy-driving, indicating either a low efficiency of the coupling and/or in-situ formation of molecular gas depending on the particular environment in each galaxy (Feruglio et al. 2015, 2017; Mizumoto et al. 2019; Bischetti et al. 2019. In Fig. 1 we show the recent results on another ULIRG, IRAS F05189−2524, along with a collection of literature results (Smith et al. 2019).

Observational evidence for the physical interaction between the putative X-ray UFO and the shocked ISM was hard to obtain. However, an intersting result was found by our group while studying the X-ray spectrum of the quasar PG 1114+445 (Serafinelli et al. 2019). This source was observed 12 times with *XMM-Newton* and once by *ASCA*. We found three distinct X-ray absorbers, variable but persistent over 15 years. The average parameters of the absorbers over the whole X-ray monitoring are: Abs1 with log$N_H = 21.88 \pm 0.05$ cm^{-2}, log$\xi = 0.35 \pm 0.04$ erg s^{-1} cm, and $v_{out} \simeq 530$ km s^{-1}; Abs2 with log$N_H = 21.5 \pm 0.2$ cm^{-2}, log$\xi = 0.50 \pm 0.36$ erg s^{-1} cm, and $v_{out} = 0.120 \pm 0.029$ c; Abs3 with log$N_H = 22.9 \pm 0.3$ cm^{-2}, log$\xi = 4.04 \pm 0.29$ erg s^{-1} cm, and $v_{out} = 0.145 \pm 0.035$ c.

The first absorber is consistent with the typical parameters observed for X-ray warm absorbers (WAs) and it has a counterpart observed in the UV (Mathur et al. 1998). The third absorber is consistent with the typical highly ionized and mildly relativistic UFOs observed in Seyfert galaxies (e.g., Tombesi et al. 2011). Instead, the second absorber seems to have parameters which are intermediate between the two, with column and ionization

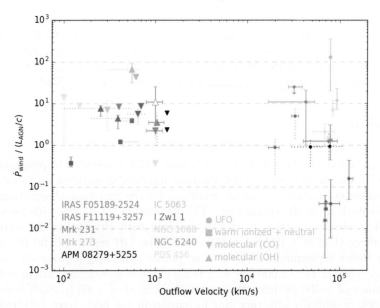

Figure 1. The momentum rate ($\dot{P}_{wind}$) normalized by the momentum of the radiation (L_{AGN}/c) is plotted against the wind outflow velocity for ten objects with observed UFOs and large-scale galactic outflows with good constraints on their spatial scales. Solid error bars indicate that upper and lower errors were calculated whereas dotted error bars indicate that only a range of values was provided. Arrows indicate limits. UFO measurements are plotted as circles, warm ionized and neutral gas as squares, the molecular (CO) as downward triangles, and the molecular (OH) as upward triangles. For molecular measurements, filled symbols indicate a time-averaged momentum rate whereas an open symbol is an "instantaneous" or local momentum rate. Figure adapted from Smith et al. (2019).

of WAs and velocity of UFOs. We interpret this intermediate absorber as an "Entrained ultra-fast outflow" (E-UFO) indicating the interaction region between the nuclear UFO and the ambient medium in this galaxy at a distance of ∼100 pc from the central AGN. The shock or interaction region is likely clumpy and unstable due to Rayleigh-Taylor and Kelvin-Helmholtz instabilities. Absorbers with similar characteristics to the E-UFO discussed in Serafinelli et al. (2019) and complex multi-phase Fe K outflows are being found in an incresing number of sourcers (e.g., Gupta et al. 2013, 2015; Longinotti et al. 2015; Pounds et al. 2016; Reeves et al. 2016; Middei et al. 2020).

2.2. From nuclear to cluster scales feedback

It is widely known that AGN in galaxy clusters preferentially reside in the central brightest cluster galaxy (BCG) and can even influence the cluster gas through powerful winds or jets (e.g., Fabian 2012). However, it is still hard to observationally constrain and quantify such effect.

In Tümer et al. (2019) we recently reported a detailed X-ray spectral/imaging analysis of XMM-Newton and Chandra observations of the non-cool core (NCC) galaxy cluster MKW 08 and its central brightest cluster galaxy (BCG), namely NGC 5718. As shown in Fig. 2, we have been able to trace a multi-phase medium extending from within the interstellar medium/corona of the central BCG at scales of less than ∼3 kpc with temperature of $kT \simeq 1$ keV, up to the hot intracluster medium (ICM) with temperature of $kT \simeq 4$ keV for scales larger than ∼10 kpc up to ∼500 kpc. The cooling time of the 1 keV gas at scales of less than 3 kpc is estimated to be of ∼65 Myrs. This is much

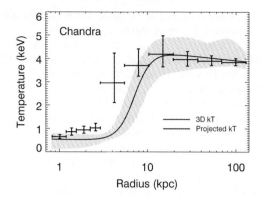

Figure 2. Projected radial ICM temperature values plotted over the deprojected temperature profiles using *Chandra* X-ray observations of the galaxy cluster MKW 08 centered on the BCG. Confidence contours (blue) were propagated from Monte Carlo realizations of the fit analytical functions. Figure adapted from Tümer et al. (2019).

shorter compared to the $>$ Gyrs timescales for the ICM and it requires a heating source in order to counterbalance catastrophic cooling. We find that AGN feedback may provide the main source of heating.

3. Implications

Our recent observational results support the picture of a mutual evolution between SMBHs and their host galaxies. This is in accordance with recent theoretical works and simulations exploring the SMBH and galaxy environments from micro (sub-pc) to meso (kpc) and macro (Mpc) scales. The scenario that is emerging is one of a self-regulation of SMBH growth, with the feeding by chaotic cold accretion (CCA) raining down from the macro and meso environment mediated by the feedback from AGN winds and jets launched at micro scales (e.g., Gaspari & Sadowski 2017). Upcoming large multi-wavelength observatories in the next decade, from radio to X-rays, in synergy with the most advanced computer simulations, will provide unprecedented advances in this field of research.

References

Bischetti, M., Piconcelli, E., Feruglio, C., et al. 2019, *A&A*, 628, A118
Bower, R. G., Benson, A. J., & Crain, R. A. 2012, *MNRAS*, 422, 2816
Fabian, A. C. 2012, *ARA&A*, 50, 455
Faucher-Giguère, C.-A. & Quataert, E. 2012, *MNRAS*, 425, 605
Feruglio, C., Fiore, F., Carniani, S., et al. 2015, *A&A*, 583, A99
Feruglio, C., Ferrara, A., Bischetti, M., et al. 2017, *A&A*, 608, A30
Gaspari, M. & Sadowski, A. 2017, *ApJ*, 837, 149
Gofford, J., Reeves, J. N., Tombesi, F., et al. 2013, *MNRAS*, 430, 60
Gofford, J., Reeves, J. N., McLaughlin, D. E., et al. 2015, *MNRAS*, 451, 4169
Gupta, A., Mathur, S., Krongold, Y., et al. 2013, *ApJ*, 772, 66
Gupta, A., Mathur, S., & Krongold, Y. 2015, *ApJ*, 798, 4
Häring, N. & Rix, H.-W. 2004, *ApJ*, 604, L89
Hopkins, P. F. & Elvis, M. 2010, *MNRAS*, 401, 7
Kormendy, J. & Ho, L. C. 2013, *ARA&A*, 51, 511
Longinotti, A. L., Krongold, Y., Guainazzi, M., et al. 2015, *ApJ*, 813, L39
Magorrian, J., Tremaine, S., Richstone, D., et al. 1998, *AJ*, 115, 2285
Middei, R., Tombesi, F., Vagnetti, F., et al. 2020, accepted in *A&A*, arXiv:2001.03979
Mizumoto, M., Izumi, T., & Kohno, K. 2019, *ApJ*, 871, 156

Pounds, K. A., Lobban, A., Reeves, J. N., *et al.* 2016, *MNRAS*, 459, 4389
Reeves, J. N., Braito, V., Nardini, E., *et al.* 2016, *ApJ*, 824, 20
Serafinelli, R., Tombesi, F., Vagnetti, F., *et al.* 2019, *A&A*, 627, A121
Shankar, F., Bernardi, M., Richardson, K., *et al.* 2019, *MNRAS*, 485, 1278
Smith, R. N., Tombesi, F., Veilleux, S., *et al.* 2019, *ApJ*, 887, 69
Sturm, E., González-Alfonso, E., Veilleux, S., *et al.* 2011, ApJ, 733, L16
Tombesi, F., Cappi, M., Reeves, J. N., *et al.* 2010, *A&A*, 521, A57
Tombesi, F., Cappi, M., Reeves, J. N., *et al.* 2011, *ApJ*, 742, 44
Tombesi, F., Cappi, M., Reeves, J. N., *et al.* 2012, *MNRAS*, 422, L1
Tombesi, F., Cappi, M., Reeves, J. N., *et al.* 2013, *MNRAS*, 430, 1102
Tombesi, F., Tazaki, F., Mushotzky, R. F., *et al.* 2014, *MNRAS*, 443, 2154
Tombesi, F., Meléndez, M., Veilleux, S., *et al.* 2015, *Nature*, 519, 436
Tombesi, F., Veilleux, S., Meléndez, M., *et al.* 2017, *ApJ*, 850, 151
Tümer, A., Tombesi, F., Bourdin, H., *et al.* 2019, *A&A*, 629, A82
Veilleux, S., Meléndez, M., Sturm, E., *et al.* 2013, *ApJ*, 776, 27
Veilleux, S., Bolatto, A., Tombesi, F., *et al.* 2017, *ApJ*, 843, 18
Zubovas, K. & King, A. R. 2014, *MNRAS*, 439, 400

Feedback and star formation in AGNs

Luis C. Ho[1,2]

[1]Kavli Institute for Astronomy and Astrophysics, Peking University, Beijing 100871, China
[2]Department of Astronomy, School of Physics, Peking University, Beijing 100871, China

Abstract. I will describe a series of experiments designed to use measurements of the cold ISM content of quasars to constrain the effectiveness of AGN feedback. I will propose new star formation rate indicators and apply them to constrain the star formation rate in quasars.

Keywords. galaxies: active, active: quasars, galaxies: ISM

The role of AGN in galaxy star formation: A case study of a radio galaxy at z = 2.6

Allison Man

Dunlap Institute for Astronomy and Astrophysics, University of Toronto, Toronto, ON, Canada

Abstract. Radio galaxies are ideal sites to scrutinize AGN feedback physics, as they are massive galaxies with jets that interact with the surrounding ISM. I will present a detailed analysis of the recent star formation history and conditions of a starbursting, massive radio galaxy at z = 2.6, PKS 0529-549. In the 8.5-hour VLT/X-Shooter spectrum, we detect unambiguous signatures of stellar photospheric absorption lines originating from OB-stars. Comparison with model spectra shows that more than one burst took place in its recent past: the most recent one at $4-7$ Myr, and another aged $\geqslant 20$ Myr. ALMA observations of the [CI] atomic carbon emission line indicates that it has a low molecular gas fraction ($\sim 13\%$) and short depletion time (~ 40 Myr). Most intriguing is the modest velocity dispersion ($\leqslant 50$ km/s) of these photospheric lines and the ALMA [CI] cold gas. We attribute its efficient star formation to compressive gas motions, induced by radio jets and/or interaction. Star formation works in concert with the AGN to remove any residual molecular gas and eventually leads to quenching.

Keywords. galaxies: active, galaxies: star formation, methods: spectroscopy

Ionized AGN outflows are less powerful than assumed: A multi-wavelength census of outflows in type II AGN

Dalya Baron

School of Physics and Astronomy, Tel-Aviv University, Tel Aviv 69978, Israel

Abstract. In this talk I will show that multi-wavelength observations can provide novel constraints on the properties of ionized gas outflows in AGN. I will present evidence that the infrared emission in active galaxies includes a contribution from dust which is mixed with the outflow and is heated by the AGN. We detect this infrared component in thousands of AGN for the first time, and use it to constrain the outflow location. By combining this with optical emission lines, we constrain the mass outflow rates and energetics in a sample of 234 type II AGN, the largest such sample to date. The key ingredient of our new outflow measurements is a novel method to estimate the electron density using the ionization parameter and location of the flow. The inferred electron densities, $\sim 10^{4.5}$ cm^{-3}, are two orders of magnitude larger than found in most other cases of ionized outflows. We argue that the discrepancy is due to the fact that the commonly-used [SII]-based method underestimates the true density by a large factor. As a result, the inferred mass outflow rates and kinetic coupling efficiencies are 1–2 orders of magnitude lower than previous estimates, and 3–4 orders of magnitude lower than the typical requirement in hydrodynamic cosmological simulations. These results have significant implications for the relative importance of ionized outflows feedback in this population.

Keywords. galaxies: active, active: outflows, galaxies: properties

Clustering dependence of Chandra COSMOS Legacy AGN on host galaxy properties

Viola Allevato[1,2]

[1]Scuola Normale Superiore, Piazza dei Cavalieri 7, I-56126 Pisa, Italy
[2]Department of Physics and Helsinki Institute of Physics, Gustaf Hällströmin katu 2a, 00014 University of Helsinki, Finland

Abstract. The presence of a super massive BH in almost all galaxies in the Universe is an accepted paradigm in astronomy. How these BHs form and how they co-evolve with the host galaxy is one of the most intriguing unanswered problems in modern Cosmology and of extreme relevance to understand the issue of galaxy formation. Clustering measurements can powerfully test theoretical model predictions of BH triggering scenarios and put constraints on the typical environment where AGN live in, through the connection with their host dark matter halos. In this talk, I will present some recent results on the AGN clustering dependence on host galaxy properties, such as galaxy stellar mass, star formation rate and specific BH accretion rate, based on X-ray selected Chandra COSMOS Legacy Type 2 AGN. We found no significant AGN clustering dependence on galaxy stellar mass and specif BHAR for Type 2 COSMOS AGN at mean $z \sim 1.1$, with a stellar - halo mass relation flatter than predicted for non active galaxies in the Mstar range probed by our sample. We also observed a negative clustering dependence on SFR, with AGN hosting halo mass increasing with decreasing SFR. Mock catalogs of active galaxies in hosting dark matter halos with logMh[Msun] > 12.5, matched to have the same X-ray luminosity, stellar mass and BHAR of COSMOS AGN predict the observed Mstar - Mh, BHAR - Mh and SFR-Mh relations, at $z \sim 1$.

Keywords. galaxies: active, galaxies: environment, galaxies: properties

CHAPTER VI. Jets and environment

Radio jets: Properties, life and impact

Raffaella Morganti[1,2]

[1]ASTRON, the Netherlands Institute for Radio Astronomy, Oude Hoogeveensedijk 4,
7991 PD Dwingeloo, The Netherlands

[2]Kapteyn Astronomical Institute, University of Groningen, P.O. Box 800,
9700 AV Groningen, The Netherlands
email: `morganti@astron.nl`

Abstract. Our view of the properties of extragalactic radio jets and the impact they have on the host galaxy has expanded in the recent years. This has been possible thanks to the data from new or upgraded radio telescopes. This review briefly summarises the current status of the field and describes some of the exciting recent results and the surprises they have brought. In particular, the physical properties of radio jets as function of their radio power will be discussed together with the advance made in understanding the life-cycle of radio sources. The evolutionary stage (e.g. newly born, dying, restarted) of the radio AGN can be derived from their morphology and properties of the radio spectra. The possibilities offered by the new generation of low-frequency radio telescopes make it possible to derive (at least to first order) the time-scale spent in each phase. The presence of a cycle of activity ensures a recurrent impact of the radio jets on their surrounding inter-stellar and inter-galactic medium and, therefore, their relevance for AGN feedback. The last part is dedicated to the recent results showing the effect of jets on the surrounding galactic medium. The predictions made by numerical simulations on the impact of a radio jet (and in particular a newly born jet) on a clumpy medium describe well what is seen by the observations. The high resolution studies of jet-driven outflows of cold gas (H I and molecular) has provided new important addition both in term of quantifying the impact of the outflows and their relevance for feedback as well as for providing an unexpected view of the physical conditions of the gas under these extreme conditions.

Keywords. galaxies: active, galaxies: jets, radio continuum: galaxies, ISM: jets and outflows

1. Introduction

Radio jets are one spectacular manifestation of nuclear activity in galaxies. They are launched by a super massive black hole (SBMH) and collimated very close to it thanks to the combined effects of black hole spin and magnetic field (see Blandford *et al.* 2019, for a review and Fig. 1 for an example). This short review will discuss some of the latest results on their energetics, life-cycle and impact on the surrounding medium.

Extragalactic radio jets span over a huge range of sizes (from pc to Mpc) and show a variety of structures and physical properties (see Sec. 2). They are also known to be recurrent during the life of a massive galaxy (see Sec. 3). All these properties have implications for the impact the jets have on the surrounding medium.

Especially on cluster scale, they have been identified as responsible for preventing gas from cooling, therefore representing one of the clearest case of feedback in action driven by Active Galactic Nuclei (AGN) (see e.g. review by McNamara & Nulsen 2012). However, their impact can also be visible on galaxy scales (see Sec. 4) and can be considered complementary to the effects of radiation and AGN-driven winds.

New results improving our understanding of radio jets and of their impact, have been made possible thanks to the upgrade of a number of radio telescopes, like JVLA, GMRT,

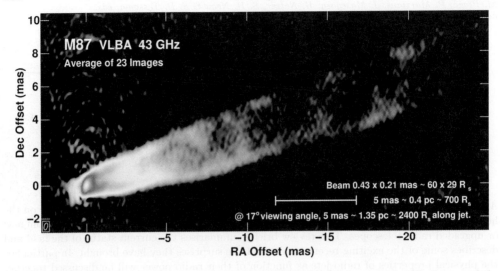

Figure 1. Very Long Baseline Array (VLBA) 43 GHz, 23-epoch average radio image of the jet and counterjet in M87 based on data from 2007 and 2008. Image taken from Walker et al. (2018); © 2018 The American Astronomical Society.

WSRT-Apertif, eVLBI and the coming in operation of new radio telescopes, like the Low Frequency Array (LOFAR; van Haarlem et al. 2013), the Murchison Widefield Array (MWA; Tingay et al. 2013) and ALMA. More radio telescopes are starting their operation, e.g. MeerKAT (Jonas & MeerKAT Team 2016), and ASKAP (Johnston et al. 2007) or are planned for the coming years, i.e. SKA and SKA-VLBI (Paragi et al. 2015), the latter particularly relevant for the connection to many countries in Africa. Thus, these are extremely interesting times for the study of radio jets. Especially relevant are the increased sensitivity and increased field-of-view that some of these telescopes have brought. Also important are the new possibilities offered by the low frequencies surveys, now reaching spatial resolutions comparable to what has been achieved so far only at GHz wavelengths. This is making possible the study of jets in this previously unexplored window. Particularly interesting are the results obtained by the LOFAR Two Meter Survey (LoTSS): an overview can be found in a special issue of A&A (February 2019) and in Shimwell et al. (2019).

It is worth noting that the occurrence of radio emission and of radio jets is a strong function of the mass of the host galaxy and of the radio power of the source, see Sabater et al. (2019) and refs therein. Thus, in massive galaxies ($M_* \gtrsim 10^{11}$ $M_\odot$) the presence of a radio AGN with radio power $< 10^{23}$ W Hz^{-1} can reach above 80%, while the fraction remains at most 30% for radio sources above 10^{24} W Hz^{-1} (Sabater et al. 2019).

2. Jets as function of the radio power

Figure 2 shows examples of radio jets observed in sources of different radio power. The figure illustrates the first-order dependence of the morphology of the radio jet on this parameter. In the most powerful sources ($P_{1.4 \text{ GHz}} \gtrsim 10^{24}$ W Hz^{-1}), typically hosted by early-type galaxies, the jets can reach many hundred of kpc to above Mpc in size, something that does not happen among the low power radio sources. The lower power sources (with power $P_{1.4 \text{ GHz}} \lesssim 10^{24}$ W Hz^{-1}), are often referred to as *radio-quiet* (see Kellermann et al. 1989, and discussion below).

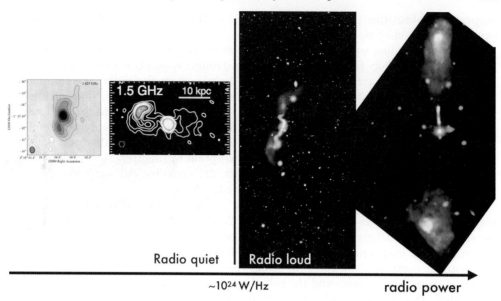

Figure 2. Examples of morphology of radio jets observed in sources with different radio power. On the left two examples of radio-quiet sources. NGC 1266 with radio power $P_{1.4\ \mathrm{GHz}} = 9.3 \times 10^{20}$ W Hz^{-1}(Alatalo et al. 2011), and the so-called "Tea cup" source with radio power $P_{1.4\ \mathrm{GHz}} = 5 \times 10^{23}$ W Hz^{-1}(Harrison et al. 2015) representing radio-quiet sources. On the right, two examples of radio loud sources. The giant FRI radio galaxy 3C 31 as observed by LOFAR at 150 MHz, from Heesen et al. (2018) and the FRII radio galaxy 3C 219 from Clarke et al. (1992) and Bridle Picture gallery; © NRAO/AUI 1999.

Following the seminal work of Fanaroff & Riley (1974), the radio-loud group of sources has been historically separated into two morphological classes, the so-called Fanaroff-Riley I and II (FRI and FRII). The separation is considered to be a function of radio power, with FRII becoming prominent for power above $P_{1.4\ \mathrm{GHz}} \sim 10^{25.5}$ W Hz^{-1}. However, a sharp separation in term of radio power has been recently questioned by Mingo et al. (2019), as result of the automatic classification performed on a large sample of radio galaxies produced by the LOFAR surveys. The Fanaroff-Riley division dependents on the properties of the jets and is, therefore, connected to the properties of the central engine (e.g. accretion rate, BH spin etc.) and to the environment into which the jet is expanding. Differences in the physical parameters of jets have been suggested and observed for FRI and FRII, e.g. differences in the composition (i.e. fraction of thermal component), speed of the jet and spectral properties. In particular, the spectral properties represent a powerful tool to trace the ageing of the relativistic electrons and test for its acceleration.

For FRI, a detailed description of the observed properties can be found in Laing & Bridle (2012, 2014); Laing (2015) and refs therein. In general, the jets in these sources widen rapidly. They decelerate from relativistic to sub-relativistic speeds on scales of 1-10 kpc (Laing & Bridle 2012). The deceleration and the turbulence in the flow result in strong entrainment of external thermal medium. This helps balancing the pressure between the jet and the external medium, preventing the jet to be under-pressured, as otherwise suggested by X-ray observations (Morganti et al. 1987; Worrall & Birkinshaw 2000). Thus, jets in FRI sources can contain a large fraction of non-radiating, thermal medium.

The situation is different in FRII. Hot spots are observed at the end of the jets indicating high Mach number jets and overpressure with respect to the medium. The spectral properties confirm this (Harwood et al. 2016). In these jets, no strong entrainment is expected (Croston et al. 2018).

Although according to the work of Sabater et al. (2019) and refs therein, the stellar mass of the host galaxy (M_*) appears to have a stronger connection to the radio emission than the mass of the central BH (M_{BH}), the link between the radio emission and the central engine is suggested by the properties of the optical emission lines of the host galaxy. In radio-loud galaxies the strength of these lines correlates, to first order, with the power of the radio source, i.e. the host of FRI radio galaxies show only weak emission lines while FRII tend to show strong emission lines (see Tadhunter 2016, for a review on the optical properties of radio galaxies). The exception to this trend is represented by a group of FRII showing only weak emission lines, i.e. weak-line radio galaxies (WLRG). The nature of this group is still unclear (Tadhunter 2016).

Despite the fact that most of the attention is given to extended sources, the great majority of radio sources are actually small (< 20 kpc) and appear unresolved in most of the surveys. Interestingly, in the past years more attention has been given to these small sources. Indeed, even a new class in the Fanaroff-Riley classification has been introduced (not without controversies): the FR0 group (Sadler et al. 2014; Baldi et al. 2015). The nature of these radio sources and the role they play is still a matter of discussion (Baldi et al. 2019). However, their relevance is that they release their energy on galaxy-scale.

The so-called *radio-quiet* sources (following Kellerman et al. 1989, these are sources with a relatively low radio-to-optical flux density ratios, $R < 10$, and low radio powers, below 10^{24} W Hz^{-1}), represent a mix bag of objects, that includes Seyfert galaxies, quasars and low luminosity AGN (LLAGN). The definition of radio-quiet can be misleading and it should be kept in mind that these sources are not *radio-silent* and in many cases they show radio jets (see also discussion in Padovani 2017).

Although as the radio power decreases it is more difficult to separate radio emission from AGN and starformation, a large number of radio-quiet sources have jets and should be treated as proper radio AGN (see examples in samples studied by Gallimore et al. 1999; Morganti et al. 1999; Jarvis et al. 2019). Despite the diversity observed in their host galaxy, all radio sources classified as radio-quiet have a number of properties in common. For example, they are typically of small sizes (up to at most a few tens of kpc), they have more complex morphologies and they are dominated by entrainment (due to the low velocity of their jets), therefore their jets can have a large fraction of thermal component.

Interestingly, the most recent work on a sample of radio-quiet, obscured quasars by Jarvis et al. (2019) has shown that the majority of the targets exhibit extended radio structures on 1 to 25 kpc scales. These radio features are associated with morphologically and kinematically distinct components in the ionised gas. This is also confirmed by detailed studies of single, radio-quiet objects e.g. Seyfert galaxies like NGC 1068, IC 5063, Mrk 6 (see García-Burillo et al. 2014; Morganti et al. 2015; Kharb et al. 2014, and refs therein) and LLAGN (Alatalo et al. 2011; Combes et al. 2013; Riffel et al. 2014; Rodríguez-Ardila et al. 2017; Fabbiano et al. 2018; Maksym et al. 2019; Murthy et al. 2019; Husemann et al. 2019a). Thus, the radio jets observed in these objects, despite their low power, can have an impact on the surrounding gas, disturb the kinematics and, in some cases, influence the physical properties (e.g. ionisation via shocks). This is further discussed in Sec 4.

Regardless the morphology and classification of the radio jets, one of the parameters that is key for quantifying their impact on the surrounding interstellar medium (ISM) is the *jet power*. This gives the actual energetics available to the jet, combining radiative and non-radiative components, i.e. taking into account the thermal component. Because of this, measuring the jet power is not trivial. Although other methods have also been proposed, see e.g. Willott et al. (1999), the studies of the X-ray cavities have provided a powerful way to estimate the jet power. By measuring the work needed to inflate the cavities, the power required from the radio plasma can be derived and then related to the radio luminosity (see Cavagnolo et al. 2010; McNamara & Nulsen 2012, for a review). The relation derived in Cavagnolo et al. (2010), $P_{jet} \approx 5.8 \times 10^{43}(P_{radio}/10^{40})^{0.70}$ erg s^{-1}, is often used for this purpose. However, it is important to keep in mind the limitations of this approach, see discussion in Croston et al. (2018); Shabala & Godfrey (2013). In low power jets, the jet power is more uncertain because of the large fraction of thermal component coming from the entrainment. Because of this, the power of these jets can be larger than expected from their radio luminosity, see Bicknell et al. (1998) and Mukherjee et al. (2018a) for a recent example. Thus, given that these jets release their energy on galactic scales, they can play a relevant role in affecting the medium of their host galaxy and a better understanding of their jet power is important (see below).

3. Life-cycle of radio jets

The recurrence of the active phase of the super massive BH (SMBH) is a key ingredient in cosmological simulations of galaxy evolution, because it ensures that the energy released by the SMBH impacts the host galaxy multiple times as needed in the feedback cycle (see e.g. Ciotti et al. 2010; Novak et al. 2011; Gaspari et al. 2017). For radio AGN, we know that the period of activity is followed by a remnant phase when the nuclear activity switches off or drastically decreases. We also know that this cycle can repeat. The availability of deep (and high spatial resolution) surveys at low frequencies is a major steps forward for quantifying the time-scale of their cycle of activity (see also Morganti 2017a, for an overview). This is because the emission at low frequencies remains, for the longest time, unaffected by energy losses, thereby acting as a "fossil record".

In radio AGN, we can identify newly started jets by the combination of small size and peaked radio spectra. Because of these properties, these young sources are usually called Compact Steep Spectrum (CSS) and GigaHertz Spectrum (GPS), see O'Dea (1998); Orienti (2016) for overviews of their properties. Their ages are typically $< 10^6$ yrs, and they have not emerged yet from the galactic medium (i.e. their size is $\lesssim 10$ kpc). Most of these sources (albeit not all, see Kunert-Bajraszewska et al. 2010, for exceptions) are believed to expand and break out from the galactic ISM, and grow to evolved radiogalaxies. Most of the life of a radio galaxy is spent in this phase (lasting roughly few $\times 10^7$ up to a few $\times 10^8$ yrs). Modelling of this active phase, in particular for FRII radio galaxies, have been presented in a number of studies, starting with the seminal work of Scheuer (1974), to more recent studies, e.g. Kaiser & Best (2007); Hardcastle (2018) and refs therein. After this active phase, the jet can switch off. In the dying phase, known as *remnant phase*, the radio emission quickly fades away due to radiative and adiabatic losses, see e.g. Parma et al. (2007); Brienza et al. (2017) and refs therein. Interestingly, radio AGN can also have a *restart phase*. The best examples of this are the so-called "double-double" radio galaxies. An example can be seen in Fig. 3 (left) and more details can be found in Schoenmakers et al. (2000); Konar et al. (2012); Mahatma et al. (2019) and refs therein. In these sources two sets of lobes are observed, resulting from two well separate phases of activity. The remnant phase in these objects has a duration comparable or shorter than their active phase, ranging from a few Myr to a few tens of Myr (e.g. Konar et al. 2012). Other type of restarted radio galaxies are known, for example those showing a bright inner, newly born radio source embedded in a low surface brightness emission, reminiscent of a remnant structure, see Fig. 3 (right) for an example. In these sources, the time scales of the active and remnant phases can only be derived from a detailed analysis of the radio spectrum (see e.g. Brienza et al. 2018).

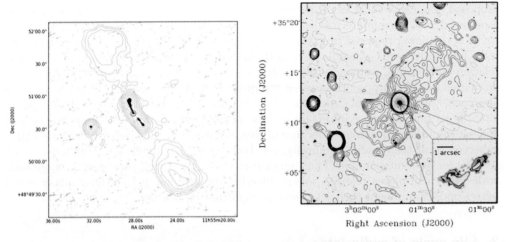

Figure 3. Examples of restarted radio galaxies. *Left:* double-double radio galaxy from the study of Mahatma *et al.* (2019). The image shows the emission at 1.4 GHz from the in greyscale, overlaid with the 144 MHz LOFAR contours from the LoTSS DR1 in orange. *Right:* Contours of the diffuse radio emission around B2 0258+35 over-laid on a DSS2 image. Figure taken from Shulevski *et al.* (2012). The inset at the bottom right shows the young, restarted source (image taken from Giroletti *et al.* 2005). A full discussion of the properties of this radiogalaxy can be found in Brienza *et al.* (2018).

Thanks to the low frequency radio data which are coming on-line (and in particular the data from the LOFAR surveys), we are getting a better view of the life cycle of radio galaxies. Brienza *et al.* (2017) and Hardcastle (2018) have suggested that remnant sources fade rapidly, thus most of the observed remnant radio galaxies are relatively young, with total ages between 5×10^7 and 10^8 yr. The majority of the remnant sources would be observed soon after the switch-off of the radio source and they are expected to evolve quickly due to dynamic expansion. The restarted phase can also follow shortly after, in a similar way as found for the group of double-double radio galaxies (Jurlin *et al.* 2020). It is interesting to note that these findings are consistent with what derived from the study of X-ray cavities in radio galaxies in clusters and groups, see e.g. Randall *et al.* (2011); Vantyghem *et al.* (2014). A full overview of the life cycle can only be obtained by including these findings in the context of modelling the evolution with respect to the luminosity function of radio galaxies (e.g. Shabala *et al.* 2020).

4. Jets and their impact

Radio jets are recognised to play a role in feedback by preventing the cooling of hot (X-ray) gas surrounding central galaxies in clusters. Extensive work has been done on this, see e.g. McNamara & Nulsen (2012). This role has been identified as *jet-mode* feedback and associated to radio sources characterised by radiatively-inefficient accretion, i.e. FRI. However, radio jets can also drive massive gas outflows on galactic scales, an other signature of AGN feedback. We will focus here on this role of the radio jets. An overview of the relevance of outflows for feedback and galaxies evolution can also be found in a number of reviews, e.g. King & Pounds (2015); Morganti (2017b); Harrison *et al.* (2018) and refs therein.

Radio jets have been known a long time to be able to drive gas outflows (see e.g. Axon *et al.* 1998; Capetti *et al.* 1999). Most of these studies focused on ionised gas (warm and hot). In the last years, however, the study of outflows has expanded also to the cold component of the gas: H I and molecular (see Morganti *et al.* 2005a; Feruglio *et al.* 2010, as examples of earlier studies). The finding of cold gas associated with AGN-drive outflows has been quite surprising. The origin of this gas associated with fast outflows is still a matter of debate. The most likely explanation is that the component of cold gas is due to fast cooling after the gas is shocked by the interaction with the jet. Interestingly, in most cases studied so far, the cold phase of the gas appears to carry the larger mass of the outflow, i.e. higher compared to what associated to the warm ionised component of the outflow.

In addition to this, the improvement of the numerical simulations describing the impact of jets on the surrounding medium has further helped in the intepretation of the results from the observations. The new generation of numerical simulations assume more realistic conditions for the gas the jet expands into. They are finding that radio jets couple strongly to the *clumpy* ISM of the host galaxy, see Wagner *et al.* (2012); Mukherjee *et al.* (2016); Cielo *et al.* (2018); Mukherjee *et al.* (2018a,b) for the details. According to these numerical simulations, a clumpy ISM, instead of a smooth one (as was assumed so far in the simulations) is key in making the impact of the jet much larger than previously found: because of the clumpiness of the gaseous medium, the progress of the jet can be temporarily halted when hitting a dense gas cloud. Thus, the jet is meandering through the ISM to find the path of minimum resistance and, doing so, creating a cocoon of shocked gas driving an outflow in all directions (Wagner *et al.* 2012; Mukherjee *et al.* 2016, 2018a). The jet power, the distribution of the surrounding medium and the orientation at which the jet enters the medium are all important parameters which determine the final impact of the jet-ISM interaction (Mukherjee *et al.* 2018a).

4.1. *Impact traced by the ionised gas*

As mentioned, the capability of radio jets to drive outflows is known since long time thanks to the studies of the optical emission lines of the warm ionised gas and of X-ray emission tracing the presence of gas heated by shocks. Evidence of jet-ISM interaction and jet-driven outflows have been found in a variety of objects. For example, Seyfert galaxies often show morphological association between the ionised gas and the radio emission. Furthermore, kinematical disturbance - traced by broad (often blueshifted) components of the emission lines - have been observed from the gas co-spatial with the radio (see e.g. Capetti *et al.* 1999; Axon *et al.* 1998; Morganti *et al.* 2007). All these signatures supports and highlight the role of the jets as responsible for the interaction. Interestingly, also low-luminosity AGN show gas outflows attributed to radio jets as show in, e.g. Riffel *et al.* (2014); Rodríguez-Ardila *et al.* (2017); May *et al.* (2018) and refs therein.

In radio galaxies the presence of kinematically disturbed gas at the location of the radio emission is a relatively common features. Figure 4 (left) shows the example of Coma A and the spatial coincidence between the distribution of the ionised gas and the radio emission, strongly suggestive of a role of the radio plasma in shaping the distribution of the gas (Tadhunter *et al.* 2000). The radio can also play a role in the ionisation of the gas but assessing this requires a detailed analysis of the line ratios and the comparison with shocks models. Particularly interesting is the result that young radio galaxies, i.e. sources hosting a newly formed jet, show these kinematically disturbance more often and of larger amplitude compared to large radio sources (Holt *et al.* 2008, 2009). The effect of jets in high-z radio galaxies has been also studied in detail by e.g. Nesvadba *et al.* (2008).

While outflows of ionised gas are very common in radio sources, the mass outflow rate associated with them is typically not high ($\lesssim 1 M_\odot$ yr^{-1}), regardless whether they are driven by jets or by radiation/winds. Thus, the actual impact of these outflows for galaxy evolution is still an open question.

Signatures of shocked-heated gas resulting from the interaction between the radio plasma and the ISM are also seen by X-ray observations. The case of NGC 3801

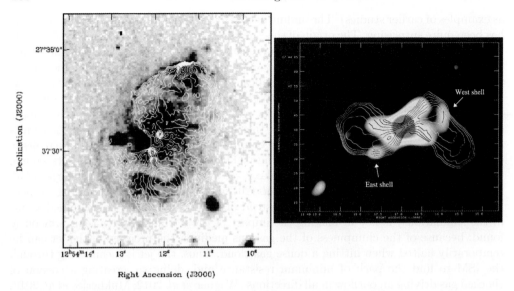

Figure 4. Left: Overlay of the Hα+continuum image for Coma A (grey-scale) with the 6-cm radio map of van Breugel *et al.* (1985) (contours). Figure taken from Tadhunter *et al.* (2000). The coincidence between the radio emission from the lobes and the ionised gas is clearly seen suggesting that the distribution of the gas is shaped by the interaction with the radio plasma. **Right** Chandra image with 1.4 GHz radio contours overlaid to illustrate the relationship between the X-ray shells and radio morphology in NGC 3801, figure taken from Croston *et al.* (2007).

(Fig. 4, right) nicely illustrates this process by showing shells of enhanced X-ray emission at the edge of the radio lobes (Croston *et al.* 2007). It is interesting to note that this object is a low-power radio galaxy. Croston *et al.* (2007) estimate an expansion speed of the shells of 850 km s^{-1}, corresponding to a Mach number of 4 and this allows to measure directly the contribution of shock heating of the jet to the total energetic input to the ISM. Similar conclusion about the relevance of shocks heating of the gas were derived for Centaurus A (Croston *et al.* 2008). In this object, a shell of X-ray emitting gas is observed, tracing the effect of the southern radio lobe expanding with a velocity of ~2600 kms^{-1}, roughly Mach 8 relative to the ambient medium.

Signatures in the form of X-ray emitting gas heated by shocks from jet-ISM interaction have been reported also for the radio galaxy 3C 305, (Hardcastle *et al.* 2012). The authors find that the X-ray emission is consistent with being shock-heated material and can be described by standard collisionally ionised models. Albeit with a number of assumptions, in this source the X-ray-emitting gas could dominate the other phases of the gas outflow (Morganti *et al.* 2005b) in terms of its energy content. This may be the case in more objects, but the lack of deep enough X-ray data prevent to draw strong conclusions.

4.2. *Jet-driven outflows of cold gas: HI component*

A number of cases have been found where a component of H I (identify by blueshifted wings seen in H I absorption) is associated with jet-driven outflows, see e.g. Morganti *et al.* (2003, 2005a); Aditya & Kanekar (2018); Morganti & Oosterloo (2018) and refs therein.

In a shallow H I survey of about 250 sources, at least 15% of H I detections show blueshifted components of H I absorption (Geréb *et al.* 2015; Maccagni *et al.* 2017). Interestingly, the large majority of these outflows are observed in young (or restarted) radio galaxies (see also Aditya & Kanekar 2018). The H I outflows can extend up to a few hundred pc to kpc in radius and are characterised by velocities between a few

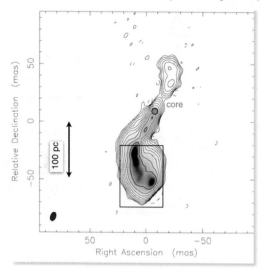 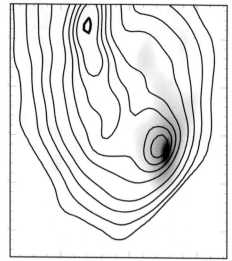

Figure 5. **Left:** radio continuum image of the young radio galaxy 4C 12.50. **Right:** in orange is shown the distribution of the outflowing H I (about 1000 km s^{-1} blueshifted compared to the systemic velocity). The spatial coincidence between the radio lobe and the gas, with a bright H I cloud at the location of the hot spot and more diffuse gas wrapping around the lobe, suggests that the outflow could be driven by the radio jet. Adapted from Morganti et al. (2013).

hundred and ∼1300 km s^{-1}, masses raging from a few ×10^6 to 10^7 M$_\odot$ and mass outflow rates up to 20–50 M$_\odot$ yr^{-1}. Their kinetic energies can be derived and compared to the Eddington (or bolometric) luminosities resulting in $\dot{E}_{kin}/L_{edd} \sim 10^{-4}$ (few ×10^{-3} bolometric luminosity).

In a few cases, the H I outflow has been located and spatially resolved using high resolution observations and VLBI. The best examples are 3C 293 (Mahony et al. 2016) and 4C 12.50 (Morganti et al. 2013). Other examples can be found in Schulz et al. (2018). Figure 5 shows the case of 4C 12.50, where the interaction jet-ISM is caught in the act. The outflowing H I gas (shown in orange in Fig. 5 right) is distributed around the head of the jet and on the side of the lobes, very suggestive of an on-going interaction. From the VLBI observations it is also interesting to see that outflowing gas is present very close to the AGN (∼40 pc, Schulz et al. 2018). Furthermore, these studies also suggest that the structure of the outflow changes with time, with an increasing amount of diffuse gas - with respect to gas in clumps - as the radio source grows (Schulz et al. in prep).

4.3. Jet-driven outflows of cold gas: molecular component

The study of molecular gas in radio galaxies in a variety of environments is rapidly expanding, in particular thanks to the capabilities of ALMA (see e.g. Ruffa et al. 2019; Russell et al. 2019).

The sensitivity of ALMA has also made possible to derive the best evidence of fast and massive outflows driven by radio jets. One of the objects where the effects of the radio jet on the molecular gas has been studied in details is the Seyfert 2 galaxy IC 5063. This object is also one of the clearest examples of low-power jet (this source would be classified as radio-quiet) disturbing the kinematics of *all the phases of the gas* (see Tadhunter et al. 2014; Morganti et al. 2015, for an overview).

The ALMA observations have confirmed the presence of molecular gas with disturbed kinematics across the entire region co-spatial with the radio emission (Morganti et al. 2015). The observations of a number of CO transitions have shown that in the immediate

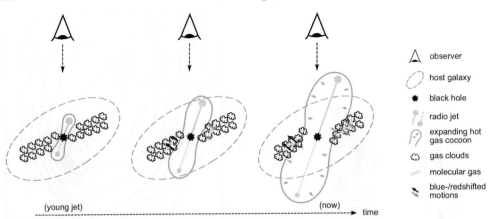

Figure 6. Cartoon of the jet-ISM interaction in the case of 3C 273 as proposed by Husemann et al. (2019b). The sketch illustrates the dependence of the impact from the geometry of the jet-disk system.This has been suggested to occur also in other objects (see text) and is predicted by the numerical simulations of Mukherjee et al. (2018b).

vicinity of the radio jet, a fast outflow, with velocities up to 800 km s^{-1}, is occurring. In addition to this, the interaction is also affecting the physical conditions of the gas (Dasyra et al. 2016; Oosterloo et al. 2017). The gas involved in the outflow has high excitation temperatures (in the range 30−55 K) and, based on the relative brightness of the ^{12}CO lines and of ^{13}CO(2-1) vs ^{12}CO(2-1), the outflow must be optically thin (see Oosterloo et al. 2017, for details). The mass of the molecular outflow is estimated to be at least 1.2×10^6 M$_\odot$ and the mass outflow rate is $\sim 12 M_\odot$ yr^{-1}. Although not extremely high, it is much higher than the one derived for the warm ionised gas. Interestingly, the kinematics of the gas can be well reproduced by the hydrodynamic simulations described above, which model the effect of the radio jets on the multiphase, clumpy interstellar medium. The detail of the simulation and the results of the comparison are described in Mukherjee et al. (2018a).

In addition to IC 5063, in a steadily growing number of radio galaxies (both radio-quiet and radio-loud) the interaction of the jets with dense clumps producing molecular outflows has been observed. This is thanks not only to the depth but also to the high spatial resolution provided by ALMA. Some examples can be found in e.g. NGC 1068 (García-Burillo et al. 2014), HE 1353-1917 (Husemann et al. 2019a), NGC 613 (Audibert et al. 2019), 4C12.50 (Fotopoulou et al. 2019), PKS1549-69 (Oosterloo et al. 2019), 3C 273 (Husemann et al. 2019b). The actual impact of the jet-ISM interaction will depend on the distribution of the gas and the orientation of the jet compared to it. Interestingly, this dependence is seen both in low and high radio power jets. This is seen very clearly in IC 5063 but it is supported by other cases as extensively discussed in (Mukherjee et al. 2018b). Other examples are e.g. the low-power AGN HE 1353-1917 (Husemann et al. 2019a) and the high-power jet in 3C 273 (Husemann et al. 2019b). The cartoon shown in Fig. 6 illustrates the scenario proposed by the authors for this object but also representing what predicted by the numerical simulations: the radio jets is creating a pressurised expanding hot gas cocoon which is impacting on the inclined gas disk.

The presence of radio jets does not preclude other forms of AGN activity (like radiation and winds) impacting the ISM. In some cases, it may even be difficult to disentangle the effect of these different phenomena. The case of PKS 1549-69, an object hosting an obscured quasars and a young radio source, is particularly interesting. The ALMA high resolution observations show the presence of three gas structures, which can be seen in

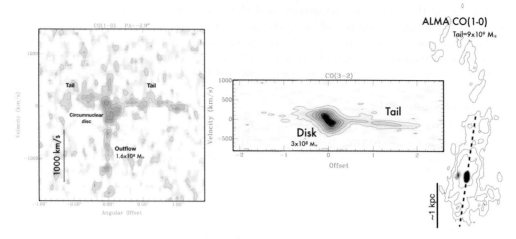

Figure 7. The complex view of the central few hundred pc of PKS 1549-69 ($z = 0.150$), see Oosterloo et al. (2019) for details. ALMA CO(1-0) and CO(3-2) detected in emission with spatial resolution ranging from 0.05 arcsec ($\sim$100 pc) to 0.2 arcsec.

Fig. 7, tracing the accretion and the outflowing of molecular gas. Kiloparsec-scale tails are observed, resulting from an on-going merger and providing gas which accretes onto the centre of PKS 1549-79. At the same time, a circum-nuclear disc has formed in the inner few hundred parsec, and a very broad (> 2300 km s^{-1}) component associated with fast outflowing gas is detected in CO(1-0) at the position of the AGN. As expected, the outflow is massive (up to 600 $M_\odot$ yr^{-1}) but, despite the fact that PKS 1549-79 should represent an ideal case of feedback in action, it is limited to the inner 200 pc. Both the jet and the wind/radiation can drive the outflow (and perhaps they both do!). Only circumstantial evidence suggest the jet may play a prominent role, as the jet appears to be affect by strong interaction with the ISM, possibly providing the driving mechanism for the massive outflow. These results illustrates that the impact on the surrounding medium of the energy released by the AGN is not always as expected from the feedback scenario.

5. Conclusions

In summary, radio jets show a variety of structures, physical conditions (velocity, composition etc.) and energetics. The importance of radio jets goes beyond the radio-loud objects and the (much more common) radio-quiet sources should also be considered. Their radio emission shows often jet-like structures and the power of these jets can have a relevant impact on the host galaxy. They can interact with the ISM of the host galaxy for a longer period and, integrated over time, they can dump large amounts of energy in the ISM.

Radio jets are also a recurrent phenomenon. From their properties (e.g. morphology and radio spectra) we can identify young, dying and restarted radio sources. The recurrence of the radio AGN ensures that they impact the host galaxy multiple times as needed in the feedback cycle. Low frequency radio surveys are now helping to derive more reliable time-scales of their life-cycle.

Jets can drive (massive) multi-phase outflows. Based on new numerical simulations, the predictions are that jets couple well with clumpy ISM and that they can produce cocoons of shocked gas expanding across the surrounding gas. These effects are already observed in a growing number of objects. Young (or restarted) jets are those showing the most clear signs of affecting the surrounding medium. This is also consistent with the predictions from the simulations.

Thanks to the increasing sensitivity and the capabilities offered by the new radio telescopes, our understanding of both the physical properties as well as the impact of radio jets, is continuously expanding. These progresses, combined with the study of AGN at other wavelengths, will hopefully help building a more complete picture of the interplay between AGN and host galaxy, a problem that still has many open questions.

Acknowledgements

I would like to thanks the organisers and, in particular, Mirjana Pović for putting together such an interesting (not only scientifically!) conference. Some of the results presented here would not have been obtained without the help of some key collaborators. In particular I would like to thank Tom Oosterloo, Robert Schulz and Clive Tadhunter.

References

Alatalo, K., Blitz, L., Young, L. M., et al. 2011, ApJ, 735, 88
Aditya, J. N. H. S. & Kanekar, N. 2018, MNRAS, 473, 59
Audibert, A., Combes, F., García-Burillo, S., et al. 2019, A&A, 632, A33
Axon, D. J., Marconi, A., Capetti, A., et al. 1998, ApJL, 496, L75
Baldi, R. D., Capetti, A., & Giovannini, G. 2015, A&A, 576, A38
Baldi, R. D., Capetti, A., & Giovannini, G. 2019, MNRAS, 482, 2294
Blandford, R., Meier, D., & Readhead, A. 2019, ARA&A, 57, 467
Bicknell, G. V., Dopita, M. A., Tsvetanov, Z. I., et al. 1998, ApJ, 495, 680
Brienza, M., Godfrey, L., Morganti, R., et al. 2017, A&A, 606, A98
Brienza, M., Morganti, R., Murgia, M., et al. 2018, A&A, 618, A45
Capetti, A., Axon, D. J., Macchetto, F. D., et al. 1999, ApJ, 516, 187
Cavagnolo, K. W., McNamara, B. R., Nulsen, P. E. J., et al. 2010, ApJ, 720, 1066
Cielo, S., Bieri, R., Volonteri, M., Wagner, A. Y., & Dubois, Y. 2018, MNRAS, 477, 1336
Ciotti, L., Ostriker, J. P., & Proga, D. 2010, ApJ, 717, 708
Clarke, D. A., Bridle, A. H., Burns, J. O., et al. 1992, ApJ, 385, 173.
Combes, F., García-Burillo, S., Casasola, V., et al. 2013, A&A, 558, A124
Croston, J. H., Kraft, R. P., & Hardcastle, M. J. 2007, ApJ, 660, 191
Croston, J. H., Hardcastle, M. J., Kharb, P., Kraft, R. P., & Hota, A. 2008, ApJ, 688, 190
Croston, J. H., Ineson, J., & Hardcastle, M. J. 2018, MNRAS 476, 1614
Dasyra, K. M., Combes, F., Oosterloo, T., et al. 2016, A&A, 595, L7
Fabbiano, G., Paggi, A., Karovska, M., et al. 2018, ApJ, 865, 83
Fanaroff, B. L. & Riley, J. M. 1974, MNRAS, 167, 31P
Feruglio, C., Maiolino, R., Piconcelli, E., et al. 2010, A&A, 518, L155
Fotopoulou, C. M., Dasyra, K. M., Combes, F., et al. 2019, A&A, 629, A30
Gallimore, J. F., Baum, S. A., O'Dea, C. P., et al. 1999, ApJ, 524, 684
García-Burillo, S., Combes, F., Usero, A., et al. 2014, A&A, 567, A125
Gaspari, M., Temi, P., & Brighenti, F. 2017, MNRAS, 466, 677
Geréb, K., Maccagni, F. M., Morganti, R., & Oosterloo, T. A. 2015, A&A, 575, A44
Giroletti, M., Giovannini, G., & Taylor, G. B. 2005, A&A, 441, 89
Hardcastle, M. J., Massaro, F., Harris, D. E., et al. 2012, MNRAS, 424, 1774
Hardcastle, M. J. 2018, MNRAS, 475, 2768
Harrison, C. M., Thomson, A. P., Alexander, D. M., et al. 2015, ApJ, 800, 45
Harrison, C. M., Costa, T., Tadhunter, C. N., et al. 2018, Nature Astronomy, 2, 198
Harwood, J. J., Croston, J. H., Intema, H. T., et al. 2016, MNRAS, 458, 4443
Heesen, V., Croston, J. H., Morganti, R., et al. 2018, MNRAS, 474, 5049
Holt, J., Tadhunter, C. N., & Morganti, R. 2008, MNRAS, 387, 639
Holt, J., Tadhunter, C. N., & Morganti, R. 2009, MNRAS, 400, 589
Husemann, B., Scharwächter, J., Davis, T. A., et al. 2019a, A&A, 627, A53
Husemann, B., Bennert, V. N., Jahnke, K., et al. 2019b, ApJ, 879, 75
Jarvis, M. E., Harrison, C. M., Thomson, A. P., et al. 2019, MNRAS, 485, 2710

Johnston, S., Bailes, M., Bartel, N., et al. 2007, *PASA*, 24, 174
Jonas, J. & MeerKAT Team 2016, Meerkat Science: On the Pathway to the SKA, 1
Jurlin, N., Morganti, R., Brienza, M., et al. 2020, *A&A*, 638, A34
Kaiser, C. R. & Best, P. N. 2007, *MNRAS*, 381, 1548
Kharb, P., O'Dea, C. P., Baum, S. A., et al. 2014, *MNRAS*, 440, 2976
Kellermann, K. I., Sramek, R., Schmidt, M., et al. 1989, *ApJ*, 98, 1195
King, A. & Pounds, K. 2015, *ARA&A*, 53, 115
Konar, C., Hardcastle, M. J., Jamrozy, M., et al. 2012, *MNRAS*, 424, 1061
Kunert-Bajraszewska, M., Gawroński, M. P., Labiano, A., et al. 2010, *MNRAS*, 408, 2261
Laing, R. 2015, Proceedings of "the Many Facets of Extragalactic Radio Surveys: Towards New Scientific Challenges" (EXTRA-RADSUR2015). 20–23 October 2015. Bologna, 67
Laing, R. A. & Bridle, A. H. 2014, *MNRAS*, 437, 3405
Laing, R. A. & Bridle, A. H. 2012, *MNRAS*, 424, 1149
Maccagni F. M., Morganti R., Oosterloo T. A., Geréb K., Maddox N., 2017, *A&A*, 604, A43
Mahatma, V. H., Hardcastle, M. J., Williams, W. L., et al. 2019, *A&A*, 622, A13
Mahony, E. K., Oonk, J. B. R., Morganti, R., et al. 2016, *MNRAS*, 455, 2453
Maksym, W. P., Fabbiano, G., Elvis, M., et al. 2019, *ApJ*, 872, 94
May, D., Rodríguez-Ardila, A., Prieto, M. A., et al. 2018, *MNRAS*, 481, L105
McNamara, B. R. & Nulsen, P. E. J. 2012, *New Journal of Physics*, 14, 055023
Mingo, B., Croston, J. H., Hardcastle, M. J., et al. 2019, *MNRAS*, 488, 2701
Morganti, R., Fanti, C., Fanti, R., et al. 1987, *A&A*, 183, 203
Morganti, R., Tsvetanov, Z. I., Gallimore, J., et al. 1999, *A&AS*, 137, 457
Morganti, R., Oosterloo, T. A., Emonts, B. H. C., et al. 2003, *ApJL*, 593, L69
Morganti, R., Tadhunter, C. N., & Oosterloo, T. A. 2005a, *A&A*, 444, L9
Morganti, R., Oosterloo, T. A., Tadhunter, C. N., et al. 2005b, *A&A*, 439, 521
Morganti, R., Holt, J., Saripalli, L., et al. 2007, *A&A*, 476, 735
Morganti, R., Fogasy, J., Paragi, Z., Oosterloo, T., & Orienti, M. 2013, *Science*, 341, 1082
Morganti, R., Oosterloo, T., Oonk, J. B. R., Frieswijk, W., Tadhunter, C. 2015, *A&A*, 580, A1
Morganti, R. 2017a, *Nature Astronomy*, 1, 596
Morganti, R. 2017b, *Frontiers in Astronomy and Space Sciences*, 4, 42
Morganti, R. & Oosterloo, T. 2018, *A&ARew*, 26, 4
Mukherjee, D., Bicknell, G. V., Wagner, A. Y., et al. 2018a, *MNRAS*, 479, 5544
Mukherjee, D., Wagner, A. Y., Bicknell, G. V., et al. 2018b, *MNRAS*, 476, 80
Mukherjee, D., Bicknell, G. V., Sutherland, R., & Wagner, A. 2016, *MNRAS*, 461, 967
Murthy, S., Morganti, R., Oosterloo, T., et al. 2019, *A&A*, 629, A58
Nesvadba, N. P. H., Lehnert, M. D., De Breuck, C., Gilbert, A. M., & van Breugel, W. 2008, *A&A*, 491, 407
Novak, G. S., Ostriker, J. P., & Ciotti, L. 2011, *ApJ*, 737, 26
O'Dea, C. P. 1998, IAU Colloq. 164: Radio Emission from Galactic and Extragalactic Compact Sources, 291
Oosterloo, T., Raymond Oonk, J. B., Morganti, R., et al. 2017, *A&A*, 608, A38
Oosterloo, T., Morganti, R., Tadhunter, C., et al. 2019, *A&A*, 632, A66
Orienti, M. 2016, *Astronomische Nachrichten*, 337, 9
Padovani, P. 2017, *Nature Astronomy*, 1, 0194
Paragi, Z., Godfrey, L., Reynolds, C., et al. 2015, *Advancing Astrophysics with the Square Kilometre Array (AASKA14)*, 143
Parma, P., Murgia, M., de Ruiter, H. R., et al. 2007, *A&A*, 470, 875
Randall, S. W., Forman, W. R., Giacintucci, S., et al. 2011, *ApJ*, 726, 86
Riffel, R. A., Storchi-Bergmann, T., & Riffel, R. 2014, *ApJL*, 780, L24
Rodríguez-Ardila, A., Prieto, M. A., Mazzalay, X., et al. 2017, *MNRAS*, 470, 2845
Ruffa, I., Prandoni, I., Laing, R. A., et al. 2019, *MNRAS*, 484, 4239
Russell, H. R., McNamara, B. R., Fabian, A. C., et al. 2019, *MNRAS*, 490, 3025
Sabater, J., Best, P. N., Hardcastle, M. J., et al. 2019, *A&A*, 622, A17
Sadler, E. M., Ekers, R. D., Mahony, E. K., et al. 2014, *MNRAS*, 438, 796

Scheuer, P. A. G. 1974, *MNRAS*, 166, 513
Shabala, S. S., & Godfrey, L. E. H. 2013, *ApJ*, 769, 129
Shabala, S. S., Jurlin, N., Morganti, R., et al. 2020, *MNRAS*, 496, 1706
Shimwell, T. W., Tasse, C., Hardcastle, M. J., et al. 2019, *A&A*, 622, A1
Shulevski, A., Morganti, R., Oosterloo, T., et al. 2012, *A&A*, 545, A91
Schoenmakers, A. P., de Bruyn, A. G., Röttgering, H. J. A., et al. 2000, *MNRAS*, 315, 371
Schulz, R., Morganti, R., Nyland, K., et al. 2018, *A&A*, 617, 38
Tadhunter, C. N., Villar-Martin, M., Morganti, R., et al. 2000, *MNRAS*, 314, 849
Tadhunter, C., Morganti, R., Rose, M., et al. 2014, *Nature*, 511, 440
Tadhunter, C. 2016, *A&A Rew*, 24, 10
Tingay, S. J., Goeke, R., Bowman, J. D., et al. 2013, *PASA*, 30, e007
van Breugel, W., Miley, G., Heckman, T., et al. 1985, *ApJ*, 290, 496
van Haarlem, M. P., Wise, M. W., Gunst, A. W., et al. 2013, *A&A*, 556, A2
Vantyghem, A. N., McNamara, B. R., Russell, H. R., et al. 2014, *MNRAS*, 442, 3192
Wagner A. Y., Bicknell G. V., Umemura M. 2012, *ApJ*, 757, 136
Walker, R. C., Hardee, P. E., Davies, F. B., et al. 2018, *ApJ*, 855, 128
Willott, C. J., Rawlings, S., Blundell, K. M., et al. 1999, *MNRAS*, 309, 1017
Worrall, D. M., & Birkinshaw, M. 2000, *ApJ*, 530, 719

The origin of X-ray emission in Low-Excitation Radio Galaxies

Shuang-Liang Li

Key Laboratory for Research in Galaxies and Cosmology, Shanghai Astronomical Observatory, Chinese Academy of Sciences, 80 Nandan Road, Shanghai 200030, China
email: lisl@shao.ac.cn

Abstract. In previous works, the radio-X-ray slope in FRI radio galaxies is found to be steeper compared with that in low-luminosity AGNs, indicating different origin of the X-ray emission. Here we reinvestigate this point by compiling a sample of 13 low-excitation radio galaxies (LERG) from 3CR radio catalog of galaxies, where the central engine in LERG is accepted to be a radiatively inefficient accretion flow (RIAF). The core radio and X-ray emissions in all the objects of our sample are detected by VLA/VLBI/VLBA and Chandra/XMM-Newton, respectively. Surprisingly, a shallower slope of $L_\mathrm{R} - L_\mathrm{X}$ relation ($L_\mathrm{R} \sim L_\mathrm{X}^{0.63}$) is given by our sample, which demonstrates that the X-ray emission in LERG may come from accretion disk rather than a jet as suggested by previous works. In addition, the slope in the fundamental plane ($\log L_\mathrm{R} = 0.52 \log L_\mathrm{X} + 0.84 \log M_\mathrm{BH} + 10.84$) of LERG is found to be well consistent with that reported by Merloni *et al.* (2003).

Keywords. accretion, accretion disks - black hole physics - galaxies: active

1. Introduction

In black hole systems with relativistic jets, a tight correlation between the X-ray and radio emissions ($L_\mathrm{R} \propto L_\mathrm{X}^{0.7}$) has been reported by numerous authors, with the X-ray and radio emissions believed to come from accretion disk and jet, respectively. Coupled with black hole mass, a so-called fundamental plane ($\log L_\mathrm{R} = 0.6 \log L_\mathrm{X} + 0.78 \log M_\mathrm{BH} + 7.33$, where $\log L_\mathrm{R}$ and $\log L_\mathrm{X}$ are the nuclear luminosity at 5 GHz and the intrinsic rest-frame luminosity in 2–10 keV band, respectively.) was developed by Merloni *et al.* (2003) (hereinafter M03). However, it is found that the slope between the radio and X-ray in radio-loud AGNs appears to be much steeper compared with that of low-luminosity AGNs (e.g., de Gasperin *et al.* 2011), possibly owing to the domination of strong jet emissions in the radio and/or X-ray bands.

In this work, we reinvestigate the origin of X-ray in LERG by constructing a sample satisfying the following conditions: (1) All the sources in the sample should be LERG in order to ensure that the accretion flow is a RIAF; (2) Radio emission comes mainly from the jet in LERG, where both the core and lobe can play important roles. In order to avoid the influence of the surrounding medium, we require that sources must be core-dominated only.

2. The LLAGN sample

Our parent sample comprises 113 3CR radio sources with redshift $z < 0.3$, in which all the emission lines necessary to identify LERG are detected in 83 sources (Buttiglione *et al.* 2009). We first exclude 43 HERG with excitation index (EI) larger than 0.95 (see Buttiglione *et al.* 2009 for details) because their accretion flows may be radiatively efficient, leaving us with 40 LERG. Radio flux in LERG is dominated by the jet synchrotron

© The Author(s), 2021. Published by Cambridge University Press on behalf of International Astronomical Union

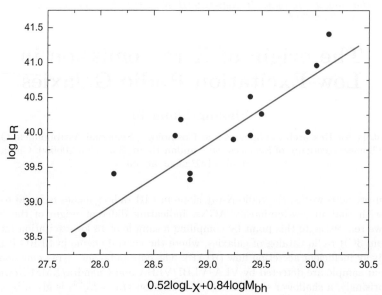

Figure 1. The fundamental plane of black hole activity in LERG, where the solid line shows the best fitting function 3.1.

emission, based on the truncated disk-jet model (Narayan & Yi 1994, 1995), which has been successfully applied to the M03 fundamental plane. In order to prevent contamination from the lobe, only sources with radio core emission detected by VLBA/VLBE/VLA are included in this work. For the X-ray, we consider only the sources with X-ray core flux detected by Chandra/XMM-Newton only, in order to maintain a high precision. The final sample includes 13 LERG (see Li & Gu 2018 for details).

3. Results

Following M03, we adopt a least χ^2 method to fit the multivariate relation coefficients in black hole fundamental plane. The best fitting result reads,

$$\log L_R = 0.52^{+0.16}_{-0.16} \log L_X + 0.84^{+0.50}_{-0.50} \log M_{BH} + 10.84^{+5.95}_{-5.95}, \quad (3.1)$$

with an intrinsic scatter of $\sigma_{\rm int} = 0.38$ dex (Figure 1). While the radio-X-ray slope is consistent with M03, it is found that the normalization of our sample is larger than that in M03 by about 0.7 dex (Figure 2). We suggest that this shift can be due to the difference in magnetic field strength (see Li & Gu 2018 for details) because X-ray emission in RIAF has a complicated relationship with the magnetic field strength (e.g., Bu *et al.* 2013, 2016).

We also reinvestigate the slope of L_R and L_X relation in our LERG sample (Figure 3). A linear fit we found can be written as,

$$\log L_R = (0.63 \pm 0.11) \log L_X + 13.78 \pm 4.71, \quad (3.2)$$

with a strong confidence level larger than 99.9% based on a Pearson test. This slope is considerably flatter than the one found by de Gasperin *et al.* (2011).

4. Discussion

In this work, we compiled a sample of 13 LERG from 3CR catalog with optical spectroscopic information (Buttiglione *et al.* 2009). Surprisingly, we discovered a radio-X-ray slope similar to that of the M03 fundamental plane, which suggests that LERG still follow the original M03 fundamental plane. We note that de Gasperin *et al.* (2011) investigated

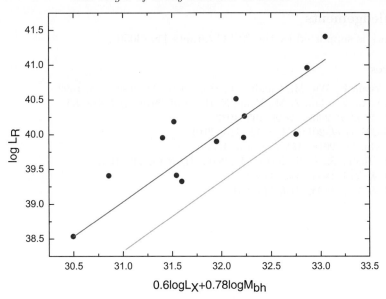

Figure 2. The fundamental plane of black hole activity in LERG, where the green line shows the M03 fundamental plane relation. The red line indicates the movement of red line by 0.7 dex.

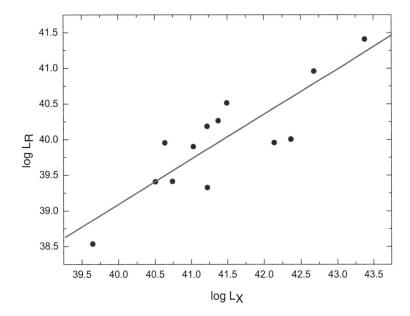

Figure 3. The relationship between the radio luminosity L_R and X-ray luminosity L_X in the present LERG sample.

the fundamental plane in a LERG sample too. They discovered a steeper radio-X-ray slope and suggested that the X-ray emission in LERG may originate from jets. The reason for this discrepancy may be that their sample included some steep-spectrum LERG in addition to the core-dominated flat-spectrum LERG. The shallower radio-X-ray slope in LERG found here suggests that the X-ray probably originate in an accretion disk rather than jet.

Acknowledgements

This work is supported by the NSFC (grants 11773056).

References

Bu, D. F., Yuan, F., Wu, M., Cuadra, J., *et al.* 2013, *MNRAS*, 434, 1692
Bu, D. F., Yuan, F., Gan, Z. M., Yang, X. H., *et al.* 2016, *ApJ*, 818, 83
Buttiglione, S. *et al.* 2009, *A&A*, 495, 1033
de Gasperin, F. *et al.* 2011, *MNRAS*, 415, 2910
Li, S.-L. & Gu, M., 2018, *MNRAS*, 481, L45
Merloni, A., Heinz, S., & di Matteo, T., 2003, *MNRAS*, 345, 1057
Narayan R. & Yi I. 1994, *ApJ*, 428, L13
Narayan R. & Yi I. 1995, *ApJ*, 444, 231

Understanding the origin of radio outflows in Seyfert galaxies using radio polarimetry

Biny Sebastian[1], Preeti Kharb[1], Christopher P. O' Dea[2,3], Jack F. Gallimore[4], Stefi A. Baum[3,5] and Edward J. M. Colbert[6,7]

[1] National Centre for Radio Astrophysics (NCRA) - Tata Institute of Fundamental Research (TIFR), S. P. Pune University Campus, Ganeshkhind, Pune 411007, India
email: biny@ncra.tifr.res.in

[2] School of Physics & Astronomy, Rochester Institute of Technology, Rochester, NY, USA

[3] Physics and Astronomy, University of Manitoba, Winnipeg, Canada

[4] Department of Physics and Astronomy, Bucknell University, Lewisburg, PA, USA

[5] Carlson Center of Imaging Science, Rochester Institute of Technology, Rochester, NY, USA

[6] Hume Center for National Security and Technology, 900 N. Glebe Rd, Arlington, VA, USA

[7] U.S. Army Research Laboratory Adelphi, MD, USA

Abstract. The role of starburst winds versus active galactic nuclei (AGN) jets/winds in the formation of the kiloparsec scale radio emission seen in Seyferts is not yet well understood. In order to be able to disentangle the role of various components, we have observed a sample of Seyfert galaxies exhibiting kpc-scale radio emission suggesting outflows, along with a comparison sample of starburst galaxies, with the EVLA B-array in polarimetric mode at 1.4 GHz and 5 GHz. The Seyfert galaxy NGC 2639, shows highly polarized secondary radio lobes, not observed before, which are aligned perpendicular to the known pair of radio lobes. The additional pair of lobes represent an older epoch of emission. A multi-epoch multi-frequency study of the starburst-Seyfert composite galaxy NGC 3079, reveals that the jet together with the starburst superwind and the galactic magnetic fields might be responsible for the well-known 8-shaped radio lobes observed in this galaxy. We find that many of the Seyfert galaxies in our sample show bubble-shaped lobes, which are absent in the starburst galaxies that do not host an AGN.

Keywords. Seyferts; Jets; Radio Continuum; Polarimetry

1. Introduction

It is known that many active galactic nuclei (AGN) like radio galaxies and quasars host very powerful radio jets, which can extend up to several hundreds of kpc or sometimes even up to a few Mpcs. However, rarely do Seyfert galaxies host such powerful jets, although several high angular resolution studies have revealed that many of them possessed double or triple components similar to radio galaxies (Ulvestad *et al.* 1981), which are sometimes distorted into S-shapes, or are loop-like, etc. Lower angular resolution studies on the other hand, have shown that most of these Seyfert galaxies possess radio emission that sometimes extends up to several kpcs. However, the origin of the kiloparsec-scaled radio structure (KSR) emission is debated in the literature. For example, Baum *et al.* 1993 have concluded that the KSR was a consequence of starburst winds, since most of the emission was aligned along the minor axis of the host galaxy. Colbert *et al.* 1996, on the other hand carried out a similar study of a sample of Seyfert galaxies but with a comparison sample of starburst galaxies. They found that while the radio powers are comparable, the morphologies of the kilo-parsec scale emission are different for a sample of starburst galaxies which are more diffuse versus that of the Seyfert galaxies which are more lobe-like and oriented at skewed angles to the minor axis. Gallimore et al. 2006 have carried out a more complete survey of a sample of 43 Seyfert galaxies, and they find that at least 44% of these galaxies host kilo-parsec scale radio emission. While they favored an AGN driven origin for these outflows, they did not rule out the role of starburst superwinds. It was demonstraed by Irwin et al. 2017 that polarization can prove to be an effective tool in distinguishing the AGN-jet related emission from that of the galactic disk radio continuum emission. The radio lobes, which were embedded in the disk emission in total intensity images, were revealed in the polarization intensity image due to the higher degrees of linear polarization of the lobes compared to the rest of the host galaxy disk emission. In this paper, we present results from an ongoing work, where we are trying to understand if polarization can be used to distinguish outflows which have a jet related origin, versus those with a starburst wind driven origin. We are trying to address questions like (i) whether we will find higher degrees of polarization in a sample of Seyfert galaxies compared to starburst galaxies, (ii) if we will find signatures of ordered magnetic fields in Seyfert galaxy outflows more often than in starburst galaxy outflows, and (iii) whether the alignment of the magnetic fields in Seyfert galaxies will be different from that expected to be generated by host galaxy dynamo mechanism as observed in typical starburst galaxies. We have chosen a sample of starburst galaxies along with a sample of Seyfert galaxies, which we describe below.

2. Sample and Observations

We chose 10 Seyferts from the CfA (Huchra & Burg 1992) and 12 μm sample (Rush et al. 1993), which possessed lobe to lobe extents $\geqslant 20''$ from the study of Gallimore et al. 2006. We chose a comparison sample of 8 edge-on starburst galaxies from Colbert et al. 1996 and Dahlem et al. 1998. The sample was observed using the EVLA in the B-array configuration. Four of the starburst sources were observed using L-band (1.5 GHz), and the rest of the sources were observed using the C-band (5.5 GHz). Additional data are being obtained at X-band (10.0 GHz) using the D-array configuration under the project code (19B-198), to decrease the effect of wavelength dependant depolarization effects dominant at lower wavelengths.

3. Results

We present the results of the detailed analysis of two Seyfert galaxies where we detected polarization, in Sections 3.1 and 3.2, and some early results from the larger sample in Section 3.3.

3.1. *Filamentary lobes in the starburst-Seyfert Composite Galaxy NGC 3079*

NGC 3079 is a Seyfert galaxy with a prominent starburst component (Dahlem et al. 1998). Duric & Seaquist 1988 identified the "eight" shaped morphology of the lobes in NGC 3079. They suggested that these lobes which are aligned along the minor axis of the galaxy are generated as a result of winds that are either powered by a starburst or by the AGN. On the other hand, Irwin & Seaquist 1988 who studied the VLBI jet argued that the jet can essentially power the entire lobes. X-ray and emission line imaging showed the presence of superwinds present in NGC 3079. We have carried out a detailed study of NGC 3079 using multi-frequency legacy VLA observations which complemented our current EVLA observations (Sebastian et al. 2019a). Our sensitive high resolution images revealed the complex filamentary morphology of the lobes. Morphologically, the radio lobes in NGC 3079 do not resemble those seen in powerful radio galaxies, which either show a hotspot and a backflow (in the case of FR II type radio galaxies) or diffuse

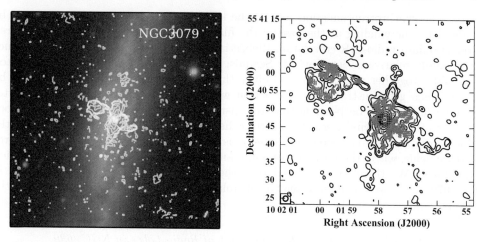

Figure 1. (Left) 1.5 GHz EVLA A-array image contours with levels 145 × (−1, 1, 2, 4, 8, 16, 64, 32, 64, 128) μJy beam^{-1} overlaid on the optical SDSS *gri* color composite image, and (right) 5 GHz VLA B-array image with contour levels 0.80 × (−0.085, 0.085, 0.17, 0.35, 0.70, 1.40, 2.80, 5.60, 11.25, 22.50, 45, 90) mJy beam^{-1} of NGC 3079 with polarization electric vectors whose length is proportional to the fractional polarization (1 arcsec length corresponds to 40% fractional polarization), superimposed in red.

plumes or tails which become more diffuse and uncollimated with distance from the core (in the case of FR I type radio galaxies). We investigated other possible mechanisms that might be influencing the formation of the lobes with bubble like morphology. For example, shocks are a feasible mechanism which can explain many of the observed features of the "ring" or loop-like like structure that are observed inside the northern lobe. Shocks, can give rise to synchrotron emission via diffusive shock acceleration (Blandford & Eichler 1987). Shock compression at the boundaries of the expanding bubble and the "ring" can also lead to the amplification of magnetic fields, which ultimately result in the high fractions of polarization (∼33%) along the "ring" (see Figure 1). The flatter spectral indices observed along the edges of the ring can also be explained as a result of re-acceleration of relativistic particles along the edges.

We also investigated how well the radio emission correlated with the thermal components of the superwind using the Chandra ACIS-S X-ray observations and HST WFPC2 Hα+ [NII] line emission data. We noticed that the thermal emission peaked to the south of the region where the radio emission peaked. Moreover, X-ray and the emission line filaments did not appear very well correlated spatially, as can be seen in Figure 9 in Sebastian et al. 2019a. Adebahr et al. 2013 found that the thermal emission traced by emission lines is correlated with the radio continuum emission in M82, a prototypical starburst galaxy hosting superwinds, which is suggestive of frozen-in magnetic fields. Brandenburg et al. 1995 suggested that disk material and frozen in magnetic fields are expected to be entrained along with winds. However, the lack of a correlation between radio and X-ray/emission line filaments in NGC 3079, argue against this scenario. Another interesting aspect about this source is that the filamentary structure in the northern lobes shows a rotation measure inversion (see Cecil et al. 2001 & Sebastian et al. 2019a). It is hard to explain such an inversion only by using shock acceleration. Organized magnetic fields on the scales of the kpc-scale ring are required to explain such a geometry.

Cecil et al. 2001 have suggested that the RM inversion seen in NGC 3079 can be explained as a result of loop-like and twisted magnetic field structures similar to solar prominences. The dome-like structures that are observed in NGC 3079 and active regions can be explained to have been generated as a result of the α^2 dynamo mechanism in the galaxy. The synchrotron emission seen in the loop, however is higher than that can be produced by the winding at the base of the filaments. The radio-excess seen compared to the standard radio-FIR correlation (Sebastian et al. 2019a), along with the presence of the mildly relativistic jet (Irwin & Seaquist 1988; Trotter et al. 1998; Sawada-Satoh et al. 2000) together points to a possibility that the relativistic plasma is supplied by the jet which gets frustrated by interaction with the surrounding medium on shorter length scales. It may be possible that the disrupted jet material gets transported out along with the superwinds, near the centre, where the thermal pressure is dominant (the region where X-ray or emission line emission is dominant) and later along the force-free magnetic fields, where the magnetic pressure and consequently the synchrotron emission is dominant. We therefore concluded that the peculiar lobe morphology observed in NGC 3079 was caused by an interplay of the relativistic jet, the galactic dynamo and the starburst superwinds in the galaxy.

3.2. Discovery of an Additional Pair of Radio Lobes in NGC 2639

With our polarization sensitive observations using the EVLA we were able to discover a pair of outer lobes in the north-south direction in addition to the already known pair of lobes which was aligned in the east-west direction (Sebastian et al. 2019b). The north-south lobes were highly polarized with fractional polarizations $\sim 21\pm 4\%$. There are several possible explanations like environmental influence, slow precession of the jets, the presence of a binary black hole at the core, and episodic activity. However, it is unlikely that precession is the leading cause because of the lack of a bridge like feature connecting both pairs of lobes, similar to that seen in typical X/S-shaped galaxies (Lal et al. 2019). Also the position angle of the host galaxy minor axis is located almost midway between the north-south and east-west lobes, disfavoring any pressure gradients due to the galaxy itself as a reason for the misalignment. The probability of the existence of binary black holes both launching kpc-scale jets simultaneously or within the gap of the typical synchrotron loss time-scale is low and hence may not be the reason for the double pair of lobes. We favor the scenario where the origin of the two pair of lobes is due to the switching off and restarting of the central engine. The optical host galaxy shows a rather settled morphology. Hence, such a restarting might have been induced probably by a minor merger which changed the direction of the spin of the central black hole, but did not disturb the host galaxy morphology. The spectral age of the secondary lobes was ~ 16 Myr. There have therefore been at least two AGN jet episodes in the past 16 Myrs.

3.3. Starburst versus Seyfert Sample Properties

We detect polarization in five out of the sixteen sources in our combined sample which includes four Seyfert galaxies, viz., NGC 3079 (Sebastian et al. 2019a), NGC 2639 (Sebastian et al. 2019b), NGC 4388 and NGC 5506 and one starburst galaxy, NGC 253 (Sebastian et al. 2020). We need to explain the following morphological differences that are observed in the sample of starbursts versus Seyferts.

• Edge-brightened bubble-like radio lobes similar to those found in NGC 3079, which make them distinct from the typical large-scale jets/lobes observed in radio galaxies are common in Seyfert galaxies but are absent in starbursts, without an AGN. These lobes are frequently roughly aligned with the minor axis, especially on several kpc-scales, but not always. We are exploring the role of the relativistic AGN jet in conjunction with the galactic dynamo in the formation of these lobes in the Seyfert sample.

• We find that the total intensity distribution of the lobes are rotationally symmetric on both sides, which is indicative of the role played by the central engine rather than being purely environment related.

- We find that many of the Seyfert galaxies (5/9) show multiple radio structures which are misaligned to each other on varying scales. However, in starburst galaxies, the distribution of radio continuum emission at different scales is often continuous and follows the star forming disk morphology or are aligned along the minor axis. We will explore if these misaligned structures represent multiple epochs of activity in our upcoming paper.
- For the sources in which we detected polarization, the Seyfert galaxies seem to show higher degrees of polarization compared to the starburst galaxies, at the scales probed by our observations. This is a tentative result and needs to be confirmed with deeper observations and larger samples.

4. Summary

We are trying to use polarization to distinguish between starburst/ AGN driven radiative winds versus the jet driven outflows in Seyfert galaxies. We have detected polarization in four Seyfert galaxies and one starburst galaxy. The peculiar morphology of the lobes in NGC 3079, which is a complex starburst galaxy also hosting a Seyfert nuclei, is a result of the interplay of the magnetised jet, the starburst wind and the galactic magnetic fields (Sebastian et al. 2019a). We have discovered a new pair of radio lobes in the north-south direction in the Seyfert galaxy NGC 2639; these lobes are aligned in a direction almost perpendicular to the previously known east-west lobes. The radio bubble-like lobes which are absent in starburst galaxies without an AGN are very common in Seyfert galaxies, and many a times not aligned with the minor axis, indicating the role of AGN in the formation of these lobes. We expect that our new observations using the EVLA at 10 GHz will enable us to nail down the reasons behind their origin and the differences between starburst and Seyfert galaxy radio emission.

Acknowledgements

The author is thankful to the organisers of the IAU Symposium 356 and the IAU for the opportunity and the financial support to attend the conference respectively. Baum and O'Dea are grateful to the Natural Sciences and Engineering Research Council of Canada (NSERC) for support.

References

Adebahr, B., Krause, M., Klein, U., et al. 2013, A&A, 555, A23
Baum, S. A., O'Dea, C. P., Dallacassa, D., de Bruyn, A. G., & Pedlar, A. 1993, ApJ, 419, 553
Blandford, R. & Eichler, D. 1987, Phys. Rep., 154, 1
Cecil, G., Bland-Hawthorn, J., Veilleux, S., & Filippenko, A. V. 2001, ApJ, 555, 338
Colbert, E. J. M., Baum, S. A., Gallimore, J. F., O'Dea, C. P., et al. 1996, ApJ, 467, 551
Dahlem, M., Weaver, K. A., Heckman, T. M., et al. 1998, ApJS, 118, 453
Duric, N. & Seaquist, E. R. 1988, ApJ, 326, 574
Gallimore, J. F., Axon, D. J., O'Dea, C. P., Baum, S. A., & Pedlar, A. 2006, AJ, 132, 546
Huchra, J. & Burg, R. 1992, ApJ, 393, 90
Irwin, J. A. & Seaquist, E. R. 1988, ApJ, 335, 658
Irwin, J. A., Schmidt, P., Damas-Segovia, A., et al. 2017, MNRAS, 464, 1333
Lal, D. V., Sebastian, B., Cheung, C. C., & Pramesh Rao, A. 2019, AJ, 157, 195
Rush, B., Malkan, M. A., & Spinoglio, L. 1993, ApJS, 89, 1
Sawada-Satoh, S., Inoue, M., Shibata, K. M., et al. 2000, PASJ, 52, 421
Sebastian, B., Kharb, P., O'Dea, C. P., Colbert, E. J. M., & Baum, S. A. 2019a, ApJ, 883, 189
Sebastian, B., Kharb, P., O'Dea, C. P., Gallimore, J. F., et al. 2019b, MNRAS, 490, L26
Sebastian, B., Kharb, P, O'Dea, C. P., Gallimore, J. C., Baum, S. A., et al. 2020, MNRAS, 499, 334
Trotter, A. S., Greenhill, L. J., Moran, J. M., et al. 1998, ApJ, 495, 740
Ulvestad, J. S., Wilson, A. S., & Sramek, R. A. 1981, ApJ, 247, 419

Role of active galactic nuclei and flow of relativistic jets

Abdissa Tassama and Tolu Biressa

Department of Physics, Jimma University, Jimma, Ethiopia

Abstract. The astrophysics of Active Galactic Nuclei (AGN) is one of the long outstanding issues in searches among the scientific communities raised with diverse perspectives like nebula, quasars, etc some decades ago. Currently, this exotic system is at least understood as the center of an active galaxy. Thus, the consensus of this recent theory has opened up a number of research issues for the progress of astrophysical science including how the hosting galaxy evolves with the AGN, how matter and energy flow towards and outwards, etc. Moreover, most of the AGNs possess Supermassive Black Holes (SMBHs) and accrete matter at a very high rate as current observations report. Consequently, both observations of electromagnetic (EM) spectrum and Gravity Waves (GWs) will considered to provide complementary information about the AGNs and the roles in their environments including black holes in their centers, outflow and inflow of matter-energy. Interested with this background rationale, we study the mechanisms of AGN interaction with its environment and flow of relativistic jets where General Relativistic (GR) Magneto-Hydrodynamic (MHD) equations are being considered. The solutions of the field equations are treated with a metric that involves charged systems for the possible relativistic jets including accretions. Then, numerical data is being generated using the latest version Mathematic software. Finally, the theoretical data is being compared with that of observation for validation of the model.

Keywords. galaxies: active, galaxies: properties

Feedback from quasars: The prevalence and impact of radio jets

Miranda Jarvis[1,2,3]

[1]Max-Planck Institut für Astrophysik, Karl-Schwarzschild-Str. 1, 85748 Garching, Germany
[2]European Southern Observatory, Karl-Schwarzschild-Str. 2, 85748 Garching, Germany
[3]Ludwig Maximilian Universität, Professor-Huber-Platz 2, 80539 Munich, Germany

Abstract. I will present our ongoing multi-wavelength study on the prevalence and impact of radio jets in a sample of $z < 0.2$ type 2 'obscured' quasars who's high bolometric luminosities make them ideal local analogues of distant, more common, quasars. Despite being classified as 'radio quiet' ($\log L[1.4\text{GHz}] = 23.3\text{-}24.4\,\text{W/Hz}$), our high spatial resolution ($\sim 0.25"$) radio observations (VLA and eMERLIN) reveal jet like structures on 1–25kpc scales in $\sim 80\%$ of the sample. Our integral field spectroscopy reveals jet-ISM interaction and outflows in all cases. Our work suggests that radio jets are an important feedback mechanism even during a typical 'quasar' phase. Using ALMA and APEX we are now investigating the impact of these jets and outflows on the molecular, star forming, gas; looking for signs of depletion and excitation. Preliminary results suggest a depleted molecular gas supply in these sources. I will present all of these results, focused on our pilot study of 10 targets and then introduce our on-going work on an expanded sample of 42 low-redshift quasars. Our latest results come from MUSE/AO and ALMA from which we are carefully characterising the properties of the ionised and molecular outflows at sub-kpc resolution.

Keywords. galaxies: active, active: quasars, active: relativistic jets

Physics of SMBH in nearby AGNs

Venkatessh Ramakrishnan

Astronomy Department, Universidad de Concepción, Casilla 160-C, Concepción, Chile

Abstract. We aim to leverage the transformational science enabled by the Event Horizon Telescope (EHT) to study the physics of, and near, the black holes in a sample of galaxies covering a large parameter space in SMBH mass, accretion rate, and jet power. To this end, we work on a sample of nearby galaxies whose directly measured black hole masses and distances imply that 40 micro-arcsec EHT observations will resolve the central engine at <100 Schwarzschild radius resolution. As an EHT member, I will present the results from the study of M87 and will discuss the impact of this finding on the study of nearby AGNs. The study of the SMBHs in these systems using molecular and ionised gas kinematics will also be presented.

Keywords. galaxies: active, galaxies: properties, galaxies: M87

The spins of supermassive black holes

Ranga-Ram Chary

Division of Physics, Math & Astronomy, California Institute of Technology, Pasadena, CA 91125, USA

Abstract. We present 1-second cadence, precise optical observations from SOFIA and Palomar of a sample of nearby supermassive black holes. The observations were taken to identify the shortest timescale variability in the nuclear photometry which may be associated with instabilities in the accretion flow in the immediate vicinity of the black hole. The shortest timescale variability, if associated with the radius of the innermost stable circular orbit (ISCO), can then be used to estimate the spin of the black hole. Despite 1% precision photometry, we obtained a non-detection of any significant variability in the nucleus of M32 ($M_{BH} \sim 2.5 \times 10^6\ M_\odot$). Given the density of the stellar cusp, this argues for a scenario where 1000 Msun seed black holes formed from the coalescence of less massive black holes, which then accrete the gas produced by stellar interactions/winds. In more luminous systems however, we find a significant deection of variability and present hypotheses to explain the signal and thereby the origin of supermassive black holes.

Keywords. galaxies: active, galaxies: properties, active: supermassive black holes

The space VLBI mission RadioAstron: AGN results

Yuri Kovalev[1,2]

[1]Astro Space Center of Lebedev Physical Institute, Profsoyuznaya 86/32, 117997 Moscow, Russia

[2]Moscow Institute of Physics and Technology, Institutsky per. 9, Dolgoprudny 141700, Russia

Abstract. The RadioAstron Space VLBI mission utilized the 10-m radio telescope on board the dedicated Spektr-R spacecraft to observe cosmic radio sources with an unprecedented angular resolution at centimeter wave lengths in total and polarized light. The longest baseline of the space-ground interferometer is about 350000 km. It operated in 2011–2019 together with 58 largest ground radio telescopes. Resolution as high as 10 microarcsec has been achieved. An overview of its AGN science results will be presented in the talk. It includes a probe of jet emission mechanism through brightness temperature measurements, reconstruction of magnetic field structure close to the jet origin using polarization data, jet formation and collimation study for well resolved nearby AGN, as well as observations and analysis of jet precession and plasma instabilities. We will also discuss a new scattering effect which was discovered by RadioAstron to affect high resolution radio measurements of AGN and SgrA*.

Keywords. galaxies: active, galaxies: properties, active: relativistic jets

Observing AGN sources with the Event Horizon Telescope

Maciek Wielgus[1,2]

[1]Black Hole Initiative at Harvard University, 20 Garden St., Cambridge, MA 02138, USA
[2]Center for Astrophysics—Harvard & Smithsonian, 60 Garden Street, Cambridge, MA 02138, USA

Abstract. In April 2017 Event Horizon Telescope (EHT) has delivered first resolved images of a shadow of a supermassive black hole. Apart from black hole sources in M87 and in the Galactic Center, observed with resolution comparable to the Schwarzschild radius scale, EHT observed multiple AGN sources during the 2017 campaign. These include 3C279, Centaurus A, OJ287 and more. For most of the considered sources EHT 2017 data set should allow to reconstruct images with highest angular resolution in the history of their observations, approaching 20 uas. While the analysis of these data is still ongoing, I will talk about the scientific opportunities related to observing AGN sources with the extreme resolution of the EHT as well as about the astrophysical questions that these observations may help answering.

Keywords. galaxies: active, galaxies: properties, active: supermassive black hole

CHAPTER VII. The youngest AGN and AGN evolution

What do observations tell us about the highest-redshift supermassive black holes?

Benny Trakhtenbrot

School of Physics and Astronomy, Tel Aviv University, Tel Aviv 69978, Israel
email: benny@astro.tau.ac.il

Abstract. I review the current understanding of some key properties of the earliest growing supermassive black holes (SMBHs), as determined from the most up-to-date observations of $z \gtrsim 5$ quasars. This includes their accretion rates and growth history, their host galaxies, and the large-scale environments that enabled their emergence less than a billion years after the Big Bang. The available multi-wavelength data show that these SMBHs are consistent with Eddington-limited, radiatively efficient accretion that had to proceed almost continuously since very early epochs. ALMA observations of the hosts' ISM reveal gas-rich, well developed galaxies, with a wide range of SFRs that may exceed ~ 1000 $M_\odot$ yr^{-1}. Moreover, ALMA uncovers a high fraction of companion, interacting galaxies, separated by <100 kpc (projected). This supports the idea that the first generation of high-mass, luminous SMBHs grew in over-dense environments, and that major mergers may be important drivers for rapid SMBH and host galaxy growth. Current X-ray surveys cannot access the lower-mass, supposedly more abundant counterparts of these rare $z \gtrsim 5$ massive quasars, which should be able to elucidate the earliest stages of BH formation and growth. Such lower-mass nuclear BHs will be the prime targets of the deepest surveys planned for the next generation of facilities, such as the upcoming *Athena* mission and the future *Lynx* mission concept.

Keywords. quasars: general, black hole physics, galaxies: active, galaxies: high-redshift, galaxies: interactions

1. Introduction and Overview

For decades, quasars have been detected and studied at increasingly high redshifts, owing to their extremely high luminosities. Wide-area surveys, required by the rarity of luminous, high-redshift quasars, provided an almost continuous progress of breaking redshift records. Since the first few $z \gtrsim 6$ quasars were detected through (early) SDSS observations (e.g., Fan et al. 2003), the trickle has turned into a steady stream of detections. The current redshift record holder is the quasar ULAS J1342+0928 at $z = 7.54$ (Bañados et al. 2018a), and there are over 170 quasars known at $z \gtrsim 6$, in addition to over 300 at $z \sim 5-6$ (see Ross & Cross 2020). The vast majority of these systems have been selected in various wide-field, multi-band optical-IR surveys, through elaborate colour-based criteria. Indeed, essentially every imaging survey with sufficient area and depth has identified samples of $z \gtrsim 5$ quasars (for some of the largest relevant samples see, e.g., Willott et al. 2010a; Bañados et al. 2016; Jiang et al. 2016; Reed et al. 2017; Matsuoka et al. 2019; Yang et al. 2019; Wang et al. 2019a). The quasar selection criteria are constantly improving, allowing to recover highly complete (or, at least, well-understood) samples that cover an ever expanding range in flux, redshift, and/or colour (e.g., Carnall et al. 2015; Wang et al. 2016; Reed et al. 2017). The recent, publicly accessible compilation of Ross & Cross (2020) provides an impressive and up-to-date status report on this still-growing population of quasars, as well as references to some of the important follow-up observations. Most importantly, several teams have been accumulating a rich collection of multi-wavelength data for these systems, which allow to study a multitude of phenomena related to the quasars and their central engines, to their host galaxies, and indeed to their large-scale environments.

A key motivation for studying the nature of the highest-redshift quasars, their hosts and their environments, is the very existence of such extreme systems, which are powered by super-massive black holes (SMBHs) with masses of $M_{\rm BH} \gtrsim 10^8\,M_\odot$. How could they grow to such high masses in less than a Gyr after the Big Bang (or seed BH formation)? Could they have grown through extremely fast, super-Eddington accretion? What do their host galaxies tell us about the availability of the (cold) gas that is needed for this fast, early growth? What sort of large-scale environments are needed to form the earliest nuclear, massive BHs and to power their fast growth? What are the effects that this fast SMBH growth exerts back on the host galaxies and/or the larger-scale environments, through radiative and/or mechanical energy output?

In this contribution, I review some of the main results related to the highest-redshift quasars, as inferred from *observations* across the electromagnetic spectrum, and over a wide range of physical scales. I also discuss the whereabouts of the lower-luminosity, lower-mass counterparts of these quasars, and outline how upcoming and future facilities and surveys will further extend our understanding of the first generation of SMBHs. For obvious reasons, I cannot fairly present *all* the recent results on this exciting topic. Detailed discussions of some of the theoretical aspects relevant to this topic can be found in several reviews including, among others, those by Volonteri (2010), Natarajan (2011), Valiante *et al.* (2017), and Inayoshi *et al.* (2019). In addition, Fan *et al.* (2006) reviews the usage of high-redshift quasars for probing the (re-)ionization state of their cosmic environments (and indeed of the Universe) – a topic that is not discussed here.

2. Observed Properties of the Highest-Redshift Quasars

2.1. *Basic spectral properties*

The first and perhaps most striking observation related to the highest-redshift quasars we know, is how "normal" their basic emission properties appear to be.

The (rest-frame) UV spectra of the highest-redshift quasars are remarkably similar to those of lower-redshift ones (matched in luminosity), including in particular the broad emission lines of C IV $\lambda1549$, C III] $\lambda1909$, and Mg II $\lambda2798$ (Fig. 1, top). Such comparisons have to account for the tendency of highly accreting quasars to show blue-shifted (UV) broad lines (e.g., Shen *et al.* 2016; Martínez-Aldama *et al.* 2018, and references therein). The relative intensities of these lines do not exhibit significant evolution out to $z \sim 7$, suggestive of early metal enrichment in the dense circumnuclear gas that constitutes the broad line region (BLR; e.g., Jiang *et al.* 2007; Kurk *et al.* 2007; De Rosa *et al.* 2011, 2014; Mazzucchelli *et al.* 2017). The same can be said about the presence of hot, circumnuclear dust, which was detected through *Spitzer* mid-IR observations in many systems (Jiang *et al.* 2006), although in this case there is evidence that some systems are "hot dust poor" (Jiang *et al.* 2010).

An increasing number of $z \gtrsim 6$ quasars have now been detected in the X-rays (e.g., Shemmer *et al.* 2006; Bañados *et al.* 2018b; Vito *et al.* 2019; Pons *et al.* 2020; see compilation by Nanni *et al.* 2017). The strength of their X-ray emission follows the expectations from lower-redshift quasars, namely the close relation between the (rest-frame) UV luminosity and UV-to-X-ray spectral slope ($\alpha_{\rm ox}$; e.g., Lusso & Risaliti 2016; Fig. 1, bottom).

A few of the highest-redshift quasars have been also detected in radio bands, implying extremely powerful radio emission, thought to originate from a relativistic jet

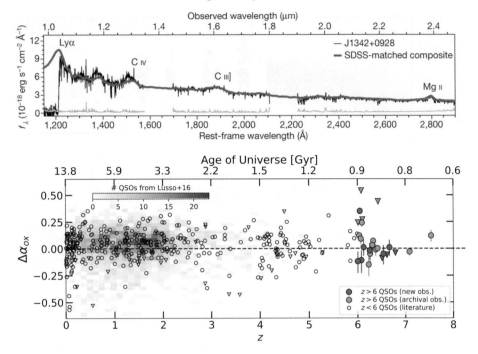

Figure 1. High-redshift quasars show "normal" UV-optical continuum and line emission, and broad-band (UV-to-X-ray) SEDs. *Top:* the (rest-frame) near-UV spectrum of ULAS J1342+0928, the highest redshift quasar known to date ($z = 7.54$), compared with a composite spectrum of lower-z SDSS quasars (taken from Bañados et al. 2018a). *Bottom:* the deviation in the optical-to-X-ray spectral slope, $\alpha_{\rm ox}$, compared with the $\alpha_{\rm ox} - L_{\rm UV}$ relation calibrated at lower z AGN (e.g., Lusso & Risaliti 2016), vs. redshift (figure taken from Vito et al. 2019).

(e.g., Bañados et al. 2018c; Belladitta et al. 2019). If (some of) these systems are interpreted as beamed AGN (i.e., blazars; e.g., Sbarrato et al. 2012; Ghisellini et al. 2015), they may potentially allow us to further constrain the accretion efficiencies and space densities of *all* (actively) accreting SMBHs at $z \gtrsim 5$ – including those which are too faint to be picked up by current optical/NIR surveys (Volonteri et al. 2011; Ghisellini et al. 2013).

2.2. Mass and accretion rates

The high luminosities of the first quasars to be identified at $z \gtrsim 5$, of $L_{\rm bol} \gtrsim 10^{47}$ erg s^{-1}, coupled with the Eddington argument (i.e., $L_{\rm bol} \lesssim L_{\rm Edd} \simeq 1.5 \times 10^{38} \, [M_{\rm BH}/M_\odot]$ erg s^{-1}), immediately indicate that these sources are powered by SMBHs with masses of at least $M_{\rm BH} \gtrsim 10^9 \, M_\odot$.

More accurate estimates of $M_{\rm BH}$ were obtained by using dedicated, intensive near-IR spectroscopy, which probes the rest-frame UV broad Mg II emission line, and by relying on so-called "virial" (or "single-epoch") $M_{\rm BH}$ prescriptions (e.g., Trakhtenbrot & Netzer 2012; Shen 2013). These reveal that the highest redshift quasars are indeed powered by SMBHs with $M_{\rm BH} \sim 10^{8-10} \, M_\odot$, accreting at rates $L/L_{\rm Edd} \sim 0.1 - 1$ (e.g., Willott et al. 2010b; Trakhtenbrot et al. 2011; De Rosa et al. 2014; Onoue et al. 2019; Shen et al. 2019; see Wang et al. 2015, Wu et al. 2015, and Kim et al. 2018 for examples of "extreme" values). Here, too, the number of sources with reliable determinations of $M_{\rm BH}$ and $L/L_{\rm Edd}$ is steadily growing. For example, the recent study by Shen et al. (2019) reported new measurements of this sort for about 30 $z \gtrsim 5.7$ quasars.

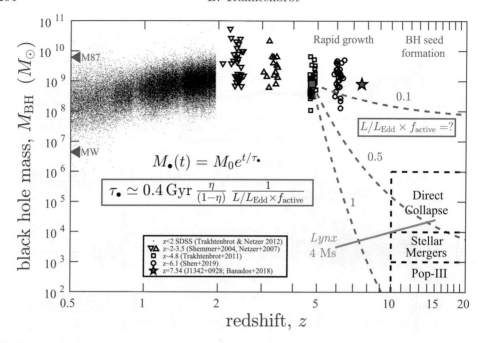

Figure 2. An overview of the emergence of the first generation of SMBHs. The plot shows various samples of highly luminous quasars with reliable $M_{\rm BH}$ estimates (black symbols, see legend; only a subset of available measurements is shown). Given the range of possible BH seed masses available at $z \sim 10-20$ (black-dashed boxes), the $z \gtrsim 5$ quasars require nearly-continuous mass accretion, at high accretion rates – as illustrated by growth tracks with various combinations of $L/L_{\rm Edd}$ and duty cycle (red-dashed lines). Future facilities, and particularly the *Lynx* mission concept, may be able to directly probe the epoch of massive seed BH formation and fast, initial BH growth – the magenta line illustrates the expected sensitivity of a deep *Lynx* survey.

With $L_{\rm bol}$ (and $L/L_{\rm Edd}$) estimates in hand, one still has to assume a certain radiative efficiency ($\eta \equiv L_{\rm bol}/\dot{M}c^2$) in order to deduce the physical accretion rates, and thus to address the dramatic growth history, of the earliest SMBHs. A few studies used simple thin-disk models to deduce $\dot{M}$ (e.g., Sbarrato *et al.* 2012; Trakhtenbrot *et al.* 2017a), showing that the (rest-frame) UV-optical SED measurements in hand can be indeed explained by Eddington-limited, radiatively efficient, thin-disk accretion. However, yet earlier episodes of super-Eddington accretion cannot be ruled out.

Comparing the known, luminous $z \gtrsim 5$ quasars to the most luminous, highest-$M_{\rm BH}$ quasars at $z \sim 2-4$, we finally see a *decrease* in the maximal $M_{\rm BH}$ with increasing redshift (Fig. 2; see also Marziani & Sulentic 2012 and Trakhtenbrot & Netzer 2012). The increase in typical $L/L_{\rm Edd}$ means that we are witnessing the earliest epoch during which the most massive BHs known ($M_{\rm BH} \sim 10^{10}\,M_\odot$) have been growing at their fastest, Eddington-limited rate (see discussion in, e.g., Trakhtenbrot *et al.* 2011).

Most importantly, given the measured masses and accretion rates (and thin-disk values for η), the main challenge concerning the highest-redshift quasars still stands: these SMBHs had to grow nearly continuously, at high accretion rates, to reach their observed masses – even if one allows for massive BH seeds. This is illustrated in Fig. 2 using several simple, exponential mass growth tracks (i.e., constant $L/L_{\rm Edd}$), scaled to match a typical $z \simeq 5$ quasar. Even the most massive BH seed ($M_{\rm seed} \sim 10^6\,M_\odot$; e.g., Volonteri 2010; Natarajan 2011) *cannot* explain the observed mass, if the growth proceeded through a combination of Eddington ratio and duty cycle of order $L/L_{\rm Edd} \times f_{\rm active} = 0.1$.

2.3. Host galaxies

Measuring the host galaxies of the earliest quasars is key to understanding their early emergence and fast growth, and to testing for any early signs of the links we may expect between SMBH and host growth (i.e., their co-evolution). In particular, one would naively expect the earliest SMBHs to grow in unstable, gas-rich systems, which would also allow for intense SF. Major galaxy mergers, which are thought to be common at early cosmic epochs, could expedite SMBH and host growth. There are even some suggestions that the most massive, rarest luminous quasars, may have out-grown (the stellar populations of) their hosts, thus predicting the SMBHs to be "over-massive", compared with the locally observed $M_{\rm BH} - M_{\rm host}$ relations (see, e.g., Agarwal et al. 2013).

All these ideas are extremely hard to address with observations. Indeed, to date there is no direct detection of the stellar component in the hosts of the highest-redshift quasars – not even with intense, high-resolution *HST* NIR imaging. ALMA and other sub-mm facilities are currently revolutionising our understanding of the host galaxies, probing directly the (cold) ISM gas, and allowing us to determine the hosts' gas content, SF activity, and dynamics. Moreover, after a few earlier efforts focusing on particular systems, and thanks to the high sensitivity of ALMA, we are now seeing a shift towards spatially-resolved studies of increasingly large and systematically defined samples.

The vast majority of $z \gtrsim 5$ quasar hosts are robustly detected, and often spatially resolved, in (rest-frame) FIR continuum emission. This implies considerable amounts of cold ($\sim 30-60$ K) gas on galaxy-wide scales ($\gtrsim 1$ kpc), and indeed that the highest redshift quasars are mostly hosted in well-developed galaxies, enriched in metals and dust.

The quasar hosts show a wide range of SFRs, reaching extremely high values. The higher-SFR systems, with $\sim 1000-3000\,M_\odot\,{\rm yr}^{-1}$, were already clearly detected with *Herschel* (e.g., Mor et al. 2012; Leipski et al. 2014; Netzer et al. 2014; Lyu et al. 2016). ALMA now allows us to probe much lower SFRs. For example, Decarli et al. (2018) studied a sample of 27 quasars at $z \gtrsim 6$, and found that their (rest-frame) FIR emission can be accounted for with SFR $\sim 30-2000\,M_\odot\,{\rm yr}^{-1}$. Other ALMA-based studies report an even wider distribution, with SFRs as low as $\sim 10\,M_\odot\,{\rm yr}^{-1}$ (Willott et al. 2017; Izumi et al. 2018), or as high as $\gtrsim 3000\,M_\odot\,{\rm yr}^{-1}$ (e.g., Trakhtenbrot et al. 2017b). The extremely high SFRs found for some $z \gtrsim 5$ quasar hosts are consistent with the highest values known, typically measured in dusty, SF galaxies (i.e., "sub-mm galaxies", or SMGs; see Casey et al. 2014 for a review). This is in line with the expectation for rapid, co-evolutionary growth of the BH and stellar components. On the other hand, the lower end of the SFR distribution overlaps with what is known about the much more abundant, "normal" SF galaxies at $z \sim 5-7$ (i.e., "Lyman break galaxies" or LBGs; see, e.g., Stark 2016). Hopefully, we will soon be able to robustly test for possible links between the host SFR and SMBH-related properties, namely $L_{\rm AGN}$, using complete samples – as is done at lower redshifts (see initial attempts to populate the SFR-$L_{\rm AGN}$ parameter space in, e.g., Netzer et al. 2014; Lyu et al. 2016; Venemans et al. 2018, and Izumi et al. 2019).

Importantly, sub-mm observations also probe several ISM emission lines, most notably [C II] $\lambda 157.74\,\mu$m, which allow to study the dynamics of the quasar hosts. Since the first detection of [C II] in the $z \simeq 6.4$ quasar J1148+5251 (Maiolino et al. 2005), such measurements have grown to become a booming "industry", with several teams pursuing increasingly better data for ever growing samples (for an impression of this progress, see, e.g., Venemans et al. 2012, 2016, 2017, 2019; Wang et al. 2013, 2019b; Willott et al. 2013, 2015, 2017; Trakhtenbrot et al. 2017b; Decarli et al. 2018; Izumi et al. 2018, 2019). In many cases, the spatially resolved [C II] line velocity maps show signs of ordered, rotation-dominated gas dynamics (Fig. 3, left; see, e.g., Trakhtenbrot et al. 2017b; Shao et al. 2017). This, in turn, provides further motivation to use the velocities and spatial extent of the line emitting regions as probes of the *dynamical* host masses. It's important to note, however, that this common practice necessitates several crucial assumptions, mainly

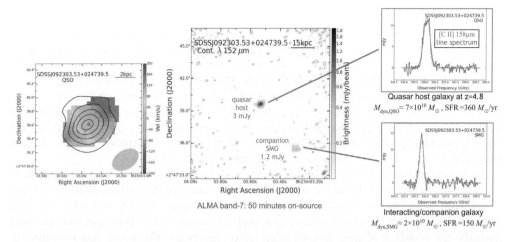

Figure 3. ALMA allows to study the host galaxies and larger-scale environments of $z \gtrsim 5$ quasars (taken from Trakhtenbrot et al. 2017b). *Centre:* A 50-minute integration with ALMA band-7 allows to robustly detect the cold ISM, and determine the SFR, in the host galaxy of a $z \simeq 5$ quasar. *Left:* The spatially-resolved [C II] $\lambda 157.74$ μm emission suggests rotation-dominated dynamics, and allows to estimate the host dynamical mass. *Right:* The same data cube reveals a dusty, star-forming companion galaxy, separated from the quasar by ~ 40 kpc and ~ 250 km s^{-1}.

that of a rotating cold gas disk, and thus carries significant uncertainties. Indeed, deeper and higher resolution ALMA data obtained for some systems show no (or very limited) evidence for rotation-dominated gas structures, thus highlighting the limitations of such assumptions. Examples of such cases, and of more elaborate analyses, can be found in, e.g., Neeleman et al. (2019), Venemans et al. (2019), and Wang et al. (2019b).

The rough dynamical host mass estimates derived from [C II] line measurements are typically of order $M_{\rm dyn} \sim 10^{10-11}$ $M_{\odot}$ (e.g., Willott et al. 2015; Venemans et al. 2016; Shao et al. 2017; Trakhtenbrot et al. 2017b; Decarli et al. 2018; Izumi et al. 2019). The *stellar* masses are naturally somewhat lower, although they would most likely remain inaccessible at least until *JWST* is operating.

Compared with the SMBH-host relations seen in the local Universe, and based on the highly uncertain $M_{\rm dyn}$ estimates, the highest-redshift quasars tend to have somewhat over-massive SMBHs, with SMBH-to-host mass ratios of $M_{\rm BH}/M_{\rm host} \sim 1/100-1/30$ (Fig. 4). This provides some evidence in support of the idea that the first generation of SMBHs emerged through some sort of preferentially efficient BH fuelling mechanism. However, as with lower-redshift systems, it's important to keep in mind that the most luminous $z \gtrsim 5$ quasars currently studied may be biased towards high $M_{\rm BH}/M_{\rm host}$, while lower-mass systems could be closer to the local $M_{\rm BH}/M_{\rm host}$ ratio (or even below it). Indeed, new and deep ALMA data obtained for (relatively) lower-luminosity $z \sim 6$ quasars from the SHELLQs sample revealed $M_{\rm BH}/M_{\rm host}$ ratios consistent with the local ones (Izumi et al. 2019). Some studies have tried to go further, and estimated the (short-term) evolution of $M_{\rm BH}$ and $M_{\rm host}$ in some $z \gtrsim 5$ quasars, based on the measured $L_{\rm AGN}$ and SFRs. Most systems, and particularly the highest $M_{\rm BH}/M_{\rm host}$ ones, seem to be able to get closer to the locally-observed mass relation (see arrows in Fig. 4).

To summarise, the best data available for the highest-redshift quasars currently known do *not* show overwhelming evidence for large, systematic and robust deviations from the locally-observed $M_{\rm BH} - M_{\rm host}$ relations.

2.4. Outflows

The radiative and/or mechanical energy output of accreting SMBHs onto their host galaxies, so-called "AGN feedback", is a widely-acknowledged cornerstone of the

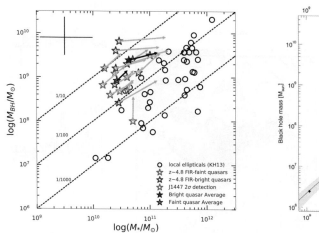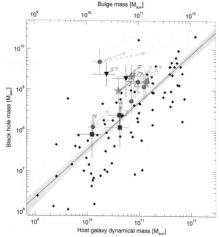

Figure 4. The SMBH-vs-host mass plane for $z \gtrsim 5$ quasars, with ALMA-enabled measurements of host *dynamical* (gas+stellar) masses, $M_{\rm dyn}$. The two panels show measurements for quasars at $z \simeq 5$ (left; taken from Nguyen *et al.* 2020) and at $z \sim 6$ (right; measurements compiled by Venemans *et al.* (2016). In both panels, the coloured symbols mark the high-z quasars, which have large systematic uncertainties (large cross in left panel), and the black symbols mark local, inactive systems (taken from Kormendy & Ho 2013). For each high-z quasar, an arrow illustrates the expected evolution of the system within 50 Myr, assuming the measured $L_{\rm AGN}$ ($\propto \dot{M}_{\rm BH}$) and SFR ($\approx \dot{M}_{\rm host}$). The highest redshift quasars appear to have higher $M_{\rm BH}/M_{\rm host}$ than what is seen in the local universe, with $M_{\rm BH}/M_{\rm host} \sim 1/100 - 1/30$. Their short-term subsequent evolution may bring them somewhat closer to the locally observed SMBH-host relation.

co-evolutionary paradigm. Specifically, given the requirement for (nearly) continuous BH growth at high rates, one could expect the highest-redshift quasars to showcase such feedback processes. Combined with the short time available for host evolution, these systems may be considered as optimal test-beds for any viable feedback scenario.

Currently, the only evidence for anything that may be interpreted as AGN feedback comes in the form of extended, high-velocity outflows of cold gas, seen in some of the highest-redshift quasars. In particular, Maiolino *et al.* (2012), and later Cicone *et al.* (2015), have identified & resolved such an outflow, extending out to >15 kpc and reaching >1,000 km s^{-1}, using PdBI observations of the [C II] line in a $z = 6.4$ quasar (Fig. 5).

More recently, two different studies used *stacking* analysis of [C II] data for dozens of $z \gtrsim 5$ quasars to try and identify such outflow signatures, through broad (but weak) [C II] emission components. Bischetti *et al.* (2019) studied 48 quasars at $4.5 < z < 7.1$ and found broad [C II] wings that are interpreted to trace outflowing gas with velocities $v \gtrsim 1,000$ km s^{-1}, an extent of $\gtrsim 5$ kpc, and a deduced a mass outflow rate of order $\sim 100\, M_\odot$ yr^{-1}. On the other hand, Stanley *et al.* (2019) found only marginal evidence ($< 3\sigma$) for such outflow signatures in their analysis of 26 $z \sim 6$ quasars (the outflow signatures may be more prominent in a subset of about half of these quasars).

Given the high SFRs and rich gas content of the highest-redshift quasar hosts, it may not seem straightforward to link the large-scale, energetic outflows probed in broad [C II] emission directly to AGN activity. However, there is strong evidence, from low-redshift systems, that the extremely high velocity regime ($\gtrsim 1,000$ km s^{-1}) tends to be indeed directly linked to AGN-driven outflows (see, e.g., Veilleux *et al.* 2013; Janssen *et al.* 2016; Stone *et al.* 2016). Future studies of high-ionization (rest-frame) optical emission lines (e.g., [O III] $\lambda 5007$, [N II] $\lambda 6584$), using *JWST*/NIRSpec 3D spectroscopy on ~ 1 kpc scales, would allow us to search for more direct signatures of AGN-driven outflows.

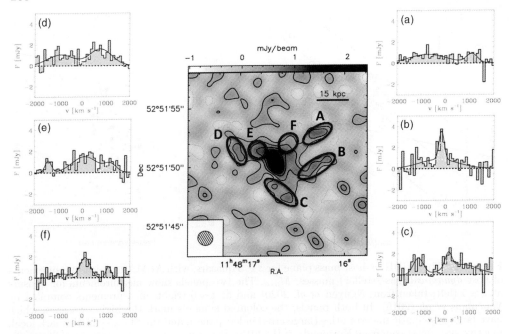

Figure 5. A large-scale outflow of cold seen in a $z \simeq 6.4$ quasar (taken from Cicone et al. 2015). The outflow, identified through a Plateau de Bure Interferometer observation of broad [C II] $\lambda 157.74\,\mu$m emission components, extends out to >20 kpc and $>1000\,{\rm km\,s}^{-1}$.

2.5. Large-scale environments

Several models suggest that the first generation of SMBHs, and indeed the most luminous quasars at $z \gtrsim 5$, would be found in over-dense large-scale environments (see, e.g., Overzier et al. 2009; Costa et al. 2014; Habouzit et al. 2019). The general idea is that in such regions, which could later evolve into rich galaxy (proto-)clusters, IGM gas can be efficiently funnelled into the central galaxies, and thus to the SMBHs at their hearts.

Attempts to address these ideas observationally, using multi-band optical imaging designed to identify LBGs at redshifts comparable to those of the quasars themselves, have so far resulted in rather ambiguous results. While some studies claimed to identify over-dense environments (e.g., Overzier et al. 2006; Kim et al. 2009; Utsumi et al. 2010; Husband et al. 2013), other systems show no evidence for such over-densities (e.g., Willott et al. 2005; Bañados et al. 2013; Simpson et al. 2014). Such observational efforts are limited by their focus on (rest-frame) UV bright LBGs; by the coarse redshift resolution they provide for any nearby source ($\Delta z \sim 0.5$); by cosmic variance; and by the relatively long exposure times they require. Linking observations to models is further complicated by the wide range in BH seeding scenarios; by the possible effects of AGN feedback; and other details (see discussion in Buchner et al. 2019 and Habouzit et al. 2019).

As with host galaxy studies, here too ALMA is revolutionising our understanding of the highest-redshift quasars. The same ALMA observations that probe the quasar hosts can also be used to search for real "companion" galaxies (i.e., not just projected neighbouring sources), separated from the quasars by up to ~ 50 kpc and/or $\sim 500\,{\rm km\,s}^{-1}$. The occurrence rate of such companion galaxies is surprisingly high. Trakhtenbrot et al. (2017b) found three such companions among a sample of six $z \simeq 4.8$ quasars, and a follow-up study of 12 additional systems from the same parent sample brings the total to 5 companions among 18 quasars (i.e., 28%; Nguyen et al. 2020). Decarli et al. (2017) identified 4 companion galaxies among their sample of 25 quasars at $z \sim 6$ (i.e., 16%).

Izumi et al. (2019) identified one companion galaxy, and another candidate companion, among a sample of merely three $z \sim 6$ quasars drawn from the SHELLQs project. Fig. 3 illustrates a quasar+companion system, from the Trakhtenbrot et al. (2017b) study. These occurrence rates are far higher than what is seen among normal, SF galaxies at $z \sim 5-7$ observed in deep extragalactic surveys (i.e., LBGs; e.g., Aravena et al. 2016).

The companion galaxies detected so far share a few interesting properties. First, their detection with ALMA immediately implies that they are themselves gas-rich and have intense SF activity, with SFRs of order 100 $M_\odot$ yr^{-1}. Second, their continuum emission and (rough) dynamical masses are within a factor of ~ 3 from those of the quasar hosts. This, along with other observed similarities (e.g., Walter et al. 2018), suggests that these systems may be considered as major galaxy mergers between rather similar galaxies. Finally, and crucially, the companion galaxies are not detected in (deep) NIR imaging, including with *HST* (Decarli et al. 2017; Mazzucchelli et al. 2019), suggesting they are indeed extremely dusty, perhaps similar to well-known merging galaxies in the local Universe (e.g., Arp 220; see Mazzucchelli et al. 2019). This latter point may also explain why the aforementioned searches for neighbouring LBGs have not identified a similarly high occurrence rate of close companions and/or signs of over-dense environments.

Linking the ALMA-detected companions to any detailed scenario for the emergence of the earliest quasars remains challenging. First, our understanding of [C II]-based selection of (inactive) $z \gtrsim 5$ galaxies – that is, the nature and abundance of galaxies comparable to the quasar companions – is itself limited, and fast-changing (see, e.g., Capak et al. 2015; Aravena et al. 2016; Faisst et al. 2020). Second, we cannot know how long would any of the observed galaxy-galaxy interactions last. Moreover, while the galaxy interactions we see may perhaps explain the triggering of the concurrent quasar phase, they could *not* have powered the entire period of nearly continuous SMBH growth, required to account for the high $M_{\rm BH}$ we measure. Given the typical timescales of galaxy-galaxy interactions, it is not unreasonable to think that the highest-redshift quasars experienced a sequence of major mergers, and we are able to witness the early, rather prolonged phases of some of these mergers, in some of the systems. In this context, perhaps the highest-SFR quasar hosts, which are generally *not* those with >10 kpc companions, may be powered by advanced (late-stage), unresolved major mergers (see also Izumi et al. 2019 for a candidate advanced merger system). Trakhtenbrot et al. (2017b) offers a rather detailed discussion of both the prospects, and limitations, of linking the (ALMA-detected) quasar companions to the scenario of merger-driven growth for the first generation of SMBHs.

Given this significant progress, as well as other exciting results at somewhat lower redshifts (e.g., Banerji et al. 2017; Bischetti et al. 2018; Miller et al. 2018), ALMA will likely continue to be the main surveyor of the environments of high-redshift quasars (together with other large sub-mm facilities).

3. Where are the Lower-Luminosity, Lower-Mass Counterparts?

We are clearly gaining various new insights into the nature of luminous, unobscured quasars at $z \gtrsim 5$, powered by highly accreting SMBHs with $M_{\rm BH} \gtrsim 10^8\ M_\odot$. However, one must recall that such systems are extremely rare, with space densities are of order of $\Phi \simeq 10^{-7}$ Mpc^{-3} (for $L_{\rm AGN} \gtrsim 10^{46}$ erg s^{-1}; e.g., Kulkarni et al. 2019; Wang et al. 2019a; Shen et al. 2020). As such, they are expected to represent only the tip of the iceberg of the entire population of (active) SMBHs at these early cosmic epochs. What do we know, observationally, about the lower-$L_{\rm AGN}$, lower-$M_{\rm BH}$ AGN at $z \gtrsim 5$? The answer is, perhaps surprisingly, "very little".

First, there is clear evidence for a drastic decline in the space densities of AGN at $z \sim 3-5$, and particularly for the lower-luminosity ones, as traced by the deepest, highly

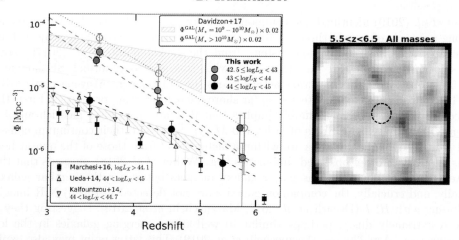

Figure 6. The "missing" population of $z \gtrsim 5$ low-luminosity, low-$M_{\rm BH}$ AGN. *Left:* the decreasing space density of X-ray selected, medium- and low-L AGN, towards high redshifts, as seen in deep multi-wavelength surveys (mainly CDF-S, Luo et al. 2017, and COSMOS, Marchesi et al. 2016). Note the lack of robustly identified, spectroscopically confirmed, $z \gtrsim 5-6$ AGN (figure taken from Vito et al. 2018). *Right:* the lack of AGN-related X-ray emission in the deepest available *stack* of X-ray data for abundant, (rest-frame) UV-bright SF galaxies at $z \sim 6$ (LBGs; figure taken from Vito et al. 2016). This X-ray stack reaches a depth of >1 *Giga*-second, and shows that the typical AGN luminosity in such galaxies is $L_{\rm AGN} \lesssim 5 \times 10^{43}$ erg s^{-1}.

complete X-ray and multi-wavelength surveys (Fig. 6, left; see, e.g., Marchesi et al. 2016, Luo et al. 2017, and Vito et al. 2018 for the most up-to-date deep surveys).

Second, these same surveys have not (yet) provided robust identifications of $z \gtrsim 5.5$ AGN, despite intensive, dedicated efforts (see, e.g., Weigel et al. 2015 and Cappelluti et al. 2016, and the compilation in Vito et al. 2018; these contrast the findings of Giallongo et al. 2015). Specifically, there is currently only one spectroscopically confirmed $z>5$ AGN in the *Chandra* Deep Fields (CDFs; that source is at $z = 5.19$, see Vito et al. 2018), and no AGN beyond $z \simeq 5.2$ in neither the two CDFs nor in the COSMOS field (Marchesi et al. 2016). Attempts to corroborate additional candidates (i.e., with high *photometric* redshifts) are still on-going, using exceedingly deep spectroscopic obseravations on large telescopes. It should be noted that the dearth of detections at $z \gtrsim 5$ is *not* due to insufficient survey volume or insufficient sensitivity (see Fig. 6 in Trakhtenbrot et al. 2016 and the discussion in Weigel et al. 2015).

Finally, *stacking* analysis of the deepest X-ray data available for hundreds of $z \gtrsim 5$ galaxies, using the CDF-S 7 Ms data – and reaching the remarkable effective depth of ~ 1 *Giga*-second – resulted in no robust detection of X-ray emission (Fig. 6, right; Vito et al. 2016). This implies that the typical AGN-related emission in abundant $z \gtrsim 5$ galaxies has $L_{\rm AGN} \lesssim 5 \times 10^{43}$ erg s^{-1}.

The reason behind this apparent dearth of lower-luminosity AGN signatures at $z \gtrsim 5$ is unclear. It could be driven by either (1) an exceptionally high fraction of highly obscured systems; (2) exceptionally low radiative efficiencies (i.e., ADAF/RIAF-like accretion); (3) extremely low duty cycles (i.e., due to "flickering"); (4) a low BH occupation fraction in such galaxies; and/or a combination of these scenarios. Given this wide range of possibilities, any attempt to compare evolutionary models with observations (including by future facilities; see §4 below) is complicated by degeneracies in the relevant model parameters.

It should finally be noted that the apparent lack of lower-luminosity AGN emission at $z \gtrsim 5$ is not (yet) in too great of a tension with neither *all* models, nor with *all* sensible extrapolations of the quasar luminosity function (compare, e.g., Fig. 16 in Vito *et al.* 2016 with Fig. 6 in Kulkarni *et al.* 2019 or Fig. 5 in Shen *et al.* 2020).

Clearly, yet deeper (X-ray) surveys are needed to detect low-luminosity, low-mass $z \gtrsim 5$ SMBHs, or otherwise to constrain their abundance, and to directly confront the wide-ranging models for their emergence. The data in hand already motivates us to consider some intriguing scenarios.

4. Future Prospects

The near future holds several opportunities to greatly improve our understanding of the currently-known $z \gtrsim 5$ quasar population and, importantly, to reveal new and much larger populations of early SMBHs.

Once operating, *JWST* will be able to directly probe the stellar and ionized gas components in the hosts of the already-known quasars. This can be done through NIR and MIR imaging, and NIR integral field spectroscopy. Several theoretical studies have already provided predictions for what *JWST* will be able to see, particularly in terms of the stellar emission (e.g., Natarajan *et al.* 2017; Volonteri *et al.* 2017). This will hopefully allow us to better understand early SMBH-host co-evolution, and perhaps even constrain high-mass BH seeding scenarios. Another improvement will come from the ability to observe the broad Balmer emission lines, which provide the most reliable kind of "virial" M_{BH} estimates. It is important to keep in mind that *JWST* is not designed to be a survey/discovery facility (at least in the context of high-z quasars), thus careful consideration of the kind of targets and samples to be observed, and the possible biases they entail, is key.

There are several upcoming and future facilities that will survey and discover *new* accreting SMBHs at $z \gtrsim 5$. In the optical/NIR regime, *Euclid* is predicted to be able to detect dozens of highly (UV) luminous quasars, with $L_{bol} \gtrsim 10^{47}$ erg s^{-1} – similar to SDSS $z \sim 6$ quasars, but out to $z \sim 8-10$ (i.e., ~ 0.5 Gyr after the Big Bang). The wide-field surveys planned to be conducted with the LSST and with *WFIRST* will be able to uncover somewhat lower-luminosity quasars at comparably early epochs, reaching $L_{bol} \simeq 10^{46}$ erg s^{-1} (i.e., driven by Eddington-limited SMBHs with $M_{BH} \simeq 10^8\ M_\odot$, or more massive, slowly accreting systems). Future X-ray missions hold the key to revealing *all* accreting SMBHs at these epochs, regardless of obscuration. *Athena* is planned to discover *hundreds* of X-ray emitting AGN beyond $z \sim 6$ with yet lower luminosities, reaching $L_{bol} \simeq 10^{45}$ erg s^{-1}. Ultimately, one has to pursue much deeper, high resolution X-ray observations to directly probe the fast growth of newly formed SMBHs. The future *Lynx* X-ray mission concept is envisioned to have about $\times 100$ the sensitivity of *Chandra*, allowing to discover AGN as faint as $L_{bol} \simeq 10^{42}$ erg s^{-1} out to $z \sim 15$, in a 4 Ms "deep drill" survey. This has the potential of detecting SMBHs with masses as low as $M_{BH} \simeq 10^4\ M_\odot$ at the epoch of massive BH seeding (see magenta line in Fig. 2), or otherwise put strong constraints on their existence.

Future facilities and surveys will thus be able to directly probe the emergence of the first SMBHs in the Universe, within the first hundreds of Myr after the Big Bang. It is important, however, to keep in mind that we will also need extensive, multi-wavelength follow-up capabilities to maximize the science return from these newly-discovered SMBH populations, and to more critically test the relevant models.

5. Summary

Table 1 summarises the key observed and derived properties of the known highest redshift quasars, at $z \sim 5-7$, as well as what would be (naively) expected for their lower-luminosity, lower-M_{BH} counterparts, which are not yet robustly identified.

Table 1. Overview of current observed and derived properties of the highest-redshift quasars, and expectations for their yet-to-be established lower-mass counterparts.

Property	Known $z \sim 5-7$ quasars	"Typical" AGN/galaxies
Luminosity, $L_{\rm bol}$	$\gtrsim 10^{46}$ erg s^{-1}	$\lesssim 10^{45}$ erg s^{-1}
Obscuration / selection	un-obscured / UV-opt.	$\sim$50% obscured / X-ray
SMBH mass, $M_{\rm BH}$	$\sim 10^9\ M_\odot$	$\sim 10^7\ M_\odot$
Accretion rate, $L/L_{\rm Edd}$	~ 1	$\sim 0.01-1$
Accretion mode	thin disk, $\eta \gtrsim 0.1$	(who knows, really?)
Implied BH seeds	massive, $M_{\rm seed} \sim 10^{4-6}\ M_\odot$	stellar (pop-III), $M_{\rm seed} < 10^3\ M_\odot$
Host mass, $M_{\rm host}$	$\sim 10^{10-11}\ M_\odot$	$\sim 10^{9-10}\ M_\odot$
Host SFR	$\sim 100-3000\ M_\odot$ yr^{-1}	$<100\ M_\odot$ yr^{-1}
Large-scale env.	over-dense, mergers, outflows	"normal"?
Demographics	rare! $\Phi \lesssim 10^{-7}$ Mpc^{-3}	common? $\Phi \gtrsim 10^{-5}$ Mpc^{-3} ($\sim$10% of galaxies? less?)
Future prospects	*Euclid, Athena, WFIRST*	*Lynx*

The key take-away points from this short overview can be summarized as follows:

• By now, there are hundreds of highly luminous, unobscured AGN (quasars) known at $z \gtrsim 5$. Their basic observed properties in the AGN-dominated, X-ray-to-NIR regime do not differ from their lower-z counterparts. We may, however, expect these early systems to stand out in *some* way, which can be linked to the fast growth of the SMBHs that power them, within the first Gyr after the Big Bang.

• ALMA is revolutionising our ability to study the host galaxies, and larger-scale environments, of the highest-redshift quasars. The hosts are massive, gas-rich galaxies with a wide range of SFRs, reaching the highest levels known ($\sim$3000 $M_\odot$ yr^{-1}).

• ALMA allowed us to uncover dusty, star-forming galaxies accompanying, and indeed interacting with, the quasar hosts. The occurrence rate of such companions is much higher than what is seen in inactive high-z galaxies, suggesting that early, fast SMBH growth may be linked to over-dense cosmic environments and/or galaxy mergers.

• We have yet to robustly identify the lower-luminosity, lower-$M_{\rm BH}$ AGN population at $z \gtrsim 5$, despite the remarkably deep X-ray survey data currently in hand. Such systems are expected to be associated with (a fraction of) the much more common, "normal" SF galaxies at these early epochs. This dearth of AGN signatures may be driven by a high fraction of highly obscured systems; by low radiative efficiencies; by a low duty cycle; and/or a low BH occupation fraction.

• Upcoming and future facilities and surveys will allow us to directly probe the stars and gas in the hosts of the currently known $z \gtrsim 5$ quasars (with *JWST*); to detect many more highly luminous quasars out to $z \sim 10$ (with *LSST* & *WFIRST*), as well as hundreds of obscured, lower-luminosity systems (with *Athena*). The *Lynx* future mission concept paves the way to directly probe the epoch of massive BH seed formation.

With growing samples and ever-improving data, our understanding of the highest-redshift quasars is now better than ever. However, to fully address the sophisticated models that were developed to explain the first generation of massive BHs in the universe, we have to strive for deeper, more detailed observations with new facilities and surveys.

I thank the organisers of the meeting for their kind invitation to present this overview, for the opportunity to take part in this exciting meeting, and indeed to visit Ethiopia and to get familiar with the local scientific community. I also thank the meeting participants, particularly N. Brandt and H. Netzer, for their insightful comments following my presentation, which helped me to improve this written contribution.

References

Agarwal, B., Davis, A. J., Khochfar, S., et al. 2013, *MNRAS*, 432, 3438
Aravena, M., Decarli, R., Walter, F., et al. 2016, *ApJ*, 833, 71
Banerji, M., Carilli, C. L., Jones, G., et al. 2017, *MNRAS*, 465, 4390
Belladitta, S., Moretti, A., Caccianiga, A., et al. 2019, *A&A*, 629, A68
Bañados, E., Venemans, B., Walter, F., et al. 2013, *ApJ*, 773, 178
Bañados, E., Venemans, B. P., Decarli, R., et al. 2016, *ApJS*, 227, 11
Bañados, E., Venemans, B. P., Mazzucchelli, C., et al. 2018a, *Nature*, 553, 473
Bañados, E., Connor, T., Stern, D., et al. 2018b, *ApJL*, 856, L25
Bañados, E., Carilli, C., Walter, F., et al. 2018c, *ApJL*, 861, L14
Bischetti, M., Piconcelli, E., Feruglio, C., et al. 2018, *A&A*, 617, A82
Bischetti, M., Maiolino, R., Carniani, S., et al. 2019, *A&A*, 630, A59
Buchner, J., Treister, E., Bauer, F. E., et al. 2019, *ApJ*, 874, 117
Capak, P. L., Carilli, C., Jones, G., et al. 2015, *Nature*, 522, 455
Cappelluti, N., Comastri, A., Fontana, A., et al. 2016, *ApJ*, 823, 95
Carnall, A. C., Shanks, T., Chehade, B., et al. 2015, *MNRAS*, 451, L16
Casey, C. M., Narayanan, D., & Cooray, A. R. 2014, *Phys. Rep.*, 541, 45
Cicone, C., Maiolino, R., Gallerani, S., et al. 2015, *A&A*, 574, A14
Costa, T., Sijacki, D., Trenti, M., & Haehnelt, M. G. 2014, *MNRAS*, 439, 2146
De Rosa, G., Decarli, R., Walter, F., et al. 2011, *ApJ*, 739, 56
De Rosa, G., Venemans, B. P., Decarli, R., et al. 2014, *ApJ*, 790, 145
Decarli, R., Walter, F., Venemans, B. P., et al. 2017, *Nature*, 545, 457
Decarli, R., Walter, F., Venemans, B. P., et al. 2018, *ApJ*, 854, 97
Faisst, A. L., Schaerer, D., Lemaux, B. C., et al. 2020, *ApJS*, 247, 61
Fan, X., Strauss, M. A., Schneider, D. P., et al. 2003, *AJ*, 125, 1649
Fan, X., Carilli, C. L., & Keating, B. 2006, *ARA&A*, 44, 415
Ghisellini, G., Haardt, F., Della Ceca, R., et al. 2013, *MNRAS*, 432, 2818
Ghisellini, G., Tagliaferri, G., Sbarrato, T., et al. 2015, *MNRAS*, 450, L34
Giallongo, E., Grazian, A., Fiore, F., et al. 2015, *A&A*, 578, A83
Habouzit, M., Volonteri, M., Somerville, R. S., et al. 2019, *MNRAS*, 489, 1206
Husband, K., Bremer, M. N., Stanway, E. R., et al. 2013, *MNRAS*, 432, 2869
Inayoshi, K., Visbal, E., & Haiman, Z. 2020, *ARA&A*, in press, arXiv:1911.05791, DOI: 10.1146/annurev-astro-120419-014455
Izumi, T., Onoue, M., Shirakata, H., et al. 2018, *PASJ*, 70, 36
Izumi, T., Onoue, M., Matsuoka, Y., et al. 2019, *PASJ*, 71, 111
Janssen, A. W., Christopher, N., Sturm, E., et al. 2016, *ApJ*, 822, 43
Jiang, L., Fan, X., Hines, D. C., et al. 2006, *AJ*, 132, 2127
Jiang, L., Fan, X., Vestergaard, M., et al. 2007, *AJ*, 134, 1150
Jiang, L., Fan, X., Brandt, W. N., et al. 2010, *Nature*, 464, 380
Jiang, L., McGreer, I. D., Fan, X., et al. 2016, *ApJ*, 833, 222
Kim, S., Stiavelli, M., Trenti, M., et al. 2009, *ApJ*, 695, 809
Kim, Y., Im, M., Jeon, Y., et al. 2018, *ApJ*, 855, 138
Kormendy, J. & Ho, L. C. 2013, *ARA&A*, 51, 511
Kulkarni, G., Worseck, G., & Hennawi, J. F. 2019, *MNRAS*, 488, 1035
Kurk, J. D., Walter, F., Fan, X., et al. 2007, *ApJ*, 669, 32
Leipski, C., Meisenheimer, K., Walter, F., et al. 2014, *ApJ*, 785, 154
Luo, B., Brandt, W. N., Xue, Y. Q., et al. 2017, *ApJS*, 228, 2
Lyu, J. et al. 2016, *ApJ*, 816, 85
Lusso, E. & Risaliti, G. 2016, *ApJ*, 819, 154
Maiolino, R., Cox, P., Caselli, P., et al. 2005, *A&A*, 440, L51
Maiolino, R., Gallerani, S., Neri, R., et al. 2012, *MNRAS*, 425, L66
Marchesi, S., Civano, F., Salvato, M., et al. 2016, *ApJ*, 827, 150
Marziani, P. & Sulentic, J. W. 2012, *NewAR*, 56, 49
Martínez-Aldama, M. L., del Olmo, A., Marziani, P., et al. 2018, *A&A*, 618, A179

Matsuoka, Y., Iwasawa, K., Onoue, M., et al. 2019, ApJ, 883, 183
Mazzucchelli, C., Bañados, E., Venemans, B. P., et al. 2017, ApJ, 849, 91
Mazzucchelli, C., Decarli, R., Farina, E. P., et al. 2019, ApJ, 881, 163
Miller, T. B., Chapman, S. C., Aravena, M., et al. 2018, Nature, 556, 469
Mor, R., Netzer, H., Trakhtenbrot, B., et al. 2012, ApJL, 749, L25
Nanni, R., Vignali, C., Gilli, R., et al. 2017, A&A, 603, A128
Natarajan, P. 2011, BASI, 39, 145
Natarajan, P., Pacucci, F., Ferrara, A., et al. 2017, ApJ, 838, 117
Neeleman, M., Bañados, E., Walter, F., et al. 2019, ApJ, 882, 10
Netzer, H., Mor, R., Trakhtenbrot, B., et al. 2014, ApJ, 791, 34
Nguyen, H. N., Lira, P., Trakhtenbrot, B., et al. 2020, ApJ, 895, 74
Onoue, M., Kashikawa, N., Matsuoka, Y., et al. 2019, ApJ, 880, 77
Overzier, R. A., Miley, G. K., Bouwens, R. J., et al. 2006, ApJ, 637, 58
Overzier, R. A., Guo, Q., Kauffmann, G., et al. 2009, MNRAS, 394, 577
Pons, E., McMahon, R. G., Banerji, M., et al. 2020, MNRAS, 491, 3884
Reed, S. L., McMahon, R. G., Martini, P., et al. 2017, MNRAS, 468, 4702
Ross, N. P. & Cross, N. J. G. 2020, MNRAS, 494, 789
Sbarrato, T., Ghisellini, G., Nardini, M., et al. 2012, MNRAS, 426, L91
Shao, Y., Wang, R., Jones, G. C., et al. 2017, ApJ, 845, 138
Shemmer, O., Brandt, W. N., Schneider, D. P., et al. 2006, ApJ, 644, 86
Shen, Y. 2013, BASI, 41, 61
Shen, Y., Brandt, W. N., Richards, G. T., et al. 2016, ApJ, 831, 7
Shen, Y., Wu, J., Jiang, L., et al. 2019, ApJ, 873, 35
Shen, X., Hopkins, P. F., Faucher-Giguère, C.-A., et al. 2020, MNRAS, 495, 3252
Simpson, C., Mortlock, D., Warren, S., et al. 2014, MNRAS, 442, 3454
Stanley, F., Jolly, J. B., König, S., et al. 2019, A&A, 631, A78
Stark, D. P. 2016, ARA&A, 54, 761
Stone, M., Veilleux, S., Meléndez, M., et al. 2016, ApJ, 826, 111
Trakhtenbrot, B., Netzer, H., Lira, P., et al. 2011, ApJ, 730, 7
Trakhtenbrot, B. & Netzer, H. 2012, MNRAS, 427, 3081
Trakhtenbrot, B., Civano, F., Urry, C. M., et al. 2016, ApJ, 825, 4
Trakhtenbrot, B., Volonteri, M., & Natarajan, P. 2017a, ApJL, 836, L1
Trakhtenbrot, B., Lira, P., Netzer, H., et al. 2017b, ApJ, 836, 8
Utsumi, Y., Goto, T., Kashikawa, N., et al. 2010, ApJ, 721, 1680
Valiante, R., Agarwal, B., Habouzit, M., et al. 2017, PASA, 34, e031
Veilleux, S., Meléndez, M., Sturm, E., et al. 2013, ApJ, 776, 27
Venemans, B. P., McMahon, R. G., Walter, F., et al. 2012, ApJL, 751, L25
Venemans, B. P., Bañados, E., Decarli, R., et al. 2015, ApJL, 801, L11
Venemans, B. P., Walter, F., Zschaechner, L., et al. 2016, ApJ, 816, 37
Venemans, B. P., Walter, F., Decarli, R., et al. 2017, ApJL, 851, L8
Venemans, B. P., Decarli, R., Walter, F., et al. 2018, ApJ, 866, 159
Venemans, B. P., Neeleman, M., Walter, F., et al. 2019, ApJL, 874, L30
Vito, F., Gilli, R., Vignali, C., et al. 2016, MNRAS, 463, 348
Vito, F., Brandt, W. N., Yang, G., et al. 2018, MNRAS, 473, 2378
Vito, F., Brandt, W. N., Bauer, F. E., et al. 2019, A&A, 630, A118
Volonteri, M. 2010, A&ARv, 18, 279
Volonteri, M., Haardt, F., Ghisellini, G., et al. 2011, MNRAS, 416, 216
Volonteri, M., Silk, J., & Dubus, G. 2015, ApJ, 804, 148
Volonteri, M., Reines, A. E., Atek, H., et al. 2017, ApJ, 849, 155
Walter, F., Riechers, D., Novak, M., et al. 2018, ApJL, 869, L22
Wang, F., Wu, X.-B., Fan, X., et al. 2015, ApJL, 807, L9
Wang, F., Wu, X.-B., Fan, X., et al. 2016, ApJ, 819, 24
Wang, R., Wagg, J., Carilli, C. L., et al. 2013, ApJ, 773, 44
Wang, F., Yang, J., Fan, X., et al. 2019a, ApJ, 884, 30

Wang, R., Shao, Y., Carilli, C. L., et al. 2019b, ApJ, 887, 40
Weigel, A. K., Schawinski, K., Treister, E., et al. 2015, MNRAS, 448, 3167
Willott, C. J., Percival, W. J., McLure, R. J., et al. 2005, ApJ, 626, 657
Willott, C. J., Delorme, P., Reylé, C., et al. 2010a, AJ, 139, 906
Willott, C. J., Albert, L., Arzoumanian, D., et al. 2010b, AJ, 140, 546
Willott, C. J., Omont, A., & Bergeron, J. 2013, ApJ, 770, 13
Willott, C. J., Bergeron, J., & Omont, A. 2015, ApJ, 801, 123
Willott, C. J., Bergeron, J., & Omont, A. 2017, ApJ, 850, 108
Wu, X.-B., Wang, F., Fan, X., et al. 2015, Nature, 518, 512
Yang, J., Wang, F., Fan, X., et al. 2019, ApJ, 871, 199

AGN astrometry: A powerful tool for galaxy kinematic studies

Naftali Kimani[1], Andreas Brunthaler[2] and Karl M. Menten[2]

[1]Kenyatta University, P.O. Box 43844, 00100 Nairobi, Kenya
email: `kimani.naftali@ku.ac.ke`

[2]Max-Planck-Institut für Radioastronomie, Auf dem Hügel 69, D-53121 Bonn, Germany

Abstract. This article highlights the successes of the high resolution astrometric VLBI observations used for measuring proper motion of galaxies in the Local group. The required, high accuracies, often in the μas yr^{-1} regime, are only attainable through the use of the phase-referencing technique. These require either a compact radio source (AGN) or strong maser emission in the target galaxy and, additionally, some compact extra-galactic radio sources (quasars) to serve as ideal background reference source. The derived proper motions can lead to lower limits on the orbital lower estimates to the mass of the host galaxy, promise a new handle on dynamical models of interacting galaxy systems and offer insights on the spatial distribution of dark matter in the near universe.

Keywords. AGN, VLBI, proper motion, Galaxy evolution

1. Introduction

Galaxies within a group or a cluster have their radial velocities and positions known to a high degree of accuracy. This information is sufficient to estimate the group mass from the observed deviations from the external Hubble field of velocities by making critical assumptions on eccentricities and equipartition (Kulessa & Lynden-Bell 1992; Karachentsev & Kashibadze 2006). For a full understanding of the evolution of a galaxy group, we require accurate knowledge of the distances and the three-dimensional velocity vectors. Knowing the 3-D velocity structure within a group requires knowledge of a combination of precise radial and proper motions. Even without accurate distances, proper motions should yield lower limits to the mass of the host galaxy, merger speed in merging clusters of galaxies, and/or provide information whether galaxies in clusters are on their first infalls (Maccarone & Gonzalez 2018). Observationally derived proper motions also provide important constraints on a dynamical model of an interacting galaxy system by establishing both the past history and the fate of such a system. Moreover, the 3-D velocity structure helps our understanding of the environment of galaxy groups and clusters, shedding light on the spatial distribution of the dark matter associated with them (Yepes et al. 2013).

In recent years, especially due to the high angular resolution and stability of the Hubble Space Telescope (HST) and other ground based optical telescopes (van der Marel et al. 2014), proper motions have now been reliably measured for a number of galaxies in the Local Group i.e. the LMC (Kallivayalil et al. 2006; Pedreros et al. 2006) the SMC (Kallivayalil et al. 2006) the Sculptor dwarf spheroidal galaxy (dShp) (Piatek et al. 2006) the Canis Major dwarf galaxy (Dinescu et al. 2005), the Ursa Minor dSph (Piatek et al. 2005), the Sagittarius dSph (Dinescu et al. 2005), the Fornax dSph (Piatek et al. 2002; Dinescu et al. 2004), and the Carina dSph (Piatek et al. 2005). On the other hand, the expected proper motion for galaxies within the Local Group and neighboring groups, which are $\ll 1\,\mathrm{mas\,yr^{-1}}$, are detectable with the Very Long Baseline Interferometry (VLBI) using the phase-referencing technique (Brunthaler et al. 2005a).

2. Overview

Astrometric VLBI phase-referencing observations require either a compact radio source or strong maser emission in the target source and, additionally, some compact extra-galactic radio sources such as quasars serving as background reference source. The technique is based on the assumption that the phase errors of two sources with a small angular separation are similar. Such an observation therefore, involves observing a target source and its adjacent calibrator ($\sim 1°$) in a fast-switching mode (a $\sim$ minute cycle). In the following we summarize successful astrometric VLBI observations done using the phase-referencing technique to detect $\mu\mathrm{as\,yr^{-1}}$ extra-galactic proper motions as part of a campaign to measure the 3-D velocity structure of the Local Universe out to the Virgo Cluster.

Milky Way (Sgr A*): Very Long Baseline Array (VLBA) observations were conducted for about 8 years, between 1995 and 2003, involving rapid switching between compact extra-galactic sources, J1745-283 and J1748-291, and Sgr A* as the phase-referencing source. The apparent proper motion of Sgr A* relative to J1745-283 was obtained as $6.379 \pm 0.024\,\mathrm{mas\,yr^{-1}}$, almost entirely in the plane of the Galaxy. Assuming a distance to the Galactic center of $8.0 \pm 0.5\,\mathrm{kpc}$, the apparent angular motion of Sgr A* in the plane of the Galaxy was obtained to be $-241 \pm 15\,\mathrm{km\,s^{-1}}$ (Reid & Brunthaler 2004).

IC 10: Using the VLBA, the proper motion of the Local Group galaxy IC10 was determined by measuring the position of a H_2O maser in IC10 relative to two background quasars (VCS1 J0027+5958 and NVSS J002108+591132) over a period of 4.3 years. The derived motion was $-39 \pm 9\,\mu\mathrm{as\,yr^{-1}}$ toward the East and $31 \pm 8\,\mu\mathrm{as\,yr^{-1}}$ toward the North. Assuming a distance to the IC10 of $660 \pm 66\,\mathrm{kpc}$, it's total space velocity of $215 \pm 43\,\mathrm{km\,s^{-1}}$ relative to the Milky Way was obtained (Brunthaler et al. 2007).

Triangulum Galaxy (M33): The proper motions of the two H_2O maser regions relative to the quasar J0137+312 were determined. The masers are associated with star-forming regions located on opposite sides of the disk of M33. The measured angular motion of the two masers allowed for an independent geometric distance estimation to M33 of $730 \pm 168\,\mathrm{kpc}$, consistent with standard candle estimates. The derived proper motion was $-29 \pm 7\,\mu\mathrm{as\,yr^{-1}}$ toward the East and $45 \pm 9\,\mu\mathrm{as\,yr^{-1}}$ toward the North. The derived total velocity of M33 relative to the Milky Way is $190 \pm 59\,\mathrm{km\,s^{-1}}$ (Brunthaler et al. 2005b).

M81 group: The proper motion of M81's central compact radio source (M81*) relative to the background quasars (0945+6924, 1004+6936) over 11 years was determined. M81* was used as the phase-referencing source. A preliminary analysis indicates that M81 is moving with a total velocity of about $500\,\mathrm{km\,s^{-1}}$ relative to the Milky Way (Kimani 2016).

3. Implications

Lower limits to the galaxy mass. The most reliable way of deriving masses is by determining orbits, which requires the knowledge of three-dimensional velocity vectors. For instance, assuming the Andromeda (M31) satellite galaxies M33 and IC10 are gravitationally bound to it, their 3-D velocities yields a lower estimate to the mass of M31 of $7.5 \times 10^{11}\,M_\odot$ (Brunthaler et al. 2007). Similarly, in the M81 group, the separation between the centers of M81 and M82 is $38\,\mathrm{kpc}$, implying that M82 is deeply embedded in the dark matter halo of M81 which spreads out to $140\,\mathrm{kpc}$. A measurement of the proper motion of M82 will allow an estimate of the lower limit on the mass of M81 in the near future (Oehm et al. 2017).

Evolution of interacting galaxy systems. Measuring proper motions of galaxies is of great value for understanding a variety of other issues related to galaxy and cluster evolution. For instance, due to its proximity, the M81 group features a fascinating interacting galaxy system comprising of M81, M82 and NGC3077. To understand the dynamics of this system, several numerical simulation studies have been performed (e.g. Sofue 1998; Yun 1999 and Gomez *et al.* 2004). The challenge that was encountered was getting the right initial conditions for the model and getting the right interaction time that results in the currently observed configuration. Key parameters for modeling of the history, evolution and fate of the group, are present spatial velocity vectors of its members which have otherwise been statistically estimated. With these velocities known, it is now possible to reconstruct and predict the past and future of the system, providing crucial information on the consequences of tidal interactions in a group environment. In the near future, with the current attempts to derive the proper motion of Andromeda (using M31*) relative to some background quasars, it will be possible to model the history and fate of the local group (Reid & Honma 2014).

Dark matter distribution. There is now an immense amount of observational evidence that firmly supports the idea that there exists much more matter in the Universe than just the luminous matter (Yepes *et al.* 2013). The non-visible mass is often referred to as Dark Matter. The Local Universe becomes the best place for the observational studies of dark matter distribution and for testing the predictions of the standard λ Cold Dark Matter (λCDM) model of cosmological structure formation, which describes very well the observations of the large scale structure (Yepes *et al.* 2013; Frenk & White 2012). Observational data for galaxies in the nearby universe such as proper motions, masses and distances are necessary for constraining initial conditions and placing constrains on simulations that reproduce the observed large scale structure and give insight on the future of the present structures.

References

Brunthaler, A., Reid, M. J., & Falcke, H. 2005a, *Astronomical Society of the Pacific Conference Series*, 340, 455
Brunthaler, A., Reid, M. J., Falcke, H., Greenhill, L. J., & Henkel, C. 2005b, *Science*, 307, 1440
Brunthaler, A., Reid, M. J., Falcke, H., Henkel, C., & Menten, K. M. 2007, *A&A*, 462, 101
Dinescu, D. I., Keeney, B. A., Majewski, S. R., & Girard, T. M. 2004, *AJ*, 128, 687
Dinescu, D. I., Martnez-Delgado, D., Girard, T. M., Peñarrubia, J., Rix, H.-W., Butler, D., & van Altena, W. F. 2005, *ApJL*, 631, L49
Frenk, C. S. & White, S. D. M. 2012, *Annalen der Physik*, 524, 507
Gomez, J. C., Athanassoula, L., Fuentes, O., & Bosma, A. 2004, *Astronomical Society of the Pacific Conference Series*, 314, 629
Kallivayalil, N., van der Marel, R. P., Alcock, C., Axelrod, T., Cook, K. H., Drake, A. J., & Geha, M. 2006, *ApJ*, 638, 772
Karachentsev, I .D. & Kashibadze, O. G. 2006, *Astrophysics*, 49, 3
Kimani, N. 2016, *Kinematics study of M81 and M82 galaxies*, Doctoral thesis, Max-Planck Institute for Radioastronomy
Kulessa, A. S. & Lynden-Bell, D. 1992, *MNRAS*, 255, 105
Reid, M. J. & Brunthaler, A. 2004, *ApJ*, 616, 872
Reid, M. J. & Honma, M. 2014, *Annu. Rev. Astronomy Astrophysics* 52, 339
Maccarone, T. J. & Gonzalez, A. H. 2018, *ASP Conference Series*, 517, 663
Pedreros, M. H., Costa, E., & Méndez, R. A. 2006, *AJ*, 131, 1461
Piatek, S., Pryor, C., Olszewski, E. W., Harris, H. C., Mateo, M., Minniti, D., Monet, D. G., Morrison, H., & Tinney, C. G. 2002, *AJ*, 124, 3198
Piatek, S., Pryor, C., Olszewski, E. W., Harris, H. C., Mateo, M., Minniti, D., & Tinney, C. G. 2003, *AJ*, 126, 2346

Piatek, S., Pryor, C., Bristow, P., Olszewski, E. W., Harris, H. C., Mateo, M., Minniti, D., & Tinney, C. G. 2005, *AJ*, 130, 95

Piatek, S., Pryor, C., Bristow, P., Olszewski, E. W., Harris, H. C., Mateo, M., Minniti, D., & Tinney, C. G. 2006, *AJ*, 131, 1445

Oehm, W., Thies, I., Kroupa, P. 2017, *MNRAS*, 467, 273

Sofue, Y. 1998, *PASJ*, 50, 227

van der Marel, R. P., Anderson, J., Bellini, A., *et al.* 2014, *Astronomical Society of the Pacific Conference Series*, 480, 43

Yepes, G., Gottlöber, S., & Hoffman, Y. 2013, *New Astronomy Reviews*, 58, 18

Yun, M. S. 1999, IAU Symposium 186, *Galaxy Interactions at Low and High Redshift*, 81

The role of AGN activity in the building up of the BCG at z ~ 1.6

Angela Bongiorno and Andrea Travascio

INAF-Observatory of Rome, via di Frascati 33, 00074, Monteporzio Catone, Rome, Italy
email: angela.bongiorno@inaf.it, andrea.travascio@inaf.it

Abstract. XDCPJ0044.0-2033 is one of the most massive galaxy cluster at $z \sim 1.6$, for which a wealth of multi-wavelength photometric and spectroscopic data have been collected during the last years. I have reported on the properties of the galaxy members in the very central region ($\sim 70 kpc \times 70 kpc$) of the cluster, derived through deep HST photometry, SINFONI and KMOS IFU spectroscopy, together with Chandra X-ray, ALMA and JVLA radio data.

In the core of the cluster, we have identified two groups of galaxies (Complex A and Complex B), seven of them confirmed to be cluster members, with signatures of ongoing merging. These galaxies show perturbed morphologies and, three of them show signs of AGN activity. In particular, two of them, located at the center of each complex, have been found to host luminous, obscured and highly accreting AGN ($\lambda = 0.4 - 0.6$) exhibiting broad Hα line. Moreover, a third optically obscured type-2 AGN, has been discovered through BPT diagram in Complex A. The AGN at the center of Complex B is detected in X-ray while the other two, and their companions, are spatially related to radio emission. The three AGN provide one of the closest AGN triple at $z > 1$ revealed so far with a minimum (maximum) projected distance of 10 kpc (40 kpc). The discovery of multiple AGN activity in a highly star-forming region associated to the crowded core of a galaxy cluster at $z \sim 1.6$, suggests that these processes have a key role in shaping the nascent Brightest Cluster Galaxy, observed at the center of local clusters. According to our data, all galaxies in the core of XDCPJ0044.0-2033 could form a BCG of $M_\star \sim 10^{12} M_\odot$ hosting a BH of $2 \times 10^8 - 10^9 M_\odot$, in a time scale of the order of 2.5 Gyrs.

Keywords. galaxy cluster, BCG formation, active galaxies, galaxy formation

1. Introduction

Relaxed, virialized and undisturbed galaxy cluster in the local Universe are characterised by a bright, massive and large elliptical galaxy at their center, the so called Brightest Cluster Galaxy (BCG). The BCG is usually located at the center of the cluster potential well, close to the peak of X-ray emission. How these galaxies form is still a matter of study. According to most models, the epoch of their assembly is mostly between z=1 and z=2, where the mass of the BCG goes from 10% to 50% of the final mass. For this reason, this redshift range is crucial to observe the BCG progenitors and to witness its assembly. Indeed, the cores of galaxy clusters at $z = 1 - 2$ show a different picture compared to the local Universe, i.e. in most cases there is no single BCG, and the core is characterised by several galaxies which are typically blue, star forming and with disturbed morphology. This implies the existence of a mechanism able to drive such transformation. Interestingly, also looking at the properties of the whole galaxy cluster population, there are evidences that z=1.5 is a crucial epoch. Indeed, while at $z < 1.4$ the number of star forming galaxies (SFGs) increases towards the cluster outskirt, at $z > 1.4$ the SF activity is higher in the core of the cluster (Brodwin *et al.* 2013). For all these reasons, we studied the core of the X-ray detected galaxy cluster XDCP J0044.0-2033 (hereafter

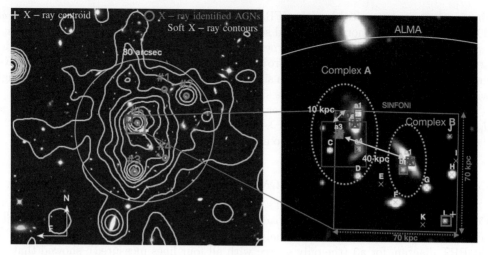

Figure 1. [Adapted from Travascio et al. 2020] *Left Panel:*: HST RGB (F105W + F140W + F160W) image of the galaxy cluster XDCP0044. The yellow cross indicates the centroid of the extended X-ray emission while the yellow circle is $R_{2500} = 250 kpc$. The cyan contours are the Chandra soft $[0.5 - 2] keV$ X-ray emissions while the red circles mark the unresolved X-ray sources as identified by Tozzi et al. (2015). Finally, the red and blue squares delimit the region analyzed in this paper, corresponding to the SINFONI and the KMOS FOVs. *Right Panel*: zoom-in of the analyzed central region, where two complexes (A and B) are highlighted with the orange dashed line. Green crosses mark the 16 HST identified photometric sources while magenta squares mark the sources for which a spectroscopic redshift has been determined. AGN are indicated with blue (type-1) and red (type-2) stars.

XDCP0044; Santos et al. 2011; Fassbender et al. 2011) at $z \sim 1.6$. XDCP0044 is the most massive galaxy clusters ($M_{200} = 4 \times 10^{14} M_\odot$) discovered in the XDCP project through XMM archival data (Fassbender et al. 2011) and one of the most massive at $z > 1.5$. It is in a quite advanced state of dynamical relaxation, and shows a reversal SF-density relation (Santos et al. 2015). XDCP0044 is a unique laboratory to study the building-up of the BCG and the interplay between galaxies, nuclear activity, and the inter-galactic gas in the core of massive high redshift galaxy clusters.

Fig. 1 (left panel) shows the HST RGB (F105W+F140W+F160W) image of XDCP0044 with overlaid the soft ($[0.5 - 2] keV$, cyan) band Xray Chandra contours. Red circles indicate the 5 point-like sources (AGN) identified by Tozzi et al. (2015) within $30''$ ($250 kpc$) from the cluster center.

In this work, we focused on a very small region ($70 kpc \times 70 kpc$) of the cluster core (right panel of Fig. 1) for which a detailed multiwavelength study has been conducted by combining the information derived from X-ray to optical, near-infrared (NIR) and radio bands, both photometrically and spectroscopilly (see Travascio et al. 2020 for more details).

2. Results

As visible in the right panel of Fig. 1, the core of XDCP0044, a very small region of 70kpc ×70kpc (slightly more than twice the milky way in size), appears very crowded. Through a SExtractor analysis (Bertin & Arnouts 1996), we indeed distinguished 16 photometric sources (green crosses in Fig. 1, right panel) and thanks to the spectroscopic SINFONI and KMOS data, we confirmed that at least seven of them are cluster members (magenta squares in Fig. 1, right panel), with redshifts ranging from $z = 1.5567$ to $z = 1.5904$ ($\Delta z \simeq 0.0337$), consistently with the redshift of the cluster. As visible in Fig. 1, these sources appear quite blue (the SFR of the entire region is $\sim 500 M_\odot/yr$, Santos et al. 2015) with disturbed morphology, and seems to be clustered in two galaxy complexes:

- **Complex A**, in the top-left corner of the central region of XDCP0044, includes at least 4 galaxies within 20 kpc in projected distance. It was detected in the HAWK-I image by Fassbender et al. (2014) as a single source and identified as the BCG, although with several extensions, interpreted as sign of ongoing or recent mergers.
- **Complex B**, located at the center of the analyzed field, is made of two sources at $d_{proj} \sim 5kpc$. One of the sources is an X-ray AGN discovered by Tozzi et al. (2013)

The two complexes are very close to each other, i.e. $d_{proj} \sim 35 kpc$.

2.1. AGN and SF activity

Interestingly, in two of the analyzed sources, a2 at the center of Complex A and b1 in Complex B, the Hα emission line is broad (FWHM > 1500 km/s), indicating galaxies with active nuclei (AGN). Moreover, the analysis of the line ratios ([NII]/Hα vs [OIII]/Hβ) in the BPT diagram for a3 (the only source with all four lines measured), showed that a3 is indeed an AGN. Three AGN have been thus discovered in the central (very small, i.e. $\sim 70 kpc \times 70 kpc$) region of the cluster (red and blue stars in the right panel of Fig. 1). Source b1, at the center of Complex B, is an X-ray point like source. From the analysis of its Chandra X-ray spectrum, we found that it is a luminous ($L_{[2-10keV]} \sim 10^{44} erg/s$) and moderately obscured ($log(N_H/cm^2) = 22.7$) AGN. From the broad (FWHM $\sim$ 2200 km/s) Hα line, we estimated its BH mass using the virial formula by Greene & Ho (2005), finding $M_{BH} = 7.2 \times 10^7 M_\odot$. Moreover, the bolometric luminosity has been computed by applying the bolometric correction by Runnoe et al. (2012) to the 5100Å luminosity, estimated from the linear interpolation of the F105W and F140W HST magnitudes. Combining the derived parameters, we found that b1 is accreting at a high rate, i.e. $\lambda_{Edd} = 0.46$. Source b1 has also been detected in ALMA continuum at 230 GHz which revealed cold dust emission from the host galaxy, detected at 5σ significance. Assuming different QSO SEDs and normalizying them at the observed ALMA flux, we derived a SFR in the range $[150-490]M_\odot/yr$, consistent with the SFR derived from Herschel for the entire central region ($452 \pm 58 M_\odot/yr$, Santos et al. 2015), suggesting that most of the IR emission, and therefore of the SF, might be associated to b1. No radio emission is associated to this source. Finally, we derived the stellar mass of its host galaxy assuming a constant M/L_K ratio (Madau et al. 1998) and a Chabrier (2003) initial mass function, finding $log M_\star \sim 11.8 M_\odot$.

Sources a2 and a3, in the very crowded Complex A, have not been detected in the Chandra X-ray data. A 3σ upper limit on the X-ray luminosity has been derived to be of the order of $L_{[2-10keV]} < 10^{43} erg/s$, assuming an unabsorbed power-law with Γ=1.9. For a2, from the bolometric luminosity, as derived from $L_{5100Å}$ (i.e. $log(L_{bol}/[erg/s]) \sim$ 45.4), and by applying $k_{bol}[2-10keV] \sim 18.96$ (Duras et al. 2020), we expect an intrinsic luminosity $L_{[2-10keV]} \sim 10^{44} erg/s$. Such value is more than 1dex higher compared to the derived X-ray luminosity upper limit, thus suggesting a high level of X-ray obscuration ($log(N_H/cm^2) > 23.8$). From the broad ($FWHM \sim 1900 km/s$) Hα line, we estimated that the BH of a2 has a mass of $M_{BH} = 3.2 \times 10^7 M_\odot$, and accretes at $\lambda_{Edd} \sim 0.51$. While no emission has been found in ALMA corresponding to a2, JVLA 1.5 GHz extended radio emission has been detected spatially correlated to it. Under the assumption that such radio signal is produced by a single source, its power (logP[1.5 GHz] = 23.45) would suggest a likely (60 to 80% of probability) AGN powered radio emission, according to the relation introduced by Magliocchetti et al. (2014, 2018). However, there is a not negligible probability that such emission is on the contrary due to SF processes. In this case the measured radio luminosity would translate into a $SFR \sim 100~M_\odot/yr$ according

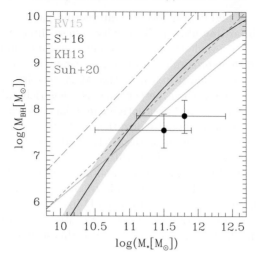

Figure 2. Correlations between central black hole mass and host galaxy total stellar mass for the two broad line AGN discovered in the core of XDCP004 (black circles with errors). As reference, the dashed red line is a linear fit to the sample of Kormendy & Ho (2013, KH13) by Shankar et al. (2019), while the solid green line is the fit to local AGN by Reines & Volonteri (2015, RV15) and the short-dashed blu line is the fit by Suh et al. (2020) to local + high-z AGN. Finally the solid black line is the de-biased $M_{BH} - M_\star$ relation derived by Shankar et al. (2016, S+16) with its scatter (yellow area).

to the relation by Brown et al. (2017). Finally, as for b1, we derived the stellar mass of its host galaxy, finding $log M_\star \sim 11.5 M_\odot$.

Summarizing, in the core of XDCP0044 we discovered three AGN hosted in massive and star-forming galaxies, i.e. two luminous, highly accreting and obscured/moderately obscured type-1 AGN and one X-ray and optically obscured type-2 AGN.

2.2. The $M_{BH} - M_\star$ plane

We studied the location of the two discovered type-1 AGN with broad Hα in the $M_{BH} - M_\star$ plane (Fig. 2). Both of them lie below the Kormendy & Ho (2013) relation for local inactive galaxies with $\Delta log(M_{BH}/M_\star)$, computed perpendicular to the local relation, at $\sim 2\sigma$ from it. Their location is more consistent, within the large errors, with the more recent determination of the local scaling relation for active galaxies at $z < 0.055$ computed by Reines & Volonteri (2015) (green line in Fig. 2). Moreover, our data points lie close to the fit recently found by Suh et al. (2020) by including local plus high-z (up to z=2.5) AGN and to the unbiased $M_{BH} - M_\star$ relation computed by Shankar et al. (2016), who interpreted the discrepancy between the observed location of quiescent and active galaxies in the $M_{BH} - M_\star$ plane as an observational bias (Shankar et al. 2019). Indeed, our newly discovered $z \sim 1.6$ AGN have $M_{BH}/M_\star$ ratio consistent with local active galaxies, thus showing no or negligible evolution in the intrinsic $M_{BH} - M_\star$ relation, in agreement with most recent works (Shankar et al. 2019; Suh et al. 2020).

3. Conclusions

XDCP0044, a massive galaxy cluster at $z \sim 1.6$, allowed us to study the processes responsible for the BCG formation in the epoch when both SF and nuclear activity are at their peak (Madau & Dickinson 2014). We confirm that high-z galaxy cluster cores show different properties compared to the z=0 ones. Indeed, no single, early-type BCG has been detected in the core of XDCP0044, which is found to host a large number (at least 7 confirmed) of highly star-forming interacting galaxies, grouped in two main merging systems, both hosting also AGN activity. These three AGN provide one of the closest AGN triple at z > 1 revealed so far with a minimum (maximum) projected distance of 10 kpc (40 kpc) and their proximity implies a future merger between them.

These results lead to a scenario in which the AGN activity is triggered during the formation of the cluster BCG, when mergers between gas-rich galaxies provide the fuel for the AGN and for triggering starburst activity in galactic nuclei. Assuming that the 7 confirmed cluster members will merge to form the local BCG, we find that in a time scale of a couple of Gyrs, all galaxies in the core of XDCP0044 will experience several major mergers, forming a massive central galaxy with a final stellar mass of $1.0 \times 10^{12} M_\odot$ at $z \sim 1$, in agreement with what predicted by semi-analytic models (De Lucia & Blaizot 2007). In fact, we considered the case in which Complex A and B are clusters sub-clumps, each of which will aggregate to form a cD-like galaxy through a gravitational phase transition and then move towards the X-ray centroid in a dynamical friction time to finally merge to form the final BCG. According to this scenario, all these galaxies will merge in $\sim$2.5 Gyrs.

References

Bertin, E. & Arnouts, S. 1996, A&As, 117, 393
Binney, J. & Tremaine, S. 1987, *Galactic dynamics*
Brodwin, M., Stanford, S. A., Gonzalez, A. H., et al. 2013, ApJ, 779, 138
Brown, M. J. I., Moustakas, J., Kennicutt, R. C., et al. 2017, ApJ, 847, 136
Cavaliere, A., Colafrancesco, S., & Menci, N. 1992, ApJ, 392, 41
Chabrier, G. 2003, PASP, 115, 763
De Lucia, G. & Blaizot, J. 2007, MNRAS, 375, 2
Duras, F., Bongiorno, A., Ricci, F.,et al. 2020, A&A, 636, 73
Fassbender, R., Böhringer, H., Nastasi, A., et al. 2011, New Journal of Physics, 13, 125014
Fassbender, R., Nastasi, A., Santos, J. S., et al. 2014, A&A, 568, A5
Greene, J. E. & Ho, L. C. 2005, ApJ, 630, 122
Kormendy, J. & Ho, L. C. 2013, ARA&A, 51, 511
Madau, P. & Dickinson, M. 2014, ARA&A, 52, 415
Madau, P., Pozzetti, L., & Dickinson, M. 1998, ApJ, 498, 106
Magliocchetti, M., Lutz, D., Rosario, D., et al. 2014, MNRAS, 442, 682
Magliocchetti, M., Popesso, P., Brusa, M., & Salvato, M. 2018, MNRAS, 473, 2493
Reines, A. E. & Volonteri, M. 2015, ApJ, 813, 82
Runnoe, J. C., Brotherton, M. S., & Shang, Z. 2012, MNRAS, 422, 478
Santos, J. S., Altieri, B., Valtchanov, I., et al. 2015, MNRAS, 447, L65
Santos, J. S., Fassbender, R., Nastasi, A., et al. 2011, A&A, 531, L15
Shankar, F., Allevato, V., Bernardi, M., et al. 2019, Nature Astronomy, 4, 282
Shankar, F., Bernardi, M., Sheth, R. K., et al. 2016, MNRAS, 460, 3119
Suh, H., Civano, F., Trakhtenbrot, B., et al. 2020, ApJ, 889, 32
Tozzi, P., Santos, J. S., Jee, M. J., et al. 2015, ApJ, 799, 93
Tozzi, P., Santos, J. S., Nonino, M., et al. 2013, A&A, 551, A45
Travascio, A., Bongiorno, A., Tozzi, P., et al. 2020, arXiv e-prints, arXiv:2008.11132
Zhao, D., Aragón-Salamanca, A., & Conselice, C. J. 2015, MNRAS, 453, 4444

The role of LoBALs in quasar evolution

Clare Wethers[1], Jari Kotilainen[1,2], Malte Schramm[3] and Andreas Schulze[3]

[1]Finnish Centre for Astronomy with ESO (FINCA) Vesilinnantie 5, FI-20014, University of Turku, Finland
email: clare.wethers@utu.fi

[2]Department of Physics and Astronomy Vesilinnantie 5, FI-20014, University of Turku, Finland

[3]National Astronomical Observatory of Japan Mitaka, Tokyo 181-8588, Japan

Abstract. Broad absorption line quasars (BALs) represent an interesting yet poorly understood population of quasars showing direct evidence for feedback processes via powerful outflows. Whilst an orientation model appears sufficient in explaining the sub-population of high-ionisation BALs (HiBALs), low-ionisation BALs (LoBALs) may instead represent an evolutionary phase, in which LoBALs exist in a short-lived phase following a merger-driven starburst. Throughout this work, we test this evolutionary picture of LoBALs by comparing the FIR detection rates, SFRs and environments for a sample of 12 LoBALs to other quasar populations at $2.0 < z < 2.5$, making use of archival *Herschel* SPIRE data. We find the LoBAL detection rate to exceed that of both HiBALs and non-BALs, indicating a potential enhancement in their SFRs. Indeed, we also find direct evidence for high SFRs ($>750\,M_\odot\,\mathrm{yr}^{-1}$) within our sample which may be consistent with an evolutionary paradigm.

Keywords. quasars: general, galaxies: general, galaxies: evolution, galaxies: active

1. Introduction

Tight correlations have long been observed between the mass of super-massive black holes (M_{BH}) and various properties of their host galaxies (e.g. Magorrian *et al.* 1998; Kormendy & Ho 2013), yet the mechanisms by which these black holes seemingly shape regions of the galaxy beyond their sphere of influence remain poorly understood. For the most massive and luminous black holes (or *quasars*), it has been proposed that energetic mass outflows may be responsible for both quenching star formation in the galaxy and self-regulating black hole growth (e.g. Silk & Rees 1998; Di Matteo *et al.* 2005; Fabian 2012), but direct observations of galaxies hosting these outflows are sparse. Broad absorption line quasars (BALs) are an important class of quasars that show direct evidence of these mass outflows, likely launched as radiation-driven disc winds. They are thought to comprise anywhere between ∼15 (e.g. Hewett & Foltz 2003; Gibson *et al.* 2009) and ∼40 per cent (Allen *et al.* 2011) of the total quasar population and are generally classified into two types: high-ionisation BALs (HiBALs) and low-ionisation BALs (LoBALs). HiBALs make up ∼85 per cent of BALs and denote those objects containing only high-ionisation absorption features in their spectra, whereas LoBALs - accounting for the remaining ∼15 per cent of BALs - contain both high- and low-ionisation lines in their spectra.

In general, the BAL population remains poorly understood and the nature of BALs is still widely debated throughout the literature. To date, two main interpretations of the BAL phenomenon exist: orientation and evolution. In the orientation scenario, BALs are said to exist in most (if not all) quasars, but can only be viewed as such along specific sight-lines due to the high covering factor of the broad absorption line region. This model is not only consistent with a unified model of quasars, but also explains the similarities between the spectra of HiBALs and non-BALs (e.g. Weymann et al. 1991; Reichard et al. 2003) and the lack of enhancement in the millimetre detection rates of HiBALs (e.g. Priddey et al. 2007; Willott et al. 2003; Lewis et al. 2003), making it a popular model for the BAL phenomenon among HiBALs. On the other hand, BALs have been observed at a wide range of inclinations (Ogle et al. 1999; Di Pompeo et al. 2011), which directly contradicts this orientation scenario. However, this observation is consistent with an evolutionary interpretation of BALs, in which BALs - particularly LoBALs - exist in a short-lived transition period between a merger-induced starburst galaxy and an UV-luminous (non-BAL) quasar (e.g. Boroson & Meyers 1992). A key prediction of this scenario is the enhancement of star formation in LoBALs. Indeed, Canalizo & Stockton 2001 find evidence for this among LoBALs at $z < 0.4$, in which star formation also appears directly linked with tidal interactions in the galaxy. At higher redshifts however, Schulze et al. 2017 find no statistical differences in the distributions of either M_{BH} or Eddington ratios of LoBALs at $z \sim 2.0$ compared to a matched sample of non-BAL quasars, implying that LoBALs do not comprise a distinct population of objects. It remains to be seen however, whether these LoBALs exhibit a similar enhancement in star formation to their low redshift counterparts (Canalizo et al. 2001).

Here, we seek to directly test the evolutionary picture of LoBALs by answering the following questions. Firstly, do we see an enhancement in the FIR detection rate of LoBALs compared to populations of HiBALs and non-BALs? Secondly, is there evidence for prolific star formation in LoBALs consistent with a remnant starburst? And finally, do LoBALs exist in overdense environments, in which we may expect more frequent galaxy-galaxy interactions? The work presented here is a summary based on an ongoing project exploring the nature of LoBALs at $z > 2$. Full details of the LoBAL sample, data reduction and methodology can therefore be found in Wethers et al. 2019, *submitted*, along with a more thorough analysis of the results outlined here.

2. FIR Detection Rates

One prediction of the LoBAL evolutionary paradigm is the enhancement in the detection rate of LoBALs with regards to other quasar populations. If LoBALs mark a post-starburst phase in the lifetime of a quasar, we would expect them to appear bright at FIR wavelengths tracing the peak of the cool dust emission from star formation. As such, the FIR detection rates of LoBALs are expected to be higher than for populations of both HiBALs and non-BAL quasars, neither of which are typically associated with starburst activity in the galaxy. To this end, we make use of targeted *Herschel* SPIRE imaging at 250, 350 and 500 μm for 12 LoBALs at $2.0 < z < 2.5$. The initial selection criteria for this sample is outlined in Schulze et al. 2017, with full details of the observations used given in Wethers et al. 2019, *submitted*. We compare the detection rate of our LoBAL sample to a population 49 HiBALs at similar redshifts (Cao Orjales et al. 2012). In each case, we classify a detection as a source appearing in every band with a flux greater than the nominal 5σ limits outlined in Cao Orjales et al. 2012 (>33.5 mJy at 250 μm; >37.7 mJy at 350 μm; >44.0 mJy at 500 μm). Fig. 1 shows the fraction of LoBALs detected in all SPIRE bands compared to the corresponding HiBAL fraction, from which the detection rate of LoBALs is found to be a factor of ~ 8.5 greater than that of HiBALs. Similarly, we compare the detection rate of our sample to a sample of non-BALs outlined in Netzer et al. 2016 based on their nominal detection thresholds of >17.4, >18.9 and >20.4 mJy at 250, 350 and 500 μm respectively (Fig. 1). Again, we find evidence for an enhancement in the FIR detection rate of LoBALs by a factor of ~ 1.6.

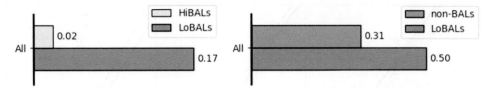

Figure 1. Figure adapted from Wethers et al. 2019. *Left:* FIR detection rate of our LoBAL sample compared to that of the HiBAL sample in Cao Orjales et al. 2012. *Right:* LoBAL detection rates compared to that of the non-BAL quasar sample in Netzer et al. 2016.

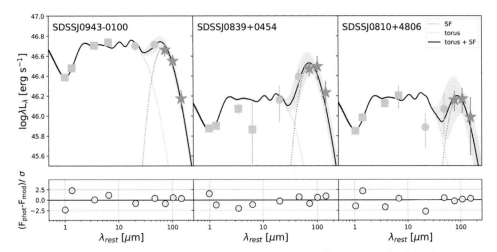

Figure 2. Figure from Wethers et al. 2019. *Upper:* Best-fit SED template based on the combined WISE (*blue squares*) + PACS (*orange circles*) + SPIRE (*pink stars*) photometry. The total model (*black solid line*) is comprised of contributions from a hot torus (*cyan dotted line*) and a star forming galaxy (*pink dotted line*). *Lower:* Error weighted residuals of the best-fit model.

3. LoBAL SFRs

Another key prediction of the LoBAL evolutionary paradigm is the enhancement in LoBAL SFRs. Of the 12 LoBALs in our sample, three are detected at $>5\sigma$ in all SPIRE bands, tracing the peak of the cold dust emission at $2.0 < z < 2.5$. Here, we present the SED fitting for the photometry of these three detected targets, from which we derive SFR estimates. We note that although any emission caused by quasar heating is thought to rapidly drop off at FIR wavelengths, some studies suggest that emission from hot dust in the torus may still contribute significantly in the *Herschel* SPIRE bands, particularly when considering bright quasars (e.g. Symeonidis et al. 2016). To this end, we therefore combine the *Herschel* SPIRE observations for our LoBAL sample with NIR photometry from both the Wide-field Infrared Survey Explorer (WISE) and *Herschel* PACS to estimate the effects of the potential quasar heating to the inferred SFRs. We simultaneously fit this combined photometry with two models: a template accounting for the expected quasar contribution (Mor & Netzer 2012) and a modified black body (or *greybody*) denoting the star forming component of the model. Fig. 2 shows the results of the SED-fitting, from which we find the quasar contamination at $\lambda > 250\,\mu$m to be negligible for all of the detected LoBAL targets, accounting for < 10 per cent of the total flux in the *Herschel* SPIRE bands.

To estimate the SFRs, we integrate over the FIR region (8-1000μm rest frame) of the star-forming component in the best-fit model to get the FIR luminosity ($L_{\rm FIR}$) and convert this to a SFR based on the relation outlined in Kennicutt & Evans 2012. i.e. SFR

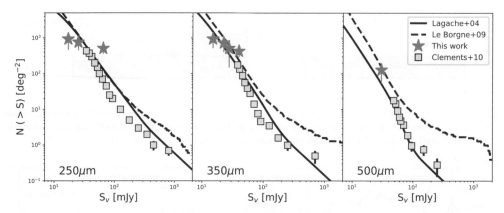

Figure 3. Figure from (Wethers *et al.* 2019, *submitted*). Number counts for the >5σ detections within 1.5 arcmin of our LoBAL targets compared to those of the H-ATLAS field given in Clements *et al.* 2010 (*grey squares*). Model predictions from Lagache *et al.* 2004 (*solid line*) and Le Borgne *et al.* 2009 (*dashed line*) are included for reference.

$= 4.5 \times 10^{-44} \times L_{FIR}$. Based on this relation, we derive SFRs of 740^{+220}_{-170}, 1610^{+280}_{-260} and 2380^{+220}_{-210} $M_\odot yr^{-1}$ for the three detected LoBALs in our sample, providing evidence for prolific star formation in LoBALs. To determine whether this phenomenon is applicable to the full LoBAL sample (not just the detections), we create a mean-weighted stack of the remaining undetected LoBALs and fit the stacked photometry with a greybody template. Again, we integrate over the FIR component of the model and estimate the SFR, deriving a 3σ upper limit on the SFR for the non-detected LoBALs of $< 440 \, M_\odot yr^{-1}$. Even in cases where our targets are not detected by *Herschel*, we therefore cannot rule out moderate to high levels of star formation in LoBALs, proving consistent with the predictions of an evolutionary scenario, in which these systems succeed starburst activity in the host.

4. LoBAL Environments

The final test of the LoBAL evolutionary paradigm employed here is to look for overdensities on the local environments of LoBALs, which may be conducive to frequent galaxy-galaxy interactions. As such, we compare the number of sources detected within 1.5arcmin of our LoBAL targets - corresponding to ∼1 Mpc at z ∼ 2 - to the blank field counts from the *Herschel* Astrophysical Tera-herz Large-Area Survey (H-ATLAS) in each of the SPIRE bands (Clements *et al.* 2010). Serendipitous sources in the image are required to lie above >5σ and fluxes for each are measured within apertures of 22 arsec (250 μm), 30 arcsec (350 μm) and 40 arcsec (500 μm). Fig. 3 shows the number counts of sources within a 1.5 arcmin projected distance from our LoBAL targets compared to the galaxy number counts of the H-ATLAS fields. All number counts have been normalised to an area of $1.0 \, deg^2$. From Fig. 3, we find no difference in the local (∼1 Mpc) environments of LoBALs compared to the the H-ATLAS field and thus conclude that LoBALs do not exist in a special environment, but rather that their FIR environments are entirely consistent with the general galaxy population at 2.0 < z < 2.5.

5. Summary

Overall, we find tentative evidence that LoBALs may exist in a distinct evolutionary phase, but cannot rule out an orientation scenario. On the one hand, we find evidence for an enhancement in the FIR detection rate of LoBALs compared to populations of both HiBAL and non-BAL quasars, indicating a likely enhancement in the SFRs in LoBALs. Indeed, SED fitting returns prolific SFRs ($>750 \, M_\odot yr^{-1}$) for the sample of LoBALs detected with *Herschel* SPIRE and similarly high SFRs ($\sim$440 $M_\odot yr^{-1}$) cannot be ruled out for the undetected sample. This observed enhancement in the SFRs of LoBALs directly supports one of the key predictions of an evolutionary explanation for the LoBAL phenomenon. On the other hand, we find no evidence that LoBAls exist in any kind of special environment. We find no statistical differences between the FIR environment of our targets compared to the H-ATLAS blank fields in any of the SPIRE bands and thus conclude LoBALs to reside in environments typical of the general galaxy population at $2.0 < z < 2.5$, potentially favouring an orientation interpretation of LoBALs.

References

Allen, J. T., Hewett, P. C., Maddox, N., et al. 2011, *VizieR Online Data Catalog*, 741
Boroson, T. A. & Meyers, K. A. 1992, *ApJ*, 397, 442
Canalizo, G. & Stockton, A. 2001, *ApJ*, 555, 719
Cao Orjales, J. M., Stevens, J. A., Jarvis, M. J., et al. 2012, *MNRAS*, 427(2), 1209–1218
Clements, D. L., Rigby, E., Maddox, S., Dunne, L., Mortier, A., et al. 2010, *A&A*, 518, L8
Di Matteo, T., Springel, V., & Hernquist, L. 2005, *Nature*, 433(7026), 604
Di Pompeo, M. A., Brotherton, M. S., et al. 2011, *ApJ*, 743(1), 71
Fabian, A. C. 2012, *A&AR*, 50, 455–489
Gibson, R. R., Brandt, W. N., Gallagher, S. C., & Schneider, D. P. 2009, *ApJ*, 696(1), 924
Hewett, P. C. & Foltz, C. B. 2003, *AJ*, 125(4), 1784
Kennicutt Jr, R. C. & Evans, N. J. 2012 *A&AR*, 50, 531–608
Kormendy, J. & Ho, L. C. 2013, *A&AR*, 51, 511–653
Lagache, G., Dole, H., Puget, J. L., Pérez-González, P. G., et al. 2004, *ApJS*, 154(1), 112
Le Borgne, D., Elbaz, D., Ocvirk, P., & Pichon, C. 2009, *A&A*, 504(3), 727–740
Lewis, G. F., Chapman, S. C., & Kuncic, Z. 2003, *ApJ Letters*, 596(1), L35
Magorrian, J., Tremaine, S., Richstone, D., Bender, R.et al. 1998,*AJ*, 115(6), 2285
Mor, R. & Netzer, H. 2012, *MNRAS*, 420(1), 526–541
Netzer, H., Lani, C., Nordon, R., Trakhtenbrot, B., et al. 2016, *ApJ*, 819(2), 123
Ogle, P. M., Cohen, M. H., Miller, J. S., Tran, H. D., et al. 1999, *ApJS*, 125(1), 1
Priddey, R. S., et al. 2007, *MNRAS*, 374, 867
Reichard, T. A., Richards, G. T., Hall, P. B., Schneider, D. P., et al. 2003, *AJ*, 126(6), 2594
Schulze, A., et al. 2017, *ApJ*, 848, 104
Silk, J. & Rees, M. J. 1998, *A&A*, 331, L1-L4
Symeonidis, M., Giblin, B. M., Page, M. J., et al. 2016, *MNRAS*, 459(1), 257–276
Wethers, C. F., Kotilainen, J., Schramm, M. & Schulze, A. 2019, *MNRAS, submitted*
Weymann, R. J., Morris, S. L., Foltz, C. B., & Hewett, P. C. 1991, *ApJ*, 373, 23–53
Willott, C. J., Rawlings, S., & Grimes, J. A. 2003, *ApJ*, 598(2), 909

Luminosity functions and quasar lifetimes in a sample of mid-IR selected quasars

Susan E. Ridgway

National Optical Astronomy Observatory, 950 North Cherry Avenue, Tucson, AZ, 85719, USA

Abstract. We have made a spectroscopic survey of luminous AGNs and quasars selected in the mid-infrared from Spitzer IRAC surveys. Mid-infrared selection is less affected by dust obscuration, and we find more high redshift quasars than are found in optical or hard X-ray surveys. We have derived luminosity functions for obscured and unobscured quasar populations, and we use these and spectral energy distribution fits to place constraints on host galaxy properties and quasar lifetimes.

Keywords. galaxies: active, galaxies: properties, active: quasars, surveys: infrared

Galactic Mergers at redshift $z \sim 5$, a sample of fast growing QSOs

Nathen Nguyen

Departamento de Astronomía, Universidad de Chile, Camino el Observatorio 1515, Las Condes, Santiago, Casilla 36-D, Chile

Abstract. In order to construct accurate Galaxy Evolution models, a more thorough understanding of the high SFRs seen at $z > 2$ is needed. To better understand AGNs at higher redshifts, we have conducted a multi-wavelength of 38 of the most luminous AGNs found in the SDSS catalogue at redshift $z \sim 4.8$, powered by fast-growing supermassive black holes (SMBHs). Using Herschel/SPIRE observations, we found star formation rates (SFRs) of up to 4000 Solar masses per year. We believe that both the AGN and star formation of these objects are fed by a common reservoir of cold gas, and that this cold gas is due to in-falling matter from major mergers. In this talk, we present ALMA band-7 data of the [CII] $\lambda 157.74$ m emission line and underlying far-infrared (FIR) continuum of twelve luminous quasars at $z \sim 4.8$ in our search for dynamically interacting companions.

Keywords. galaxies: active, active: quasars, galaxies: properties, galaxies: star formation

Supermassive black hole seed formation and the impact on black hole populations across cosmic time

Colin DeGraf

Institute of Astronomy and Kavli Institute for Cosmology, University of Cambridge,
Madingley Road, Cambridge CB3 0HA, UK

Abstract. Although it is well understood that supermassive black holes are found in essentially all galaxies, the mechanisms by which they initially form remain highly uncertain, despite the importance that the formation pathway can have on AGN and quasar behaviour at all redshifts. Using a post-processing analysis method combining cosmological simulations and analytic modeling, I will discuss how varying the conditions for formation of supermassive black hole seeds leads to changes in AGN populations. Looking at formation via direct collapse or from PopIII remnants, I will discuss the impact on black hole mass and luminosity functions, scaling relations, and black hole mergers, which each have effects at both high- and low-redshifts. In addition to demonstrating the importance of initial seed formation on our understanding of long-term black hole evolution, I will also show that the signatures of seed formation suggest multiple means by which upcoming electromagnetic and GW surveys (at both high- and low-z) can provide the data required to constrain initial supermassive black hole formation.

Keywords. galaxies: active, active: quasars, galaxies: formation

Chapter VIII. AGN posters

Multiwavelength morphological study of active galaxies

Betelehem Bilata-Woldeyes[1,2], Mirjana Población[2,3], Zeleke Beyoro-Amado[2], Tilahun Getachew-Woreta[2] and Shimeles Terefe[2]

[1]Debre Berhan University (DBU), Debre Berhan, Ethiopia
[2]Ethiopian Space Science and Technology Institute (ESSTI), Addis Ababa, Ethiopia
[3]Instituto de Astrofísica de Andalucía (IAA-CSIC), Granada, Spain

Abstract. Studying the morphology of a large sample of active galaxies at different wavelengths and comparing it with active galactic nuclei (AGN) properties, such as black hole mass (M_{BH}) and Eddington ratio (λ_{Edd}), can help us in understanding better the connection between AGN and their host galaxies and the role of nuclear activity in galaxy formation and evolution. By using the BAT-SWIFT hard X-ray public data and by extracting those parameters measured for AGN and by using other public catalogues for parameters such as stellar mass (M_*), star formation rate (SFR), bolometric luminosity (L_{bol}), etc., we studied the multiwavelength morphological properties of host galaxies of ultra-hard X-ray detected AGN and their correlation with other AGN properties. We found that ultra hard X-ray detected AGN can be hosted by all morphological types, but in larger fractions (42%) they seem to be hosted by spirals in optical, to be quiet in radio, and to have compact morphologies in X-rays. When comparing morphologies with other galaxy properties, we found that ultra hard X-ray detected AGN follow previously obtained relations. On the SFR vs. stellar mass diagram, we found that although the majority of sources are located below the main sequence (MS) of star formation (SF), still non-negligible number of sources, with diverse morphologies, is located on and/or above the MS, suggesting that AGN feedback might have more complex influence on the SF in galaxies than simply quenching it, as it was suggested in some of previous studies.

Keywords. active galaxies, morphology, multiwavelength study, star formation rate

1. Introduction

Morphology is one of the key parameters for understanding the whole picture of galaxy formation and evolution along cosmic time (Conselice 2014). It gives us an important information about galaxy structure and how other parameters such as environment, interactions and mergers, nuclear activity in galaxies, interstellar medium (ISM), etc., affect morphology and vice-versa. It also has a connection with other galaxy properties such as M_* (Brinchmann & Ellis 2000), SFR (Poggianti et al. 2008), M_{BH} (Schawinski et al. 2010), etc.

In addition, active galaxies, having active galactic nuclei (AGN) in their centers, emit radiation at different parts of electromagnetic spectrum (EMS). Different studies have been done regarding the connection between AGN and their host galaxies leading to different results (very often not very consistent). Taking into account X-ray data it has been suggested that majority of AGN reside in the green valley and could be a transitional population of galaxies (Población et al. 2009a,b; Población, et al. 2012), moving from disk-dominated to bulge-dominated (Gabor et al. (2009)). Optical spectroscopic results suggested the same (e.g., Leslie et al. (2016)). However, other works in mid-infrared

(MIR) and radio found AGN to be located in the blue cloud and red sequence, respectively (e.g., Hickox et al. 2009). Most of morphological studies, at both lower and higher redshifts suggested that AGN tend to be hosted by massive elliptical or bulge-dominated galaxies (Kauffmann et al. 2003, Pović, et al. 2012), although the nearby Seyfert galaxies were found predominately in spirals (Ho 2008). Being one of the key parameters, morphological properties of AGN are still not well known, and in particular there is a lack of multiwavelength morphological studies of active galaxies.

We went for the first time through the multiwavelength morphological analysis in optical, radio, and X-rays of the ultra-hard X-ray AGN host galaxies in the BAT AGN spectroscopic survey (BASS). We tried to understand better the morphological properties of active galaxies at different wavelengths, and also their connection with supermassive black hole (SMBH) mass, Eddington ratio (λ_{Edd}), bolometric luminosity (L_{bol}), SFR, and stellar mass (M_*).

2. Data

Swift/BAT is an all-sky ultra-hard X-ray survey in 14–195 keV energy range. It is the only ultra hard X-ray survey of the whole sky (Ricci et al. 2017, Koss et al. 2017). We used the BASS database 70-month Swift BAT all-sky catalogue with 1210 sources (Baumgartner et al. 2013). Out of these, we found optical photometric data for 640, 468, and 317 galaxies in optical, radio, and X-rays, using the Sloan Digital Sky Survey (SDSS) Data release 14 (DR14), FIRST and NVSS, and XMM-Newton, respectively.

3. Analysis and Results

We classified galaxies visually in optical, radio and X-rays. We classified 45% (290/640) of galaxies in optical (as elliptical, spiral, irregular, or peculiar), while other 55% (350/640) of galaxies stayed unclassified due to the poor resolution data and/or edge-on sources. In radio we classified 84% (394/468) of galaxies as radio-loud (RL) or radio-quiet (RQ), while other 16% (74/468) are uncertain. Finally, we classified 99% of X-ray sources as either compact or extended, while only 1% (3/317) stayed uncertain. We found that ultra-hard X-ray AGN can be hosted by different morphologies, but mainly by spiral galaxies in optical, RQ in radio, and compact in X-rays.

We compared our morphological classification obtained at different wavelengths. We found 48% (174/361) of sources with both optical and radio classifications, where the majority of sources (52%) are spiral and RQ. With both optical and X-ray morphologies we found 52% (139/265) of sources, where again 55% of sources are spiral in optical and compact in X-rays. Having both radio and X-ray classifications 85% (172/202) of sources are found, with majority of sources being RQ (85%) and compact in X-rays. We observed that there are very rare RL as well as extended X-ray galaxies in the BASS survey. 55% (97/197) of sources have classifications in all three parts of EMS, of those 42% are spiral in optical, RQ in radio, and compact in X-rays.

By studying the correlation between multiwavelength morphology and other AGN properties we found early-type (ET) galaxies to have more massive M_{BH}, higher L_{bol}, and slightly higher M_* than late-type (LT) galaxies, while galaxies classified as LT have slightly higher accretion rate as shown in Figure 1 (as raised in some of previous studies, e.g., Pović et al. 2009a, Fanidakis et al. 2010, Koss et al. 2011). For the radio and X-ray sources, galaxies classified as RL have a bit larger M_{BH}, higher luminosity, and higher M_* than RQ sources and extended X-ray sources have higher λ_{Edd}, higher M_* and lower luminosity than compact X-ray sources.

We studied the location of our sources on the M_* - SFR diagram, where for the main sequence (MS) of star formation (SF) we used results from Elbaz et al. 2007. In general

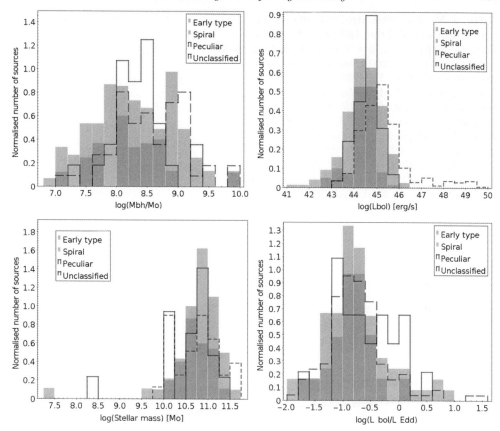

Figure 1. From top to bottom and from left to right: Distribution of M_{BH}, L_{bol}, M_*, and λ_{Edd} of optically classified sources. Different color of the histogram is related with different morphological type, as indicated on each plot.

we found that ultra-hard X-ray detected AGN can be on, below, and above the MS of SF. Although we found higher population to be located below the MS, having galaxies on and above the MS might suggest that not necessarily AGN are responsible for SF quenching.

Acknowledgements

We thank the financial support from the DBU, Ethiopian Ministry of Higher Education, EORC under the ESSTI, and Ethiopian Ministry of Innovation and Technology. MP also acknowledges the support from the Spanish Ministry of Science, Innovation and Universities (MICIU) through project AYA2016-76682C3-1-P and the State Agency for Research of the Spanish MCIU through the Center of Excellence Severo Ochoa award to the IAA-CSIC (SEV-2017-0709). Finally, we thank to the BASS collaboration for making there data available to public.

References

Baumgartner, W. H. *et al.* 2013, *ApJS*, 207, 19
Brinchmann, J. & Ellis, R. S. 2000, *ApJ*, 536, L77
Conselice C. J., 2014, *Annu. Rev. Astron.* Astrophys, 52, 291
Elbaz, D. *et al.* 2007, *A&A*, 468, 33
Fanidakis, N. *et al.* 2010, *MNRAS*, 410, 53

Gabor, J. M. et al. 2009, *ApJ*, 691, 705
Hickox, R. C. et al. 2009, *ApJ*, 696, 891
Ho L. C., 2008, *Annu. Rev. Astron. Astrophys.*, 46, 475
Kauffmann, G. et al. 2003, *MNRAS*, 346, 1055
Koss, M. et al. 2011, *ApJ*, 739, 57
Koss, M. et al. 2017, *ApJ*, 850, 74
Leslie, S. K. et al. 2016, *MNRAS*, 455, L82
Poggianti, B. M. et al. 2008, *ApJ*, 684, 888
Pović, M., et al. 2009a, *ApJ*, 706, 810
Pović, M. et al. 2009b, *ApJ* , 702, L51
Pović, M. et al. 2012, *A&A* , 541, A118
Ricci, C. et al. 2017, *arXiv preprint* arXiv:1709.03989
Schawinski, K. et al. 2010, *ApJ*, 711, 284

The role of active galactic nuclei in galaxy evolution in terms of radial pressure

Biressa Tolu[1] and Abate Feyissa[2]

[1]Department of Physics, Astronomy and Space Stream, Jimma University, Jimma, Ethiopia
email: tolu_biressa@yahoo.com

[2]Department of Physics, Mada Walabu University, Robe, Ethiopia

Abstract. Irrespective of whether Active Galactic Nuclei (AGN) is cored with Supermassive Black Holes (SMBH) or not, there is a general consensus that observations indicate that the AGN plays fundamental role in galaxy evolution. The accretion disc powered fueling of the AGN and counter-feedback on its environment in the form of stress-energy-momentum along the radial component and an associated polodial jets seems viable model. On the theoretical ground there is no unified theory that compromise the observations. But there are pull of such diverse physics simulated to describe the observational works. So, there is unsettled theoretical framework how the activity of the AGN plays role in the evolution of host galaxy. Motivated by this we studied the role of AGN on its host galaxy evolution where General relativistic (GR) Magnetohydrodynamics (MHD) equation is considered to derive radial pressure that invokes star forming cold gases. Methodologically the central engine of the AGN is considered with SMBH/pseudo-SMBH. Locally, around the AGN, Reissner-Nordstrom-de Sitter metric is considered that reduces to the Schwarzschoild-de Sitter (SdS) background. Geometrically, a simple spherical geometry is superimposed with central disc structure assumed by cored void mass ablating model. The results of the work indicates that the AGN plays role in galaxy evolution, especially in the nearby environment. Also we report that the adjacent envelope to the AGN seems quiet with no activity in formation.

Keywords. AGN, GR, MHD, Reissner-Nordstrom-de Sitter metric, SMBH, Galaxy evolution

---

1. Introduction

Regardless of whether Active Galactic Nuclei (AGN) is cored with Supermassive Blackholes (SMBH) or not, there is a general consensus that observations indicate that the AGN plays role in hosting galaxy evolution. The accretion disc powered fueling of the AGN and counter-feedback on its environment in the form of stress-energy-momentum along the radial component and an associated polodial jets seems a viable model (Yuan et al. 2018, Camera et al. 2018, McKinnon et al. 2018). On the theoretical ground there is no unified theory that compromise the observations; but there are pull of such diverse physics simulated to describe the observational works. Thus, we develop an extended General relativistic Tolman-Oppenheimer-Volkoff (TOV) equations of AGN with such geometry and additional void cores to study the role of AGN on its host galaxy evolution where the equation is considered to derive radial pressure that invokes star forming cold gases.

2. Method

Methodologically the central engine of the AGN is considered with SMBH/pseudo-SMBH. Locally, spacetime around the AGN is considered with Reissner-Nordstrom-de Sitter metric that reduces to the Schwarzschoild-de Sitter (SdS) background retaining

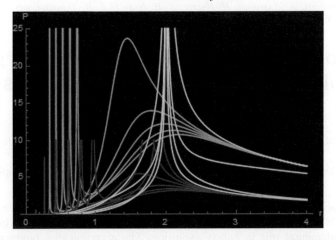

Figure 1. In the plot the green spectrum represents charge free AGN. The orange one represents low mass, low mean density galaxy where charge is being varied. Accordingly the higher peaks stand for higher charge value and so on. The blue spectrum represents high mass with reasonably higher mean density galaxy where charge is being varied.

the current standard ΛCDM cosmology that best fits to a number of various observations across a wide range of physical scales and cosmic time (Camera et al. 2018, Pellegrini et al. 2018, and the references therein). Moreover, a simple spherical geometry being superimposed with central disc structure is assumed by the cored void mass ablating model which still holds for the spherical symmetry.

3. Results and discussion

3.1. Radial pressure

With simplifying boundary conditions, we derived a conservation like generalized TOV equation that has been integrated to give the total radial pressure that depends on the parameters: q (the charge of the AGN, $M(r)$ (effective mass of the central AGN at r), $\rho(r)$ (density of the AGN surrounding fluid as a function of r), Λ (cosmological constant) and the core parameters (void mass, density, radius, height) given by:

$$P = \frac{[(\alpha\rho) + (q^2 + 16\pi r^4)\beta] \pm \sqrt{[\alpha\rho + (q^2 + 16\pi r^4)\beta]^2 - 512\pi^3\rho r^6 \gamma \beta^2}}{16\pi r^2 \gamma \beta} \quad (3.1)$$

where, α, β and γ are expressed in terms of the parameters mentioned earlier.

3.2. Numerical results

Using *Mathematica 11* we have generated a numerical result of the spectrum of pressure for some selected theoretically acceptable mass, mean density of the galaxy and void core parameters, as shown in Figure 1.

4. Conclusions

As we learn from the pressure spectrum plot and its equation, we draw the following points and comments.

a) The model developed here appears to be in agreement with what observations relatively tell us in terms of pressure distribution from the center of the hosting galaxy.

b) The AGN plays a role on hosting galaxy evolution, especially closer to it. Because as pressure creates turbulence to trigger rotations that can possibly enhance star formation where there is sufficient cold gas cloud system. In fact, the effect decreases with distance.

c) In confirmation with observations, we draw a conclusion that just next to the AGN the environment is quiet and hence no star formation expected.

d) The existence of net charge on the central gravitating system plays significant role in star formation.

e) We have also learned that Λ has no significant effect in AGN role against the host galaxy.

f) The void parameters of the AGN plays significant role in (its host) galaxy evolution.

g) The developed model shall be considered to describe AGN role in its host galaxy evolution. Moreover, it is our belief that the model also needs further developments to enrich for being more complete and comprehensive theory.

References

Camera, S., Fonseca, J., Maartens, R., & Santos, M. G. 2018, *MNRAS*, **481**(1), 1251

McKinnon, R., Vogelsberger, M., Torrey, P., Marinacci, F., & Kannan, R. 2018, *MNRAS*, 478(3), 2851

Pellegrini, S., Ciotti, L., Negri, A., & Ostriker, J. P. 2018, *ApJ*, **856**(2), 115

Yuan, F., Yoon, D., Li, Y.-P., Gan, Z.-M., Ho, L. C., & Guo, F. 2018, *ApJ*, **857**(2), 121

Optical variations in changing-look AGNs selected at X-rays

Sara Cazzoli[1], Josefa Masegosa[1], Isabel Márquez[1],
Lorena Hernández-García[2], A. Álvarez-Hernández[3,1]
and Laura Hermosa-Muñoz[1]

[1]IAA - Instituto de Astrofísica de Andalucía (CSIC), Apdo. 3004, 18008, Granada, Spain
email: sara@iaa.es

[2]Instituto de Física y Astronomía, Facultad de Ciencias, Universidad de Valparaíso, Gran Bretaña 1111, Playa Ancha, Chile

[3]Departamento de Astrofísica, Universidad de La Laguna, E-38205 La Laguna, Tenerife, Spain

Abstract. Recent observations of local AGNs have revealed that many of them show a 'changing look' behavior at optical and X-rays wavelengths in the sense of transiting between different AGNs families (e.g. from type-1 to type-2 or vice-versa). In order to pinpoint the possible relation of the changes, we performed optical spectroscopic observations (with CAFOS/CAHA) of 15 changing look AGNs selected at X-rays. Highlights from our spectroscopic study are presented.

Keywords. galaxies: active, galaxies: kinematics and dynamics, techniques: spectroscopic.

1. Introduction

Under the AGN unification scheme (see Padovani *et al.* 2017 for a review) the different optical classifications are explained in terms of orientation effects. The torus, is responsible for obscuring the broad line region (BLR) where broad Balmer lines, as H$\beta\lambda$4861 and H$\alpha\lambda$6563, are originated. This way, type 2 AGNs are those observed throughout the torus (only narrow lines) whereas type 1s have a direct view onto the BLR (broad Balmer lines and narrow forbidden lines). Intermediate classes from 1.2 to 1.9 are those in between, with type 1.2 having the largest contribution of broad lines, and 1.9 having only a broad Hα. The classification at X-rays is done by measuring the column density (N_H): for N_H below $\sim 10^{22}$ cm^{-2} the source is unobscured (i.e. type 1) whereas for larger values the source is obscured (i.e. type 2). For N_H larger than 1.5×10^{24} cm^{-2}, the sources are classified as Compton-thick (Maiolino *et al.* 1998).

In the last decades some AGNs have been observed to transit (e.g. from Compton-thick to Compton-thin in X-rays, or vice versa) between type 1 and type 2 AGNs (e.g. Matt *et al.* 2003). These objects are called "changing-look" (CL) AGN. At optical wavelengths, sources showing variations in the broad lines have also been observed, changing from one type to another type (e.g. Lamassa *et al.* 2015).

The physical mechanism of the changes is still under debate and two main different scenarios have been proposed to explain the CL behavior. On the one hand, the variation can be caused by a dramatic change of the obscuration along the line of sight which results in a variation of the column density inferred both from X-rays and optical data (e.g. Risaliti *et al.* 2005). On the other hand, the variations of the accretion onto the super-massive black hole may result in a disappearance of the BLR as below a certain accretion rate no type-1 AGN should be observed (e.g. Nicastro *et al.* 2000).

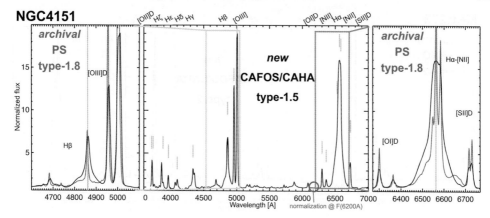

Figure 1. Example of the CL-behavior for NGC4151: comparison between new CAFOS/CAHA (Dic 2018, black) and archival Palomar Survey (PS, red, Ho *et al.* 1995).

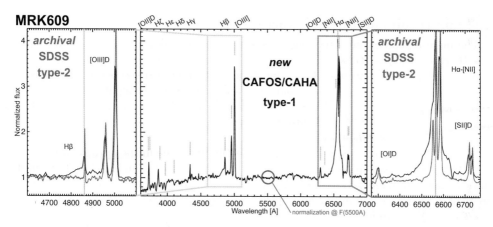

Figure 2. Comparison between new CAFOS/CAHA (Nov 2018, black) and archival SDSS-DR7 (data gathered in 2008, Abazajian *et al.* 2009, red) spectra for MRK609.

Many individual and serendipitous discoveries of CL candidates (Ricci *et al.* 2016 and reference therein) can be found in the literature. Surveys of CL-AGNs are rare and have been only lately carried out. For example, in the recent work by Yang *et al.* 2018, their blind search via archival SDSS and LAMOST spectra resulted in a discovery of 21 CL-AGNs.

2. Sample and observations

The CL-AGNs in our sample have been selected in X-rays either from previous works by our team (Hernández-García *et al.* 2016, 2017) or from literature, e.g. NGC2617 (Oknyansky *et al.* 2017) and NGC4151 (Puccetti *et al.* 2007). Our total sample is composed by 15 nearby ($z < 0.04$) CL-AGNs. For these AGNs, we obtained broad band (3200-7000 Å) low resolution (~ 4 Å/pixel) optical long-slit spectra, in Nov-Dic 2018, with CAFOS mounted at the 2.2m telescope of the Calar Alto Observatory (CAHA). Examples of our new optical spectra and their comparison with archival data are shown in Figures 1, 2 and 3.

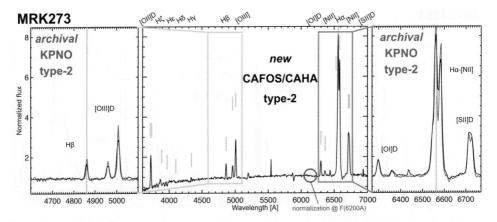

Figure 3. New CAFOS/CAHA (black) and archival Kitt Peak National Observatory (KPNO, red, Kim *et al.* 1995) optical spectra of the CL-candidate MRK273 (Hernández-García *et al.* 2016)

3. Main results and future works

The main results of our study can be summarized as follows:
- *Detection of CL-AGNs.*

Among the X-rays selected AGNs, the 30 percent (5 out 15) shows the CL-behavior at optical wavelengths. Among these, four known CL-AGNs were caught varying again: two were selected on the basis of N_H-variations (e.g. NGC4151, Puccetti *et al.* 2007, Fig. 1) and two from switching from Compton-thin to Compton-thick regime.
- *MRK609, a new CL-AGN.*

This AGN is found to show CL-behavior for the first time in the optical band (Fig. 2). MRK609 has been selected on the basis of the transition from Compton-thin to Compton-thick regime (Hernández-García *et al.* 2017). The AGN-component is absent in the SDSS archival data but visible in our new optical spectra, with FWHM (Hα) ~ 5150 km/s.
- *CL-AGNs candidates.*

We did not find variations in the optical spectra of 10 (70 percent) AGNs (an example in Fig. 3). N_H variations or changes in the Compton-regime do not guarantee to catch CL-phenomenon in action.

We plan to measure emission lines in both new and archival optical spectra of the selected AGNs fitting both narrow and broad lines, as in Cazzoli *et al.* (2018). The measurements will be compared to those in the literature and to the optical/X-ray properties.

Acknowledgements

We acknowledge financial support by the Spanish MCIU, grants SEV-2017-0709 and AYA2016-76682-C3. LHM acknowledge the financial support by MCIU, grant BES-2017-082471. AAH acknowledges CSIC, grant JAE-intro. LHG thanks FONDECYT, grant 3170527 and Conicyt PIA ACT 172033. SC and LHM thank the IAU for the travel grants.

References

Abazajian, K. N., Adelman-McCarthy, J. K,. *et al.* 2009, *ApJS*, 182, 543
Cazzoli S., Marquez I., Masegosa, J., *et al.* 2018, *MNRAS*, 480, 1106
Hernández-García, L., Masegosa, J., Gonzalez-Martín O., *et al.* 2016, *ApJ*, ApJ, 824, 7
Hernández-García, L., Masegosa, J., Gonzalez-Martín, O., *et al.* 2017, *A&A*, 602, A65
Ho, L., Filippenko, A. V., & Sargent, W. L. W. 1995, *ApJS*, 98, 477

Lamassa, S., Cales, S., Moran, C.E., et al. 2015, *ApJ*, 800, 2
Kim, D., Sanders, D. B., & Veilleux, S. 1995, *ApJs*, 98,129
Maiolino, R., Salvati, M., Bassani, L., et al. 1998, *A&A*, 338, 781
Matt, G., Guainazzi, M., & Maiolino, R. 2003, *MNRAS*, 342, 422
Nicastro, F., Piro, L., De Rosa, A., et al. 2000, *ApJ*, 530, L65-L6
Oknyansky, V., Gaskell, C. M., Huseynov, N. A., et al. 2017, *MNRAS*, 467, 1496-1504
Puccetti, S., Fiore, F., Risaliti, G., et al. 2007, *MNRAS*, 377, 607
Ricci, C., Bauer, F. E., Arevalo, P., et al. 2016, *ApJ*, 802, 5
Risaliti, G., Elvis, M., Fabbiano, G., et al. 2005, *ApJ*, 623, L93-L96
Yang, Q., Xue-Bing, W., Fan, X., et al. 2018, *ApJ*, 862, 109

NGC7469 as seen by MEGARA at the GTC

Sara Cazzoli[1], Armando Gil de Paz[2,3], Isabel Márquez[1],
Josefa Masegosa[1], Jorge Iglesias[1], Jesus Gallego[2,3],
Esperanza Carrasco[4], Raquel Cedazo[5], María Luisa García-Vargas[6],
África Castillo-Morales[2,3], Sergio Pascual[2,3], Nicolás Cardiel[2,3],
Ana Pérez-Calpena[6], Pedro Gómez-Alvarez[6],
Ismael Martínez-Delgado[6] and Laura Hermosa-Muñoz[1]

[1]IAA - Instituto de Astrofísica de Andalucía (CSIC), Apdo. 3004, 18008, Granada, Spain
email: sara@iaa.es

[2]Departamento de Física de la Tierra y Astrofísica, Universidad Complutense de Madrid,
E-28040 Madrid, Spain

[3]Instituto de Física de Partículas y del Cosmos IPARCOS, Facultad de Ciencias Físicas,
Universidad Complutense de Madrid, E-28040 Madrid, Spain

[4]INAOE - Instituto Nacional de Astrofísica, Óptica y Electrónica, Luis Enrique Erro No.1,
C.P. 72840, Tonantzintla, Puebla, Mexico

[5]Universidad Politécnica de Madrid, Madrid, Spain,

[6]Fractal, S.L.N.E., Madrid, Spain

Abstract. We present the main results from the analysis of the Hα-[NII] emission lines with integral field spectroscopy observations gathered with MEGARA at the GTC of the nearby Seyfert 1.5 galaxy NGC7469. We obtained maps of the ionised gas in the inner 12.5 arcsec × 11.3 arcsec, at spatial scales of 0.62 arcsec, with an unprecedented spectral resolution (R ∼ 20 000). We characterized the kinematics and ionisation mechanism of the distinct kinematic components (Cazzoli *et al.* 2019).

Keywords. galaxies: active, galaxies: kinematics and dynamics, techniques: spectroscopic.

1. Introduction

With the advent of MEGARA (*Multi-Espectrógrafo en GTC de Alta Resolución para Astronomía*, Gil de Paz *et al.* 2018) the new integral field unit at the 10.4m Gran Telescopio Canarias (GTC) high spectral resolution (from 6 000 up to 20 000) integral field spectroscopy observations have become available. MEGARA started the operation in 2017 and during the commissioning run the NGC7469 galaxy was observed.

NGC7469 is a nearby ($z \sim 0.016$), grand-design spiral galaxy hosting a Seyfert 1.5 AGN (Landt *et al.* 2008). Powerful star formation activity, SFR = 48 M$_\odot$/yr (Pereira-Santaella *et al.* 2011), is mainly occurring in its circumnuclear ring bright at various wavelengths (e.g. Davies *et al.* 2004). Features of non-rotational motions, such as outflows, have been found in the infrared and X-rays bands (Müller Sánchez *et al.* 2011; Blustin *et al.* 2007).

2. Main Results

After data reduction (as in Pascual *et al.* 2019), we modeled the H$\alpha\lambda$6563-[NII]$\lambda\lambda$6548,6584 emission lines with four kinematic components that can be discriminated according to their width. Figures 1, 2 and 3 show the kinematic maps (velocity

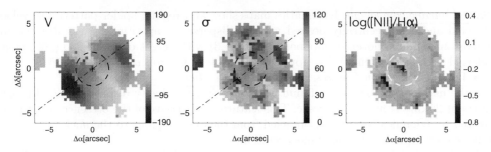

Figure 1. Ionised gas maps derived from the fit to the Hα-[NII] emission lines for the narrow component: velocity field (V), velocity dispersion map (σ) and log ([NII]/Hα) maps. The cross marks the photometric center, the dashed circle indicates the nuclear region. Kinematic maps are in km s^{-1} units. The photometric major axis is marked with a dot-dashed line.

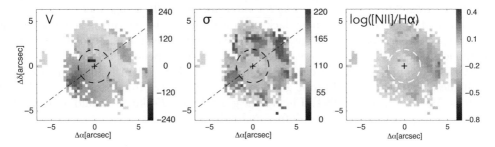

Figure 2. The same as Fig. 1 but for the second (narrow) component.

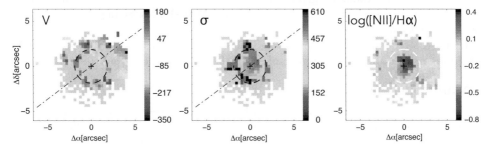

Figure 3. The same as Fig. 1 but for the third (intermediate-width) component.

and velocity dispersion, i.e. V and σ) and the log ([NII]/Hα) map, for the narrow, second and intermediate-width line-components. An unresolved broad Hα component, from the AGN's broad line region (BLR), is present in the nuclear region (within the area of the PSF).

• *Kinematics, dynamical support and disc-height for the two narrow components.*
Narrow component. The velocity field shows a pattern consistent with kpc-scale ordered rotational motions (Fig. 1, left). The peak-to-peak semi-amplitude is $163 \pm 1 \,\mathrm{km\,s^{-1}}$, with the maximum velocity gradient oriented as the photometric major axis (∼125°). The σ-map is not centrally-peaked (Fig. 2, middle), contrary to what is expected for a rotating disc. The average velocity dispersion inside the nuclear region ($40 \pm 1 \,\mathrm{km\,s^{-1}}$) is similar to that in the disc ($38 \pm 1 \,\mathrm{km\,s^{-1}}$) that presents some perturbations. The dynamical ratio (V/σ) is 4.3 indicating a rotation-dominated kinematics. By using a thin-disc approximation (as in Cazzoli et al. 2014, 2016) we calculate that the disc height is ∼20 pc.

Second narrow component. Kinematics maps show an irregular spider-pattern-like velocity field and a non-centrally peaked velocity dispersion map (Fig. 2). The disc has an important random-motion component ($V/\sigma = 1.3$) with height of about 200 (500) pc as inferred with a thin (thick) disc approximation (as in Cazzoli et al. 2014, 2016).

- *Velocity dispersion features: σ-drop and anomalies.*

For the narrow component, we found a decrease in the velocity dispersion radial profile of about $25\,\mathrm{km\,s^{-1}}$ at $r \leq 1.5$ arcsec. This feature is suggestive of the presence of a σ-drop related to dynamically cold gas funneled from the outer regions to the nucleus by a bar during a fast episode of central gas accretion.

We detect two kinds of 'velocity dispersion anomalies'. On the one hand, the velocity dispersion maps of both narrow components reveal a number of clumps with low-σ. On the other hand, a velocity dispersion enhancement along the minor photometric axis is found only in the σ-map of the narrow component. While the former could be potentially associated to star forming clumps, the origin of the latter is unclear.

- *Very turbulent emission probed by the intermediate-width component.*

The kinematics of this component lack of any rotating disc features, being irregular, with no peculiar morphology/orientation (Fig. 3). On the one hand, a broad blue-shifted component is indicative of outflows, even if not oriented perpendicular to the disc (e.g. as in NGC1068, García-Burillo et al. 2014). On the other hand, the ring-like emission with the highest turbulence could probe gas flows at the Inner Lindblad Resonance radius of the lens (Márquez & Moles 1994). Finally, the gas with $\sigma \sim 250\,\mathrm{km\,s^{-1}}$ could be associated to either disc-perturbations or to diffuse and not virialised gas.

- *Ionisation mechanisms.*

For the narrow (second) component, the $\log([\mathrm{NII}]/\mathrm{H}\alpha)$ values (Figures 1 and 2, right) are indicative of star-formation as the unique (dominant) mechanism of ionisation. For the intermediate component, given the observed widths ($> 300\,\mathrm{km\,s^{-1}}$ typically), the $\log([\mathrm{NII}]/\mathrm{H}\alpha)$ values are indicative of shock-ionisation (Fig. 3, right).

- *BLR properties.*

In the unresolved nuclear region of NGC7469 ($r \leq 1.85$ arcsec) the broad (FWHM $\sim 2590\,\mathrm{km\,s^{-1}}$) H$\alpha$ component from the BLR is dominating the global Hα-[NII] profile.

3. Main conclusions

For the two narrow components, we found that a thicker disc nearly co-rotates with the thinner one, with similar velocity amplitudes (137 vs. 163 $\mathrm{km\,s^{-1}}$), but different velocity dispersions (108 vs. 38 $\mathrm{km\,s^{-1}}$) and dynamical status (4.3 vs. 1.3 in V/σ). The morphology and the kinematics of the intermediate-width component is suggestive of the presence of turbulent non-circular motions, possibly associated either to a wide angle outflow or to gas flows related to the lens. Part of the ionised gas traced by this component could also be due to turbulent motions outside the plane of the disc related to disc perturbations, and/or with diffuse gas gravitationally bound to the host galaxy, but not virialised.

Acknowledgements

We thank the financial support by the Spanish MCIU and MEC, grants SEV-2017-0709, AYA 2016-76682-C3 and AYA2016-75808-R. LHM acknowledge the financial support by MCIU, grant BES-2017-082471. SC and LHM thanks the IAU for the travel grants.

References

Blustin, A.J., Kriss, G.A., Holczer, T., et al. 2007, A&A, 466, 107
Cazzoli, S., Gil de Paz, A., Márquez, I., et al. 2019, MNRAS, submitted
Cazzoli, S., Arribas, S., Maiolino, R., & Colina, L. 2016, A&A, 590, A125
Cazzoli, S., Arribas, S., Colina, L., et al. 2014, A&A, 569, A14

Davies, R.I., Tacconi, L. J., & Genzel, R. 2004, *ApJ*, 602, 148
García-Burillo, S., Combes, F., Usero, A., *et al.* 2014, *A&A*, 567, A125
Gil de Paz, A., Carrasco, E., Gallego, J., *et al.* 2018, *SPIE*, 10702, 17
Landt, H., Bentz, M. C., Ward, M. J., *et al.* 2008, *ApJS*, 174, 282
Márquez, I., & Moles, M. 1994, *AJ*, 108 , 90
Müller Sánchez, F., Prieto, M. A., Hicks, E. K. S., *et al.* 2011, *ApJ*, 739, 69
Pascual, S., Gil de Paz, A., *et al.* 2019, *Proceedings of Spanish Astrophysics meeting*, 227
Pereira-Santaella, M., Alonso-Herrero, A., Santos-Lleo, M., *et al.* 2011, *A&A*, 535, A93

Optical spectral properties of radio loud quasars along the main sequence

Ascensión del Olmo[1], Paola Marziani[2], Valerio Ganci[2,3]†, Mauro D'Onofrio[3], Edi Bon[4], Natasa Bon[4] and Alenka C. Negrete[5]

[1]Instituto de Astrofísica de Andalucía, IAA-CSIC, Granada, Spain,
email: chony@iaa.es

[2]INAF, Astronomical Observatory of Padova, Padova, Italy

[3]Dipartimento di Fisica e Astronomia, University of Padova, Padova, Italy

[4]Astronomical Observatory Belgrade, Belgrade, Serbia

[5]Instituto de Astronomía - UNAM, CDMX, México

Abstract. We analyze the optical properties of Radio-Loud quasars along the Main Sequence (MS) of quasars. A sample of 355 quasars selected on the basis of radio detection was obtained by cross-matching the FIRST survey at 20cm and the SDSS DR12 spectroscopic survey. We consider the nature of powerful emission at the high-FeII end of the MS. At variance with the classical radio-loud sources which are located in the Population B domain of the MS optical plane, we found evidence indicating a thermal origin of the radio emission of the highly accreting quasars of Population A.

Keywords. quasars, radio-loud, star formation

1. Introduction

The 4D Eigenvector 1 (E1) is a powerful tool to contextualize the diversity of observational properties in type-1 AGN(see e.g., Marziani et al. 2018 for a recent review). The distribution of the quasars in the E1 optical plane, defined by the FWHM of the Hβ broad component vs. the FeII strength (parametrized by the ratio $R_{\rm FeII} = I({\rm FeII}\lambda 4570)/I({\rm H}\beta)$), outlines the quasar Main Sequence (MS). The shape of the MS (Figure 1, left plot) allow us for the subdivision of quasars in two Populations (A and B) and in bins of FWHM(Hβ) and FeII which define a sequence of spectral types (STs):

1) Pop. A with FWHM(Hβ) $\leqslant 4000\,{\rm km s}^{-1}$, and with STs defined by increasing $R_{\rm FeII}$ from A1 with $R_{\rm FeII} < 0.5$ to A4 with $1.5 \leqslant R_{\rm FeII} \leqslant 2$. STs A3 and A4 encompass the extreme Pop. A of highly accreting quasars radiating near the Eddington limit.

2) Pop. B with FWHM(Hβ) $> 4000\,{\rm km s}^{-1}$, and STs bins (B1, B1$^+$, B1^{++},...) defined in terms of increasing ΔFWHM(Hβ) $= 4000\,{\rm km s}^{-1}$ (see sketch in Figure 1, left plot).

Eddington ratio $\lambda_{\rm E} = L_{\rm Bol}/L_{\rm Edd}$ and orientation are thought to be the main physical drivers of the MS (Marziani et al. 2001; Sulentic et al. 2017). Eddington ratio changes along the $R_{\rm FeII}$ axis. Pop. B quasars are the ones with high black hole mass ($M_{\rm BH}$) and low $\lambda_{\rm E}$ and Pop. A are fast-accreting sources with relatively small $M_{\rm BH}$. Radio-Loud (RL) quasars, defined as relativistic jetted sources, are not distributed uniformly along the MS (Zamfir et al. 2008). They are predominantly found in the Pop. B domain, having $R_{\rm FeII} < 0.5$ and FWHM(Hβ) $> 4000\,{\rm km s}^{-1}$. The extreme broad Pop. B^{++} quasar bin (FWHM(Hβ) $> 12000\,{\rm km s}^{-1}$) contains about 30% of the RLs but only ~ 3% of quasars (Marziani et al. 2013).

† Present address: Institute of Physics, University of Cologne, Germany

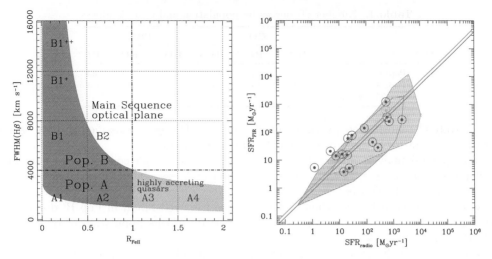

Figure 1. Left: sketch of the optical plane of the MS of quasars and the STs. Green area traces the location of the extreme Pop. A quasars (xAs); Right: FIR vs. radio SFR for the xAs with FIR data. The shaded areas trace the loci of star-forming galaxies (grey) and RQ quasars (pink).

2. Sample

We selected quasars from the SDSS-BOSS DR12 Quasar Catalog (Pâris et al. 2017) with $m_i \leqslant 19.5$ and redshift $\leqslant 1$ and cross-matched with the VLA FIRST radio survey at 20cm (Becker et al. 1995). A detailed description of the sample and the performed analysis can be found in Ganci et al. (2019). The classification of the quasars was carried out according to three criteria:

— Radio power, based on the $R_K = f_{rad}/f_{opt}$ parameter (Kellermann et al. 1989), defined on the basis of the 1.4 GHz and g magnitude estimates. Sources with $\log R_K < 1.0$ are classified as Radio-Detected (RD); those with $1.0 \leqslant \log R_K < 1.8$ as radio-intermediate (RI) and for $\log R_K \geqslant 1.8$ as RL.

— Radio morphology as Core Dominated (CD) and FRII sources, selected using the statistical procedure defined by de Vries et al. (2006).

— Optical ST classification in terms of the MS.

The final sample consist of 355 objects, 289 CD and 66 FRII. There are a total of 38 RD, 139 RI and 178 RL objects.

3. Results

We have analyzed the source distribution along the MS. Table 1 shows a summary of the number of sources for both Pop. A and B and for the particular case of the extreme accreting Pop. A quasars (xAs, STs A3-A4), according to its radio loudness and radio morphology. From this analysis we found:

• The most populated STs are the Pop. B bins (72% of the sources), especially B1 (36%) and B1$^+$ (22%) at variance with optically selected samples (Marziani et al. 2013) where B1 and A2 are the most populated STs.

• RD sources are only present in CD radio morphology (21 Pop. A, and 17 Pop. B).

• RIs sources have also only CD morphology, but are more numerous in Pop. B (70%).

• RLs are significantly more numerous in Pop. B for the CD morphology and FRII sources appear only in RL class, and are almost exclusively Pop. B (95%).

The most powerful radio quasars are located in Pop. B both for FRII and CD radio morphologies and they are most likely jetted RL quasars, meanwhile Pop. A shows almost

Table 1. Number of sources by population and radio-loudness.

	CD			FRII		
	RD	RI	RL	RD	RI	RL
Pop. B	17	96	81	-	-	63
Pop. A	21	43	31	-	-	3
xA (STs A3-A4)	5	17	5	-	-	-

exclusively CD morphology, with about 28% of the CD objects classified as Pop. A but with significant lower radio power that CD Pop. B sources.

The spectral analysis of the SDSS-BOSS spectra of the sample confirms the systematic differences between Pop A and Pop. B, and the trends associated with the quasar MS, similar for all radio classes.

- The centroid at half maximum of Hβ passes from redshifted in Pop B, due to the presence of the very broad component characteristic of this population of quasars, to clearly blueshifted in xAs.
- Equivalent width of Hβ remains roughly constant in Pop. B quasars, but dramatically decreases in Pop. A.
- From the B1^{++} to the A4 STs, λ_E increases and M_BH decreases, independently of the radio classes.

A very interesting result from our analysis is that about 80% (22/27) of the xA CD have a radio emission in the loud range (RI and RL). Are these RL and RI xAs truly jetted sources?, or is its radio power dominated by thermal emission? We have investigated the possible origin of their radio emission by analyzing its FIR luminosity and through the FIR-radio star formation rate (SFR) relation (Bonzini et al. 2015), since RL jetted sources are expected to have FIR luminosity significantly lower than the one expected for their radio power. We show in Figure 1 (right) the FIR and radio power SFR for the xAs sources with FIR data available from both this sample and the xAs in Bonzini et al. (2015) sample (see Ganci et al. 2019), together with the location of the star forming galaxies (gray shaded area) and radio-quiet (RQ) quasars (pink area). xAs sources have both high radio power and SFR. They are placed in the RQ and star-forming galaxies regions, suggesting that their radio power is due to the thermal emission. And none of the other sources, included in Table 1, that fall in the RL region (outside and to the rigth of the shaded areas) is an xA source.

It is therefore reasonable to suggest that most of the xAs that are RI and RL are non-jetted sources and they might be truly "thermal sources". Our study (Ganci et al. 2019) supports the suggestion that RI and RL A2, A3, A4 sources could be thermal in origin.

Acknowledgements

AdO acknowledges financial support from the Spanish grants MEC AYA2016-76682-C3-1-P and the State Agency for Research of the Spanish MCIU through the "Center of Excellence Severo Ochoa" award for the IAA (SEV-2017-0709).

References

Becker R. H., White R. L. & Helfand D. J. 1995, *ApJ*, 450, 559
Bonzini M., Mainieri V., Padovani P., et al. 2015, *MNRAS*, 453, 1079
de Vries W. H., Becker R. H., & White R. L. 2006, *AJ*, 131, 666
Ganci V., Marziani P., D'Onofrio M., del Olmo A., et al. 2019, *A&A*, 630, A110
Kellermann K. I., Sramek R., Schmidt M., et al. 1989, *AJ*, 98, 1195
Marziani P., Sulentic J. W., Zwitter T., et al. 2001, *ApJ*, 558, 553.
Marziani P., Sulentic J. W., Plauchu-Frayn I., del Olmo A., 2013, *A&A*, 555, A89

Marziani P., Dultzin D., Sulentic J. W., del Olmo A., *et al.* 2018, *Frontiers in A&SS*, 5, 6
Pâris I., Petitjean P., Ross N. P., *et al.* 2017, *A&A*, 597, A79
Sulentic J. W., del Olmo A., Marziani P., *et al.* 2017, *A&A*, 608, A122
Zamfir S., Sulentic J. W. & Marziani P. 2008, *MNRAS*, 387, 856

Accretion rate in AGN and X-ray-to-optical flux ratio at z ⩽ 0.2

Asrate Gaulle[1,2], Mirjana Pović[2,3] and Dejene Zewdie[4,5]

[1]Dilla University, Department of Physics, Dilla, Ethiopia
email: asrieguale@gmail.com

[2]Astronomy and Astrophysics Research and Development Division, Ethiopian Space Science and Technology Institute, Addis Ababa, Ethiopia

[3]Institute of Astrophysics of Andalucía (IAA-CISC), Department of Extragalactic Astronomy, Granada, Spain

[4]Núcleo de Astronomía, Universidad Diego Portales, Santiago, Chile

[5]Department of Physics, Debre Berhan University, Debre Berhan, Ethiopia

Abstract. We explored a sample of 545 local galaxies using data from the 3XMM-DR7 and SDSS-DR8 surveys. We carried out all analyses up to z ∼ 0.2, and we studied the relation between X/O flux ratio and accretion rate for different classes of active galaxies such as LINERs and Seyfert 2. We obtained a slight correlation between the two parameters if the whole sample of AGN is used. However, LINERs and Sy2 galaxies show different properties, slight correlation and slight anti-correlation, respectively. This could confirm that LINERs and Sy2 galaxies have different accretion efficiencies and maybe different accretion disc properties, as has been suggested previously.

Keywords. galaxies - active; AGN - accretion rate; AGN - black hole masses; AGN - X-ray properties; AGN - optical properties.

---

1. Introduction

Active galactic nuclei (AGN) are powerful sources of radiation in a wide spectral range, from gamma-rays to radio waves (Netzer 2015). In particular, AGN are strong X-ray sources. X-ray emission is shown to be a powerful tool of AGN detection and a study of the growth of supermassive black holes (SMBHs) and AGN properties (Brandt & Hasinger 2005). On the other side, optical data are very important for AGN classification and for studying the properties of AGN host galaxies (Pović et al. 2009a,b; 2012). Therefore, the combination of X-ray and optical data allows the successful study of the connection between the AGN and their host galaxies. Furthermore, the accretion rate (AR) in galaxies remains a prerequisite for understanding the physics behind SMBHs and AGN, the evolution and growth of galaxies, and the connection between active and non-active galaxies. However, AR measurements are still not easy since they mainly depend on the availability of spectroscopic data, which contain smaller data sets and poorer statistics than photometric data. In previous studies, Pović et al. (2009a,b) suggested that there might be a photometric indicator of AR in galaxies, based on the ratio between the flux in X-rays (0.5 - 4.5 keV) and optical flux. In this work, we used X-ray data, and optical spectroscopic and photometric data to measure both X/O flux ratio and AR in different types of active galaxies and to test the correlation between the two parameters.

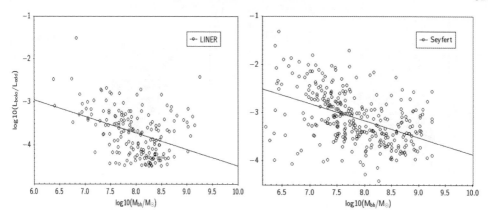

Figure 1. The correlation between BH mass and AR, left-right (LINER,Sy2)

2. Data and Sample Selection

X-ray data were obtained from the 3XMM-DR7 catalogue (Rosen et al. 2016) which contains 727,790 sources. Also, we used optical photometric and spectroscopic data from the Sloan Digital Sky Survey (SDSS) data release 8 (DR8). Spectroscopic measurements have been obtained from the MPA-JHU SDSS DR8 catalogue (Brinchmann et al. 2004) for 1,472,581 sources. Finally, morphological classification has been obtained from the Galaxy Zoo catalogue (Lintott et al. 2011). We cross-matched all three catalogues using a cross-matched radius of 3 arcsec and we selected 2151 sources in total up to redshift $z \leq 0.2$. We stick to this redshift for avoiding the K-correction in X-rays. Using the BPT-NII diagram (Baldwin, Phillips & Terlevich 1981), and signal-to-noise ratio of emission lines of $S/N > 3$, we selected 545 AGN galaxies in total, of those 209 and 336 being Seyfert 2 and LINER sources, respectively.

3. Analysis and Results

In this work, we analyzed the correlation between the X/O flux ratio and AR of spectroscopically selected AGN up to $z \leq 0.2$. After selecting LINERs and Sy2 sources, we went through the following analysis: we first measured the X/O flux ratio by using the (0.5 - 4.5keV) X-ray and optical r bands (Pović et al. 2009a,b). Secondly, we measured the mass of a black hole using the velocity dispersion method (Tremaine et al. 2002) and velocity dispersion measurements from the MPA-JHU catalogue (Brinchmann et al. 2004). Eddington luminosity was then measured as $L_{edd} = 1.5 \times 10^{38}$ (Mbh/M)erg/s. Bolometric luminosity was measured through the $H\beta$ and [OIII] emission lines using the results of Netzer (2013). Emission lines were first corrected for extinction through the $H\alpha/H\beta$ emission lines. Finally, the accretion rate was measured as AR = Lbol /Ledd. We tested the correlation between the SMBH mass and AR (see Fig. 1) and X/O flux ratio and AR (see Fig. 2) for the whole sample of AGN, and also for LINERs and Sy2 galaxies. We finally analyzed the same relations, but for different morphological types. We found anti-correlation between the SMBH mass and AR independently on AGN type, as shown in Fig. 1, confirming some of the previous results (e.g., Woo & Urry 2002). Only mild correlation has been found between the X/O flux ratio and AR in LINERs, and mild correlation in Seyfert 2 galaxies, as shown in Fig. 2. When observing the previous in relation to morphology, the same trends have been found in all cases for both elliptical and spiral galaxies at redshifts $z \leq 0.2$.

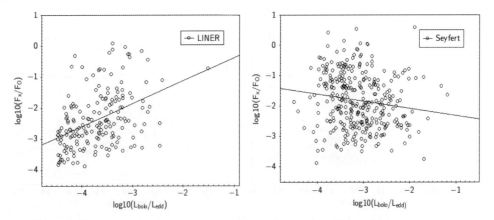

Figure 2. The correlation between X/O flux ratio & AR, left- right: LINER, Sy2.

4. Conclusion

The results obtained above could suggest previous findings that Sy2 and LINER galaxies have different accretion properties and belong to two different modes: radiatively efficient and radiatively inefficient advection dominated accretion. Preliminary results obtained in this work do not discard the possibility of using the X/O flux ratio as an AR indicator, however, more detailed studies and larger statistical samples are needed for confirming this.

Acknowledgements

AG acknowledges the support from Dilla University. AG and MP acknowledge financial support from the EORC under the ESSTI and Ethiopian Ministry of Innovation and Technology. MP acknowledges the support from the Spanish Ministry of Science, Innovation, and Universities (MICIU) through project AYA2016-76682C3-1-P and the State Agency for Research of the Spanish MCIU through the Center of Excellence Severo Ochoa award to the IAA-CSIC (SEV-2017-0709).

References

Baldwin, J. A., Phillips, M. M. & Terlevich, R. 1981, *PASP*, 93, 5
Brandt, W. N., & Hasinger, G. 2005, *ARA&A*, 43, 827
Brinchmann J., Charlot S., White S. D. M., et al. 2004, *MNRAS*, 351, 1151
Lintott, C., et al. 2011, *MNRAS*, 410, 166
Netzer H. 2013, Cambridge University Press
Netzer H. 2015, *ARA&A*, 53, 365
Pović, M., et al. 2009a, *ApJ*, 702, 51
Pović, M., et al. 2009b, *ApJ*, 706, 810
Pović, M., Sánchez-Portal, M., and Perez-García, A., et al. 2012, *A&A*, 541, 118
Rosen, S. R., et al. 2016, *A&A*, 590, 1
Tremaine, S., et al. 2002, *ApJ*, 574, 740
Woo, J.-H., & Urry, C. M., 2002, *ApJ*, 579, 530

Optical spectroscopy of type-2 LINERs

Laura Hermosa Muñoz, Sara Cazzoli, Isabel Márquez and Josefa Masegosa

Instituto de Astrofísica de Andalucía - CSIC, Glorieta de la Astronomía s/n 18008, Granada, Spain
email: lhermosa@iaa.es

Abstract. Low-Ionisation Nuclear Emission-line Regions (LINERs) are the least luminous and the most numerous among the local population of Active Galactic Nuclei (AGN). They can be classified as type-1 or type-2 if their optical spectra show or do not show, respectively, a broad component. It is associated with the presence of a Broad Line Region (BLR) in these systems. However, recent studies have proven that the classification of type-1 LINERs may be controversial, since space- and ground-based spectroscopy provide contradicting results on the presence of very broad components (Cazzoli et al. 2018). We have studied the nuclear spectra of 9 type-2 LINERs with intermediate spectral resolution HST/STIS data. We present the results on our analysis of the different spectral components, and discuss the eventual presence of BLR components in type-2 LINER galaxies, together with the possible presence of outflows, both in comparison with type-1 LINERs. We have found a BLR component in 7 out of the 9 analysed objects within the HST/STIS data.

Keywords. galaxies: active; galaxies: nuclei; galaxies: kinematics and dynamics; techniques: spectroscopic

1. Introduction

LINERs are low luminosity AGNs whose spectra are dominated by low-ionisation emission lines (Heckman 1980). Their true nature is still uncertain (Ho 2008, Márquez et al. 2017) as their spectral features can be explained with other models apart from an AGN (see e.g. for shock ionisation models: Heckman 1980, Dopita, & Sutherland 1995; and for photoionisation by post-AGB stars: Binette et al. 1994, Papaderos et al. 2013). As other AGNs, they are optically classified depending on whereas the BLR is accessible along the line of sight, and hence a very broad Balmer line is detected. Type-2 LINERs have been classified by Ho et al. (1993) as not showing a broad component in the Hα line, although it is visible in type-1 AGNs, as we have a direct view of the innermost parts of the AGN. However, it has been reported that the broad component is not visible in some optically classified type-1 LINERs, with differing results from space- to ground-based data (Cazzoli et al. 2018). Additionally, the emission line profiles may be affected by the presence of outflows. These components, whose presence could compromise the detection of weak BLR components in LINERs, can be detected in [S II]$\lambda\lambda 6716,6731$Å; H$\alpha\lambda 6564$Å; [N II]$\lambda\lambda 6548,6584$Å and [O I]$\lambda\lambda 6300,6363$Å lines.

We have analysed archival data of the 9 type-2 LINERs from the sample of 82 LINERs by González-Martín et al. (2009) with Hubble Space Telescope (HST)/Space Telescope Imaging Spectrograph (STIS) nuclear spectra. We have completed the sample with ground-based spectra from the Double Spectrograph/Palomar (Ho et al. 1993, 1995) for 8 out of 9 LINERs (NGC 4676B was not studied in the Palomar sample). The main objective is to study the presence of different kinematic components in the nuclei of these galaxies to unveil the true nature of type-2 LINERs.

2. Methodology

The stellar subtraction and spectral fitting were done as in Cazzoli et al. (2018) (hereafter C18). Spectra from both space- and ground-based data were retrieved from the archival already fully reduced. The gratings for both data-sets covered the wavelength range where [S II], Hα-[N II] and, in some cases, [O I] emission lines are available.

As the host galaxy contribution on the nuclear spectra of a low-luminosity AGN could be significant, the starlight should be subtracted before fitting the lines. However, for the HST/STIS data the wavelength range is small (572 Å) for a proper stellar continuum modelling. Nevertheless, thanks to the small aperture of the instrument, Constantin et al. (2015) proved that the correction is negligible for HST/STIS spectra. Thus the starlight was modelled and subtracted only for the Palomar data using a penalised PiXel fitting analysis (pPXF version 4.71, Cappellari, & Emsellem 2004; Cappellari 2017).

The emission lines were modelled applying a non-linear least-squares minimization and curve-fitting routine (LMFIT) implemented in Python. We used a maximum of two Gaussians per forbidden line and narrow Hα plus a broad Gaussian for Hα when needed. To prevent overfitting, we used the standard deviation of a region of the continuum without absorption or emission lines (ε_c). This value was compared to the standard deviation of the residuals under the lines after adding the Gaussian component. If it was higher than $3\varepsilon_c$ a new component was added. All the lines were tight to have the same velocity (v) and velocity dispersion (σ) as the [S II] or [O I] lines. We used them as reference for Hα-[N II] as they are usually unblended in the spectra, thus causing less uncertainties when modelling the line profiles (see C18 for more details).

3. Results

The kinematics (v and σ) of each Gaussian component used in the fit (except for the broad component originated in the BLR) can be associated to either rotational or non-rotational motions that occur in the host galaxy. The main result of this analysis is summarised in Fig. 1. The narrowest Gaussian profile is called here *narrow component*, and a broader Gaussian which is needed in all emission lines for some LINERs is called *secondary component*. The narrow component is generally associated to rotation coming from the galactic disk, whereas the secondary component could be due to rotational motions or non-rotational motions (outflows or inflows). Dividing lines in Fig. 1 have been established by C18 measuring the gas velocity field from the 2D spectra of their sample of type-1.9 LINERs, and estimating the maximum broadening of the emission lines associated to rotational motions. Their conservative upper limit for the rotational component is ~ 400 kms^{-1}, although the values of σ in their sample are well below this limits (see Sect. 5.2 in C18).

For the type-2 LINERs within our sample, all the narrow components found in the modelling are consistent with being produced by rotational motions, as initially expected. Only four galaxies needed a secondary component to reproduce the observed line profiles. We have found that this component is associated to both rotational (NGC 4552 & NGC 4374) and non-rotational motions (NGC 4594 & NGC 4486). Thus the latter objects are candidates to have outflows. As for the BLR component, we found that it is needed to reproduce the Hα profile in 7 out of the 9 galaxies from the HST/STIS data and in 2 out of 8 in the Palomar data. Three of the HST detections (NGC 3245, NGC 4594 and NGC 4736) were already reported by Constantin et al. (2015). Our results for the Full Width at Half Maximum and the flux contribution of the BLR component agree with theirs. Therefore, the true nature of most LINERs needs to be revisited with high spectral resolution.

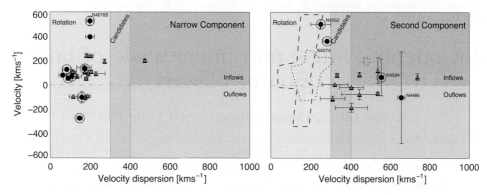

Figure 1. Kinematic classification of the components from the spectral modelling. The velocity and velocity dispersion for the narrow (left) and secondary (right) components are shown for both type-2 (black circles) and type-1 LINERs (red triangles from C18). Grey bands indicate rotational motions; pink regions are candidates to non rotational motions; yellow and blue zones indicate inflows and outflows, respectively. Black circles surround galaxies where a broad component was detected. The dashed (dotted) line indicates the region occupied by the narrow component in type-2 (type-1) LINERs.

4. Conclusions

The main conclusions of the work Hermosa-Muñoz et al. (2020) are summarised below:
- A BLR component is detected in 7 out of the 9 objects in the HST/STIS data, with 3 already reported in Constantin et al. (2015).
- Two out of 8 objects from the Palomar sample show a broad component that was not found by Ho et al. (1995).
- The detection of a broad component is favoured in the HST data as the spatial resolution is 10 times better than for the Palomar data.
- NGC 4594 is the only object with a broad detection in both ground- and space-based spectra. We propose its reclassification as a type-1 LINER.
- A secondary component consistent with outflows is detected in NGC 4594 and NGC 4486.

References

Binette, L., Magris, C. G., Stasińska, G., et al. 1994, *AAP*, 292, 13
Cappellari, M., & Emsellem, E. 2004, *PASP*, 116, 138
Cappellari, M. 2017, *MNRAS*, 466, 798
Cazzoli, S., Márquez, I., Masegosa, J., et al. 2018, *MNRAS*, 480, 1106
Constantin, A., Shields, J. C., Ho, L. C., et al. 2015, *ApJ*, 814, 149
Dopita, M. A., & Sutherland, R. S. 1995, *ApJ*, 455, 468
González-Martín, O., Masegosa, J., Márquez, I., et al. 2009, *A&A*, 506, 1107
Heckman, T. M. 1980, *A&A*, 500, 187
Hermosa-Muñoz, L., Cazzoli, S., Márquez, I., et al. 2020, *arXiv e-prints*, arXiv:2001.02955
Ho, L. C., Filippenko, A. V., & Sargent, W. L. W. 1993, *ApJ*, 417, 63
Ho, L. C., Filippenko, A. V., & Sargent, W. L. 1995, *ApJS*, 98, 477
Ho, L. C. 2008, *ARAA*, 46, 475
Márquez, I., Masegosa, J., et al. 2017, *Frontiers in Astronomy and Space Sciences*, 4, 34
Papaderos, P., Gomes, J. M., Vílchez, J. M., et al. 2013, *A&A*, 555, L1

Jet opening angle and linear scale of launch region of blazars

Xiang Liu[1,2], Pengfei Jiang[1,3] and Lang Cui[1,4]

[1] Xinjiang Astronomical Observatory, CAS, 150 Science-1 Street, Urumqi 830011, China.
Email: liux@xao.ac.cn
[2] Qiannan Normal University for Nationalities, 558000 Duyun, China
[3] Graduate University of Chinese Academy of Sciences, Beijing 100049, China
[4] Key Laboratory of Radio Astronomy, Chinese Academy of Sciences, Nanjing 210008, China

Abstract. We explore the intrinsic jet opening angle (IJOA) of blazars, from the literature, we found that the blazar number density peaks around 0.5° of IJOA and declines quickly with increasing IJOA for flat spectrum radio quasars (FSRQs), while the number density has double peaks around 0.3° and 2.0° of IJOA for BL Lacs. We assume that the black hole accretion-produced jet may have the smaller IJOA (for its larger linear scale of launch region), and the BH spin-produced jet may have the larger IJOA (for its smaller launch region), such that the FSRQs are accretion dominated for their single peaked small IJOA, while the BL Lacs are either accretion or BH spin dominated for their double peaked IJOA.

Keywords. Galaxies: jets, quasars: general, BL Lacertae objects: general

1. Introduction

There are two jet-production mechanisms of active galactic nuclei (AGNs), namely black hole (BH) accretion produced jet by Blandford & Payne (1982) and the BH spin produced jet by Blandford & Znajek (1977). These jet formation mechanisms are also useful in the study of the black hole X-ray binaries (BHXBs). The jet in the low-hard state of the BHXBs is often observed, while it is much weaker in the high-soft state of the BHXBs. It was argued whether the radio power positively correlates to the BH spin parameter. Steiner *et al.* (2013) found a positive correlation between the jet power and the BH spin in BHXBs, but Russell *et al.* (2013) argued that there is no correlation with their data (concerning different jet power estimates).

Liu *et al.* (2016) suggested that there may have been two states in both AGNs and BHXBs: a high accretion state which will lead to a linear correlation between the jet power and the accretion power, such as in quasars, their jets are predominated by the BH accretion; a lower accretion state which will lead to a flatter power-law index (<1) of the jet power to accretion power, such as in low luminosity AGNs and also in the BHXBs (with the power-law index of 0.6). The flatter power-law indices are expected in the BH spin-jet model by Liu *et al.* (2016), which can be observed when the accretion rate is not sufficiently large. For the high-soft X-ray state of the BHXBs, there is only weak or no observable jet in relatively high accretion rate, the reason might be that their accretion rate is still not high enough to produce an accretion-jet. In the low X-ray state, it is often jetted, this gentle jet could be produced by the BH spin, but it may be smeared out or quenched in the high state by the higher accretion matter.

For AGNs, it is also argued that the bright gamma-ray emitting blazars could be powered by the supermassive BH spin at their center (Ghisellini *et al.* 2014). They found

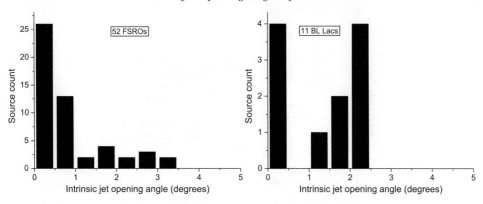

Figure 1. The intrinsic jet opening angles of 52 FSRQs (left) and 11 BL Lacs (right).

a nearly linear correlation between jet power and disk luminosity for their blazar sample. However, the linearity is also predicted in the accretion-jet model by Liu et al. (2016), as found in FRII quasars by van Velzen & Falcke (2013). Ghisellini et al. (2014) estimated the total jet power and the bolometric disk luminosity (10 times the BLR luminosity), found that the jet power is much larger than the disk accretion energy release (assuming 30% mass to energy efficiency), and suggested that the jet power is dominated by the BH spin in the blazars. As noted in Liu et al. (2016) linear correlation between jet power and disk luminosity is also a strong indicator of the accretion produced jet, and the disk luminosity in Ghisellini et al. (2014) might be underestimated (considering the uncertainty of the scaling relation of bolometric disk energy release to BLR luminosity).

2. The opening angle of jet in AGNs

Blazars are AGNs with relativistic jets oriented close to our line of sight. The very long baseline interferometry (VLBI) can resolve the inner jets or radio core at sub-milliarcsecond scales. As described in the BK model (Blandford & Konigl 1979, the core emission is the superposition of self-absorbed components which are moving in a conical channel. The core position can shift between different positions when viewed at different frequencies, so-called core-shift (e.g. Lobanov 1998; Pushkarev et al. 2012a).

The source intrinsic jet-opening angle (IJOA) can be estimated with $IJOA = JOA_{obs} \times sin\theta$, where the JOA_{obs} is observed (apparent) jet-opening angle and θ is jet viewing angle (Pushkarev et al. 2012a). The apparent opening angles have been determined by analyzing transverse jet profiles from the 2cm VLBA data in the image plane by using stacked images (Pushkarev et al. 2012b). The IJOAs were then estimated for 52 FSRQs and 11 BL Lacs (Finke 2019 and references therein). We have made a simple statistics for these opening angles in Finke (2019). The result is shown in Fig. 1, where we saw that the source count peaks around the small IJOA of 0.5° and declines quickly with increasing IJOA for FSRQs (left plot), while there are two peaks around 0.3° and 2.0° for BL Lacs (right plot). The IJOA of the BL Lacs is on average larger than that of the FSRQs. However, better statistics are needed to confirm this.

3. Possible connection between IJOA and jet formation mechanisms?

As mentioned, there are two jet-formation mechanisms, i.e. the BH accretion produced jet, and the BH spin produced jet. The former is fundamentally formed by the radiation pressure and the larger scale magnetic field from the accretion disk, in the scale of a few to tens of BH radii (Cao 2014). The latter is supposed to be formed from the BH ergosphere and its magnetosphere, in a scale of less than a few BH radii. Therefore, it is

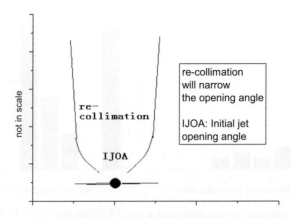

Figure 2. A cartoon of intrinsic jet opening angle and re-collimation in AGN.

expected that the linear scale of the jet launch region is much larger in the accretion-jet mechanism than that in the BH spin-jet mechanism, assuming a similar BH mass.

The jet, after formation, is also powered by the internal shocks in the shock-in-jet model (Marscher & Gear 1985; Marscher et al. 2002). The initial opening angle of jet base is usually parabolic (with larger opening angle), as shown in Fig. 2, and then re-collimated by the shocks afterwards to become a hyperbolic shape with a narrower opening angle (Hada et al. 2013). It is expected that the opening angles in the down-stream of jet for the blazars in Fig. 1, are smaller than the initial jet opening angles of launch region, but they could be positively correlated. With this idea, we expect that there is a connection between the IJOA and the jet formation mechanism, that the jet powers of FSRQs may be mainly formed by the accretion mechanism for their smaller IJOA (with the larger linear scale of jet launch region). In the meanwhile the jet powers of BL Lacs may be formed by either the accretion or the BH spin mechanism for their double peaks of the IJOA distribution. Further studies are required to test this assumption.

Acknowledgments

This work was supported by the National Key R&D Program of China under grant number 2018YFA0404602. We thank M. Pović for revision and Fu-Guo Xie for comment.

References

Blandford, R. D., & Znajek, R. L., 1977, *MNRAS*, 179, 433
Blandford, R. D., & Payne, D. G., 1982, *MNRAS*, 199, 883
Blandford, R. D., & Konigl, A., 1979, *ApJ*, 232, 34
Cao, X., 2014, *ApJ*, 783, 51
Finke, J. D., 2019, *ApJ*, 870, 28
Ghisellini, G., Tavecchio, F., Maraschi, L., Celotti, A., & Sbarrato, T., 2014, *Nature*, 515, 376
Hada, K., Kino, M., Doi, A., et al., 2013, *ApJ*, 775, 70
Liu, X., Han, Z. H., & Zhang, Z., 2016, *Astrophys. Space Sci*, 361, id9
Lobanov, A. P., 1998, *A&A*, 330, 79
Marscher, A. P., & Gear, W. K., 1985, *ApJ*, 298, 114
Marscher, A. P, Jorstad, S. G, Gómez, J.-L, et al., 2002, *Nature*, 417, 625
Pushkarev, A. B., Hovatta, T., Kovalev, Y. Y., et al., 2012a, *A&A*, 545, A113
Pushkarev, A. B., et al., 2012b, *ArXiv e-prints*, 1205.0659
Russell, D. M., Gallo, E., & Fender, R. P., 2013, *MNRAS*, 431, 405
Steiner, J. F., McClintock, J. E., & Narayan, R., 2013, *ApJ*, 762, 104
van Velzen, S., & Falcke, H., 2013, *A&A*, 557, L7

Detailed characterisation of LINERs and retired galaxies in the local universe

Daudi T. Mazengo[1], Mirjana Pović[2,3], Noorali T. Jiwaji[4] and Jefta M. Sunzu[1]

[1]The University of Dodoma (UDOM), College of Natural and Mathematical Sciences
P.O. Box 338, Dodoma, Tanzania
email: Mazengod@yahoo.co.uk

[2]Ethiopian Space Science and Technology Institute (ESSTI), Entoto Observatory and Research Center (EORC), Astronomy and Astrophysics Research and Development Division, P.O. Box 33679, Addis Ababa, Ethiopia

[3]Instituto de Astrofísica de Andalucía (IAA-CSIC), 18008 Granada, Spain

[4]The Open University of Tanzania (OUT), Faculty of Science, Technology and Environmental Studies, P.O. Box 23409, Dar es Salaam, Tanzania

Abstract. We present a detailed characterisation of physical properties of low-ionization nuclear emission-line regions (LINERs) and retired galaxies (RGs) in the local universe for redshift range $0 < z < 0.4$ and two subranges $z < 0.1$ and $0.1 < z < 0.4$. Furthermore, we test the effectiveness of WHAN diagnostic diagram in separating the two populations. We used photometric data, public spectroscopic data and morphological classification from SDSS-DR8, MPA-JHU SDSS-DR8 catalogue and Galaxy Zoo survey, respectively. We studied the distribution of LINERs, RGs and AGN-LINERs in relation to luminosity, stellar mass, star formation rate (SFR), colour, and their location on the SFR-stellar mass and colour-stellar mass diagrams. We then studied the morphologies of both populations. Results have shown that for higher redshift range, AGN-LINERs have higher apparent g magnitude, SFRs and dominate on/above the main sequence (MS) of star formation compared to RGs. However, both populations have similar stellar mass and luminosity distributions at all redshift ranges hence suggesting a significant difference in terms of star formation of RGs and AGN-LINERs with redshift. However, larger and more complete samples of LINERs are needed from the future surveys (e.g., LSST) and missions (e.g., JWST) to study in more details the properties of RGs and AGN-LINERs and find alternative methods of separating the two populations, since using simply WHAN diagram from our study we do not find it to be effective for separating the two populations.

Keywords. LINERs; retired galaxies; WHAN diagnostic method.

1. Introduction

Low-ionization nuclear emission-line regions (LINERs) dominate two third of all active galactic nuclei (AGN) and about one third of all galaxies in the local universe (Kewley et al. 2006). LINERs represent the most numerous local AGN population and could be the link between normal and active galaxies as suggested by their low luminosities (Márquez et al. 2017). Studying physical properties of LINERs helps to understand the standard model of AGN, AGN triggering mechanisms, and the role of AGN in galaxy formation and evolution. Optical classifications using BPT diagrams fail to separate LINERs and retired galaxies (RGs) since they include RGs that cover most of the region assigned as LINERs (Stasińska et al. 2008). Using WHAN diagnostic diagram, (Cid Fernandes et al. 2011) classified galaxies into SF, strong AGN, weak AGN, RGs, and passive galaxies.

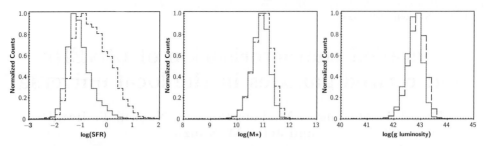

Figure 1. Distribution of SFR (*left*), stellar mass (*middle*), and g band luminosity (*right*), of RG (red solid lines) and AGN-LINER (dashed blue lines) galaxies.

However, their study did not deal with detailed characterisation of AGN-LINERs and RGs. Therefore, it is important to test the WHAN effectiveness in separating the two populations. LINERs have not been studied much at higher redshifts, and only few studies are known (Tommasin et al. 2012; Pović et al. 2016). Our study characterises in detail AGN-LINERs and RGs in the nearby universe by testing different properties, including morphology, and by testing the effectiveness of WHAN diagram in two redshift subranges.

2. Data

Photometric data were taken from SDSS-DR8†, public spectroscopic data from the MPA-JHU SDSS-DR8 catalogue (Brinchmann et al. 2004), while morphological classification was taken from Galaxy Zoo survey containing 667,944 galaxies (Lintott et al. 2008).

3. Sample selection

A total of 17,036 LINERs were separated from star forming, composite and Seyfert 2 galaxies using BPT diagrams based on [OIII]/Hβ, [NII]/Hα, [SII]/Hα, and [OI]/Hα line ratios (Kewley et al. 2006). We separated LINERs into 64% RGs and 36% AGN-LINERs using conditions EW(H$_\alpha$) $\leq$ 3 Å and EW(H$_\alpha$) $>$ 3 Å, respectively (Cid Fernandes et al. 2011). We cross-matched 17,036 LINERs with the Galaxy Zoo survey and obtained 16,032 LINERs where 44.7% were elliptical and spiral galaxies and 55.3% unclassified galaxies. We analysed the location of both RGs and AGN-LINERs at different redshifts in relation to the main-sequence (MS) of star formation (SF) by considering the results from Elbaz et al. (2007). We plotted rest-frame $u - r$ colour versus stellar mass diagram values based on Schawinski et al. (2014).

4. Analysis and results

We have found that for low redshift subrange, RGs are slightly brighter in g band than AGN-LINERs. AGN-LINERs have higher SFRs especially at higher redshift subrange compared to RGs (see Fig. 1 left plot) and hence dominate on/above the MS and in the blue cloud. Generally, RGs and AGN-LINERs have similar stellar masses and luminosities and hence difficult to separate them in terms of these properties as can be seen in (Fig. 1 middle plot) and (Fig. 1 right plot), respectively. Majority of all morphological types are early-types dominating below the MS and in the red sequence and green valley, while few sources especially AGN-LINERs suggest the presence of late-types.

† http://www.sdss3.org/dr10/spectro/galaxy_mpajhu.php

5. Conclusions

The connection between RGs and LINERs has been one of the long standing problems in astrophysics for understanding better the entire picture of galaxy formation and evolution. Our results could be important in understanding better RGs and AGN-LINERs based on physical properties and morphological classifications. Results also suggest finding an alternative method in separating the two populations.

Acknowledgments

The management of UDOM for the MSc financial support and the management of ESSTI for the supports during data analysis at the ESSTI are highly appreciated. MP acknowledges financial supports from the ESSTI under the Ethiopian Ministry of Innovation and Technology (MInT), from the Spanish Ministry of Science, Innovation and Universities (MICIU) through project AYA2016-76682C3-1-P and the State Agency for Research of the Spanish MCIU through the Center of Excellence Severo Ochoa award to the IAA-CSIC (SEV-2017-0709).

References

Brinchmann, J., *et al.*, 2004, *MNRAS*, 351, 1151
Cid Fernandes, R., *et al.*, 2011, *MNRAS*, 413, 1687
Elbaz, D., *et al.*, 2007, *A&A*, 468, 33
Kewley, L. J., *et al.*, 2006, *MNRAS*, 372, 961
Lintott, C. J., *et al.*, 2008, *MNRAS*, 389, 1179
Márquez, I., *et al.*, 2017, Frontiers in *A&S*, 4, 1570
Pović, M., *et al.*, 2016, *MNRAS*, 462, 2878
Schawinski, K., *et al.*, 2014, *MNRAS*, 440, 889
Stasińska G., *et al.*, 2008, *MNRAS*, 391, L29
Tommasin, S., *et al.*, 2012, *ApJ*, 753, 155

Multiwavelength study of potential blazar candidates among *Fermi*-LAT unidentified gamma-ray sources

Jean Damascène Mbarubucyeye[1], Felicia Krauß[2] and Pheneas Nkundabakura[3]

[1]Deutsches Elektronen-Synchrotron (DESY), Platanenallee 6, 15738 Zeuthen, Germany,
email: mbjdamas@gmail.com

[2]Department of Astronomy & Astrophysics, Pennsylvania State University, University Park, PA 16801, USA
email: Felicia.Krauss@psu.edu

[3]University of Rwanda, College of Education, P.O. Box 5039, Kigali, Rwanda
email: nkundapheneas@yahoo.fr

Abstract. Studying unidentified γ-ray sources is important as they may hide new discoveries. We conducted a multiwavelength analysis of 13 unidentified Fermi-LAT sources in the 3FGL catalogue that have no known counterparts (Unidentified Gamma-ray Sources, UnIDs). The sample was selected for sources that have a single radio and X-ray candidate counterpart in their uncertainty ellipses. The purpose of this study is to find a possible blazar signature and to model the Spectral Energy Distribution (SED) of the selected sources using an empirical log parabolic model. The results show that the synchrotron emission of all sources peaks in the infrared (IR) band and that the high-energy emission peaks in MeV to GeV bands. The SEDs of sources in our sample are all blazar like. In addition, the peak position of the sample reveals that 6 sources (46.2%) are Low Synchrotron Peaked (LSP) blazars, 4 (30.8%) of them are High Synchrotron Peaked (HSP) blazars, while 3 of them (23.0%) are Intermediate Synchrotron Peaked (ISP) blazars.

Keywords. radiation mechanism: non-thermal, gamma-rays: galaxies, galaxies: active galaxies, BL Lacertae objects: general, X-ray: general, methods: data analysis.

1. Introduction

The Large Area Telescope (LAT) on board the *Fermi* Gamma-ray Space Telescope detects high-energy photons between 20 MeV and 300 GeV (Atwood *et al.* 2009). In the 3FGL catalog among 3033 sources detected, 33% are still unassociated to any counterparts (Acero *et al.* 2015) while in the latest catalog (4FGL) more than 38% remain unassociated (The Fermi-LAT Collaboration 2019). Among 2023 sources that have been already identified in 3FGL, 1100 are Active Galactic Nuclei (AGN) of blazar type (57%). Blazars are a subclass of AGN with their relativistic jets pointing at Earth. The two main classes of blazars according to their optical properties are: (1) Flat Spectrum Radio Quasars (FSRQs) characterised by the strong emission lines and (2) BL Lacs characterised by weak or no emission lines at all. In addition, blazars are classified according to the position of their peak frequency (Massaro *et al.* 2004): Low Synchrotron-Peaked (LSP) objects with $\nu_p^{\rm synch} \leqslant 10^{14}$ Hz, Intermediate Synchrotron-Peaked (ISP) objects with 10^{14} Hz $\leqslant \nu_p^{\rm synch} \leqslant 10^{15}$ Hz, and High Synchrotron-Peaked (HSP) objects with $\nu_p^{\rm synch} \geqslant 10^{15}$ Hz. In this research, we determined the type of emissions from a sample of 13 blazar candidates and classified them according to the shape of their SEDs.

2. Sample selection

The 3FGL catalogue includes 1010 unidentified sources (Acero *et al.* 2015). The release of the 3FGL catalogue enabled the *Swift*-XRT survey of the *Fermi*-LAT unassociated sources with the purpose of performing their follow-up observations in an attempt to find their potential X-ray counterparts. Two common features that all blazars share are that they are radio emitters (radio-loud objects) and that they are highly variable. Yang & Fan (2005) showed that there is a strong correlation between radio and γ-ray emission in blazars when they are in a their flaring high state. However, the most recent studies are continuously debating about the correlation between radio to γ-ray bands (Massaro *et al.* 2017; Bruni *et al.* 2018). The selected sample takes advantage of the fact that most of the studied blazars exhibit strong radio-X-ray emission (Padovani *et al.* 2007). In addition, most of the known γ-ray bright AGN are above a Galactic latitude of $|b| > 10$ degrees, due to possible source confusion as well as higher diffuse background in the Galactic plane (Ackermann *et al.* 2015). A sample of unassociated sources was selected based on the following criteria:

(*a*) Only a single XRT detection lies in the uncertainty ellipse of the 3σ of the LAT unassociated source.

(*b*) Only a single radio source lies in the uncertainty box of *Fermi*-LAT uncertainty region which is coincident with the XRT detection.

(*c*) Only sources with $|b| > 10$ degrees in order to limit for any source confusion in the Galactic plane.

Applying all cuts to the population of unassociated sources listed in the 3FGL and observed by *Swift*, we isolated a sample of 13 unassociated sources assumed to be potential blazar candidates.

3. Data analysis

The *Swift*-XRT/UVOT data available were considered. We used standard extraction methods to produce *Swift*/XRT and *Swift*/UVOT spectra.

The *Fermi*-LAT spectral data point were extracted using Fermi Science tools (v10r0p5 released on June 24, 2015). The analysis takes into account the source region and the region of interest (ROI). We selected events within the energy range 100 MeV – 300 GeV, a maximum zenith angle of 90 degrees, a ROI of 10 degrees.

A complete multiwavelength SED was obtained by supplementing the XRT/UVOT and *Fermi*-LAT data with other data from across the whole electromagnetic spectrum. These data were obtained using SEDbuilder tool†.

In order to identify the type of emission from the source, we apply an empirical model on the data. The SED shows that the log-parabolic model fits the data well for blazars, in different energy bands (Massaro *et al.* 2004; Tramacere *et al.* 2009). The model used includes two log parabolas, one fitting the low-energy spectrum (synchrotron emission) and another one fitting the high-energy spectrum. In addition to these two log-parabolas, three other components are added: absorption, extinction, and a blackbody.

4. Results and discussions

The SEDs resulted from the fitting and reported herein show the two typical bump known as the main signature of blazars. The low-energy bump (peaking between IR and optical bands) is interpreted as synchrotron emission from highly accelerated relativistic electrons and the high-energy bump (peaking at X-ray and γ-ray energy bands) is related to high-energy emission, which could be either leptonic (Inverse Compton) or hadronic in nature. Based on the position of the $\nu_{\text{peak}}^{\text{sync}}$ (Abdo *et al.* 2010), we found that 6 ($\sim 46.2\%$)

† http://www.asdc.asi.it/

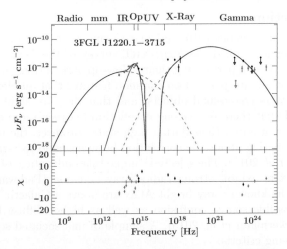

Figure 1. A sample of modeled SED of our sample with a log parabolic model. The model includes two absorbed logarithmic parabolas shown in blue, a blackbody in red dashed, and the total unabsorbed model shown in black. The extracted data are black colored while 3FGL data are in cyan and non-simultaneous archival data are in magenta.

sources of our sample are LSP objects, 3 (23.0%) sources are ISP objects and 4 (30.8%) of them are HSP objects. Figure 1 shows an example of the SED for 3FGL J1220.1−3715 where data are fitted by the model. The lower panel represents residuals. For most of the cases, our classification is in agreement with Parkinson et al. (2016); Lefaucheur & Pita (2017) and Salvetti et al. (2017), who classified the sources of our sample as AGN and blazars. All three works use Machine Learning techniques. We finally used the WISE IR colour-colour plot taking into account sources that have WISE matching sources. Only 9 sources have matching counterparts in WISE catalogue. As a results we found that only 7 sources lies in the WISE blazar region which emphasize their candidacy to be blazars.

References

Abdo, A. A., Ackermann, M., Agudo, I., et al. 2010, ApJ, 716, 30
Acero, F., Ackermann M., Ajello, M., et al. 2015, ApJS, 218, 23
Ackermann, M., Ajello, M., Atwood, W. B., et al. 2015 ApJ, 810, 14
Atwood W. B., et al. 2009, ApJ, 697, 1071
Bruni, G., et al. 2018, ApJ, 854, L23
Houck J. C., Denicola L. A. 2000, in Manset N., Veillet C., Crabtree D., 653 eds, Astronomical Society of the Pacific Conference Series Vol. 216, 654 Astronomical Data Analysis Software and Systems, IX. p. 591
Lefaucheur, J., and Pita, S. 2017, A&A, 602, A86
Massaro, E., Perri, M., Giommi, P. and Nesci, R. 2004, A&A, 413, 489
Massaro F., Marchesini E. J., D'Abrusco, R., et al. 2017, ApJ, 834, 113
Padovani, P., Giommi, P., Landt, H., et al. 2007, ApJ, 662, 182
Saz Parkinson, P. M., Xu, H., Yu, P. L. H., et al. 2016, ApJ, 820, 8
Salvetti, D., Chiaro, G., La Mura, G., Thompson, D. J. 2017, MNRAS, 470, 1291
The Fermi-LAT collaboration 2019, arXiv e-prints, p. arXiv:1902.10045
Tramacere, A., Giommi, P., Perri, M., et al. 2009, A&A, 706 501, 879
Yang J.-H., and Fan J.-H. 2005, Chinese J. Astron. Astrophys., 5, 229

A search for new γ-ray blazars from infrared selected candidates

Blessing Musiimenta[1], Bruno Sversut Arsioli[2], Edward Jurua[1] and Tom Mutabazi[1]

[1]Mbarara University of Science and Technology (MUST), Mbarara, Uganda
email: mblessing78@gmail.com

[2]Instituto de Fisica Gleb Wataghin IFGW, Unicamp, Brasil

Abstract. We present a systematic study of gamma-ray blazar candidates based on a sample of 40 objects taken from the WIBR catalogue. By using a likelihood analysis, 26 of the 40 sources showed significant gamma-ray signatures $\geqslant 3\sigma$. Using high-energy test statistics (TS) maps, we confirm 8 sources, which are completely new, and show another 15 promising γ-ray candidates. The results from this analysis show that a multi-frequency approach can help to improve the current description of the gamma-ray sky.

Keywords. Galaxies: active; gamma-rays: blazars.

1. Introduction

Blazars are Active Galactic Nuclei (AGN) with relativistic jets pointed towards Earth (Urry et al. 1995). While blazars are rare extragalactic sources, they represent the largest fraction of gamma-ray sources (Abdo et al. 2010) and contribute to the extragalactic gamma-ray background (EGB) (Mukherjee et al. 1997; Abdo et al. 2010; Ajello et al. 2015). A complete description of the EGB is still an open issue in gamma-ray astronomy. Therefore, searching for new gamma-ray sources from low energy blazar candidates can improve the description of the gamma-ray sky (Arsioli et al. 2017).

2. Methods

The WIBR catalogue (D'Abrusco et al. 2014) contains 7855 sources. Many of them happen to be part of the 2WHSP catalogue (Chang et al. 2017) and 5BZcat (Massaro et al. 2015). We selected 40 sources from the WIBR catalogue which have no counterparts in the Fermi catalogue, therefore with focus on undetected gamma-ray sources.

(i) The Spectral Energy Distributions (SEDs) were analysed using the ASDC SEDbuilder tool.

(ii) The gamma-ray analysis was performed using Fermi science tools. The significance ($\sqrt{TS}\,\sigma$) of the detections was determined using the Test Statistic (TS) parameter defined as:

$$TS = -2(\ln L_{no\text{-}source} - \ln L_{source}), \quad (2.1)$$

where $L_{no\text{-}source}$ is the null hypothesis, and L_{source} is the likelihood value for a model with the additional candidate source at the same position (Mattox et al. 1996). The gamma-ray spectrum of the sources was assumed to be described by a power law model given by:

$$\frac{dN}{dE} = N_o \left(\frac{E}{E_o}\right)^{-\Gamma}, \qquad (2.2)$$

where E_o is pivot energy, N_o is the prefactor (corresponding to the flux density in ph/cm^2/s/MeV) at E_o and Γ is the photon spectral index for a given energy range.
(iii) TS maps were obtained by performing an unbinned likelihood analysis using Fermi science tools.

3. Results and discussion

3.1. Significant detections

In 3FGL and 4FGL catalogues, sources were detected at 4σ significance (Acero *et al.* 2015; Abdollahi *et al.* 2020, respectively). However, in this analysis, sources with $\sqrt{TS} \sim 3\sigma$ are considered to be γ–ray detections with low significance following the discussion from Arsioli *et al.* (2017, 2018). All cases with TS between 9 and 25 are considered to be a relevant excess signature, given that those cases actually have a multi-frequency counterpart (radio to X-rays) as expected from blazars.

Out of the 26 sources with a significant gamma-ray signature, 10 were detected with TS > 25 and 16 were detected with 9 < TS < 25. All these sources are entirely new detections out of previous Fermi catalogues.

3.2. New WIBR gamma-ray detections

Different TS maps were built with different photon energy cuts i.e., 950 MeV, 1 GeV, 2 GeV, 2.5 GeV, and 3 GeV. The TS maps were built taking into consideration the computational time and also to obtain the TS peak with the best resolution to better solve the gamma-ray signature as a point-like source. All these sources are at high Galactic latitudes ($|b| > 10°$), avoiding the Galactic diffuse emission and also preventing spurious detections.

Out of the 26 new WIBR γ-ray detections, we built TS maps for 23 sources (both high and low significance detections). Three sources were not fully analysed due to very high photon counts that increased the computational heaviness. This is because those cases seemed more relevant at the lowest energy channel from Fermi-LAT (E < 900 MeV) and building TS maps in this energy range becomes computationally far too heavy, especially if necessary to integrate over many years of observations. Out of the 23 TS maps built, 8 show a point-like source at the positions of the target sources and 15 sources could not be confirmed with high energy TS maps, but might be relevant at lower energies and therefore are flagged as promising γ-ray candidates. Figure 1 (left panel) shows a TS map of one of the new gamma-ray sources.

3.3. Promising gamma-ray candidates

A total of 15 sources were found to be promising gamma-ray candidates. In some cases, the TS peak corresponded to closeby 4FGL sources or there was an FGL source within the region. In other cases, the signatures were identified as spurious. One of the promising gamma-ray candidates is shown in the right panel of Figure 1.

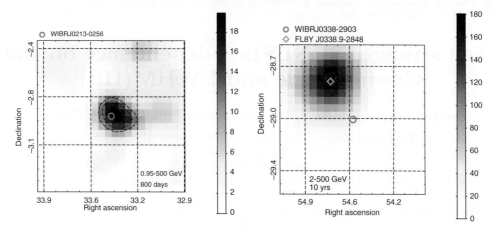

Figure 1. We show the TS map corresponding to a new detection of WIBRJ0213-0256 (left plot) and WIBRJ0338-2903 (right plot). The WIBR seed positions are shown by the magenta circles. In the right plot, the green diamond indicates the nearby Fermi source. The blue dashed contour lines correspond to 68%, 95%, and 99% containment region for the gamma-ray signature position (from inner to outer lines) in both TS maps.

References

Abdo A. A., et al., 2010, *ApJS*, 188, 405
Abdo A. A., et al., 2010, *ApJ*, 720, 435
Abdollahi S., et al., 2020, *ApJS*, 247, 33
Acero F., et al., 2015, *ApJS*, 218, 23
Ajello M., et al., 2015, *ApJ*, 800, 27
Arsioli B. & Chang Y.-L., 2017, *A&A*, 598, 134
Arsioli B. & Polenta G., 2018, *A&A*, 616, 20
Chang Y.-L., Arsioli B., Giommi P. & Padovani P., 2017, *A&A*, 598, 17
D'Abrusco R., et al., 2014, *ApJS*, 215, 14
Massaro E., et al., 2015, *Ap&SS*, 357, 75
Mattox J. R., et al., 1996, *ApJ*, 461, 396
Mukherjee R., et al., 1997, *ApJ*, 490, 116
Urry C. M., & Padovani P., 1995, *PASP*, 107, 803

FeII strength in NLS1s – dependence on the viewing angle and FWHM(Hβ)

Swayamtrupta Panda[1,2], Paola Marziani[3] and Bożena Czerny[1]

[1]Center For Theoretical Physics, Polish Academy of Sciences, Al. Lotników 32/46, 02-668 Warsaw, Poland
email: panda@cft.edu.pl

[2]Nicolaus Copernicus Astronomical Center, Polish Academy of Sciences, ul. Bartycka 18, 00-716 Warsaw, Poland

[3]INAF-Astronomical Observatory of Padova, Vicolo dell'Osservatorio, 5, 35122 Padova PD, Italy

Abstract. We address the effect of orientation of the accretion disk plane and the geometry of the broad line region (BLR) in the context of understanding the distribution of quasars along their Main Sequence. We utilize the photoionization code CLOUDY to model the BLR, incorporating the 'un-constant' virial factor. We show the preliminary results of the analysis to highlight the co-dependence of the Eigenvector 1 parameter, $R_{\rm FeII}$ on the broad Hβ FWHM (i.e. the line dispersion) and the inclination angle (θ), assuming fixed values for the Eddington ratio ($L_{\rm bol}/L_{\rm Edd}$), black hole mass ($M_{\rm BH}$), spectral energy distribution (SED) shape, cloud density ($n_{\rm H}$) and composition.†

Keywords. accretion; accretion disks; radiation mechanisms: thermal; radiative transfer; galaxies: active; (galaxies:) quasars: emission lines; galaxies: Seyfert

1. Introduction

Narrow-Line Seyfert 1 (NLS1) galaxies constitute a typical class of Type-1 active galaxies which have "narrow" broad profiles (e.g. FWHM(Hβ) < 2000 km s^{-1}) and contain supermassive black holes (BH) that have masses lower than the typical broad-line Seyfert galaxies (Marziani et al. 2014; Shen & Ho 2014). Having a lower-than typical mass is a result of how the black hole mass ($M_{\rm BH}$) is estimated. One of the methods to estimate the $M_{\rm BH}$ is the dynamical method that involves the use of the virial relation (Chen et al. 2019). According to this, the $M_{\rm BH}$ is a function of (i) the size of the emitting region, here, the broad-line region (BLR); and (ii) the FWHM(Hβ) of the virialized gas. The size of the BLR (i.e. $R_{\rm BLR}$) is measured by estimating the light-travel time from the central ionizing source to the emitting medium. This method is known as reverberation mapping (Peterson 1993). The other quantity, the line FWHM, can be measured reliably from high quality spectroscopy.

Since, the emitting line regions of quasars are extended, the energy that an observer receives from these luminous objects is also dependent on the geometry of the source with respect to the observer. By geometry, we mean the structure and how this structure is oriented to the observer's line of sight. We address this aspect of the geometry of the quasars using photoionisation modelling with CLOUDY v17.01 (Ferland et al. 2017) in

† The project was partially supported by NCN grant no. 2017/26/A/ST9/00756 (MAESTRO 9) and MNiSW grant DIR/WK/2018/12. PM acknowledges the INAF PRIN-SKA 2017 program 1.05.01.88.04.

© The Author(s), 2021. Published by Cambridge University Press on behalf of International Astronomical Union

the context of understanding better the main sequence of quasars (see Panda et al. 2019, 2020 for more information).

The eigenvector 1 of the original principal component analysis (PCA) paved way for the quasar main sequence picture as we know it today (Sulentic et al. 2000; Shen & Ho 2014). The main sequence connects the velocity profile of 'broad' Hβ with the strength of the FeII emission (R$_{\rm FeII}$), i.e., the intensity of the FeII blend within 4434-4684 Å normalized with the 'broad' Hβ intensity.

2. Method

We assume a single cloud model where the density ($n_{\rm H}$) of the ionized gas cloud is varied from 10^9 cm^{-3} to 10^{13} cm^{-3} with a step-size of 0.25 (in log-scale). We utilize the *GASS10* model (Grevesse et al. 2010) to recover the solar-like abundances and vary the metallicity within the gas cloud, going from a sub-solar type (0.1 Z$_\odot$) to super-solar (100 Z$_\odot$) with a step-size of 0.25 (in log-scale). The total luminosity of the ionizing continuum is derived assuming a value of the Eddington ratio ($L_{\rm bol}/L_{\rm Edd} = 0.25$) and the respective value for the black hole mass ($M_{\rm BH} = 10^8$ M$_\odot$). The Eddington ratio is appropriate for sources of Population A (Sulentic et al. 2000; Marziani et al. 2014). The $M_{\rm BH}$ is representative of optically-selected low-z quasar samples. The shape of the ionizing continuum used here is taken from Korista et al. (1997). The size of the BLR is estimated from the virial relation, assuming a black hole mass, a distribution in the viewing angle [0–90 degrees] and FWHM of Hβ which is given as:

$$R_{BLR} = \frac{GM_{BH}}{f * FWHM(H\beta)^2}, \quad (2.1)$$

where G is the gravitational constant. The f factor, which contains the information about the geometry of the source, can be expressed as:

$$f = \frac{1}{4\left[\kappa^2 + sin^2\theta\right]}, \quad (2.2)$$

where, θ is the angle of inclination with respect to the observer and κ is the ratio between $v_{\rm iso}$ and $v_{\rm K}$, which decides how isotropic the gas distribution is around the central potential. If the value is close to zero, it represents a flat disk with thickness almost zero. On the other hand, if the value of κ is close to unity, it represents an almost spherical distribution of the gas. Here, we assume $\kappa = 0.1$ that is consistent with a flat, Keplerian-like gas distribution.

Substituting the values for the $M_{\rm BH}$ and κ, we have

$$R_{BLR} \approx 5.31 \times 10^{24} \left[\frac{0.01 + sin^2\theta}{FWHM(H\beta)^2}\right] \quad \text{(in cm)} \quad (2.3)$$

3. Results and Conclusions

Figure 1 shows dependence of the R$_{\rm FeII}$ both on FWHM(Hβ) and θ. R$_{\rm FeII}$ values are anti-correlated with the FWHM(Hβ). This is directly coming from the Eq. 2.3, where the $R_{\rm BLR}$ is anti-correlated with the square of the FWHM(Hβ), but also correlates with θ which shows that if $R_{\rm BLR}$ is too low, the ionizing radiation is too high for the FeII species to exist and the emitting zones of the FeII only begins to materialize deeper in the cloud (see Panda et al. 2018). This effect reduces the R$_{\rm FeII}$ for lower $R_{\rm BLR}$ values. θ is $< 60°$ for Type-1 quasars. Within this limit, we find an increasing trend in R$_{\rm FeII}$ with respect to θ. This trend is more substantial for lower values of the FWHM(Hβ) that are consistent with NLS1 population, i.e. $\lesssim$ 3000 km s^{-1}. These results are described and analyzed in Panda et al. (2019b).

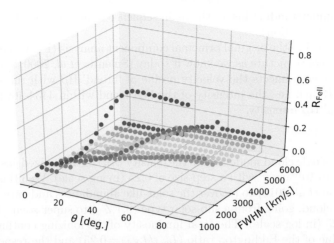

Figure 1. 3D scatter plot showing the dependence of the parameter $R_{\rm FeII}$ on the Hβ FWHM and the inclination angle (θ). The Hβ FWHM ranges from 1000 km s^{-1} to 6000 km s^{-1} with a step size of 500 km s^{-1}. Similarly, the θ values range from 0°-90° with a step size of 3°. The black hole mass is assumed to be $10^8 M_\odot$ and the Eddington ratio, $L_{\rm bol}/L_{\rm Edd} = 0.25$. The value of the $\kappa = 0.1$ consistent with a flat, Keplerian-like gas distribution around the central supermassive black hole.

References

Boroson, T.A. & Green, R.F., 1992, *ApJS*, 80, 109
Chen, Z.-F., Yi, S.-X., Pang, T.-T., *et al.*, 2019, *ApJS*, 244, 36
Collin, S., Kawaguchi, T., Peterson, B. M., & Vestergaard, M., 2006, *A&A*, 456, 75
Ferland, G. J., Chatzikos, M., Guzmán, F., *et al.*, 2017, *RMxAA*, 53, 385
Grevesse, N., Asplund, M., Sauval, A. J., & Scott, P., 2010, *ApSS*, 328, 179
Korista, K., Baldwin, J., Ferland, G., & Verner, D., 1997, *ApJS*, 108, 401
Marziani, P., Sulentic, J. W., Negrete, C. A., *et al.*, 2014, *AstRv*, 9, 6
Panda, S., Czerny, B., Adhikari, *et al.*, 2018, *ApJ*, 866, 115
Panda, S., Marziani, P., & Czerny, B., 2019a, *ApJ*, 882, 2
Panda, S., Marziani, P., & Czerny, B., 2019b, *Proceedings of the International Astronomical Union (IAU)*, 356, 2
Panda, S., Marziani, P., & Czerny, B., 2020, *Contributions of the Astronomical Observatory Skalnaté Pleso*, 50, 293
Peterson, B. M., 1993, *PASP*, 105, 247
Shen, Y., & Ho, L.C., 2014, *Nature*, 513, 210
Sulentic, J. W., Zwitter, T., Marziani, P. & Dultzin-Hacyan, D., 2000, *ApJL*, 536, L5

Discovering exotic AGN behind the Magellanic Clouds

Clara M. Pennock[1], Jacco Th. van Loon[1], Cameron P. M. Bell[2], Miroslav D. Filipović[3], Tana D. Joseph[4] and Eleni Vardoulaki[5]

[1]Lennard-Jones Laboratories, Keele University, Keele, ST5 5BG, UK
email: c.m.pennock@keele.ac.uk
[2]Leibniz Institute for Astrophysics Potsdam, Potsdam, Germany
[3]Western Sydney University, Sydney, Australia
[4]University of Manchester, Manchester, UK
[5]Max-Planck-Institut für Radioastronomie, Bonn, Germany

Abstract. The nearby Magellanic Clouds system covers more than 200 square degrees on the sky. Much of it has been mapped across the electromagnetic spectrum at high angular resolution and sensitivity X-ray (XMM-Newton), UV (UVIT), optical (SMASH), IR (VISTA, WISE, Spitzer, Herschel), radio (ATCA, ASKAP, MeerKAT). This provides us with an excellent dataset to explore the galaxy populations behind the stellar-rich Magellanic Clouds. We seek to identify and characterise AGN via machine learning algorithms on this exquisite data set. Our project focuses not on establishing sequences and distributions of common types of galaxies and active galactic nuclei (AGN), but seeks to identify extreme examples, building on the recent accidental discoveries of unique AGN behind the Magellanic Clouds.

Keywords. AGN; Machine-learning; Magellanic Clouds; Multi-wavelength

1. Newest Data

The Vista survey of the Magellanic Clouds (VMC; Cioni et al. 2011) is a recently completed deep-field multi-epoch, near-IR survey taken with the 4m VISual and Infrared Telescope for Astronomy (VISTA) at the European Southern Observatory at Cerro Paranal, Chile. It observed in the Y, J and K_s bands, going to a depth of $K_s = 22.2$ mag (AB). It covers the Large Magellanic Cloud (LMC) area (116 $\deg^2$), the Small Magellanic Cloud (SMC) area (45 $\deg^2$), the Bridge (20 $\deg^2$) and two tiles in the Stream (3 $\deg^2$).

The Evolutionary Map of the Universe (EMU; Norris et al. 2011) is a wide-field radio continuum survey which uses the Australian Square Kilometre Array Pathfinder (ASKAP). It has a frequency range from 700 MHz to 1.8 GHz and is capable of a 30 $\deg^2$ field-of-view. EMU's primary goal is to make a deep (rms $\sim$ 10 μJy/beam) radio continuum survey of the Southern sky, with 10 arcsec resolution. The SMC survey (Joseph et al. 2019) was taken as part of ASKAP early science verification. The LMC was also observed and is currently in the data processing stage.

The combination of these two wavelength ranges will provide a powerful tool for searching for AGN, in particular towards higher redshifts and/or obscured AGN.

2. The Exotic

SAGE1C J053634.78−722658.5 (SAGE0536AGN) is an abnormal AGN discovered behind the LMC in the Spitzer Space Telescope "Surveying the Agents of Galaxy

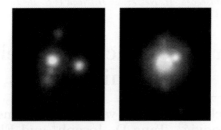

Figure 1. (Left) Original X-shaped radio AGN candidate and (right) near-IR look-a-like. Colours represent Y (blue), J (green), K_s (red) from the VMC.

Evolution" Spectroscopy survey (Hony et al. 2011) and has since been better characterised with the Southern African Large Telescope (SALT) (van Loon & Sansom 2015). Sitting at redshift $z = 0.1428 \pm 0.0001$, the striking feature of this AGN is its uniquely strong silicate emission at 10 μm and its lack of star-formation activity. From our use of Galfit (Peng et al. 2002) it has been found that this is a disc galaxy instead of an early-type galaxy as previously thought.

Another interesting source is a striking X-shaped radio AGN candidate (Joseph et al. 2019). Predicted to be a binary AGN, seen just before the AGN merge and consisting of two jets emitting in parallel and opposite directions from the central source. The third and middle jet is relativistically boosted towards us. We also found a near-IR look-a-like using WISE colour-colour diagrams cross-matched with EMU radio sources and VMC images. These sources need to be disentangled and we are using the combination of the VMC and optical spectroscopy from SALT to ascertain the true natures of each of the radio counterparts. See Figure 1 for these two radio sources.

3. Random Forest Basics

Supervised machine learning algorithms predict classifications/values based off example data with known classifications/values. It does this by analysing a known sample, the training set, and producing a model, which can then be used to make predictions of the output of an unseen dataset.

Decision trees (Breiman 2001) predict the value/class of a target variable by creating a model that has learnt simple decision rules inferred directly from the data features it is trained upon. It consists of nodes where a condition is given that is either true or false. The answer to this condition leads down a branch to the next node and condition, where either another split happens or the output variable is given.

An ensemble of decision trees is called the random forest algorithm (Breiman 2001), which can be used for both classification and regression problems. The algorithm builds several decision trees independently and then averages the predictions of these to obtain the final prediction. This reduces variance over using a single estimator and creates an overall more stable model. Randomness is injected into the training process of each individual tree via a method called 'bagging'. This method splits up the training set into randomly selected subsets, and each decision tree is then trained on one of those subsets.

Advantages of the random forest include: they require very little data preparation; it constructs a non-linear model during training, which is advantageous as most problems require non-linear solutions. It can handle numerous features and objects. It can produce probabilities of each class for each prediction (fuzzy logic) and produce feature importance, showcasing which features have more bearing on the classifications it had made. Lastly, it generalises well to unseen datasets due to the inclusion of randomness.

4. The Training and Results

The features used in the random forest are magnitudes/fluxes and colours from survey catalogues: EMU (Norris et al. 2011), unWISE (Schlafly et al. 2019), SAGE (LMC — Meixner et al. 2006, SMC/Bridge — Gordon et al. 2011), VMC (Cioni et al. 2011), Gaia DR2 (Gaia Collaboration et al. 2018), GALEX (Bianchi et al. 2017), XMM-Newton (Sturm et al. 2013) and including features calculated from VMC data such as source density and sharpness. It is trained on 75% of the training data and tested on remaining 25%.

The output classifications are: AGN (spectroscopically confirmed), Galaxies (no AGN, spectroscopically confirmed), Stars - Proper Motion/Emission/RGB/Carbon/Red Supergiant (SG), Young Stellar Object (YSO).

The resulting trained random forest produced AGN accuracies ranging from 80–90%. While AGN can be misclassified as other sources, the predicted AGN tend to be correct, with a false positive rate of 1–2%. The current limitations are caused by limited coverage of X-ray and UV catalogues that avoid the central parts of the Magellanic Clouds. Furthermore, only 26.7% and 7.5% of the known AGN sources are X-ray and radio sources, respectively. As these emission types are often indications of an AGN, their lack of representation in the training set lessens the likelihood the random forest will see these emissions as clear indications of AGN.

5. Future Work

Future investigations will look into the difference between the correctly classified and the misclassified AGN. Optical spectroscopic observations using SAAO's (South African Astronomical Observatory) SALT and 1.9 m telescopes to increase the number of AGN in the training set from behind the Magellanic Clouds are currently underway. This is, however, biased towards the more unobscured sources. We have proposed to use the ESO NTT to observe spectroscopically in the infrared the higher redshift sources that, while many of the radio sources are detected in the VMC, their optical counterparts are all invisible except for nearby, unobscured and/or extreme cases. We will also incorporate optical SMASH data that has just become publicly available into the machine learning as an improvement over Gaia data and augment the UV GALEX features with UVIT data. In addition, we will include variability data from VMC into machine learning as well as features based on the environment around the sources and account for reddening (by dust) and redshift. Lastly, we will employ the use of a probabilistic random forest, which allows the addition of uncertainty in the features and labels of the sources and missing data. It has been shown to provide greater accuracy in predictions (Reis et al. 2018). Current tests of using it, however, have yielded lower accuracy's in the prediction of AGN and this needs to be improved upon.

References

Bianchi, L., Shiao, B., & Thilker, D., 2017, *APJS*, 230, 24
Breiman, L., 2001, *Machine Learning*, 45, 1
Cioni, M.-R. L., Clementini, G., Girardi, L., et al., 2011, *AAP*, 527, A116
Gaia Collaboration, et al. 2018, *AAP*, 616, A1
Gordon, K. D., Meixner, M., Meade, M. R., et al., 2011, *AJ*, 142, 102
Hony, S., Kemper, F., Woods, P. M., et al., 2011, *AAP*, 531, A137
Joseph, T. D., Filipović, M. D., et al., 2019, *MNRAS*, 490, 1202
van Loon, J. T., & Sansom, A. E., 2015, *MNRAS*, 453, 2341
Meixner, M., Gordon, K. D., Indebetouw, R., et al., 2006, *AJ*, 132, 2268

Norris, R. P., Hopkins, A. M., Afonso, J., et al., 2011, *PASA*, 28, 215
Peng, C.Y., Ho, L.C., Impey, C.D., & Rix, H.-W., 2002, *AJ*, 124, 266
Reis, I., & Baron, D. & Shahaf, S., 2018, *AJ*, 157, 16
Schlafly, E. F., Meisner, A. M., & Green, G. M., 2019, *APJS*, 240, 30
Sturm, R., Haberl, F., Pietsch, W., et al., 2013, *AAP*, 558, A3

Environmental effects on star formation main sequence in the COSMOS field

Solohery M. Randriamampandry[1,2], Mattia Vaccari[3,4] and Kelley M. Hess[5,6]

[1]South African Astronomical Observatory, P.O. Box 9, Observatory 7935, Cape Town, South Africa, email: solohery@saao.ac.za

[2]A&A, Department of Physics, Faculty of Sciences, University of Antananarivo, B.P. 906, Antananarivo 101, Madagascar

[3]Department of Physics and Astronomy, University of the Western Cape, Robert Sobukwe Road, Bellville 7535, South Africa

[4]INAF - Istituto di Radioastronomia, via Gobetti 101, I-40129 Bologna, Italy

[5]ASTRON, the Netherlands Institute for Radio Astronomy, PO Box 2, 7990 AA Dwingeloo, The Netherlands

[6]Kapteyn Astronomical Institute, University of Groningen, PO Box 800, 9700 AV Groningen, The Netherlands

Abstract. We investigate the relationship between environment and star formation main sequence (the relationship between stellar mass and star formation rate) to shed new light on the effects of the environments on star-forming galaxies. We use the large VLA-COSMOS 3 GHz catalogue that consist of star-forming galaxies (SFGs) and active galactic nuclei (AGN) in three different environments (field, filament, cluster) and for different galaxy types. We examine for the first time a comparative analysis for the distribution of SFGs with respect to the star formation main sequence (MS) consensus region from the literature, taking into account galaxy environment and using radio selected sample at $0.1 \leqslant z \leqslant 1.2$ drawn from one of the deepest COSMOS radio surveys. We find that, as observed previously, SFRs increase with redshift independent on the environments. Furthermore, we observe that SFRs versus M_* relation is flat in all cases, irrespective of the redshift and environments.

Keywords. Galaxies: evolution, galaxies: star-forming, galaxies: environment, galaxies: stellar mass, galaxies: star formation rate

1. Overview

We use a multi-wavelength counterpart catalogue based on the VLA-COSMOS 3 GHz Large Project compiled by Smolčić et al. (2017). We matched this catalogue to the cosmic web environment catalogue of Darvish et al. (2015, 2017) in aiming to study the effects of the environment (field, filament, cluster) on star formation main sequence for various type of galaxies (satellite, central, isolated). After that we then matched the later catalogue to the COSMOS2015 of Laigle et al. (2016) to obtain stellar mass (M_*) and photometric redshifts (z) for each galaxy (bad data, i.e. FLAG = 1 as in COSMOS2015, were all discarded).

We apply the SFGs/AGN separation criteria of Smolčić et al. (2017) to further select AGN. The sample utilises the classified AGN/SFGs of Smolčić et al. (2017) which consists of X-ray, MIR, and SED-based and plus a combined rest-frame colour with radio excess diagnostics where it is known as "clean" classification.

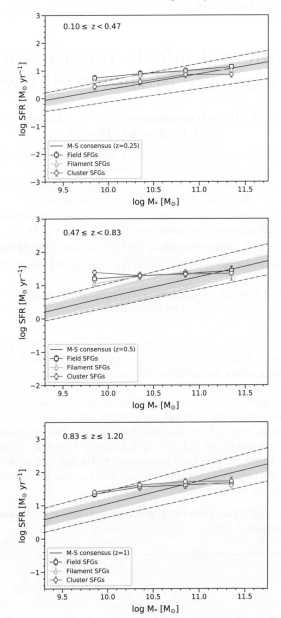

Figure 1. Star formation rate (SFR) against stellar mass (M_*) of SFGs for the three different environments for the lower redshift (*top panel*), intermediate (*middle panel*), and higher (*bottom panel*) redshift bins. The black solid line of each panel indicates the best-fit MS consensus of Speagle *et al.* (2014) at z=0.25 (*top*), z=0.5 (*middle*), and z=1 (*bottom*) while the shaded yellow region indicates a scatter of ±0.2 dex consensus dispersion. The black dashed-dotted lines indicate MS consensus of Speagle *et al.* (2014) at z=0 and 0.5 (*top*), z=0.25 and 1 (*middle*), and z=0.5 and 2 (*bottom*), respectively, which are shown as benchmark for comparison. Each panel presents the behaviour of the average M_* of the four M_* bins for the three environments. Blue square, teal triangle, and red circle represent field, filament, and cluster SFGs. Error bars correspond to average 1σ errors based on the standard error of the mean.

The final number of sources out of these selection procedures resulted to 2568 galaxies where 1836 are SFGs and 732 are AGN. We indicate that we entirely focus on studying for the SFGs only throughout the paper (i.e. we removed AGN from our sample).

2. Properties of galaxies

The stellar mass (M_*) were estimated based on spectral energy distribution (SED) fitting using the stellar population synthesis model of Bruzual & Charlot (2003) templates by assuming an initial mass function (IMF) of Chabrier (2003). The full details of the method for measuring the (M_*) is presented in Laigle et al. (2016).

The star formation rates (SFRs) were inferred from the total infrared luminosity by using the Kennicutt (1998) conversion factor that were scaled to a Chabrier (2003) IMF. The full details of the method for measuring the total infrared luminosity and SFR is presented in Smolčić et al. (2017).

We note that measurements of galaxy environments and types are based on density field Hessian matrix and we refer the readers to the work of Darvish et al. (2015, 2017) for full details.

3. Results

Figure 1 shows the behaviour of the MS as a function of the average M_* for four M_* bins. Each panel presents the behaviour of the average M_* of the four M_* bins for the three environments. Blue square, teal triangle, and red circle represent field, filament, and cluster SFGs. Error bars correspond to average 1σ errors based on the standard error of the mean.

We observe that the MS for all SFGs do not agree with the yellow region from the MS consensus and plus they all have shallower slopes in all environments.

4. Summary

In this paper, we present a study of the relationship between environment and star formation main sequence to shed new light on the effects of the environments (field, filament, cluster) on galaxies.

We summarise our main results as follows: (i) as observed previously, we find that SFRs increase with redshift independent on the environments; (ii) Furthermore, we observe that SFRs versus M_* relation is flat in all cases, irrespective of the redshift and environments.

References

Bruzual G., Charlot S., 2003, *MNRAS*, 344, 1000
Chabrier G., 2003, *PASP*, 115, 763
Darvish B., Mobasher B., Sobral D., Scoville N., & Aragon-Calvo M., 2015, *ApJ*, 805, 121
Darvish B., Mobasher B., Martin D. C., Sobral D., Scoville N., et al., 2017, *ApJ*, 837, 16
Kennicutt R. C. J., 1998, *ApJ*, 498, 541
Laigle C., McCracken H. J., Ilbert O., Hsieh B. C., & Davidzon I., 2016, *ApJS*, 224, 24
Smolčić V., Delvecchio I., Zamorani G., Baran N., Novak M., et al., 2017, *A&A*, 602, A2
Speagle J. S., Steinhardt C. L., Capak P. L., & Silverman J. D., 2014, *ApJS*, 214, 15

A dying radio AGN in the ELAIS-N1 field

Zara Randriamanakoto[1]

[1]South African Astronomical Observatory
P.O Box 9, Observatory 7935, South Africa
email: zara@saao.ac.za

Abstract. We use low-frequency GMRT observations and 1.4 GHz VLA archival data to study the radio spectrum of a dying radio galaxy discovered in the field of ELAIS-N1. With a linear size of ~ 100 kpc at a redshift $z \sim 0.33$, the diffuse source J1615+5452 exhibits a steep spectral index $\alpha_{612}^{1400} < -1.5$ and a convex radio spectrum. Its radio morphology also seems to lack compact features such as a nuclear core, relativistic jets and hotspots. We record a spectral curvature $\Delta \alpha \approx -1$ and a synchrotron age estimated between 34 - 70 Myr. These characteristics suggest that J1615+5452 is most likely a remnant radio AGN that has spent more than half of its total lifetime in the quiescence phase. The detection of such an elusive source is important since it represents the final phase in the evolution of a radio galaxy unless the nuclear core gets replenished with fresh particles and undergoes a restarting activity.

Keywords. galaxies: active - radio continuum: galaxies - galaxies: individual: J1615+5452

1. Introduction

Through radiation and/or jets of relativistic particles, active galactic nuclei (AGN) release large amounts of energy affecting the dynamical evolution of the surrounding interstellar and intergalactic medium (e.g. Brüggen, & Kaiser 2002; McNamara, & Nulsen 2012). Such a mechanism also plays an important role in regulating the star formation of the host galaxy. Determining the duty cycle of the nuclear activity (active vs. dormant phase) is therefore crucial for understanding the co-evolution process between the accreting supermassive black hole (SMBH) in AGN and its host galaxy (Kormendy & Ho 2013).

The active phase is typically in the range of 10 - 100 Myr (Cordey 1987). During that period, the radio galaxy morphology is characterized by the presence of a compact core, relativistic jets and hotspots besides the extended radio lobes inflated by the jets. A single power-law distribution best describes the radio spectrum over a wide range of frequencies.

On the other hand, once the quiescence phase kicks in, as the nuclear engine switches off due to a shortage of fresh particles accreting the SMBH, the compact structures fade away. This will result to a steepening of the spectral index and the appearance of a break frequency in the convex radio spectrum (Komissarov & Gubanov 1994; Murgia et al. 2011; Morganti 2017).

Because of the fast spectral evolution throughout the dormant phase, remnant radio AGNs remain elusive, especially in the cm wavelength regime. Low frequency observations with a relatively high sensitivity are thus required to increase their detection which are often observed as relatively bright and diffuse extended lobe emission.

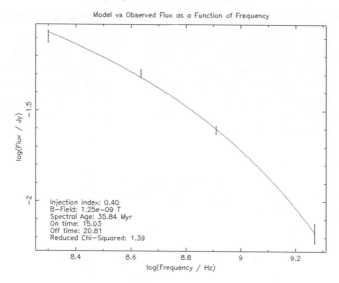

Figure 1. The radio spectrum of J1615+5452 at rest-frame frequencies fitted with a KGJP model developed by Komissarov & Gubanov (1994).

2. J1615+5452: a dying radio galaxy in EN1 field

A thorough visual inspection of the GMRT 612 MHz image of the ELAIS-N1 field (EN1, RA = 16^h10^m, DEC = $+54°35'$) led to the discovery of J1615+5452: a candidate remnant radio AGN. The source is likely to be hosted by an early-type elliptical galaxy, with a spectroscopic redshift $z \sim 0.32936 \pm 0.00005$ (SDSS-DR14, Abolfathi et al. 2018).

Observed as a fuzzy source in the radio image with a sensitivity of $\sim 40\,\mu$Jy/beam, the flux density is ~ 25 mJy and the linear size in the order of ~ 100 kpc. Although J1615+5452 is also detected at 150 MHz and 325 MHz, it is however invisible in the FIRST 1.4 GHz VLA survey. There are no signs of compact features.

An upper limit of the radio core prominence (CP $\lesssim 3.3 \times 10^{-3}$), a steep spectral index of $\alpha^{1400}_{612} < -1.5$ at high frequencies, and an estimate of the spectral curvature parameter ($\Delta\alpha = \alpha^{1400}_{612} - \alpha^{325}_{150} = -0.97 \pm 0.19$) were derived to investigate the remnant nature of J1615+5452. All the values are consistent with the radio source being dominated by non-thermal synchrotron emission. This is also reflected in the shape of the radio spectrum (Figure 1) which exhibits a break frequency between 325 and 612 MHz.

Assuming minimum energy conditions between particles and magnetic field, an estimate of the equipartition magnetic field $B_{eq} \sim 7.5\,\mu$G was used to derive a first order approximation of the synchrotron age $t_s \sim 34$ Myr. This value considers the 612 MHz radio emission. The age of the radio plasma is around 70 Myr for a reference frequency at 325 MHz.

We also fit a KGJP model (also known as CI_{off} model, Komissarov & Gubanov 1994) to the data while running BRATS (Harwood et al. 2013, 2015). The software returns a characteristic age of 36 Myr and a spectral break $\nu_b = 465$ MHz. The active and quiescent phases are equal to $t_{on} \sim 15$ Myr and $t_{off} \sim 21$ Myr, respectively.

The radio morphology, the synchrotron age as well as the spectral properties of the diffuse radio plasma are comparable to those of other remnant radio AGNs found in the literature (e.g. Jamrozy et al. 2004; Parma et al. 2007; Murgia et al. 2011; Brienza et al. 2016). All these factors led us to classify J1615+5456 as a dying radio galaxy.

3. Way forward: searching for dying radio AGNs with MeerKAT

This paper briefly summarizes the discovery of a dying radio galaxy in the ELAIS-N1 field (a comprehensive report on this work will be available in Randriamanakoto et al. 2020). Thanks to the sensitivity of low frequency GMRT observations, the diffuse radio emission from the remnant AGN could be observed. Large survey projects such as MeerKAT/MIGHTEE (rms noise level down to $\sim 10\mu$Jy at 1.4 GHz) are expected to provide an unprecedented detailed study of these elusive radio sources that are key to understanding the radio galaxy life cycle.

Acknowledgements

ZR acknowledges financial support from the South African Astronomical Observatory and the South African Radio Astronomical Observatory which are facilities of the National Research Foundation.

References

Abolfathi, B., Aguado, D. S., Aguilar, G., et al., 2018, ApJS, 235, 42
Brienza, M., Godfrey, L., Morganti, R., et al., 2016, A&A, 585, A29
Brüggen, M., & Kaiser, C. R., 2002, Nature, 418, 301
Cordey, R. A., 1987, MNRAS, 227, 695
Harwood, J. J., Hardcastle, M. J., Croston, J. H., et al., 2013, MNRAS, 435, 3353
Harwood, J. J., Hardcastle, M. J., & Croston, J. H., 2015, MNRAS, 454, 3403
Jamrozy, M., Klein, U., Mack, K.-H., et al., 2004, A&A, 427, 79
Komissarov, S. S., & Gubanov, A. G., 1994, A&A, 285, 27
Kormendy, J., & Ho, L. C., 2013, ARA&A, 51, 511
McNamara, B. R., & Nulsen, P. E. J., 2012, New Journal of Physics, 14, 055023
Murgia, M., Parma, P., Mack, K.-H., et al., 2011, A&A, 526, A148
Morganti, R., 2017, Nature Astronomy, 1, 596
Parma, P., Murgia, M., de Ruiter, H. R., et al., 2007, A&A, 470, 875
Randriamanakoto, Z., Ishwara-Chandra, C. H., Taylor, A. R., 2020, submitted to MNRAS

Understanding galaxy mergers and AGN feedback with UVIT

Khatun Rubinur[1], Mousumi Das[2], Preeti Kharb[1] and P. T. Rahne[2]

[1]National Centre for Radio Astrophysics - Tata Institute of Fundamental Research, S. P. Pune University Campus, Ganeshkhind, Pune, 411007, India
email: rubinur@ncra.tifr.res.in
[2]Indian Institute of Astrophysics, 2nd Block, Koramangala, Bengaluru, 560034, India

Abstract. Simulations expect an enhanced star-formation and active galactic nuclei (AGN) activity during galaxy mergers, which can lead to formation of binary/dual AGN. AGN feedback can enhance or suppress star-formation. We have carried out a pilot study of a sample of ∼10 dual nuclei galaxies with AstroSat's Ultraviolet Imaging Telescope (UVIT). Here, we present the initial results for two sample galaxies (Mrk 739, ESO 509) and deep multi-wavelength data of another galaxy (Mrk 212). UVIT observations have revealed signatures of positive AGN feedback in Mrk 739 and Mrk 212, and negative feedback in ESO 509. Deeper UVIT observations have recently been approved; these will provide better constraints on star-formation as well as AGN feedback in these systems.

Keywords. Galaxy merger, star-formation, AGN feedback, binary/dual AGN

1. Introduction

Galaxies evolve through major and minor mergers as well as through interactions with nearby galaxies (Barnes & Hernquist 1992). Simulations show that galaxy mergers/interactions cause gas inflow onto the nuclei resulting in the accumulation of dense gas around the nuclei (Bournaud et al. 2010). This triggers vigorous star-formation around the single or double nuclei. Hence, mergers and close interactions of galaxies often fall in the category of starbursts (Bournaud et al. 2010). Models of star-formation predict an increase of the star formation rate both in the disks, nuclei and even in the outer tidal tails (e.g. Duc et al. 2000). However, it has been found that the increase in star-formation is much lower than what is considered a typical starburst (Ellison et al. 2013). As the nuclear mass concentration and gas inflow increases, active galactic nuclei (AGN) activity may be triggered in one or both nuclei. This is due to mass accretion onto the supermassive black holes (SMBHs) of the individual galaxies which can lead to binary/dual AGN (Begelman et al. 1980). Studying dual AGN is one of the ways in which we can follow the SMBHs as they sink into the centres of the merger remnants and finally coalesce in the central bulge. There are several studies which have attempted to detect these binary/dual AGN (Rubunur et al. 2017, 2019 and references therein).

Once AGN activity is triggered and the SMBHs reach a certain critical mass (Ishibashi & Fabian 2012), they give out energy to the surrounding medium via winds, jets and radiation. This can enrich the circum-galactic medium (CGM). The winds also trigger star formation beyond the AGN by shocking gas and suppress gas infall by blowing out the gas. This is collectively called AGN feedback (see Harrison et al. 2017 for review). UV emission is a good tracer of the star formation rate (SFR). It can trace recent as well as older star formation. High resolution UV observations can help us understand how

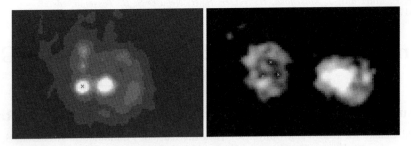

Figure 1. (left) UVIT image of Mrk 739. It shows UV emission from two nuclei and outer disk. Two star-forming knots (green cross) are detected, which may be related to one of the nuclei (red cross). (right) UVIT image of ESO 509. It shows three cavities (green cross). Red cross is the centre of the galaxy.

SFR is affected by mergers and AGN feedback. Hence, the new ultra-violet telescope UVIT on the Astrosat satellite (Kumar et al. 2012) with a spatial resolution of $1.2''$ is crucial for tracing the SFRs in nearby galaxies. We have carried out a pilot study of dual nuclei galaxies with UVIT to detect merger- and AGN- related star-formation in galactic disks and tidal tails. Here, dual nuclei can be a pair of AGN, AGN-starburst nuclei or starburst-starburst. We have selected our sample from surveys of dual core galaxies (e.g. Mezcua et al. 2014) and confirmed dual AGN or AGN-starburst nuclei (Koss et al. 2012). We had chosen 17 nearby galaxies with redshifts < 0.1, having dual nuclei with projected separations below 10 kpc and exhibiting strong UV emission in GALEX observations.

So far, we have observed ~ 10 sources with UVIT using short exposures (< 3 ks). Multi-frequency observations have also been carried out for one galaxy. Here, we present results from the initial observations of two galaxies and multi-wavelength data along with deep UVIT observation of Mrk 212 (Rubunur et al. 2019, MNRAS, submitted).

2. Results:

<u>Mrk 739:</u> is a confirmed dual AGN (Koss et al. 2011). One of the nuclei is a Seyfert type 1 while the other is a Seyfert type 2. The nuclei are at a projected separation of 3 kpc. The optical spectra show signatures of outflows from one of the nuclei, while the molecular gas profile is asymmetric (Koss et al. 2011). Our UVIT images (2.5 ks) show the star forming regions around the nuclei of Mrk 739 (Fig 1). These star-forming knots can be induced by the positive AGN feedback activity. Deeper UV observations are required to confirm direct signatures of AGN-feedback-related star-formation.

<u>ESO 509-IG 066 NED 02 (ESO 509):</u> is a confirmed dual AGN in an ongoing galaxy merger; the two nuclei are at a projected separation of 11.2 kpc (Koss et al. 2012). Our 2.5 ks UVIT image shows UV emission from the individual galaxies. Furthermore, we have found that three UV holes (Fig 1) in Source 2, surrounding the central nuclei. These could be signatures of negative AGN feedback (e.g., george et al. 2019).

<u>Mrk 212:</u> is a double-pinwheel galaxy with a companion. Mrk 212 has two radio sources (S1 and S2) with a projected separation of ~ 6 kpc (Fig 2). The 15 GHz VLA image and $1.4 - 8.5$ GHz core spectral index (-0.81 ± 0.06) supports the presence of an AGN in S1. S2 has a compact structure at 15 GHz which coincides with the optical centre of the companion galaxy and an extended structure at 8.5 GHz, which is offset by $\sim 1''$ from the optical centre. Hernandez et al. (2016) have identified S2 as an AGN, from optical spectroscopy. Our deep UVIT observations of Mrk 212 resolve the star-forming knots in S2 (Fig 2) and detect tidal tails. The star-forming knots in S2 coincide with the two sided radio structure detected at 8.5 GHz, which could be the result of positive

Figure 2. This is a deep UVIT image of MRK 212 (15 ks). It has resolved star-forming knots in S2 (green cross) and tidal arms. The red cross is the centre of S2.

AGN feedback in S2. However, the overall SFR is similar to isolated galaxies. This is not typically expected in the standard picture of galaxy merger and subsequent enhanced SFR.

3. Summary

We have carried out UVIT observations of ~ 10 dual nuclei galaxies. These observations have revealed signatures of AGN feedback in Mrk 739, ESO 509 and Mrk 212. While Mrk 739 and Mrk 212 may show positive AGN feedback, ESO 509 shows negative AGN feedback. We require high sensitivity radio and molecular gas observations in Mrk 212 to confirm this. Mrk 739 and ESO 509 will be observed for 15 ksec in the upcoming UVIT cycle, which would confirm the star-forming knots as well as UV cavities in these two systems. While some studies find that mergers can enhance star-formation, others have shown the opposite trend (see Pearson *et al.* 2019). We find that Mrk 212 has an SFR similar to isolated galaxies, consistent with the latter. Systematic multi-wavelength observations of large samples of galaxies are required to fully understand AGN feedback.

References

Barnes, J. E., and Hernquist, L., 1992, *ARAA*, 30, 705
Bournaud, F., 2010, *arXiv0909.1812*
Duc, P.-A., Brinks, E., Springel, V., et al. 2000, *AJ*, 120, 1238
Ellison, S. L., Mendel, J. T., Patton, D. R., & Scudder, J. M. 2013, *MNRAS*, 435, 3627
Begelman, M. C., Blandford, R. D., Rees, M. J., 1980, *Nature*, 287, 307
Ishibashi, W., and Fabian, A. C., 2012, *MNRAS*, 427, 2998
Harrison, C. M., 2017, *Nature Astronomy*, 1, 0165
Kumar, A., et al., 2012, *SPIE Conference Series*, Vol. 8443, procspie, p. 84431N
Koss, M., Mushotzky, R., Treister, E., et al. 2012, *ApJ*, 746L, 22
Mezcua, M., Lobanov, A. P., Mediavilla, E., and Karouzos, M., 2014, *ApJ*, 784, 16
Hernandez Ibarra F., et al., 2016, *MNRAS*, 459, 291
Pearson, W. J., et al., 2019, *AA*, 631, A51
Rubinur, K., Das M., Kharb, P., Honey, M., 2017, *MNRAS*, 465, 4772
Rubinur, K., Das, M., Kharb, P., 2019, *MNRAS*, 484, 4933
George, K., et al., 2019, *MNRAS* 487,3102

Exploring the X-ray universe via timing: mass of the active galactic nucleus black hole XMMUJ134736.6+173403

Eva Šrámková[1], K. Goluchová[1], G. Török[1], Marek A. Abramowicz[1], Z. Stuchlík[1] and Jiří Horák[2]

[1]Research Centre for Computational Physics and Data Processing, Institute of Physics, Silesian University in Opava, Bezručovo nám. 13, CZ-746 01 Opava, Czech Republic
email: eva.sramkova@fpf.slu.cz

[2]Astronomical Institute, Academy of Sciences, Boční II 1401, CZ-14131 Prague, Czech Republic

Abstract. A strong quasi-periodic modulation has recently been revealed in the X-ray flux of the X-ray source XMMUJ134736.6+173403. The two observed twin-peak quasiperiodic oscillations (QPOs) exhibit a 3:1 frequency ratio and strongly support the evidence for the presence of an active galactic nucleus black hole (AGN BH). It has been suggested that detections of twin-peak QPOs with commensurable frequency ratios and scaling of their periods with BH mass could provide the basis for a method intended to determine the mass of BH sources, such as AGNs. Assuming the orbital origin of QPOs, we calculate the upper and lower limit on the AGN BH mass M, reaching $M \approx 10^7 - 10^9 M_\odot$. Compared to mass estimates of other sources, XMMUJ134736.6+173403 appears to be the most massive source with commensurable QPO frequencies, and its mass represents the current observational upper limit on the AGN BH mass obtained from the QPO observations.

Keywords. black hole physics; accretion; accretion disks

1. Introduction

The X–ray power density spectra (PDS) of several Galactic BH systems show high-frequency (HF) QPOs. Their frequencies lie within the range of $40 - 450$Hz, which corresponds to timescales of orbital motion in the vicinity of a BH, suggesting thus that the observed signal likely originates in the innermost parts of an accretion disk (McClintock & Remillard, 2006). In Galactic microquasars, the commonly found detections of HF QPO peaks display more or less constant frequencies that are specific for a given source. The observed frequencies often come in ratios of small natural numbers, the most common one being 3:2 (Abramowicz & Kluźniak, 2001; Remillard et al., 2002; McClintock & Remillard, 2006).

The 3:2 frequencies observed in the Galactic microquasars can be matched by the following relation (McClintock & Remillard, 2006),

$$\nu_U = \frac{2.8\text{kHz}}{M^*}. \qquad (1.1)$$

Here, ν_U is the higher of the two frequencies that form the 3:2 ratio, $R = \nu_U/\nu_L = 3/2$, and $M^* = M/M_\odot$, where M is the BH gravitational mass and $M_\odot$ is the solar mass. The microquasar's 3:2 QPO frequencies and their scaling are illustrated in Figure 1. It has been proposed by Abramowicz et al. (2004) that detections of 3:2 QPOs and scaling (1.1) could provide the basis for a method intended to determine the mass of BH sources,

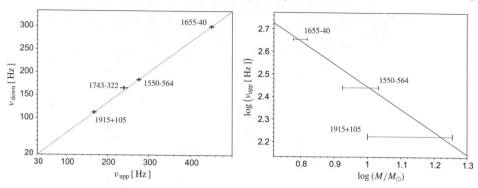

Figure 1. Left: The 3:2 HF QPO frequencies observed in Galactic microquasars. Right: The $1/M$ scaling of these frequencies, which supports the QPO orbital origin hypothesis.

such as AGNs. Here we briefly report on a recent progress in this field. A more detailed discussion along with a number of references can be found in Goluchová et al. (2019).

2. Estimation of the upper limit on the AGN mass

The two QPOs that occur on a daily timescale and exhibit a 3:1 frequency ratio have recently been observed in the X-ray source XMMUJ134736.6+173403 by Carpano & Jin (2018). This observation allows us to obtain the upper limit on the AGN BH mass in this source.

The Keplerian frequency of matter orbiting a BH monotonically increases as the orbital radius r decreases down to the inner edge of the accretion disk. Location of the inner edge depends on the disk's radiative efficiency, ranging from the innermost stable circular orbit (r_{ISCO}) to the marginally bound orbit (r_{RISCO}). At these orbits, Keplerian frequency for a Schwarzschild BH ($a \equiv cJ/(GM^2) = 0$) scales with BH mass as (e.g., Bardeen et al., 1972)

$$\nu_{ISCO} = \frac{2.20\text{kHz}}{M^*}, \quad \nu_{RISCO} = \frac{4.04\text{kHz}}{M^*}, \tag{2.1}$$

while for an extremely rotating Kerr BH ($a = 1$), one may write:

$$\nu_{ISCO} = \nu_{RISCO} = \frac{16.2\text{kHz}}{M^*}. \tag{2.2}$$

Figure 2 shows the above relations, which determine the highest allowed orbital frequencies. Assuming that the higher QPO frequency observed in XMMUJ134736.6+173403 corresponds to Keplerian frequency of matter anywhere inside the disk, we estimate the mass of the source to be no higher than $M \doteq 1.1 \times 10^9 M_\odot$.

3. Discussion

Provided that the observed BH HF QPOs can be described by a QPO model that deals with orbital motion, XMMUJ134736.6+173403 likely represents the most massive BH source with commensurable QPO frequencies. Using the ISCO frequency relation that gives the highest possible orbital frequency inside the disk, the estimation of M is found to be no higher than $M \approx 10^9 M_\odot$. This result can be viewed as the current observational upper limit on the AGN BH mass inferred from the QPOs. It is, however, valid only as long as the oscillation frequencies characteristic for a given QPO model are not much higher than the Keplerian frequency. This condition is, nevertheless, likely to be satisfied (Straub & Šrámková, 2009).

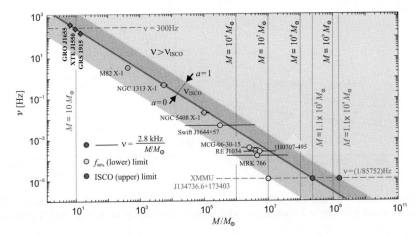

Figure 2. Large scaling of BH HF QPO frequencies vs. Keplerian frequencies in the accretion disk. The upper left corner of the plot corresponds to Galactic microquasar BHs, while the lower right corner corresponds to supermassive BHs. The light orange region denotes the ISCO frequencies in the range of $a \in [0, 1]$. The grey area indicates frequencies that are higher than the ISCO frequency. The light yellow area denotes Keplerian frequencies in the inner part of the (thin) disk that radiates more than 90% of the whole disk luminosity ($a \geq 0$). The green circles denote intermediate-mass BH sources, whose mass estimate is based either fully or in large part on the observations of HF QPOs.

For a large range of M, various orbital QPO models provide similar mass-spin relations. Following a quadratic approximation, we may write:

$$M \approx (5^{+11}_{-3} a^2 + 8^{+17}_{-4} a + 8^{+12}_{-4}) \times 10^7 M_\odot, \tag{3.1}$$

where the upper limit corresponds to models that imply a high BH mass (low QPO excitation radius) and the lower limit to models giving a low BH mass (high QPO excitation radius, Goluchová *et al.*, 2019).

Finally, we also note that, very recently, Gupta *et al.* (2019) discussed a gamma ray QPO in the high-redshift blazar B2 1520+31. Using the ISCO frequency relation, they found the upper limit on the BH mass to be $M \approx 10^{10} M_\odot$.

Acknowledgments

We acknowledge the Czech Science Foundation (GAČR) grant No. 17-16287S. We wish to thank the INTER-EXCELLENCE project No. LTI17018 that supports our international collaboration. Furthermore, we acknowledge two internal grants of the Silesian University, SGS/12,13/2019.

References

Abramowicz, M. A. & Kluźniak, W., 2001, *A&A*, 374, L19
Abramowicz, M. A., Kluźniak, W., McClintock, J. E., Remillard, R. A., 2004, *APJL*, 609, L63
Bardeen, J. M., Press, W. H., Teukolsky, S. A., 1972, *APJ*, 178, 347
Carpano, S. & Jin, C., 2019, *MNRAS*, 477, 3178
Goluchová, K., Török, G., Šrámková, E., Abramowicz, M. A., Stuchlík, Z., Horák, J., 2019, *A&A*, 622, L8
Gupta, A. C., Tripathi, A., Wiita, P. J., Kushwaha, P., Zhang, Z., Bambi, C., 2019, *MNRAS*, 484, 5785
McClintock, J. E., Remillard, R. A., 2006, Black hole binaries (Cam. Uni. Press), 157213
Remillard, R. A., Muno, M. P., McClintock, J. E., Orosz, J. A., 2002, *ApJ*, 580, 1030
Straub, O. & Šrámková, E., 2009, *CQG*, 26, 055011

Dichotomy of radio loud and radio quiet quasars in four dimensional eigenvector one (4DE1) parameter space†

Shimeles Terefe[1], Ascensión del Olmo[2], Paola Marziani[3] and Mirjana Pović[1,2]

[1] Ethiopian Space Science and Technology Institute (ESSTI), Addis Ababa, Ethiopia
email: `shimeles11@gmail.com`

[2] Instituto de Astrofisica de Andalucía (IAA-CSIC), Granada, Spain

[3] Istituto Nazionale di Astrofisica (INAF), Osservatorio Astronomico di Padova, Padova, Italy

Abstract. Recent work has shown that it is possible to systematize quasars (QSOs) spectral diversity in 4DE1 parameter space. The spectra contained in most of the surveys have low signal to noise ratio which fed the impression that all QSOs are spectroscopically similar. Exploration of 4DE1 parameter space gave rise to the concept of two populations of QSOs that present important spectroscopic differences. We aim to quantify broad emission line differences between radio quiet and radio loud sources by exploiting more complete samples of QSO with spectral coverage in Hβ, MgII and CIV emission lines. We used a high redshift sample ($0.35 < z < 1$) of strong radio emitter QSOs observations from Calar Alto Observatory in Spain.

Keywords. Active galaxies, quasars, radio loud and radio quiet quasars, four dimensional eigenvecter 1 parameter space

---

1. Introduction

Over fifty years after their discovery, people are beginning to see progress in both defining and contextualizing the properties of QSOs, some of the brightest AGNs (Sulentic *et al.* 2007). Bright type-1 AGNs show widely differing line profiles, intensity ratios and ionization levels (Marziani *et al.* 2018).

A much debated problem in AGN studies involves the possibility of a real physical dichotomy between radio loud (RL) and radio quiet (RQ) QSOs (Zamfir *et al.* 2008). Another complication is introduced by the fact that some good fraction of RQ sources share common properties with the RL QSOs as discussed in Zamfir *et al.* (2008). For instance, about 30 - 40 % of RQ QSOs are spectroscopically similar to RL (Sulentic *et al.* 2000) and with the improvement of radio interferometry techniques, it was possible to notice that both QSO types are capable of producing radio jets (Chiaberge, & Marconi 2011).

The other which remains a perplexing question 50 years after the discovery of QSO is the origin of radio loudness (Ruff 2012). From the theoretical point of view, in spite of the great advancement in the ability of collecting unbiased sets of data, most of the researchers argued, the origin of the relativistic radio jets in AGNs as an open question.

A first contextualization, RQ vs. RL (Sulentic *et al.* 1995; Corbin 1997) showed intriguing spectroscopic differences between the two types of QSOs with large blue shifts observed in the emission line profiles only among RQ sources.

† Based on observations obtained at the CAHA Observatory, Calar Alto, Spain

Recent work has shown that it is possible to systematize quasar spectral diversity in a space called 4DE1 parameter space (see Marziani et al. 2018 and references therein). As stated in Zamfir et al. (2008), the value of studying the RL phenomenon within the 4DE1 context is at least two fold: (i) the approach compares RL and RQ sources in a parameter space defined by measures with no obvious dependence on the radio properties (Marziani et al. 2003b), (ii) it allows to make predictions about the probability of radio loudness for any population of QSOs with specific optical (UV) spectroscopic properties. As the RQ/RL separation in 4DE1 is not complete (Zamfir et al. 2008), many open questions are present after 50+ years of study of QSOs.

Therefore, this work studies the properties of RL and RQ QSOs by using the 4DE1 parameter space. We specifically consider a possible dichotomy between them, and the reason behind observed low fractions of RL.

2. Data

The work focus on type 1 QSOs, as they can be unambiguously identified based on the presence of a broad component in the hydrogen Balmer emission lines (mainly in H$\beta\lambda$4861), the doublet of MgIIλ2800, or in High Ionization Lines (HILs) like CIVλ1549 in the UV. The best clues to study the RQ/RL dichotomy in QSOs lies in the optical/UV spectra and many of the brightest RL QSOs with radio coverage have no published optical spectra with S/N high enough to permit a detailed study. Data from astronomical archives were supplemented with new data obtained at the Observatory of Calar Alto in Spain. Regarding the new data, a sample of 50 strongly-RL QSOs were obtained by using the TWIN spectrograph attached at the 3.5m telescope of the Calar Alto Observatory. The TWIN spectrograph has two arms that allows to obtain (1) near UV spectra for studying the region of the MgIIλ2800 doublet, and (2) the Hβ-FeII region for the QSOs in the redshift range (0.35 < z < 1).

Radio images and the fluxes at radio continuum were obtained from the VLA Faint Images of the Radio Sky at Twenty-Centimeters (FIRST) survey and from the NVSS survey for those QSOs with no available FIRST radio measurements. This information is fundamental to identify QSOs according to their morphology at radio frequencies and to determine the power of radio emission.

The Kellermann factor R_k which is a value defined by the ratio between radio/optical flux density was used as a discriminant between RQ and RL. High S/N spectra with Hβ, CIVλ1549 and MgIIλ2800 coverage were considered.

3. Data analysis and preliminary results

Analyzing the broad lines by doing multicomponent non-linear fitting of broad emission lines in particular of Hβ, FeII, MgII and the UV lines of CIV and HeII allows us to quantify the properties and kinematics of the broad line region, to detect very broad components (observed only in population B of QSOs within the 4DE1 scheme) as well as to calculate fundamental quantities such as the mass of the super massive black hole (SMBH) and the Eddington ratio (L/L_{Edd}). In order to get spectra suitable for scientific use, we used a standard spectroscopic data reduction by using IRAF astronomical package which produces a spectrum as shown in Figure 1 left for one of our source, S5_1856+73 at z = 0.46 which incorporates most of the emission lines.

The optical (Hβ-FeII) and UV (MgII) analysis mainly focus on the spectral fitting of emission lines using the SPECFIT routine in the IRAF package. The result of the SPECFIT for Hβ emission line for S5_1856+73 is shown in Figure 1 right.

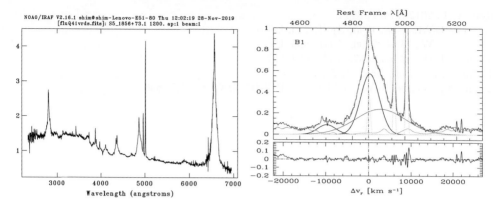

Figure 1. (Left) Rest frame spectra for S5_1856+73, abscissa corresponds to rest frame wavelength, ordinate corresponds to specific flux in units of 10^{-15}erg $s^{-1}cm^{-2}\text{Å}^{-1}$ and (right) multicomponent fitting in the Hβ region. Yellow lines represent the narrow components of [OIII] and Hβ; black line for the broad component; red line corresponds to the very broad component and the green one the FeII. Bottom plot corresponds to the residuals of the fitting.

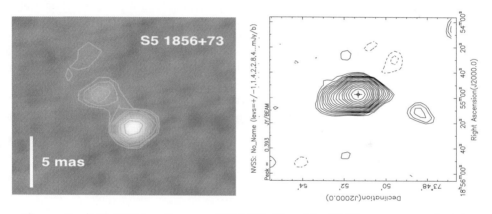

Figure 2. (left) VLBA radio map of S5_1856+73 and its NVSS contour map (right)

In radio, we searched the VLBA radio map and the NVSS contour map as shown in Figure 2 for S5_1856+73 which can be taken as an indication of very different morphologies in radio. The analysis of the connection between the optical/UV and radio measurements will be presented in a forthcoming paper.

Acknowledgements

AdO and MP acknowledges financial support from Spanish grants AYA2016-76682-C3-1-P and SEV-2017-0709. We also acknowledge ESSTI, Jimma University, Ethiopian Ministry of Innovation and Technology, and Ethiopian Ministry of Science and Higher Education for their support.

References

Chiaberge, M., & Marconi, A. 2011, *MNRAS*, 416, 917
Corbin, M. R., 1997, *ApJS*, 113, 245
Marziani, P., *et al.*, 2018, Frontiers in *A&SS*, 5, 6
Marziani, P., *et al.*, 2003, *ApJS*, 145, 199

Ruff, A. J., 2012, *PhDT*, 314T
Sulentic, J. W., *et al.*, 1995, *ApJL*, 445, L85
Sulentic, J. W., *et al.*, 2000, *ApJL*, 536, L5
Sulentic, J. W., *et al.*, 2007, *ApJL*, 666, 757
Taylor, G. B., *et al.*, 1996, *ApJS*, 107, 37
Zamfir, S., Sulentic, J. W., & Marziani, P. 2008, *MNRAS*, 387, 856

Determination of K4000 of potential blazar candidates among EGRET unidentified gamma-ray sources

Emmanuel Uwitonze[1], Pheneas Nkundabakura[2] and Tom Mutabazi[1]

[1]Physics Department, Mbarara University of Science and Technology,
P.O. Box 1410, Mbarara, Uganda
email: uwitonze_emmanuel@yahoo.com

[2]Department of Mathematics, Sciences and Physical Education, University of Rwanda-College of Education, P.O. Box 5039, Kigali, Rwanda

Abstract. Blazars are radio-loud Active Galactic Nuclei (AGN) with relativistic jets oriented towards the observer's line-of-sight. Based on their optical spectra, blazars may be classified as flat-spectrum radio quasars (FSRQs) or BL Lacs. FSRQs are more luminous blazars with both narrow and broad emission and absorption lines, while BL Lacs are less luminous and featureless. Recent studies show that blazars dominate ($\sim 93\%$) the already-identified EGRET sources (142), suggesting that among the unidentified sources (129) there could still be faint blazars. Due to the presence of a strong non-thermal component inside their jets, blazars are found to display a weaker depression at ~ 4000 Å (K4000 ≤ 0.4). In this study, we aimed at determining the K4000 break for a selected sample among the potential blazar candidates from unidentified EGRET sources to confirm their blazar nature. We used two blazar candidates, 3EG J1800-0146 and 3EG J1709-0817 associated with radio counterparts, J1802-0207 and J1713-0817, respectively. Their optical counterparts were obtained through spectroscopic observations using Robert Stobie spectrograph (RSS) at the Southern African Large Telescope (SALT) in South Africa. The observed Ca II H & K lines depression at ~ 4000 Å in spectra of these sources show a shallow depression, K4000 = 0.35 ± 0.02 and 0.24 ± 0.01, respectively, suggesting that these sources are blazar candidates. Moreover, the redshifts z = 0.165 and 0.26 measured in their spectra confirm the extragalactic nature of these sources.

Keywords. line: identification, radiation mechanisms: non-thermal, techniques: spectroscopic, galaxies: active, galaxies: jets, galaxies: BL Lacertae objects.

1. Introduction

Different studies reveal that out of 271 sources detected by the Energetic Gamma-Ray Experiment Telescope (EGRET) during its mission, ~ 142 sources were identified and $\sim 93\%$ of them were associated with blazars (e.g., Hartman *et al.* 1999; Sowards-Emmerd *et al.* 2003; Nkundabakura & Meintjes 2012). It is assumed that among 129 ($\sim 48\%$) remaining unidentified EGRET sources, there could be still faint blazars in abundance. There have been efforts to try to identify blazars from the unidentified EGRET sources via their variability and spectral energy distribution (SED). This study joined this effort and aimed at determining the extent of non-thermal emissions from two potential blazar candidates selected among the remaining unidentified EGRET sources through the analysis of the strength of their Ca II H & K break at ~ 4000 Å (K4000). Blazars are a special class of AGN with a weaker Ca II contrast, i.e., K4000 $\leqslant 0.4$ (Marcha *et al.* 1996; Caccianiga *et al.* 1999; Landt *et al.* 2002), due to the contribution of the nuclear non-thermal component produced inside their jets, i.e., for blazars,

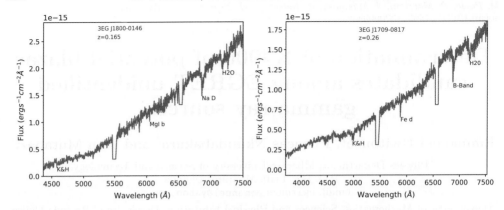

Figure 1. The left panel shows the spectrum of 3EG J1800-0146. The right panel shows the spectrum of 3EG J1709-0817.

this Ca II H & K break decreases as the non-thermal emission from jet increases and vice-versa. In fact, observations in optical, radio and X-rays show that the increase in the jet's luminosity leads to the decrease in the Ca II H & K break (Landt et al. 2002).

2. Data source and methods

The study used two blazar candidates (3EG J1800-0146 and 3EG J1709-0817) from a sample of 13 blazar candidates selected from Nkundabakura & Meintjes (2012) which were selected from the 3EG catalogue. The spectroscopic raw data were obtained at the Southern African Large Telescope (SALT) with RSS spectrograph. The raw data of the two sources were reduced using the Image Reduction and Analysis Facility (IRAF) data reduction pipeline. The presence of nuclear non-thermal component in the host galaxy of each source was confirmed by measuring the K4000 in the rest frame. For each spectrum, the contrast was calculated using the relation:

$$K4000 = \frac{f^+ - f^-}{f^+}, \quad (2.1)$$

where f^- and f^+ are average fluxes within $3750\,\text{Å}$–$3950\,\text{Å}$ and $4050\,\text{Å}$–$4250\,\text{Å}$, respectively, in rest frame (Caccianiga et al. 1999; Landt et al. 2002). The fluxes measured in each spectrum are shown in Table 2.

3. The spectra of 3EG J1800-0146 and 3EG J1709-0817

Figure 1 (left panel) represents the optical spectrum of 3EG J1800-0146. The spectrum features some absorption lines at redshift $z = 0.165$, e.g., Ca II H & K $\lambda\lambda 3969\,\text{Å}$, $3934\,\text{Å}$, Mg I b $\lambda\lambda 5169, 5175, 5184\,\text{Å}$ and Na D $\lambda\lambda\, 5890, 5896\,\text{Å}$ in the rest frame. H_2O absorption (7168–$7394\,\text{Å}$) from the Earth's atmosphere, not redshifted has been detected in this source. The line centre, continuum, flux and the equivalent width (W_λ) shown in Table 1 were measured using SPLOT command in IRAF.

It shows a shallow depression (K4000 = 0.35 ± 0.02), suggesting the presence of a strong non-thermal emission in this source, emanating from a jet oriented at a small angle. The spectrum of this source resembles that of a BL Lac (Marcha et al. 1996; Landt et al. 2002). On the other hand, Figure 1 (right panel) shows the spectrum of 3EG J1709-0817. It features some absorption lines at redshift $z = 0.26$, e.g., Ca II H & K $\lambda\lambda 3969\,\text{Å}$, $3934\,\text{Å}$ and Fe d at $\lambda 4668\,\text{Å}$ in the rest frame (Table 1). The telluric absorption lines resulting from Earth's atmosphere, not redshifted have been identified in this object, e.g., H_2O (7168–$7394\,\text{Å}$), and B-Band (6867–$6884\,\text{Å}$). The spectrum displays a shallow

Table 1. Absorption lines in 3EG J1800-0146 and 3EG J1709-0817. Each line is redshifted by $z = 0.165$ and 0.26, respectively.

Absorption line	Line centre (Å)	Continuum (erg s^{-1}cm^{-2}Å^{-1})	Flux (erg s^{-1}cm^{-2}Å^{-1})	W_λ (Å)
		3EG J1800-0146		
Ca II K	4580.18	2.55×10^{-16}	4.95×10^{-16}	1.93
Ca II H	4625.53	2.30×10^{-16}	8.00×10^{-17}	0.34
Mg I b	6046.36	1.25×10^{-15}	8.25×10^{-16}	0.65
Na D	6863.19	1.81×10^{-15}	3.25×10^{-15}	1.79
		3EG J1709-0817		
Ca II K	4973.92	3.483×10^{-16}	6.95×10^{-16}	1.996
Ca II H	4997.05	3.472×10^{-16}	6.69×10^{-16}	1.927
Fe d	5882.79	8.13×10^{-16}	2.03×10^{-15}	2.49

Table 2. Ca depression near 4000 Å in 3EG J1800-0146 and 3EG J1709-0817 spectra.

Object	Ca band (Å, rest frame)	Ca band (Å, redshifted)	z	Flux (erg s^{-1}cm^{-2}Å^{-1})	K4000	Class by K4000
3EG J1800-0146	3750–3950	4368.75–4601.75	0.165	$f^- = (2.170 \pm 0.040) \times 10^{-16}$		
	4050–4250	4718.25–4951.25		$f^+ = (3.322 \pm 0.035) \times 10^{-16}$	0.346 ± 0.02	BL Lac
3EG J1709-0817	3750–3950	4725–4977	0.26	$f^- = (4.333 \pm 0.030) \times 10^{-16}$		
	4050–4250	5103–5355		$f^+ = (5.720 \pm 0.050) \times 10^{-16}$	0.242 ± 0.01	BL Lac

depression, K4000 = 0.24 ± 0.01. This is within the range of $0.25 \leq K4000 \leq 0.4$ of typical BL Lac candidates (Marcha et al. 1996). In Table 2, we present the average fluxes, f^- and f^+ used for the measurement of K4000 in the two spectra.

4. Conclusions

From our spectroscopic data, we confirmed 3EG J1800-0146 and 3EG J1709-0817 as extragalactic objects with redshift $z = 0.165$ and 0.26, respectively, and as BL Lacs through the strength of their K4000 contrast.

References

Caccianiga, A., Maccacaro, T., Wolter, et al., 1999, ApJ, 513, 51
Hartman, C., Bertsch, L., Bloom, et al., 1999, ApJS, 123, 79
Landt, H., Padovani, P., Giommi, et al., 2002, MNRAS, 336, 945
Marcha, M., Browne, A., Impey, et al., 1996, MNRAS, 281, 425
Nkundabakura & P., Meintjes., 2012, MNRAS, 427, 859
Sowards-Emmerd, D., Romani, R. W., Michelson, et al., 2003, ApJ, 590, 109

The properties of inside-out assembled galaxies at z < 0.1

Dejene Zewdie[1,2,3], Mirjana Pović[3,4], Manuel Aravena[1], Roberto J. Assef[1] and Asrate Gaulle[5]

[1] Núcleo de Astronomía, Universidad Diego Portales, Santiago, Chile
email: dejene.woldeyes@mail.udp.cl

[2] Department of Physics, Debre Berhan University (DBU), Debre Berhan, Ethiopia

[3] Ethiopian Space Science and Technology Institute-AAU, Addis Ababa, Ethiopia

[4] Instituto de Astrofísica de Andalucía (IAA-CSIC), Granada, Spain

[5] Department of Physics, Dilla University, Dilla, Ethiopia

Abstract. In this work, we study the properties of galaxies that are showing the inside-out assembly (which we call inside-out assembled galaxies; IOAGs), with the main aim to understand better their properties and morphological transformation. We analysed a sample of galaxies from the Sloan Digital Sky Survey (SDSS) Data Release 8 (DR8), with stellar masses in the range $\log M_\star = 10.73 - 11.03$ $M_\odot$ at $z < 0.1$, and analyze their location in the stellar mass-SFR and the color-stellar mass diagram. We found that IOAGs have different spectroscopic properties, most of them being classified either as AGN or composite. We found that the majority of our sources are located below the main sequence of star formation in the SFR-stellar mass diagram, and in the green valley or red sequence in the color-stellar mass diagram. We argue that IOAGs seem to correspond to the transition area where the galaxies are moving from star-forming to quiescent, and from the blue cloud to the red sequence and/or to recently quenched galaxies.

Keywords. Galaxies: properties; galaxies: inside-out

1. Introduction

How galaxies form and evolve through cosmic time is one of the major open questions in extragalactic astronomy and modern cosmology. In particular, we need to understand what role do galaxy mergers, AGN activity and star formation feedback play in driving the observed morphological evolution and the quenching of star formation of galaxies.

Pérez et al. (2013) showed that local galaxies with stellar masses in the range $\log M = 10.73 - 11.03$ $M_\odot$ have systematically shorter assembly times within their inner regions ($< 0.5 R_{50}$) when compared to that of the galaxy as a whole, contrary to lower or higher mass galaxies which show consistent assembly times at all radii. We refer to galaxies in this stellar mass range as Inside Out Assembled Galaxies, or IOAGs.

In this work, we aim to characterize physical properties of IOAGs, such as morphology, star formation rates, and AGN activity, and to constrain the relation of these properties to the evolutionary state of these sources.

2. Data and sample selection

We select our sample of IOAGs from the Sloan Digital Sky Survey (SDSS), Data Release (DR8) MPA-JHU spectroscopic catalogue (Brinchmann et al. 2004). Following Pérez et al. (2013), we selected galaxies with the stellar mass range of

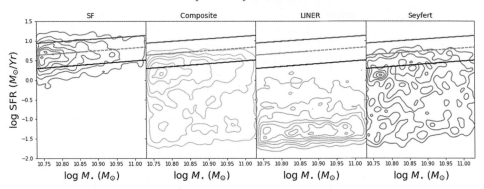

Figure 1. SFR as a function of stellar mass for the IOAGs in each spectral class. A red dashed line in each panel represents the local main-sequence of star formation from Whitaker *et al.* (2012). The blue and black solid lines indicate the typical scatter around the MS of 0.3 dex.

$\log M_\star = 10.73 - 11.03$ M$_\odot$ and redshift $z < 0.1$. We found 23816 IOAGs with spectroscopic observations from SDSS that can be classified using the NII-BPT diagram (Baldwin *et al.* 1981; Kewley *et al.* 2006). We found that 40% are classified as AGN (LINERs + Sy2), 40% as composites and only 20% as star-forming galaxies. Based on the Galaxy Zoo morphologies, 12% of IOAGs are classified as ellipticals and 37% as spirals. The rest (51%) have uncertain morphologies in this catalogue.

3. Analysis and results

Here, we have analysed the spectroscopic and morphological properties IOAGs by using the stellar masses and SFRs provided by the MPA-JHU catalogue. We found that the majority of our sources are located below the main sequence of star formation (76%) in the SFR-stellar mass diagram. We also find the majority of IOAGs to be within the green valley (27%) or red-sequence (49%) in the colour-stellar mass diagram. All the details analysis and statistical results are presented in Zewdiw *et al.* (2020).

Figure 1 shows that IOAGs spectroscopically classified as star-forming are typically consistent with the main sequence (61%), with only 11% (28%) above (below) this sequence in the SFR-stellar mass diagram. Galaxies spectroscopically classified as composite, LINER or Seyfert 2 have lower SFRs, and are found to be below the main sequence and few of them are on the main sequence. With respect to their morphological classifications as ellipticals are mostly located in the red sequence, with 88% of this population in this colour range, as expected, and 10% in the green valley.

To visualize more clearly the number of sources of each class above on, or below the MS, we compute the specific SFR (sSFR = SFR/$M_\star$), normalized by the sSFR of the main sequence at $z = 0$ at a given stellar mass. Figure 2 shows the normalized distribution of δ_{MS} for our galaxies split by spectroscopic and morphological class. We found that the majority of the samples of IOAGs are located below the MS, some of them are on the MS and very few of the sources being in the starburst regime in both classifications.

4. Conclusions

We studied the morphological properties of our galaxies using the Galaxy Zoo morphological classifications, as well as their spectroscopic properties. The main results of our study are the following:

⋆ Most IOAGs in our spectroscopic sample may have AGN activity.

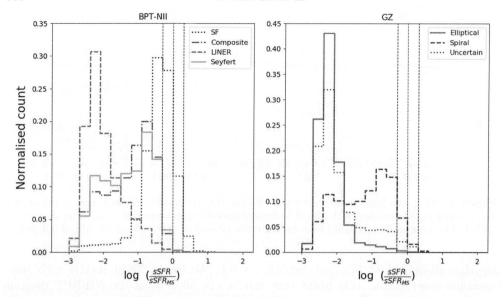

Figure 2. Distribution of the sSFR, normalized to the sSFR of the MS at a given redshift and stellar mass, for the spectroscopic BPT (left) and morphological (right) classifications. The red vertical line represents the location of the MS (at $\mathrm{sSFR}/\mathrm{sSFR}_{\mathrm{MS}}(M_*, z) = 0$). The black dotted lines show the typical spread of the MS, of ±0.3 dex.

⋆ Majority of IOAGs are located below the MS of star formation in the SFR-stellar mass diagram, and in the green valley or red sequence in the colour-stellar mass diagram.
⋆ IOAGs spectroscopically classified as SF have spiral morphologies and are in the main sequence as expected.
⋆ Seyfert 2 and composites have spiral morphologies but quiescent SFRs, which points to the idea that the AGN could be related to their evolutionary state (Zewdie et al. 2020, submitted).

Acknowledgements

DZ acknowledges support from the European Southern Observatory - Government of Chile Joint Committee through a grant awarded to Universidad Diego Portales. DZ and MP acknowledges financial supports from the Ethiopian Space Science and Technology Institute (ESSTI) under the Ethiopian Ministry of Innovation and Technology (MInT). MP acknowledges the support from the Spanish Ministry of Science, Innovation and Universities (MICIU) through project AYA2016-76682C3-1-P and the State Agency for Research of the Spanish MCIU through the Center of Excellence Severo Ochoa award to the IAA-CSIC (SEV-2017-0709). RJA was supported by FONDECYT grant number 1191124.

References

Baldwin J. A., Phillips M. M., and Terlevich R. 1981, *PASP*, 93, 5
Brinchmann, J., et al., 2004, *MNRAS*, 351, 1151
Kewley, L. J., et al., 2006, *MNRAS*, 372, 961
Pérez E., et al., 2013, *ApJ*, 764, L1
Schawinski K., et al., 2003, *MNRAS*, 440, 889
Whitaker K. E., et al., 2012, *ApJ*, 754, L29
Zewdiw, D., et al. 2020, *MNRAS*, 498, 4345

Catalogue with visual morphological classification of 32,616 radio galaxies with optical hosts

Natalia Żywucka[1], Dorota Koziel-Wierzbowska[2] and Arti Goyal[2]

[1]Centre for Space Research, North-West University,
2520, Potchefstroom, South Africa
email: n.zywucka@oa.uj.edu.pl

[2]Astronomical Observatory, Jagiellonian University,
30-244, Kraków, Poland
emails: dorota.koziel@uj.edu.pl; arti@oa.uj.edu.pl

Abstract. We present the catalogue of Radio sources associated with Optical Galaxies and having Unresolved or Extended morphologies I (ROGUE I). It was generated by cross-matching galaxies from the Sloan Digital Sky Survey Data Release 7 (SDSS DR 7) as well as radio sources from the First Images of Radio Sky at Twenty Centimetre (FIRST) and the National Radio Astronomical Observatory VLA Sky Survey (NVSS) catalogues. We created the largest handmade catalogue of visually classified radio objects and associated with them optical host galaxies, containing 32,616 galaxies with a FIRST core within 3 arcsec of the optical position. All listed objects possess the good quality SDSS DR 7 spectra with the signal-to-noise ratio > 10 and spectroscopic redshifts up to $z = 0.6$. The radio morphology classification was performed by a visual examination of the FIRST and the NVSS contour maps overlaid on a DSS image, while an optical morphology classification was based on the 120 arcsec snapshot images from SDSS DR 7.

The majority of radio galaxies in ROGUE I, i.e. $\sim 93\%$, are unresolved (compact or elongated), while the rest of them exhibit extended morphologies, such as Fanaroff-Riley (FR) type I, II, and hybrid, wide-angle tail, narrow-angle tail, head-tail sources, and sources with intermittent or reoriented jet activity, i.e. double–double, X–shaped, and Z–shaped. Most of FR IIs have low radio luminosities, comparable to the luminosities of FR Is. Moreover, due to visual check of all radio maps and optical images, we were able to discover or reclassify a number of radio objects as giant, double–double, X–shaped, and Z–shaped radio galaxies. The presented sample can serve as a database for training automatic methods of identification and classification of optical and radio galaxies.

Keywords. Radio continuum: galaxies; surveys: individual (SDSS, FIRST, NVSS); catalogues: galaxies

1. Introduction

Up to date, the majority of catalogues containing optical galaxies with radio counterparts do not include detailed morphological classification of radio structure and/or associated optical host galaxy (e.g., Lin et al. 2010, Best & Heckman 2012). We provide a catalogue of radio sources identified with optical galaxies, possessing spectroscopic redshift and good quality optical spectra from the Sloan Digital Sky Survey Data Release 7 (SDSS DR 7; Abazajian et al. 2009), measured radio flux densities of radio structures with a flux density limit of the First Images of Radio Sky at Twenty Centimetre (FIRST; White et al. 1997), visually assigned morphological classifications of the radio structure and the optical host galaxy. The catalogue of Radio sources associated with Optical

Galaxies and having Unresolved or Extended morphologies I (ROGUE I) is the largest sample of spectroscopically selected radio galaxies, covering $\sim 30\%$ of the entire sky.

As a parent sample, containing 662,531 unique SDSS galaxies, we used objects from the Red Galaxy Sample (Eisenstein et al. 2001) and the SDSS Main Galaxy Sample (Strauss et al. 2002), introducing a limit on a signal-to-noise ratio in the continuum at 4020 Å ≥ 10 (Koziel-Wierzbowska & Stasińska 2017). In order to identify the SDSS galaxies with radio sources, we cross-match the positions of the SDSS galaxies with the sources from the FIRST catalogue by applying a matching radius of 3 arcsec. Subsequently, we generated the FIRST and the NRAO VLA Sky Survey (NVSS; Condon et al. 1998) radio contour — optical gray images, centred at the host galaxy position and having an angular size equal to a linear size of 1 Mpc at the source distance. This allowed us to visualise the morphologies of small and giant radio sources as well as associate them with an SDSS galaxy. Both radio sky surveys used in this work were conducted at 1.4 GHz and have different angular resolution of the radio images and of the sensitivity for the point-like and extended/diffuse emission. FIRST provides 5.4 arcsec synthesized beam size images and is complete down to 1 mJy flux density limit for point-like sources, while for NVSS these parameters are 45 arcsec and 2.5 mJy, respectively. Using this procedure, we have found 32,616 matching sources, which we further classified morphologically. 629,815 remaining galaxies from the parent sample can still host extended radio emission without a core. This will be searched for in future work on *ROGUE II: A catalog of SDSS galaxies without FIRST cores*.

2. Classification schemes

Our radio morphological classification scheme is detailed and complex, consisting of:
- compact, i.e. point-like single-component sources,
- elongated, i.e. elliptical profile single-component sources,
- FR I, FR II, and hybrid, i.e. linear structure brighter near core, linear structure brighter near edges, and one lobe of FR I and another of FR II morphology, respectively,
- Z–shaped, i.e. sources with Z– or S–shaped radio morphology,
- X–shaped, i.e. sources with X–shaped radio morphology,
- double-double, i.e. two pairs of collinear lobes,
- narrow-angle, wide-angle, and head-tail, i.e. sources with angle between lobes $< 90°$, with angle between lobes $> 90°$, and with bright core and a tail, respectively,
- O I and O II, i.e. one-sided sources with FR I or FR II lobe,
- halo, i.e. diffuse radio emission around the core,
- star-forming region, i.e. emission from the host galaxy

We also classified some of the sources as not clear (radio source with unclear morphology), blended (radio emission blended with other source), and not detected (when an optical galaxy is not a host of the radio emission).

We used the standard Hubble classification scheme to assign a morphological type of the optical host galaxy, extending it with some additional types:
- spiral, i.e. disc galaxy with visible spiral arms,
- elliptical,
- lenticular, i.e. disc galaxy without spiral arms,
- distorted,
- ring, i.e. ring-like shape,
- merger, mainly major merger,
- star-forming region, i.e. SDSS spectrum of star-forming region, not galaxy center,
- off-center, i.e. spectrum not corresponding to star-forming region.

We also included in our scheme an interacting galaxy (with visible signs of interaction) and barred galaxy (spiral or lenticular with prominent bars). Sources with uncertain attribution of any of aforementioned radio or optical types are marked with p for *possible*.

3. Results

Our cross-matching procedure and visual classification allowed us to find that:
- unresolved (compact and elongated) radio sources dominate in ROGUE I, constituting 92%. The remaining 8% of sources show extended morphology,
- secure and possible radio sources of FR I, II, and hybrid form group containing 73% of the extended sources, bent (wide-angle, narrow-angle, head-tail) sources are 23%, and sources with intermittent or re-oriented jet activity (double–double, X–shape, Z–shape sources) — 3%,
- we discovered or reclassified 55 giant, 16 double–double, 9 X–shaped, and 25 Z–shaped radio galaxies,
- the optical morphological classification revealed that 64% of radio galaxies in ROGUE I have elliptical, 15% spiral, 12% distorted, and 7% lenticular hosts.

4. Summary and conclusions

We have created the largest handmade catalogue of 32,616 radio sources associated with optical galaxies. All ROGUE I objects have provided 1.4 GHz radio flux densities of the core from FIRST and flux densities of the total emission from FIRST or NVSS, radio and optical morphological classifications, luminosity distance, spectroscopic redshifts, good quality spectra, and apparent optical magnitudes from SDSS. The ROGUE I sample can serve as a database for training automatic methods of radio sources' identification and their morphological classification.

Acknowledgements

NŻ work is supported by the NCN through the grant DEC-2014/15/N/ST9/05171. DKW acknowledges the support of Polish National Science Centre (NCN) grant via 2016/21/B/ST9/01620, and AG acknowledges support from the Polish National Science Centre through the grant 2018/29/B/ST9/02298.

References

Abazajian, K. N., Adelman-McCarthy, J. K., Agüeros, M. A., *et al.*, 2009, *ApJS*, 182, 543
Best, P. N., & Heckman, T. M., 2012, *MNRAS*, 421, 1569
Condon, J. J., Cotton, W. D., Greisen, E. W., *et al.*, 1998, *AJ*, 115, 1693
Eisenstein, D. J., Annis, J., Gunn, J. E., *et al.*, 2001, *AJ*, 122, 2267
Kozieł-Wierzbowska, D. & Stasińska, G., 2017, *MNRAS*, 415, 1013
Lin, Y.-T., Shen, Y., Strauss, M. A., Richards,G. T., & Lunnan, R. 2010, *ApJ*, 723, 1119
Strauss, M. A., Weinberg, D. H. & Lupton, R. H., 2002, *AJ*, 124, 1810
White, R. L., Becker, R. H., Helfand, D. J., & Gregg, M. D., 1997, *ApJ*, 475, 479

Extragalactic background light inhomogeneities and Lorentz-Invariance Violation in gamma-gamma absorption and Compton scattering

Hassan Abdalla

Centre for Space Research, North-West University, Potchefstroom 2520, South Africa

Abstract. At energies approaching the Planck energy scale $10^19 GeV$, several quantum-gravity theories predict that familiar concepts such as Lorentz (LIV) symmetry can be broken. Such extreme energies are currently unreachable by experiments on Earth, but for photons traveling over cosmological distances the accumulated deviations from the Lorentz symmetry may be measurable using the Cherenkov Telescope Array (CTA). To study the spectral hardening feature observed in some VHE gamma-ray blazars, we calculate the reduction of the EBL gamma-gamma opacity due to the existence of underdense regions along the line of sight to VHE -gamma ray sources and we compared with the possibility of a LIV signature. Considering the LIV effect, we found that the cosmic opacity for VHE-gamma rays with energy more than 10 TeV can be strongly reduced. I will further discuss the impact of LIV on the Compton scattering process, and how future CTA observations may open an exciting window on studies of the fundamental physics.

Keywords. : active, active: blazars, surveys: gamma-rays

Probing black hole - host galaxy scaling relations with obscured type II AGN

Dalya Baron

School of Physics and Astronomy, Tel-Aviv University, Tel Aviv 69978, Israel

Abstract. The scaling relations between supermassive black holes and their host galaxy properties are of fundamental importance in the context black hole-host galaxy co-evolution throughout cosmic time. Beyond the local universe, such relations are based on black hole mass estimates in type I AGN. Unfortunately, for this type of objects the host galaxy properties are more difficult to obtain since the AGN dominates the observed flux in most wavelength ranges. In this poster I will present a new correlation we discovered between the narrow L([OIII])/L(Hβ) line ratio and the FWHM(broad Hα). This scaling relation ties the kinematics of the gas clouds in the broad line region to the ionization state of gas in the narrow line region, connecting the properties of gas clouds kiloparsecs away from the black hole to material gravitationally bound to it on sub-parsec scales. This relation can be used to estimate black hole masses from narrow emission lines only, and thus brings the missing piece required to estimate black hole masses in obscured type II AGN. Using this technique, we estimate the black hole mass of about 10,000 type II AGN, and present, for the first time, M(BH)-sigma and M(BH)-M(stars) scaling relations for this population. These relations are remarkably consistent with those observed for type I AGN, suggesting that this new method may perform as reliably as the classical estimate used in non-obscured type I AGN. These findings open a new window for studies of black hole-host galaxy co-evolution throughout cosmic time.

Keywords. galaxies: active, galaxies: black holes, galaxies: main relations

Peering into the heart of darkness: Radio VLBI survey of the NEP deep field

Joseph Gelfand

New York University Abu Dhabi, Abu Dhabi, United Arab Emirates

Abstract. Active Galactic Nuclei (AGN), accreting supermassive black holes at the centers of galaxies, are believed to produce powerful outflows – often observed as radio jets – which significantly influence the evolution of the surrounding galaxy and inter-galactic medium. However, how these jets – which are produced in the central parsecs of the AGN – impact gas on scales thousands to millions times larger is poorly understood. Doing so requires measuring the properties on all the relevant size scales. In this talk I will present initial results from the deepest-ever radio VLBI survey of an extragalactic field, whose milli-arcsecond angular resolution allows us to probe the central parsecs around these AGN. By comparing the radio properties of the detected radio jets with the multi-wavelength properties of their host galaxies, we are better to understand what galaxies generate powerful radio jets, and how do these outflows affect their host galaxies.

Keywords. galaxies: active, galaxies: properties, surveys: radio

X-ray variability plane revisited: Role of obscuration

Omaira González-Martín

IRyA – Instituto de Radioastronomía y Astrofíisica, 3-72 Xangari, 8701, Morelia, Mexico

Abstract. Scaling relations are the most powerful astrophysical tools to set constraints to the physical mechanisms of astronomical sources and to infer properties for objects where they cannot be accessed directly. We have re-investigated one of these scaling relations using powerful type 1 Seyferts; the so-called X-ray variability plane (or mass-luminosity-timescale relation, McHardy et al. 2006). This relation links the power-spectral density (PSD) break frequency with the SMBH mass and the bolometric luminosity. We used all available XMM-Newton observations to study the PSD and spectra in short segments within each observation. This allows us to report for the first time that the PSD break frequency varies for each object, showing variations in 19 out of the 22 AGN analyzed. Our analysis of the variability plane confirms the relation between the break frequency and the SMBH mass and finds that the obscuration along the line of sight (or the variations on the obscuration using its standard deviation) is also a required parameter. We constrain a new variability plane of the form: $log(\nu_{Break}) = -Alog(M_{BH}) + Blog(N_H) - C$ (or $log(\nu_{Break}) = -Alog(M_{BH}) + B\Delta(N_H) + C$). The X-ray variability plane found by McHardy et al. (2006) is roughly recovered when we use unobscured segments. We speculate the PSD shape is related with the outflowing wind close to the accretion disk at least for these powerful type 1 AGN (Gonzalez-Martin et al. 2018).

Keywords. galaxies: active, galaxies: scaling relations, galaxies: outflows

References

McHardy, I. M., et al. 2006, *Nature*, 444, 730
Gonzalez-Martin, O. 2018, *ApJ*, 858, 2

Optical variability of faint quasars

Endalamaw Ewnu Kassa

Department of Physics, Woldia University, Woldia, Ethiopia

Abstract. The variability properties of a quasar sample, spectroscopically complete to magnitude $J = 22.0$, are investigated on a time baseline of 2 yr, using three different photometric bands (U, J and F). The original sample was obtained using a combination of different selection criteria: colours, slitless spectroscopy and variability, based on a time baseline of 1 yr. The main goals of this work are two-fold: first, to derive the percentage of variable quasars on a relatively short time baseline; secondly, to search for new quasar candidates, missed by the other selection criteria, and thus to estimate the completeness of the spectroscopic sample. In order to achieve these goals, we have extracted all the candidate variable objects from a sample of about 1800 stellar or quasi-stellar objects with limiting magnitude $J = 22.50$ over an area of about 0.50 deg^2. We find that $> 65\%$ of all the objects selected as possibly variable are either confirmed quasars or quasar candidates, on the basis of their colours. This percentage increases even further if we exclude from our lists of variable candidates a number of objects equal to that expected on the basis of 'contamination' induced by our photometric errors. The percentage of variable quasars in the spectroscopic sample is also high, reaching about 50%. On the basis of these results, we can estimate that the incompleteness of the original spectroscopic sample is $< 12\%$. We conclude that variability analysis of data with small photometric errors can be successfully used as an efficient and independent (or at least auxiliary) selection method in quasar surveys, even when the time baseline is relatively short. Finally, when corrected for the different intrinsic time lags corresponding to a fixed observed time baseline, our data do not show a statistically significant correlation between variability and either absolute luminosity or redshift.

Keywords. galaxies: active, active: quasars, galaxies: variability

Search for binary black holes in 10 years of Fermi LAT data with information field theory

Michael Kreter

Centre for Space Research, North-West University, Private Bag X6001,
Potchefstroom 2520, South Africa

Abstract. Blazars are powered by super-massive black holes at their centers and are known for extreme variability on timescales from minutes to years. In case of a binary black hole system, this duality is traceable as periodic modulation of their MeV to GeV emission. So far, no high-significance periodicity has been found with standard approaches. We developed a method to search for periodic patterns in Fermi/LAT light curves, using information field theory (IFT). IFT is a formulation of Bayesian statistics in terms of fields. Bayesian statistics is ideal for the problem at hand since the data is incomplete, irregularly sampled and obeys non-Gaussian statistics such that common least-squares methods do not apply. We present a proof of principle of this method, analyzing a sample of promising binary black hole candidates like PG 1553+113 and Mrk 501.

Keywords. galaxies: active, active: blazars, active: properties

Stellar populations and ages of ultra-hard X-ray AGN in the BASS survey

Mirjana Pović[1,2]

[1] Astronomy and Astrophysics Research and Develoment Division, Etiopian Space Science and Technology Institute (ESSTI), Addis Ababa, Ethiopia

[2] Instituto de Astrofísica de Andalucía (IAA-CSIC), Granada, Spain

Abstract. Connection between star formation and AGN activity has been studied widely over the past years, which shown to be very important for understanding better the role of AGN in galaxy evolution. In this context, what are the stellar ages and average stellar populations of AGN host galaxies, and if there are any differences depending on AGN type, are still open questions that brought many inconsistencies, very often due to different selection criteria used. The AGN sample detected in the ultra-hard X-rays (14–195 keV) by the Swift BAT telescope is not affected by obscuration nor is it contaminated by stellar emission, and presents some of the most unbiased samples. In this talk we will present the results obtained on AGN stellar populations and ages through spectral fittings by using the Swift-BAT AGN Spectroscopic Survey (BASS) which gives us an unique opportunity to understand better the connection between AGN and their host galaxies.

Keywords. galaxies: active, galaxies: properties; surveys: ultra-hard X-rays

Adaptive optics imaging and spectroscopy of the radio galaxy 3C294

Andreas Quirrenbach

Landessternwarte, Zentrum für Astronomie der Universität Heidelberg, Königstuhl 12, 69117 Heidelberg, Germany

Abstract. 3C 294 is a powerful FR II type radio galaxy at $z = 1.786$. Due to its proximity of a bright star, it has been subject to several adaptive optics supported imaging studies. The system shows a clumpy structure indicative of a merging system. There is even tentative evidence that 3C 294 hosts a dual AGN. In order to distinguish between the various scenarios for 3C 294 we performed deep high-resolution adaptive optics imaging and optical spectroscopy of 3C 294 with the Large Binocular Telescope. We resolve the 3C 294 system in three distinct components separated by a few tenths of an arcsecond. One of them is compact, the other two are extended. The nature of the latter is unclear. They could be a single galaxy with an internal dust absorption feature, a galaxy merger, or two galaxies at different redshifts. We can now uniquely associate the radio source of 3c 294 with one of the extended components. Based on our spectroscopy, we determine a slightly different redshift of $z = 1.784$. We find, however, in addition a single emission line at a wavelength of 6745 AA, which might be identified with Lyα at $z = 4.56$. It thus appears unlikely that 3C294 hosts a dual AGN; it might rather be a pair of AGNs with very small projected separation.

Keywords. galaxies: active, galaxies: properties, methods: adaptive optics

GeMS/GSAOI near-infrared imaging of z ∼ 0.3 BL Lacs

Susan Ridgway

National Optical Astronomy Observatory, 950 North Cherry Avenue, Tucson, AZ, 85719, USA

Abstract. Bright quasars at low z have generally been found in massive, evolved host galaxies, consistent with formation at early epochs. However, deep, high resolution, multicolor imaging of some quasar hosts have found morphological evidence of tidal tails and colors indicative of active star formation. These results are consistent with theories of galaxy formation and evolution in which merger processes trigger the activation of the quasar phase, and energetic feedback is essential. Understanding the role the black hole population plays in the galaxy formation process is important, but imaging the host galaxies around bright quasars is difficult because of the contribution of the bright nuclei. Very high resolution, deep imaging is necessary to successfully remove the nuclear component. We made high-resolution near-infrared images of several bright $z \sim 0.3$ BL Lacs with the Gemini Multi-Conjugate Adaptive Optics System (GeMS)/GSAOI in order to study their host galaxies. We will present the results of this imaging with the 1 arcmin AO-corrected field provided by GeMS/GSAOI and compare with available HST imaging available in the archive.

Keywords. galaxies: active, galaxies: properties, methods: near-infrared imaging

The circum-nuclear environment and jets of active galaxies at z ∼ 0: Recent results from a multi-frequency investigation

Prajval Shastri

Indian Institute of Astrophysics, Sarjapur Road, Bengaluru 560034, India

Abstract. We report results from a multi-frequency investigation of very nearby accreting supermassive black holes. We seek to test the hypothesis that imprints of AGN feedback are present at z ∼ 0. Our sample contains about 130 AGN which were chosen to have redshifts less than 0.02, so our optical imaging typically has several spatial resolution elements across the host galaxy. In addition to optical IFU measurements for all the 130 objects, we have GMRT and archival VLA imaging for subsets of the sample, and also ATCA, ASTROSAT and Chandra observations for select objects. We present our most recent results based on the multi-frequency data, for the systematics of the sample as a whole and on individual objects, that includes a binary black hole precursor.

Keywords. galaxies: active, galaxies: black holes, surveys: multiwavelength

Spectral energy distribution of blazars

Prospery Simpemba

Department of Physics, Copperbelt University, Copperbelt, Zambia

Abstract. This study focuses on spectral energy distributions and light-curves of blazars and radio galaxies, and the testing of the existing models with a view to appropriately predict a new model that will nearly accurately present the nature of the energy outflows of these supermassive bodies. Understanding blazar emission is very important as it relates more directly to the physics of the AGN's central black hole. X-ray, radio and gamma-ray wavelength range data on blazars and radio galaxies from archived data has been collected and a detailed investigation of the spectral energy distribution patterns of the blazars and radio galaxies carried out so as to fit the data in the various models. The results of this investigation will be discussed in detail in this presentation.

Keywords. galaxies: active, galaxies: properties, active: blazars

The GLEAM 4-Jy Sample: The brightest radio-sources in the southern sky

Sarah White[1,2,3]

[1]South African Radio Astronomy Observatory (SARAO), 2 Fir Street, Observatory, Cape Town, 7925, South Africa

[2]Department of Physics and Electronics, Rhodes University, PO Box 94, Grahamstown, 6140, South Africa

[3]Astrophysics, University of Oxford, Denys Wilkinson Building, Keble Road, Oxford, OX1 3RH, UK

Abstract. Low-frequency radio emission allows powerful active galactic nuclei (AGN) to be selected in a way that is unaffected by dust obscuration and orientation of the jet axis. It also reveals past activity (e.g. radio lobes) that may not be evident at higher frequencies. Currently, there are too few "radio-loud" galaxies for robust studies in terms of redshift-evolution and/or environment. Hence our use of new observations from the Murchison Widefield Array (the SKA-Low precursor), over the southern sky, to construct the GLEAM 4-Jy Sample (1,860 sources at $S_{151MHz} > 4$ Jy). This sample is dominated by AGN and is 10 times larger than the heavily relied-upon 3CRR sample (173 sources at $S_{178MHz} > 10$ Jy) of the northern hemisphere. In order to understand how AGN influence their surroundings and the way galaxies evolve, we first need to correctly identify the galaxy hosting the radio emission. This has now been completed for the GLEAM 4-Jy Sample – through repeated visual inspection and extensive checks against the literature – forming a valuable, legacy dataset for investigating relativistic jets and their interplay with the environment.

Keywords. galaxies: active, galaxies: relativistic jets, surveys: radio

An accreting $< 10^5 M_\odot$ black hole at the center of dwarf galaxy IC750

Ingyin Zaw

New York University Abu Dhabi, Abu Dhabi, United Arab Emirates

Abstract. Nuclear black holes in dwarf galaxies are important for understanding the low end of the supermassive black hole mass distribution and the black hole-host galaxy scaling relations. IC 750 is a rare system which hosts an AGN, found in ~0.5% of dwarf galaxies, with circumnuclear 22 GHz water maser emission, found in ~3-5% of Type 2 AGNs. Water masers, the only known tracer of warm, dense gas in the center parsec of AGNs resolvable in position and velocity, provide the most precise and accurate mass measurements of SMBHs outside the local group. We have mapped the maser emission in IC 750 and find that it traces a nearly edge-on warped disk, 0.2 pc in diameter. The central black hole has an upper limit mass of $\sim 1 \times 10^5$ $M_\odot$ and a best fit mass of $\sim 8 \times 10^4$ $M_\odot$, one to two orders of magnitude below what is expected from black hole-galaxy scaling relations. This has implications for models of black hole seed formation in the early universe, the growth of black holes, and their co-evolution with their host galaxies.

Keywords. galaxies: active, galaxies: black holes, galaxies: evolution

Chapter IX. Non-AGN posters

Science strategy of the African Astronomical Society (AfAS): *An outcome of the Science Business held in synergy with the IAUS 356*

Lerothodi L. Leeuw, Kevin Govender, Charles M. Takalana, Zara Randriamanakoto, and Alemiye Mamo

College of Graduate Studies, University of South Africa, Barney Pityana: Office 01-105, Florida Science Campus, PO Box 392, UNISA, 0003, South Africa
email: lerothodi@alum.mit.edu

Abstract. Presented here, is a summary of discussions at African Astronomical Society (AfAS) Science Business Meeting, Addis Ababa, Ethiopia, 10-11 October 2019. This summary was deliberated with delegates of the International Astronomical Union (IAU) Symposium 356, during a lunch session of the meeting.

Keywords. African Astronomy, African Astronomical Society, Science Strategy

1. Introduction

The African Astronomical Society (AfAS, Buckley (2019); AfAS (2019)) is a Pan-African Professional Society of Astronomers. The Society was relaunch at the Astronomy in Africa business meeting, which was held in Cape Town at the South African Astronomical Observatory on 25-26 March 2019. AfAS held a Science Business Meeting in Addis Ababa, Ethiopia, on the 10th to 11th October 2019, which was attended by about 30 delegates from Africa and the diaspora. One of the key aims and outcomes of the meeting was a draft strategy for the Science Committee of the Society. Presented here, is extracts from the resulting draft science strategy and summary of the discussions from the meeting, that was deliberated with delegates of the International Astronomical Union (IAU) Symposium 356, during a lunch session of the meeting. The goal was to get feedback and input from the IAUS 356 delegates on such discussions, given their deep and broad backgrounds, and to introduce them to the Society.

2. AfAS Science Strategy

As a professional astronomical society, AfAS recognises science and in extension the science committee as a cornerstone of the Association. The vision of the AfAS Science Committee is to create an interlinked and world-class African astronomy community contributing to the advancement of human knowledge. AfAS envisages establishing an experienced and representative science committee, which will carry out the mission to advance astronomy through the development of strategies, facilitation of interdisciplinary collaborations, encouragement of cross border engagements, and stimulation of human capital development. The meeting proposed objectives in these five categories or goals, that are itemized and detailed further below.

© The Author(s), 2021. Published by Cambridge University Press on behalf of International Astronomical Union

2.1. To support the development and sharing of national strategies

At this early developmental stage of the Society, points of contact in each African country must be established as they will be instrumental in identifying science focus areas of research groups in Africa - these focus areas would feed science strategies and be used to guide the science discussions at upcoming AfAS conferences/workshops, including AfAS2020†. A database of national and regional strategies may be compiled and used by AfAS to identify strategic science focus areas (or big science questions) that are of interest to the African astronomy community. Workshops at AfAS2020 or other meetings will create platforms to discuss strategic science focus areas for a connected African community in addition to this a task force of experienced individuals may be established to advise on the development of strategies and identify resources available (and required) in different locations. In the long term, African countries will be encouraged to have appropriately aligned strategies which include astronomy.

2.2. To advance Astronomical Science and Technologies

To take astronomy forward on the continent, science collaborations, research exchanges, focused meetings, and workshops including observational/hands-on activities can be used to connect research groups across Africa. For this AfAS will need to identify partners to support such meetings, workshops, and any other relevant activities, including fellowships, research grants. Wherever possible AfAS can encourage streaming and/or recording of science meetings, this will ensure that those who are unable to travel to such meetings have access to the content which may be instrumental in their research. AfAS can also facilitate coordinated access to research telescopes through requesting time on instruments available on the continent e.g., time on SALT earmarked for African researchers. The community can be encouraged through awards for recognition of Excellence. AfAS can establish a prize system for excellence in various areas and launch an AfAS fellowship to nurture early career researchers. However, to implement this AfAS may be required to establish an awards committee.

The African Astronomical community is gradually growing and it is important to carry out various skills development projects. Proposal writing skills for research funding and observing time can be delivered to members of this growing community through various workshops and mentoring programmes. AfAS can contribute to the exchange of knowledge through meetings and publications. These initiatives will provide a space for student contributions, non-research publications, historical information, and proceedings. To benefit the community AfAS can house databases of astronomy expertise, active research areas, infrastructures across Africa, and the African diaspora. This can be done by the development of a web portal for the exchange of information, which will be used to disseminate (open source) resources that will benefit astronomy, including research tools e.g. software packages and a list of experts.

2.3. Human Capital Development

AfAS is well-positioned as a professional society to create the bridge between teaching and research. This means that AfAS may have to identify and share information on available teaching materials and programmes e.g. lecture notes, books, etc. In addition to this AfAS can disseminate a list of online courses that could supplement astronomy teaching and research and establish a database of experts willing to offer courses in other universities. AfAS can play a role in driving collaborations for joint supervision,

† www.africanastronomicalsociety.org/conferences-and-meetings/afas-astronomy-in-africa-meetings/afas2020

and implementation of twinning to support institutions without astronomy programmes. Part of this would be identifying sources of funding to support exchanges. AfAS may also encourage short term research visits between universities for postgraduate students and ensure that the African community is aware of training opportunities and scholarships: academic writing, schools, DARA, SKA, etc. This can be done by AfAS establishing strategic relationships with training programmes to create opportunities for Africans. AfAS may begin looking at identifying and disseminating information about scholarships available to African students and in the long term, AfAS may begin looking into the establishment of an AfAS scholarship.

In connecting the African community AfAS can play a role in identifying and sharing opportunities with robotic telescopes/remote observing to improve the quality of training e.g. symposia, workshops, etc. AfAS will need to also develop best practice guidelines for workshops including access to training materials and explore the feasibility of a certification/vetting/endorsement process. AfAS can be involved in arranging workshops for training of trainers and explore MOOCs to support trainer development.

2.4. International and national engagements

The major role of AfAS and a driving force for the Science Strategy is connecting the African community to international initiatives. This can be done through supporting and encouraging the presence of African delegates at international meetings and involvement in international science projects. Platforms may need to be established to raise awareness on international opportunities e.g. European Research funding, and support for African countries to become active participants in international fora. At present it is pivotal that AfAS focus on increasing African membership of IAU (African countries could become either full or observer IAU members), this process can be accelerated in light of the IAU General Assembly, which will be hosted for the first time on African soil in 2024. In addition to human capital efforts, it is envisaged that AfAS will engage with the relevant stakeholders for the protection of astronomy and astronomical facilities. This will involve Lobbying for dark skies and preservation and regulation of radio quiet zones, which are advantageous for astronomy. AfAS could facilitate connections for expert support for site testing and planning of new astronomical instruments and Infrastructure.

2.5. Facilitate interdisciplinary collaborations

Going forward partnering with relevant organisations outside astronomy on issues related to the AfAS vision would allow the society to bridge the gap between modern and indigenous astronomy. Processes that could be initiated may include identifying and engaging with groups that research, collect, register, preserve indigenous knowledge of astronomy, and/or translate astronomical terms into local languages. It is envisaged that future AfAS meetings may include interdisciplinary sessions.

3. Conclusion

As a result of the Astronomy in Africa Science Business meeting, AfAS has been tasked with working on items deliberated at the meeting and is expected to report back on objectives and actions at the AfAS 2020 meeting. The Society has issued an open call for Nominations for members of the Science Committee in addition to members nominated at this meeting. This committee has also been mandated with identifying overlaps/gaps between AfAS science strategy and relevant African Union (AU) strategies that would form the basis for approaches to the AU.

Acknowledgement

The authors acknowledge funding from the South African National Foundation (NRF) and Department of Science and Innovation (DSI) to the African Astronomical Society (AfAS) and the Ethiopian Space Science and Technology Institute (ESSTI) and partners for funding the meeting and its participation. Further, the authors acknowledge the organizers of this symposium for the opportunity to have the AfAs Science Business meeting in synergy with the symposium and to present and deliberate its outcome in a lunch session of the symposium. The content presented here had input from the participants of the AfAS Science Business and this record serves to further their participation, wishes, and those of AfAS.

References

Buckley, D. A. H. 2019, *Nature Astronomy*, 3, 369–373
African Astronomical Society Online Pages 2019, https://www.africanastronomicalsociety.org/

Mass-radius relation of compact objects

Seman Abaraya and Tolu Biressa

Department of Physics, Astronomy and Space Science Stream, Jimma University, Jimma, Ethiopia
emails: `semanabaraya06@gmail.com`; `tolu_biressa@yahoo.com`

Abstract. Compact objects are of great interest in astrophysical research. There are active research interests in understanding better various aspects of formation and evolution of these objects. In this paper we addressed some problems related to the compact objects mass limit. We employed Einstein field equations (EFEs) to derive the equation of state (EoS). With the assumption of high densities and low temperature of compact sources, the derived equation of state is reduced to polytropic kind. Studying the polytropic equations we obtained similar physical implications, in agreement to previous works. Using the latest version of Mathematica-11 in our numerical analysis, we also obtained similar results except slight differences in accuracy.

Keywords. Compact objects, mass limit, equation of state, mass-radius relation.

1. Introduction

Compact objects (COs) represent the final stage in the evolution of ordinary stars. They are formed when stars cease their nuclear fuel to support themselves against gravity. They are generally categorized as black holes, neutron stars, or white dwarfs depending upon the masses of their progenitors (Shapiro and Teukolsky 1986). Today, the COs are amongst important objects used to understand better the structure, formation, and evolution of stellar sources. Yet, there are issues concerning the structure and size of the COs by themselves (Cardoso and Pani 2019). Thus, there is an active research for better understanding and application of the COs. So, in this paper we worked out EFEs to study the COs mass limit in relation to their radii. The detailed method is given as in the upcoming section.

2. Methodology

EFEs were used to derive EoS with simplifying boundary conditions such as variational techniques (Weinberg 1972). With the assumption of high densities and low temperature characteristics of the COs, the state equation is further reduced to polytrope kind (Chandrasekhar 1967). Finally, the obtained polytrope is being worked out for analysis both analytically and numerically where the results are being compared with previous results.

3. Results and discussion

Working out the EFEs, (as outlined in the methodology) we obtained the Lane-Emden equation given as in the standard books (Chandrasekhar 1967; Weinberg 1972). The analytical solutions exist only for polytropes of indices n = 0, 1 and 5 as shown earlier (Chandrasekhar 1967). For the other values of n, the Lane-Emden equation is to be integrated numerically. Thus, we used the latest version of *Mathematica-11* to integrate the Lane-Emden equation numerically for polytropes of integral indices of $n = 0-6$. The results are depicted as in Figure 1. The stable solutions of the COs are determined from

Table 1. Numerical solutions of Lane-Emden equation for mass-radius relationship of compact objects with integral polytropic indices, $n = 0 - 5$. Authors' solutions vs Chandrasekhar (1967).

	ξ_1		$\left(-\xi^2 \frac{d^2\theta(\xi)}{d\xi^2}\right)\Big|_{\xi=\xi_1}$	
n	Seman & Tolu	Chandrasekhar (1967)	Seman & Tolu	Chandrasekhar (1967)
0	2.449489	2.4494	4.89898	4.8988
1	3.141581	3.14159	3.14159	3.14159
2	4.35287	4.35287	2.41104	2.41105
3	**6.89685**	**6.89685**	**2.01824**	**2.01824**
4	14.97153	14.97155	1.79723	1.79723
5	∞	∞	1.73204	1.73205

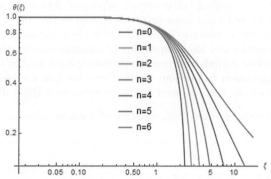

Figure 1. The LogLogPlot of polytropic solutions of Lane-Emden equation for Mass-Radius relationship of compact objects with integral polytropic indices, $n = 0 - 6$.

the intersections of θ-ξ curves with the ξ-axis. Where, θ and ξ are the parameterized dimensionless density and radius of the COs respectively. So, in our sample plots the solutions exist for integral polytropic indices $n = 0 - 5$, while $n = 6$ is non-existent. On the other hand, we compare our numerical results of the parameterized mass-radius relations of the COs with that given by (Chandrasekhar 1967) as in table 1. As we observe from the table, our numerical results are also similar except slight differences in accuracy.

4. Conclusions

The polytropic equation we have obtained has similar physical implications, in agreement to the earlier works by others. Furthermore, our numerical results where the latest version of Mathematica-11 used is similar except slight differences in accuracy.

References

Camenzind, M., 2007, *Compact objects in astrophysics*, Springer, Berlin, Heidelberg
Cardoso, V. & Pani, P., 2019, *Living Reviews in Relativity* 22(1)
Chandrasekhar, S., 1967, *An introduction to the study of stellar structure*, Dover, New York
Shapiro, S. L. & Teukolsky, S. A., 1986, *Black Holes, White Dwarfs and Neutron Stars*
Weinberg, S., 1972, *Gravitation and Cosmology*, John Wiley & Sons, Inc.

Non-thermal radio emission from dark matter annihilation processes in simulated Coma like galaxy clusters

Fitsum Woldegerima Beyene[1] and Remudin Reshid Mekuria[2]

[1]Department of Physics, Wachamo University, Hossana, Ethiopia
email: `fitsewgerima@gmail.com`

[2]Department of Physics, Addis Ababa University, Addis Ababa, Ethiopia
email: `remudin.mekuria@gmail.com`

Abstract. Taking secondary particles produced from dark matter (DM) annihilation process to the origin of the extended diffuse radio emission observed in galaxy clusters, we studied both their morphology and radio spectral profile using simulated Coma like galaxy clusters. We have considered a neutralino annihilation channel dominated by $b\bar{b}$ species with a branching ratio of 1 and neutralino mass of 35 GeV with annihilation cross-section of 1×10^{-26} cm^3 s^{-1}. The radio emission maps produced for the two simulated galaxy clusters which are based on the MUsic SImulation of galaxy Clusters (MUSIC) dataset reveal the observed radio halo morphology showing radio emission both from the central regions of the cluster and substructures lying out off cluster centre. The flux density curve is in a good agreement for $\nu \leq 2$ GHz with the obsevational values for the Coma cluster of galaxies showing a small deviation at higher frequencies.

Keywords. Dark matter (DM) annihilation, neutralino, radio emission, integrated flux

1. Introduction

Observations have revealed that there is a diffuse radio emission in cluster of galaxies which is non-thermal synchrotron in origin. The diffuse extended radio emission of the Coma cluster of galaxies was investigated by Willson (1970). Several years later, extended diffuse radio emission was also detected at the periphery of the Coma cluster and at the center of the Perseus cluster (van Weeren et al. 2019). This diffuse extended radio sources are not associated to the individual galaxies and their origin and evolution is still a matter of debate (Colafrancesco et al. 2015). It is therefore important to carry out new studies aimed at discriminating between these theoretical models. We studied the radio emission from DM annihilation/decay processes in two simulated Coma like galaxy clusters at different frequencies choosen to make comparison with the observational results of (Thierbach et al. 2003). We model the radio emission from DM annihilation processes because of the fact that the clusters are dominated by DM and since the Coma cluster is one of the best studied we choose simulated Coma like galaxy clusters. For this purpose we took two simulated Coma like galaxy clusters, named SGC280 and SGC282.

2. Methodology

A fortran code is used to allocate the DM and gas density on a three dimensional cube of side 2 Mpc which is grided into smaller cubes of side $\sim$ 10 kpc. Now knowing the DM and gas density in each cube help us to determine and allocate the magnetic

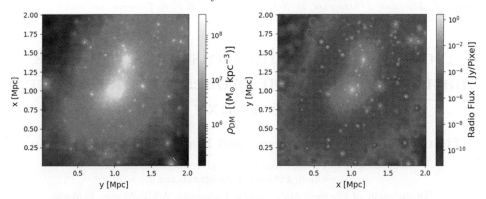

Figure 1. The DM density map of SGC280 [Left Panel] and radio emission map of SGC280 [Right Panel] at 110 MHz.

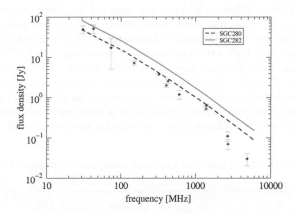

Figure 2. Integrated radio flux for SGC282 and SGC280 indicated in red (solid) and dark (broken) curves, respectively in comparison to the observational data of radio emission of Coma cluster indicated in blue dots from Thierbach et al. (2003).

field and electron density within the cluster which in turn allow us to compute the gyro- and plasma frequency. Combining this results with the electron production spectra from the DarkSUSY package (a fortran package that calculate the yield of particles per a neutralino annihilation) enables us to calculate all the necessary ingredients for the synchrotron emission. Finally, the local emissivity and the integrated flux density are computed in each cube.

3. Implications

All the densest regions on the maps which are near the center and out of the center are the sources (or amplifiers) of higher radio emission (see Fig. 1 [Left Panel]). And this is clearly depicted in Fig. 1 [Right Panel] by the radio emission map which is morphologically similar to the DM density distribution map (Giovannini et al. 2012). From the curves in Fig. 2 we can see that the flux density decreases as frequency increases in both cases. This indicates that the integrated radio spectrum of SGC280 is in a good agreement with the observed radio spectrum of the Coma cluster of galaxies (Thierbach et al. 2003).

References

Colafrancesco, S., Marchegiani, P., & Beck, G. 2015, *Evolution of dark matter halos and their radio emissions*, 32, 2015

Giovannelli, F. & Sabau-Graziati, L. 2012, *Multifrequency behaviour of high energy cosmic sources. A review*, 17, 83

Mekuria, R. & Colafrancesco, S. 2017, *Multi-wavelength emissions from dark matter annihilation processes in galaxy clusters using cosmological simulations*, 9

Thierbach M., Klein U., & Wielebinski R. 2003, *The diffuse radio emission from the Coma cluster at 2.675 GHz and 4.85 GHz*, 53, 397

van Weeren, R. J., de Gasperin, F., Akamatsu, H., Brüggen, M., Feretti, L., Kang, H., Stroe, A., & Zandanel, F. 2019, *Space Science Reviews*, 215, 1, 16

Star formation efficiency of magnetized, turbulent and rotating molecular cloud

Gemechu M. Kumssa[1,2,3] and Solomon Belay Tessema[1,2]

[1] Ethiopian Space Science and Technology Institute (ESSTI), Entoto Observatory and Research Center (EORC), Astronomy and Astrophysics Research and Development Division, P. O. Box 33679, Addis Ababa, Ethiopia

[2] Addis Ababa University, P. O. Box 1176, Addis Ababa, Ethiopia

[3] Jimma University, College of Natural Sciences, Department of Physics, Jimma, Ethiopia
emails: gemechumk@gmail.com; tessemabelay@gmail.com

Abstract. The formation of stars constitutes one of the basic problems in astrophysics. Understanding star formation efficiency of molecular clouds (MCs) of a galaxy is necessary for studying the galactic evolution. Present data and theoretical formulations show that the structure and dynamics of the interstellar medium (ISM) are extremely complex. Therefore, there is no simple model that can explain adequately the star formation efficiency of MCs because of its complex nature. The initial mass of the cloud needed for collapse varies based on the environment in which the cloud resides and the strength of its magnetic field, turbulence, as well as the speed of rotation. In this paper, we estimate the star formation efficiency by combining pre-determined models and the critical mass formulated by Kumssa & Tessema (2018).

Keywords. Stars, star formation, critical mass

1. Introduction

Understanding star formation efficiency of molecular clouds (MCs) of a galaxy is important for studying the galactic evolution as well as to build a comprehensive theory of star formation. Moreover, Star formation lies at the center of the processes that drive synthesis of elements, formation of planets, and development of life (Krumholz 2014). Due to the complexity of Galactic interstellar medium (ISM) structure and dynamics, either predicting or explaining the star formation efficiency (SFE) of MCs is not an easy task. Therefore, finding SFE and the mass of the star formed from the collapsing cloud using the critical mass, different time scales will provide an additional knowledge on how currently evolving MCs can be converted to stars. The purpose of this study is to theoretically calculate the stellar mass and SFE of magnetized, turbulent and rotating MC.

2. Critical mass and stellar mass

The critical mass theoretically modeled by Kumssa & Tessema (2018) is:

$$M_{crt} = \frac{5R_{core}}{G}\left(\frac{P_g(\rho_{core})}{\rho_{core}} + \sigma_r^2\right) + \frac{10R_{core}^3}{3G}c_{rot}\Omega_{core}^2 + \frac{5B_o^2 R_o^4}{4\pi G \rho_{core} R_{core}^3}\left(\frac{1}{3} - \frac{1}{2\pi R_{core}^2}\right) \quad (2.1)$$

The analytic model developed by Burkert, A. and Hartmann (2013) is:

$$M_\star(\tau) = M_{crt,0}(\tau=0)\left(1 - e^{-\epsilon\tau}\right) + \dot{M}_{in}t_{ff}\left[\tau - \left(1 - e^{-\epsilon\tau}\right)/\epsilon\right] \quad (2.2)$$

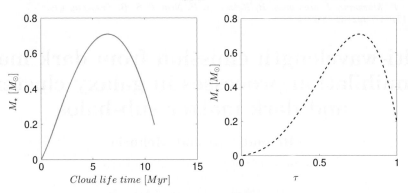

Figure 1. The left panel shows MC life-time (t) versus the stellar mass, while the right panel shows t/t_{ff} versus stellar mass, where $\epsilon = 0.1$, $R_c = 0.01$pc, $\dot{M}_{in} = 10^{-5} - 10^{-7} M_\odot yr^{-1}$, $\tau = 0.1$–10. $B_o \Rightarrow 0$, $M_{crt} = 1.502 M_\odot$, $R_{core} = 0.01$pc, and $t_{ff} \sim 10^6 - 10^{7.04} yr$.

3. Results

Using analytic stellar mass model described in previous section we define stellar mass as:

$$M_\star = \frac{5R_{core}}{G}\left(\frac{P_g(\rho_{core})}{\rho_{core}} + \sigma_r^2\right) \frac{10 R_{core}^3}{3G} c_{rot} \Omega_{core}^2$$
$$+ \frac{5B_o^2 R_o^4}{4\pi G \rho_{core} R_{core}^3}\left(\frac{1}{3} - \frac{1}{2\pi R_{core}^2}\right)(\tau = 0)\left(1 - e^{-\epsilon\tau}\right)$$
$$+ \dot{M}_{in} t_{ff}\left[\tau - \left(1 - e^{-\epsilon\tau}\right)/\epsilon\right] \qquad (3.1)$$

The time scale is defined as $\tau = t/t_{ff}$, where t is the life-time of MC, t_{ff} is its free-fall time, while ϵ is its star formation efficiency. Recent studies showed that average values of the magnetic field of MC is $\approx$ 5-15 μG, while the angular speed of MC is approximately $10^{-15} s^{-1}$ - $10^{-13} s^{-1}$ (Beck 2001; Crutcher et al. 2010). In this work we used these values.

4. Summary

By setting SFE $\simeq \frac{M_\star}{M_{crt}}$, we obtained the maximum SFE of ~ 0.6223 and the minimum of ~ 0.0233. When life time of the cloud is almost approaching to its local free-fall time we get the maximum SFE $\sim$ of 0.4660, while if the life time is shorter than t_{ff} SFE is ~ 0.

Acknowledgments

We thank Ethiopian Space Science and Technology Institute (ESSTI), Entoto Observatory & Research Center, and Astronomy and Astrophysics Research and Development Division for their support. We gratefully acknowledge the International Science Programme (ISP) from Uppsala University for their financial support. GMK thanks Jimma University for its support.

References

Beck, R. 2001, *Space Science Reviews*, 99, 243
Burkert, A. & Hartmann, L., 2013. *APJ*, 773(1), p. 48
Crutcher, R. M., Hakobian, N., & Troland, T. H. 2010, MNRAS, 402, L64
Kumssa, G. M. & Tessema, S. B. 2018. *IJAA*, 8(04), p. 347
Krumholz, M. R. 2014. *Physics Reports*, 539(2), pp. 49–134.
Richard, M. Crutcher, Nicholas H., & Thomas H. Troland, *MNRAS Letters*, 402(1):L64–L66, 2010.

Multi-wavelength emission from dark matter annihilation processes in galaxy clusters and dark matter sub-halos

Remudin Reshid Mekuria

Department of Physics, Addis Ababa University,
Addis Ababa, Ethiopia
email: remudin.mekuria@gmail.com

Abstract. Multi-wavelength emission maps from dark matter (DM) annihilation processes in galaxy clusters are produced using Marenostrum-MultiDark SImulation of galaxy Clusters (MUSIC-2) high resolution cosmological simulations. Comparison made with observational radio emission flux data (spectral shape) and the spatial distribution from the simulated emission maps show that secondary particles from DM annihilation could describe the origin of energetic particles which are the sources of the diffuse radio emission observed in large number of galaxy clusters. DM sub-halos which are dominantly composed of DM, but with very little or no gas and stellar content, are ideal objects to study the nature and properties of DM. Therefore, statistical studies of a large number of them as well the emission maps of high mass-to-light ratio DM sub-halos will not only explain the observed diffused radio emission but also provide very crucial information about the nature and properties of DM particles.

Keywords. Dark matter, galaxy cluster, sub-halo, radio emission.

1. Introduction

Large-scale diffuse radio sources have been observed by sensitive radio telescopes in many clusters (van Weeren et al. 2019). Various observations have shown that these emissions follow a power-law spectrum, suggesting that these sources are of non-thermal origin due to synchrotron emission produced by relativistic particles in a magnetic field (Colafrancsco et al. 2006). Because of the synchrotron and inverse Compton energy losses, the typical lifetime of the relativistic electrons in the intra-cluster medium (ICM) is expected to be relatively short $\sim 10^8$ yr (e.g. Sarazin 1999). As a result, the electrons suffer from difficulties to diffuse over a Mpc-scale region within their radiative lifetime (see for e.g. Feretti et al. 2012). To resolve these and more, several models have been proposed which describe the mechanism of energy transfer into the relativistic electron population as well as the origin of these electrons (Colafrancsco et al. 2006).

2. Overview

Emissions across the multi-wavelength spectrum are expected from the secondary particles, i.e., from the electrons/positrons which are by-product of the self annihilation of super-symmetric DM particles (Colafrancsco et al. 2006). Thus we have studied these emissions expected in galaxy clusters by considering the annihilation of the most viable DM particles, the neutralinos. We have considered neutralino masses of 35 and 60 GeV DM models with two magnetic field models, and averaged DM annihilation cross-section times velocity $<\sigma V>$ of 1.0×10^{-26} cm^3 s^{-1}. Applying a Smooth Particle

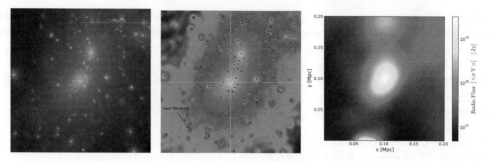

Figure 1. *Left Panel*: DM density projection maps of 2.0 Mpc diameter cluster. *Middle Panel*: Identifying DM sub-halos in the cluster. *Right Panel*: Synchrotron emission map at 110 MHz for the DM dominated sub-halo indicated by the arrow in the middle panel.

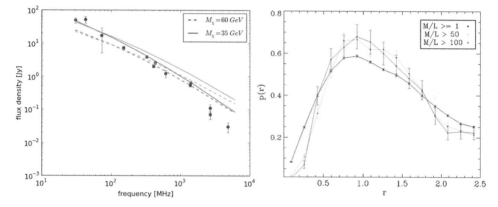

Figure 2. *Left Panel*: Flux density from synchrotron emission are compared with observations of the Coma cluster (blue data points) from Thierbach *et al.* (2003) for neutralino masses of 60 GeV (dotted curves) and 35 GeV (solid curves). Magnetic field model used are Model A, in red and Model B in cyan color. *Right Panel*: the probability distribution function of sub-halos based on their mass-to-light ratio (M/L) at a relative distance from the center of the host clusters.

Hydrodynamics (SPH) we have determined the DM densities at arbitrary locations within the cluster volume which is used to focus on the contribution of sub-halos (Fig. 1 [*Middle Panel*]) to the radio emission maps of clusters. A map showing the DM density projection is given in Fig. 1 [*Left Panel*]. A map showing the spatial distribution of the synchrotron emission of the chosen high M/L ratio DM sub-halo is given in Fig. 1 [*Right Panel*]. The flux density of synchrotron emission is compared with the observational data of Coma cluster in Fig. 2 [*Left Panel*]. The relative distance (from the center of the host clusters) of finding high M/L ratio DM sub-halos is also given in Fig. 2 [*Right Panel*].

3. Implications

We have investigated the nature and properties of DM indirectly by studying the non-thermal radio emission from their annihilation processes in simulated Coma like galaxy clusters. The DM radio flux densities for a neutralino mass of 35 GeV and the magnetic field based on Model A match best the observed diffuse radio emission of the Coma cluster (Thierbach *et al.* 2003). The fact that the flux density shows a very good agreement with the observed one without invoking a boost factor from DM sub-structures shows that the distribution of sub-structures is described well in the simulations. Even though not presented here, our model also predicts gamma-ray emissions. This multi-wavelength study will therefore have a strong implication to unveil both the source of the observed diffuse radio emission in clusters, and learn about the nature and properties of DM.

References

Colafrancesco, S., Profumo, S., & Ullio, P. 2006, *Astronomy and Astrophysics*, 455, 21
Feretti, L., Giovannini, G., Govoni, F., & Murgia, M. 2012, *The Astronomy and Astrophysics Review*, 20, 1, 54
Sarazin, C. L. 1999, *APJ*, 520, 529
Thierbach, M., Klein, U., & Wielebinski, R. 2003, *Astronomy and Astrophysics*, 397, 53
van Weeren, RJ., de Gasperin, F., Akamatsu, H., Brüggen, M., Feretti, L., Kang, H., Stroe, A. & Zandanel, F. 2019, *Space Science Reviews*, 215, 1, 16

14 years of 6.7 GHz periodic methanol maser observations towards G188.95+0.89

Martin M. Mutie[1,2], Paul Baki[1], James O. Chibueze[3,4] and Khadija El Bouchefry[5]

[1]Department of Physics, Technical University of Kenya, P. O. Box 52428-00200, Nairobi, Kenya

[2]Department of Physical Sciences, Chuka University, P. O. Box 109-60400, Chuka, Kenya

[3]Centre for Space Research, Physics Department, North-West University, Potchefstroom 2520, South Africa

[4]Department of Physics and Astronomy, University of Nigeria, Carver Building, 1 University Road, Nsukka, Nigeria

[5]South African Radio Astronomy Observatory, Rosebank, Johannesburg, South Africa

emails: martmulesh@gmail.com; paulbaki@gmail.com; james.chibueze@gmail.com and kelbouchefry@ska.ac.za

Abstract. We report the results of 14 years of monitoring of G188.95+0.89 periodic 6.7 GHz methanol masers using the Hartebeesthoek 26-m radio telescope. G188.95+0.89 (S252, AFGL5180) is a radio-quiet methanol maser site that is often interpreted as precursors of ultra-compact HII regions or massive protostar sites. At least five bright spectral components were identified. The maser feature at 11.36 km s^{-1} was found to experience an exponential decay during the monitoring period. The millimetre continuum reveals two cores associated with the source.

Keywords. ISM: individual (G188.95+0.89), ISM: molecules, stars: imaging.

1. Introduction

One property that is unique to massive young stellar objects (MYSO) is the presence of methanol masers in their early formative phases (Menten 1991; Caswell *et al.* 1995). This paper is devoted to a study of a radio-quiet methanol maser site G188.95+0.89 (S252, AFGL 5180) and its environment, which is particularly good candidate for hosting a massive protostar (Minier *et al.* 2003; Goedhart *et al.* 2004). Our aim of this study is to observe the long-term variation along 14 years (between 2003-2008 and 2010-2018) and reveal the origin of the flux variation of the 6.7-GHz methanol maser.

2. Results

Radio data were obtained from Hartebeesthoek Radio Astronomy Observatory (HartRAO) 26-m telescope for the ongoing methanol maser monitoring programme (Goedhart et al., 2004). We obtained Atacama millimeter/submillimeter Large Millimeter Array (ALMA) band 6 archival data of G188.95+0.89 (Project ID: 2015.1.01454.S) taken on 2016-April-23 (42 antennas), 2016-Sepember-17 (38 antennas) in its compact and extended configurations, respectively. The time series of the selected maser velocity channels associated with G188.95+0.89 are shown in sub-figure (a). The variations observed in five velocity channels covering the bright CH_3OH emission at 6.7 GHz, between 8 and 12 km s^{-1} are periodic (395 d) with exception of 11.36 km s^{-1} which is exponentially decaying as shown in sub-figures (a) and (b). The spectra and intensity map of

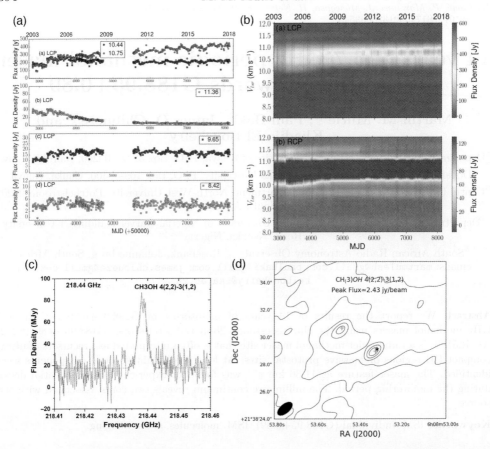

Figure 1. (a) Time series of G188.95+0.89. (b) Dynamic spectrum plot from 2003 to 2018. (c) Spectra of CH_3OH observed by ALMA. (d) CH_3OH integrated map of the emission. The star and plus signs show the position of the methanol maser and the peak positions of the 1.3 mm dust-continuum, respectively. The synthesized beam size is shown in the lower left-hand corner.

CH_3OH masers in G188.95+0.89 observed by ALMA are shown in sub-figures (c) and (d), respectively.

3. Conclusions

Observed changes in G188.95+0.89 masers are likely due to changes in the background free-free emissions which are amplified by the masers. The decay of 11.36 km s^{-1} maser can be explained in terms of the recombination of the ionized gas against which the maser is projected.

Acknowledgements

The authors would like to thank Technical University of Kenya, NRF-South Africa, HartRAO-South Africa and the Newton Fund (UK) for their support.

References

Caswell, J. L., Vaile, R. A., & Ellingsen, S. P. 1995, *PASA*, 12, 37
Goedhart S., Gaylard M. J. & van der Walt D. J. 2004, *MNRAS*, 355, 553
Menten, K. M. 1991, *ApJ*, 380, L75
Minier, V., Ellingsen, S. P., Norris, R. P. & Booth, R. S. 2003, *A&A*, 403, 1095

Testing stellar evolution models

Seblu H. Negu and Solomon Belay Tessema

Ethiopian Space Science and Technology Institute (ESSTI), Entoto Observatory and Research Center (EORC), Astronomy and Astrophysics Research and Development Division, Box 33679, Addis Ababa, Ethiopia
emails: seblu1557@gmail.com; tessemabelay@gmail.com

Abstract. In this study, we have investigated the stellar evolution models using the open-source software instrument Modules for Experiments in Stellar Astrophysics. We examine the evolution of angular momentum and the stability of mass transfer in the evolution of Algol-type binaries through the inner Lagrangian point via the Roche lobe overflow. Also, we have determined the ongoing challenge of chemical mixing and exhibit improvements that make easier the simulation of Algol-type binaries evolution.

Keywords. Binary systems, Algol-type binaries, stellar models, mass transfer, method-numerical.

1. Introduction

The development of a relatively complete and quantitative picture of evolution of binary stars is of great interest in astrophysics. The comprehensive and detailed stellar models of various processes involved in binary systems evolution are lacking and suffer many theoretical uncertainties associated with mass transfer, chemical reactions, and orbital evolution, especially in the domain of low-mass stars (Nelson & Eggleton 2001). Based on the binary stars evolution code such as Modules for Experimental Stellar Astrophysics, MESA (Paxton *et al.* 2015), we investigate the stability of mass transfer in Algol-type binaries by using the binary stars evolution models and mass-radius exponent relations. Algol-type binaries are one of the semi-detached binary systems, which are formed from detached binaries by exchanging mass between their components. In semi-detached binaries, mass transfer proceeds from low-mass subgiant to more massive main-sequence stars in circular orbit (Peters 2001). Hence, as noted by Negu & Tessema (2018), here we examine the theoretical models of mass transfer in semi-detached binaries and its stability by using the evolution of orbital and stellar parameters through the physical principle of Roche lobe overflow, RLOF (Hilditch 2001).

2. Analysis

We considered the MESA version 7624 in our simulations of semi-detached binary systems evolution taking into account different stellar masses. For numerical calculations we applied the Eggleton (1983) formula:

$$\frac{R_{Ld}(q)}{a} = \frac{0.49q^{2/3}}{0.6q^{2/3} + \log(1 + q^{1/3})}, \qquad (2.1)$$

where R_{L_d}, a, and $q = \frac{M_d}{M_a}$ are the Roche lobe radius of the donor star, semi-major axis, and mass ratio of the semi-detached binary system, respectively. This is a basic mathematical model that can be used to compute the evolutionary phases of semi-detached binary systems and its chemical composition. Hence, we determine the stability of mass

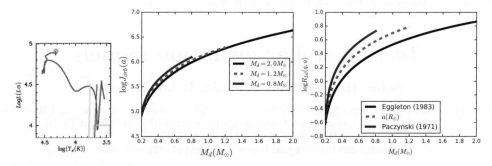

Figure 1. Left plot: relation between L and T_{eff} for different chemical composition of semi-detached binary systems. **Middle plot:** evolution of J_{orb} with M_d. **Right plot:** evolution of the R_{L_d} with Md.

transfer in these systems using the orbital and stellar parameters due to the distribution of orbital angular momentum, J_{orb}, which is given by:

$$J_{orb}(a) = \left(\frac{q\sqrt{GM^3 a}}{(1+q)^2} \right), \quad (2.2)$$

where G and $M = M_a + M_d$ are the universal gravitational constant and the sum of mass of the accretor, M_a, and donor, M_d, respectively. The components of semi-detached binary systems should also satisfy the Stefan-Boltzmann law that the stellar luminosity, L, is related to the stellar radii, R, and effective temperature, T_{eff}:

$$L = 4\pi R^2 \sigma T_{eff}^4, \quad (2.3)$$

where σ is the Stefan-Boltzmann constant. In Figure 1 (left plot) we show the core helium (red line) and hydrogen (yellow line) shell burnig for different values of L and T_{eff}. Numerical solution of the system is showed with cyan line. Middle and right plot of Figure 1 show the relation of J_{orb} and R_{L_d} with the stability of mass transfer, respectively.

3. Conclusion

In this study, we have tested the binary star evolution code MESA to determine the stable mass transfer in semi-detached binary systems with the evolution of orbital angular momentum.

Acknowledgment

This work makes use of the binary star code, MESA version 7624. We would like to thank EORC under ESSTI for the support of this study.

References

Eggleton, P. P. (1983), *ApJ*, 268, 368
Hilditch, R. W. (2001), An introduction to close binary stars, *Cambridge University Press*
Negu, S. H. & Tessema, S. B. (2018), *Astron. Nachr.*, 339, 478
Nelson, C. A. & Eggleton, P. P. (2001), *ApJ*, 552, 664–678
Paxton, B., Marchant, P., & Schwab, J. (2015), *ApJS*, 220, 15–44
Peters, G. J. (2001), *D. Vanbeveren, Proc. ASSL*, Vol. 264, Kluwer, Dordrecht, p. 79

Accelerating universe in modified teleparallel gravity theory

Shambel Sahlu[1,2], Joseph Ntahompagaze[3], Amare Abebe[4] and David F. Mota[5]

[1]Astronomy and Astrophysics Research and Development Division, Entoto Observatory and Research Center, Ethiopian Space Science and Technology Institute, Addis Ababa, Ethiopia.
[2]Department of Physics, College of Natural Science, Wolkite University, Wolkite, Ethiopia.
[3]Department of Physics, College of Science and Technology, University of Rwanda, Kigali, Rwanda.
[4]Center for Space Research, North-West University, Mafikeng, South Africa.
[5]Institute of Theoretical Astrophysics, University of Oslo, Oslo, Norway

Abstract. This paper studies the cosmology of accelerating expansion of the universe in modified teleparallel gravity theory. We discuss the cosmology of $f(T, B)$ gravity theory and its implication to the new general form of the equation of state parameter w_{TB} for explaining the late-time accelerating expansion of the universe without the need for the cosmological constant scenario. We examine the numerical value of w_{TB} in different paradigmatic $f(T, B)$ gravity models. In those models, the numerical result of w_{TB} is favored with observations in the presence of the torsion scalar T associated with a boundary term B and shows the accelerating expansion of the universe.

Keywords. $f(T, B)$ gravity theory; accelerating universe; equation of state parameter.

---

1. Introduction

To explain the cause behind the current cosmic accelerating expansion, different suggestions have been put forward. For instance, one suggestion is that the cosmological constant is the one responsible for this cosmic acceleration, as presented in Saul et al. (2003) and the second approach is the modification of General Relativity (GR) as Timothy et al. (2012). In the second approach, several extra degrees of freedom are presented through the modification of GR to account for the present cosmic accelerating expansion and to study if the cosmic history from the early universe can produce this cosmic acceleration. This paper discusses how the modified teleparallel gravity so-called, $f(T, B)$ gravity scenario, is taken as an alternative approach for the ΛCDM model to describe the late-time accelerating expansion of the universe. We obtain the new expression of the equation of state parameter w_{TB} in the effective torsion fluid.

2. The cosmology of $f(T, B)$ gravity

We consider different paradigmatic $f(T, B)$ gravity models, and in each model, we describe the accelerating universe in the late time by plotting the w_{TB} versus redshift z. It is close to the well-known equation of state parameter $w = -1$. We start by providing the action that contains the $f(T, B)$ Lagrangian:

$$I_{f(T,B)} = \frac{1}{2\kappa^2} \int d^4 x e \Big[T + f(T, B) + L_m \Big], \qquad (2.1)$$

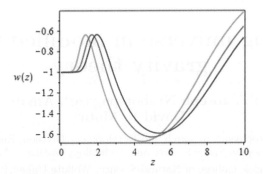

Figure 1. $w(z)$ versus redshift z for $f(T, B) = B - T + \alpha T_0 \left(1 - e^{b\frac{T}{T_0}}\right)$ gravity model with different value of b. We use $b = 0.002$, $b = 0.004$ and $b = 0.006$ for green, red and blue lines, respectively.

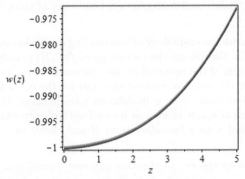

Figure 2. $w(z)$ versus redshift z for power-law $f(T, B) = B - T + \alpha(-T)^b$ gravity model with different value of b. We use $b = 0.1$ for green line, $b = 0.4$ for red line and $b = 0.7$ for blue line.

where e is the determinate of the tetrad field e^μ_A, L_m is the matter Lagrangian and $\kappa^2 = 8\pi G/c^4$ is the coupling constant†. We assume that the total cosmic medium is composed of matter ρ_m, radiation ρ_r and effective torsion fluid ρ_{TB}, and that one can directly derive the corresponding thermodynamic quantities in $f(T, B)$ gravity from eq. (2.1), such as the energy density ρ_{TB} and pressure p_{TB} of the torsion-like fluid as presented in Bahamonde et al. (2017):

$$\kappa^2 \rho_m = -3H^2(3f_B + 2f_B) + 3H\dot{f}_B - 3\dot{H}f_B + \frac{f(T, B)}{2}, \quad (2.2)$$

$$\kappa^2 p_m = 3H^2 + \dot{H})(3f_B + 2f_T)2H\dot{f}_T - \ddot{f}_B - \frac{f(T, B)}{2}. \quad (2.3)$$

Therefore, the new general form of w_{TB} can be constructed from the above equations for the effective torsion fluid defined as $w_{TB} = p_{TB}/\rho_{TB}$ and is given by:

$$w_{TB} = -1 + \frac{\ddot{f}_B - 3H\dot{f}_B - 2\dot{H}f_T - 2H\dot{f}_T}{3H^2(3f_B + 2f_T) - 3H\dot{f}_B + 3\dot{H}f_B - f(T, B)/2}. \quad (2.4)$$

In the following we consider the well known two models namely: exponential and power-law $f(T)$ gravity models associated with the boundary term as $f(T, B) = B + f(T)$, for $B = 0$ it reads $f(T, B) = f(T)$ as presented in Nayem (2017).

† We assume that $\kappa^2 = 8\pi G/c^4 = 1$.

3. Conclusions

Generally, in all models that are treated above, we numerically computed the effective equation of state parameter form of $w(z)$ through the use of eq. (2.4) and the value of $w(z)$ is favored with the observed value of the effective equation of state parameter of cosmological constant $w \approx -1$ in the present universe. So, we clearly show that all two $f(T, B)$ gravity models can be regarded as an alternative way of cosmological constant model to describe the late-time accelerating expansion of the universe. Surprisingly, in all models the value of $w(z)$ in the present and near past universe asymptotically approaches to the equation of state parameter of the cosmological constant $w = -1$. For instance in Fig. 1, we clearly observe the history of the universe phase with the phantom-like $w(z) < -1$ and quintessence-like $w(z) > -1$ phases, while the other two models in Fig. 2 show only the quintessence-like phase. GR can be recovered for $B = b = 0$ for all models.

References

Bahamonde *et al.* 2017, *EPJC*, 77, 107
Nayem, S. K. 2017, *Phy.Letters B.*, 775, 100
Saul *et al.* 2003, *Physics today*, 56, 53
Timothy *et al.* 2012, *Physics reports*, 513, 1

The impact of CMEs on the critical frequency of F2-layer ionosphere (foF2)

Alene Seyoum[1,2], Nat Gopalswamy[3], Melessew Nigussie[4] and Nigusse Mezgebe[1]

[1]Ethiopian Space Science and Technology Institute (ESSTI), Addis Ababa, Ethiopia

[2]Dire Dawa University, College of Natural and Computational Science, Physics Department, Dire Dawa, Ethiopia

[3]NASA Goddard Space Flight Center, Greenbelt, MD 20771, USA
email: nat.gopalswamy@nasa.gov.

[4]Washera Geospace and Radar Science Laboratory, College of Science, Physics Department, Bahir Dar University, Bahir Dar, Ethiopia

Abstract. The ionospheric critical frequency (foF2) from ionosonde measurements at geographic high, middle, and low latitudes are analyzed with the occurrence of coronal mass ejections (CMEs) in long term variability of the solar cycles. We observed trends of monthly maximum foF2 values and monthly averaged values of CME parameters such as speed, angular width, mass, and kinetic energy with respect to time. The impact of CMEs on foF2 is very high at high latitudes and low at low latitudes. The time series for monthly maximum foF2 and monthly-averaged CME speed are moderately correlated at high and middle latitudes.

Keywords. CME impact; critical frequency; ionosphere; maximum foF2; geographic latitudes.

1. Introduction

Active regions on the Sun contribute to the variability of Earth's ionosphere, in particular to the variability of neutral and ionized densities. The ionosphere becomes variable due to lower atmospheric internal waves, and geomagnetic and solar activity variations from the above atmosphere (Yiğit *et al.* 2016). The origin of space weather effects such as intense geomagnetic storms is due to CMEs (Gopalswamy 2009). CME parameters are thought to cause a large volume of the Earth's ionosphere to increase ionization (Farid *et al.* 2015). In the ionospheric dynamo, equatorial electrojet is produced due to Eastward electric field (Seba and Nigussie 2016). The electric field E interacts with Earth's magnetic field B causing strong vertical upward E×B drift velocity and enhanced foF2 values in low latitudes (Horvath and Essex 2003). This paper shows the impact of CMEs on foF2 from 1996 to 2018 in geographical high, middle, and low latitudes. This impact has importance for communication depending on relationship with solar activity. The parameters used in this work are monthly maximum value of foF2 and monthly-averaged values of CMEs angular width, speed, mass, and kinetic energy.

2. Data Sources

The foF2 data are obtained from ionosonde of UK Solar System Data Center CEDA-UKSSDC for three station data. These stations are SO166 (67.400°N, 26.600°E) in Finland, BP440 (40.080°N, 116.260°E), in China, and VA50L (−2.700°S, 141.300°E) in

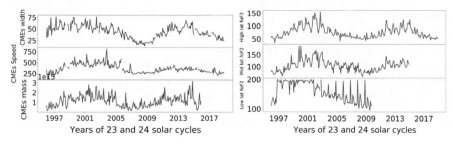

Figure 1. The relation of monthly-averaged CME parameters such as CME width (deg), CME speed (km/s), and CME mass (ergs) (left panels, from top to bottom) with monthly maximum foF2 values (in MHz) (right panels) from 1996 to 2018.

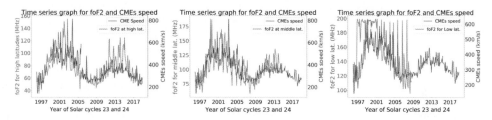

Figure 2. The time series of monthly-averaged CME speed in solar cycles of 23rd and 24th and monthly maximum values of foF2.

Papua New Guinea†. The CMEs data are found from the SOHO/LASCO ‡ (Yashiro et al. 2004; Gopalswamy et al. 2009) in 23rd and 24th solar cycles from 1996 to 2018.

3. Results

In Figure 1 we showed that the monthly maximum foF2 (right panels) tracks the monthly-averaged CME parameters (left panels) in 23rd and 24th solar cycles. In the ascending periods from 1997 to 2001, the maximum foF2 values are higher at low latitudes. In Figure 2 we showed the time series of CME speed and foF2. High and middle latitudes foF2 show similar time series to CME speed, but not for low latitudes. The foF2 at low latitudes (equatorial region) is rapidly fluctuating foF2 values, which needs further investigation.

4. Conclusions

We studied the relation between monthly-averaged values of CME parameters and the monthly maximum value of foF2 at high, middle, and low latitudes. While there is moderately good correlation between the two time series graph at high and middle latitudes, for the low-latitudes of ionosphere correlation is not clear. Because the foF2 values here are higher and have fast fluctuation at low latitudes and it is not symmetric to CME parameters at high and middle latitudes. This may be due to the $\mathbf{E}\times\mathbf{B}$ upward drift velocity of the ionospheres at equatorial region (low latitudes) that can not be easily compressed and highly influenced by CMEs at high latitudes. This indicates the impact of CMEs on foF2 is higher at high latitudes (polar region) than at the low latitudes. The coincidence relation between maximum value of foF2 and CME parameters are important for modeling of space weather prediction at high and middle latitudes.

† (https://www.ukssdc.ac.uk/wdcc1/ionosondes/secure/ion_data.shtml)
‡ (http://cdaw.gsfc.nasa.gov/CME_list/)

References

Farid, H. M., Mawad, R., Yousef, M., & Yousef, S. 2015, *Elixir Space Sci*, 80
Gopalswamy N. 2009, *Proceedings of IAU*, 5.S264
Gopalswamy N., Yashiro S., Michalek G., *et al.* 2009, *Earth, Moon, and Planets*, 104, 1
Horvath, I., and Essex, E. A. 2003, *Annales Geophysicae*, 21, 4
Seba, E. B., and Nigussie, M. 2016, *Advances in Space Res.*, 58, 9
Yashiro, S., Gopalswamy, N., Michalek, G., *et al.* 2004, *JGR*, 109, A7
Yiğit E., Knížová, P. K., Georgieva, K., & Ward, W. 2016, *JASTP*, 141, 1

Mass-loss varying luminosity and its implication to the solar evolution

Negessa Tilahun Shukure,[1,2,3] Solomon Belay Tessema[1] and Endalkachew Mengistu[1]

[1]Astronomy and Astrophysics Research and Development Division, Entoto Observatory and Research Center, Ethiopian Space Science and Technology Institute, P.O.Box 33679, Addis Ababa, Ethiopia

[2]Addis Ababa University, P.O.Box 1176, Addis Ababa, Ethiopia

[3]Dilla University, P.O.Box 419, Dilla, Ethiopia

emails: nagessa2006@gmail.com; tessemabelay@gmail.com; 2fendalk@gmail.com

Abstract. Several models of the solar luminosity, $L_\odot$, in the evolutionary timescale, have been computed as a function of time. However, the solar mass-loss, $\dot{M}$, is one of the drivers of $L_\odot$ variation in this timescale. The purpose of this study is to model mass-loss varying solar luminosity, $L_\odot(\dot{M})$, and to predict the luminosity variation before it leaves the main sequence. We numerically computed the $\dot{M}$ up to 4.9 Gyrs from now. We used the $\dot{M}$ solution to compute the modeled $L_\odot(\dot{M})$. We then validated our model with the current solar standard model (SSM). The $L_\odot(\dot{M})$ shows consistency up to 8 Gyrs. At about 8.85 Gyrs, the Sun loses 28% of its mass and its luminosity increased to $2.2L_\odot$. The model suggests that the total main sequence lifetime is nearly 9 Gyrs. The model explains well the stage at which the Sun exhausts its central supply of hydrogen and when it will be ready to leave the main sequence. It may also explain the fate of the Sun by making some improvements in comparison to previous models.

Keywords. Sun; mass-loss; solar luminosity

1. Introduction

Understanding of the red giant evolution allows us to predict the final fate of the Sun during its final nuclear burning phase. Nowadays, numerical simulations make possible to test different evolutionary scenarios. Several models of solar luminosity, $L_\odot$, in the evolutionary timescale have been computed as a function of time (Gough 1981; Bahcall et al. 2001). However, the solar mass-loss, $\dot{M}$, is one of the drivers of $L_\odot$ variation. The purpose of this study is to model mass-loss varying solar luminosity, $L_\odot(\dot{M})$, to predict the magnitude of luminosity before it leaves the main sequence, and to estimate the solar mass-loss just before the red giant branch (RGB).

2. Mathematical formulation

The base of the formulation of our model is the equation of solar standard model (SSM) given by Gough (1981):

$$L(t) = \frac{L_\odot}{1 + \frac{2}{5}\left(1 - \frac{t}{t_\odot}\right)}, \qquad (2.1)$$

where $t_\odot$ is the current solar age and $L_\odot$ is the solar luminosity at this time.

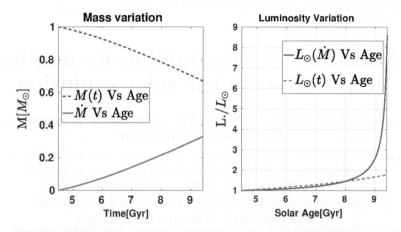

Figure 1. Panel A. The solar mass variation (dashed blue line) and mass-loss (solid red line) in relation to coming lifetime. Panel B. Solar luminosity variation in our model, using eq. 3.1 (blue solid line), and in SSM model, using eq. 2.1 (red dashed line).

3. Analysis

We modeled the solar mass variation with time as an exponential decay, $M(t) = M_\odot e^{-\delta t}$, and $\dot{M} = M_\odot - M(t)$, where $M_\odot$ is the current solar mass, δ is the reduction coefficient of mass of the Sun defined in the greatest interval $0 \leq \delta \ll 1$, and t is the age of the Sun. We modeled the $L_\odot(\dot{M})$ in the evolutionary time scale as:

$$L(\dot{M}) = \frac{L_\odot}{1 + \frac{2}{5}ln\left[1 + \frac{1}{\delta t_0}ln\left(1 - \frac{\dot{M}}{M_\odot}\right)\right]}, \quad (3.1)$$

where $L_\odot = 3.85 \times 10^{26}$ W and $t_\odot = 4.57$ Gyrs (Feulner 2012) are the current solar luminosity and solar age, respectively, $M_\odot = 1.9891 \times 10^{30}$ kg (Kaplan, 1981) is the current solar mass, and $0 \leq \delta \leq 10^{-10}$ is the input boundary condition to numerical solution.

4. Results and conclusions

We studied solar mass variation and mass-loss (Figure 1, A) and luminosity variation (Figure 1, B) along the solar coming lifetime. The $L_\odot(\dot{M})$ shows consistency up to 8 Gyrs (blue solid line in panel B) with the SSM (red dashed linein panel B). Our model suggests that Sun will enter RGB at about 8.85 Gyrs which approaches the time suggested by Kopp (2016). The Sun will lose about 28% of its mass and will have a luminosity of $2.2L_\odot$ at this age. At 9 Gyrs, the Sun will lose about 30% of its mass and the luminosity will raise by about $2.5L_\odot$. The model explains well the stage in which the Sun exhausts its central supply of hydrogen and is ready to leave the main sequence. This model contributes to previous studies and our knowledge about Solar evolution.

Acknowledgment

This research was partially supported by the International Science Program (ISP), grant code IPPS/AFRO5.

References

Bahcall, J. N., Pinsonneault, M. H., & Basu, S. 2001, *APJ*, 555(2), p. 990
Feulner, G. 2012, *Reviews of Geophysics*, 50 (2)
Gough, D. O. 1981, *Sol. Phys.*, 74, 21–34
Kaplan G. H. 1981, editor *US Naval Observatory*, No. 163
Kopp G. 2016, *Journal of space weather and space climate*, 6, p. A30

Howusu Metric Tensor - problems and prospects

Obini Ekpe Ekpe

Department of Physics, Michael Okpara University of Agriculture Umudike, Nigeria

Abstract. A new metric tensor proposed by Howusu is presented. Problems associated with the metric tensor are pointed out. Some of the good aspects of the Howusu Metric are also outlined. It is argued that Howusu Metric holds some promise in generating healthy academic debate.

Keyword. Metric tensor

Progress of astronomy in Tanzania

Noorali Jiwaji

Open University of Tanzania, Dar es Salaam, Tanzania

Abstract. A summary in a timeline of astronomy developments over the past several decades and listing major achievements and current work will be presented.

Keywords. astronomy: development.

The study of short orbital period of delta scuti pulsating variable stars

Abduselam Mohammed

Department of Physics, Woldia University, Woldia, Ethiopia

Abstract. As a pulsating star moves in its binary orbit, the path length of the light between us and the star varies, leading to the periodic variation in the arrival time of the signal from the star to us (earth). With the consideration of pulsators light arrival time delay effects several new methods which allows using Kepler photometric data (light curves) alone to find binary stars have been recently developed. Among these modern techniques we used binarogram method and we identified that several δSct pulsating stars have companions. The application of these method on detecting long periods(i.e. longer than about 50 d) δSct pulsating stars is not new, but the uniqueness of this study is we verified that it is also applicable to detect and determine the orbital elements of short periods (i.e short orbital period) δSct pulsating stars. With this investigation, we identified the possible way to overcome effects of fictious peaks, even, on the maximum peaks helpful to verify weather the star has companion or not depend up on the existence of the time-delay. Then, we applied the technique on known binary stars and their orbital elements are previously published. Finally, we identified some new short orbital period δSct pulsating stars and obtained their orbital frequency and period with the same procedures. Because of with our attempts we succeeded and verified the applicability of the method (the Binarogram method) on these stars (i.e short orbital period) for the first time, we expect that our present study will play a great role for similar study and to improve our binary statistics.

Keywords. stars: variable; stars: properties; stars: binary.

Design and development of a two-dish interferometer

Dorcus Mulumba

Kenyatta University, Nairobi, Kenya

Abstract. The angular resolution and the sensitivity of a parabolic dish telescope increase with the diameter of its aperture at a given frequency. This implies that as the telescope gets larger, its resolution becomes better. However, constructing telescopes of ever increasing size is prohibitive for both technical and financial reasons. This problem is solved by using an interferometer which consists of two or more separate telescopes that combine their signals offering a resolution equivalent to the largest separation distance between the telescopes. In this work, the electric field variations from two telescopes will be obtained. The voltage signals from the two telescopes will be coherently combined in order to derive the structure of the target source of radio emission. This combination will be done by a cross-correlator, which multiplies and averages the voltage outputs V1 and V2 of the two dishes. A major challenge to be addressed in this work is to design an instrument capable of making professional-type radio astronomy measurement in a local interference environment. In this regard, the investigative part of this work will verify whether it is possible to achieve a high sensitivity enough to detect some cosmic sources where the presence of man-made interference and cost adversely influences the system. The design of an interferometer will be presented and implemented. It may also serve as a demonstrator for engineering students to gain a working knowledge of radio interferometry.

Keywords. telescopes: radio; radio: interferometry.

Electron-proton interaction in radio sources

Halima Ugomma Obini

Department of Physics, Michael Okpara University of Agriculture Umudike, Nigeria

Abstract. A method of treating electron-proton interaction is presented. The energies involved in the interaction are estimated. Only elastic collisions are considered. The cross sections of the processes are not taken into account. Calculations are carried out in the centre of mass frame. Relevant quantities are transformed into the laboratory frame. Results indicate that the energy per collision gained by an electron ranges from 0.5 MeV to 0.6 MeV, under suitable conditions.

Keywords. Electron-proton interaction; radio sources.

Author index

Abaraya, S. – 383
Abdalla, H. – 364
Abebe, A. – 397
Abramo, R. – 12
Abramowicz, M. A. – 348
Alcaniz, J. – 12
Alexander, D. M. – 199
Allevato, V. – 226
Almudena Prieto, M. – 97
Álvarez-Hernández, A. – 302
Amiri, A. – 171
Aravena, M. – 358
Aretxaga, I. – 95
Arévalo, P. – 56
Arsioli, B. S. – 329
Asmus, D. – 50
Assef, R. J. – 358

Baki, P. – 393
Balmaverde, B. – 87
Bandyopadhyay, B. – 56
Barai, P. – 184
Baron, D. – 225, 365
Baum, S. A. – 247
Belay Tessema, S. – 163, 388, 395, 403
Bell, C. P. M. – 335
Benitez, N. – 12
Berton, M. – 94
Beswick, R. – 137
Beyene, F. W. – 385
Beyoro-Amado, Z. – 158, 163, 295
Bhattacharjee, A. – 72
Bilata-Woldeyes, B. – 295
Biressa, T. – 252, 383
Blanchard, J. – 137
Bon, E. – 66, 310
Bon, N. – 66, 310
Bongiorno, A. – 280
Bonoli, S. – 12
Brandt, W. N. – 11
Brotherton, M. S. – 143
Brunthaler, A. – 276
Bu, D.-F. – 214

Calderone, G. – 12
Calistro Rivera, G. – 26
Cao Orjales, J. M. – 204
Cardaci, M. V. – 61
Cardiel, N. – 306
Carneiro, S. – 12
Carrasco, E. – 306

Castillo-Morales, Á. – 306
Cazzoli, S. – 87, 302, 306, 317
Cedazo, R. – 306
Cenarro, J. – 12
Chakraborty, A. – 72
Chary, R.-R. – 255
Chen, H. – 143
Chibueze, J. O. – 393
Circosta, C. – 194
Colbert, E. J. M. – 247
Combes, F. – 177
Comerón, S. – 209
Cristóbal-Hornillos, D. – 12
Cui, L. – 320
Czerny, B. – 66, 77, 332

D'amour Kamanzi, J. – 132
D'Onofrio, M. – 66, 310
Das, M. – 345
de Diego, J. A. – 66
de Witt, A. – 137
DeGraf, C. – 292
del Olmo, A. – 66, 87, 310, 351
Diaz, Y. – 56
Dors, O. L. – 61
Dultzin, D. – 66
Dupke, R. – 12

Ederoclite, A. – 12
Ekpe, O. E. – 405
El Bouchefry, K. – 393

Fernández-Ontiveros, J. A. – 17, 29, 97
Feyissa, A. – 299
Filipović, M. D. – 335

Gallagher, S. C. – 143
Gallego, J. – 306
Gallimore, J. F. – 247
Ganci, V. – 310
García-Benito, R. – 61
García-Burillo, S. – 87
García-Vargas, M. L. – 306
Gaulle, A. – 314, 358
Gelfand, J. – 366
Getachew-Woreta, T. – 158, 163, 295
Gil de Paz, A. – 306
Giustini, M. – 82
Goluchová, K. – 348
González-Martín, O. – 87, 93, 367

Author index

Góomez-Alvarez, P. – 306
Gopalswamy, N. – 400
Gorbovskoy, E. S. – 127
Govender, K. – 379
Goyal, A. – 361
Gurvits, L. I. – 137

Hägele, G. F. – 61
Hardcastle, M. J. – 204
Harrison, C. M. – 199
Hermosa-Muñoz, L. – 302, 306, 317
Hernández-García, L. – 87, 302
Hess, K. M. – 339
Ho, L. C. – 223
Hoare, M. – 137
Horák, J. – 348

Ichikawa, K. – 44
Iglesias, J. – 306
Ilić, D. – 144

Järvelä, E. – 172
Jarvis, M. – 253
Jarvis, M. J. – 204
Jiang, P. – 320
Jiwaji, N. – 406
Jiwaji, N. T. – 323
Joseph, T. D. – 335
Jurua, E. – 329

Kalfountzou, E. – 204
Kaspi, S. – 116
Kassa, E. E. – 368
Kharb, P. – 247, 345
Kimani, N. – 276
Knapen, J. H. – 209
Kotilainen, J. – 170, 285
Kovalev, Y. – 256
Koziel-Wierzbowska, D. – 361
Krauß, F. – 326
Kreter, M. – 23, 369
Kumssa, G. M. – 388

Leeuw, L. L. – 379
Levenson, N. A. – 44
Li, S.-L. – 243
Li, Y.-R. – 143
Lipunov, V. M. – 127
Lira, P. – 101
Liu, X. – 320
López, E. – 56
López San Juan, C. – 12
Lopez-Rodriguez, E. – 44
Luo, B. – 143

Mahoro, A. – 147, 152
Maithil, J. – 143

Mamo, A. – 379
Man, A. – 224
Marín-Franch, A. – 12
Márquez, I. – 87, 302 306, 317
Marquez Perez, I. – 158
Martinez-Aldama, M. L. – 66
Martínez-Delgado, I. – 306
Marziani, P. – 66, 77, 310, 332, 351
Masegosa, J. – 87, 158, 302, 306, 317
Mazengo, D. T. – 323
Mbarubucyeye, J. D. – 132, 326
Mekuria, R. R. – 385, 390
Mendes de Oliviera, C. – 12
Mengistu, E. – 403
Menten, K. M. – 276
Mezgebe, N. – 400
Mohammed, A. – 407
Moles, M. – 12
Mordini, S. – 17
Morganti, R. – 229
Mota, D. F. – 397
Mulumba, D. – 408
Musiimenta, B. – 329
Mutabazi, T. – 132, 329, 355
Mutie, M. M. – 393

Nagar, N. M. – 56
Negrete, A. C. – 310
Negrete, C. A. – 66
Negu, S. H. – 395
Nguyen, N. – 291
Nigussie, M. – 400
Nikutta, R. – 44
Njeri, A. – 137
Nkundabakura, P. – 132, 147, 152, 326, 355
Ntahompagaze, J. – 397
Nyiransengiyumva, B. – 147, 152

O' Dea, C. P. – 247
Obi, I. – 25
Obini, H. U. – 409
Oknyansky, V. L. – 127

Packham, C. C. – 44
Panda, S. – 66, 77, 332
Pascual, S. – 306
Péerez-Calpena, A. – 306
Pennock, C. M. – 335
Perea, J. – 66, 158
Pérez-Montero, E. – 61
Placco, V. – 12
Povic, M. – 3, 87, 147, 152, 158, 163, 295, 314, 323, 351, 358, 370
Proga, D. – 82
Pu, D. – 143

Quirrenbach, A. – 371

Rahne, P. T. – 345
Ramakrishnan, V. – 56, 254
Randriamampandry, S. M. – 339
Randriamanakoto, Z. – 342, 379
Ridgway, S. – 372
Ridgway, S. E. – 290
Rosario, D. J. – 199
Rubinur, K. – 345

Sahlu, S. – 397
Sánchez-Portal, M. – 163
Sani, E. – 96
Schleicher, D. R. G. – 56
Scholtz, J. – 199
Schramm, M. – 285
Schulze, A. – 285
Sebastian, B. – 247
Seidel, M. K. – 209
Seyoum, A. – 400
Sharpe, C. – 137
Shastri, P. – 373
Shemmer, O. – 143
Shukure, N. T. – 403
Simpemba, P. – 374
Sodré, L. Jr. – 12
Spinoglio, L. – 17, 29
Šrámková, E. – 348
Stalevski, M. – 50
Stanley, F. – 199
Stevens, J. – 204
Stirpe, G. M. – 66
Stuchlík, Z. – 348
Sunzu, J. M. – 323

Takalana, C. M. – 379
Tassama, A. – 252

Taylor, K. – 12
Terefe, S. – 295, 351
Tiplady, A. – 137
Tolu, B. – 299
Tombesi, F. – 218
Török, G. – 348
Trakhtenbrot, B. – 261
Travascio, A. – 280
Tristram, K. R. W. – 50
Tsygankov, S. S. – 127
Tyurina, N. V. – 127

Uwitonze, E. – 355

Vaccari, M. – 339
Väisänen, P. – 147, 169
van der Heyden, K. – 147
van Loon, J. T – 335
Vardoulaki, E. – 335
Varela, J. – 12
Vázquez Ramió, H. – 12
Verma, A. – 204
Vílchez, J. M. – 61

Wang, J.-M. – 143
Wethers, C. – 285
White, S. – 375
White, S. V. – 204
Wielgus, M. – 257
Winkler, H. – 122

Xie, F.-G. – 56, 189

Yesuf, H. – 173

Zaw, I. – 24, 376
Zewdie, D. – 314, 358
Żywucka, N. – 361